A Comprehensive Approach to Evolutionary Biology

A Comprehensive Approach to Evolutionary Biology

Editor: Jesse Santos

R CALLISTO
REFERENCE

www.callistoreference.com

Callisto Reference,
118-35 Queens Blvd., Suite 400,
Forest Hills, NY 11375, USA

Visit us on the World Wide Web at:
www.callistoreference.com

ISBN: 978-1-64116-167-1 (Hardback)

Cataloging-in-Publication Data

A comprehensive approach to evolutionary biology / edited by Jesse Santos.
    p. cm.
Includes bibliographical references and index.
ISBN 978-1-64116-167-1
1. Evolution (Biology). 2. Evolutionary developmental biology. I. Santos, Jesse.
QH366.2 .C66 2019
576.8--dc23

Table of Contents

Permissions

List of Contributors

Index

Preface

The world is advancing at a fast pace like never before. Therefore, the need is to keep up with the latest developments. This book was an idea that came to fruition when the specialists in the area realized the need to coordinate together and document essential themes in the subject. That's when I was requested to be the editor. Editing this book has been an honour as it brings together diverse authors researching on different streams of the field. The book collates essential materials contributed by veterans in the area which can be utilized by students and researchers alike.

Evolutionary biology is a study of the evolutionary processes that have resulted in the biological diversity on Earth. It integrates the concepts and principles of genetics, ecology, paleontology and systematics. Evolutionary processes studied under this domain include natural selection, common descent and speciation. Modern evolutionary biology also delves into the genetic architecture of adaptation, sexual selection, genetic drift, molecular evolution and biogeography. Due to the merger of biological science with applied sciences, many domains of evolutionary biology have emerged such as evolutionary robotics, engineering, economics, architecture and algorithms. This book is compiled in such a manner, that it will provide in-depth knowledge about the theory and practice of evolutionary biology. It explores all the important aspects of this field in the present day scenario. This book is a vital tool for all researching or studying evolutionary biology as it gives incredible insights into emerging trends and concepts.

Each chapter is a sole-standing publication that reflects each author's interpretation. Thus, the book displays a multi-facetted picture of our current understanding of application, resources and aspects of the field. I would like to thank the contributors of this book and my family for their endless support.

Editor

Sexual selection and the evolution of male pheromone glands in philanthine wasps (Hymenoptera, Crabronidae)

Katharina Weiss, Gudrun Herzner and Erhard Strohm[*] ⓘ

Abstract

Background: Sexual selection is thought to promote evolutionary changes and diversification. However, the impact of sexual selection in relation to other selective forces is difficult to evaluate. Male digger wasps of the tribe Philanthini (Hymenoptera, Philanthinae) scent mark territories to attract receptive females. Consequently, the organs for production and storage of the marking secretion, the mandibular gland (MG) and the postpharyngeal gland (PPG), are subject to sexual selection. In female Philanthini, these glands are most likely solely subject to natural selection and show very little morphological diversity. According to the hypothesis that sexual selection drives interspecific diversity, we predicted that the MG and PPG show higher interspecific variation in males than in females. Using histological methods, 3D-reconstructions, and multivariate statistical analysis of morphological characters, we conducted a comparative analysis of the MG and the PPG in males of 30 species of Philanthini and three species of the Cercerini and Aphilanthopsini, two related tribes within the Philanthinae.

Results: We found substantial interspecific diversity in gland morphology with regard to gland incidence, size, shape and the type of associated secretory cells. Overall there was a phylogenetic trend: Ensuing from the large MGs and small PPGs of male Cercerini and Aphilanthopsini, the size and complexity of the MG was reduced in male Philanthini, while their PPG became considerably enlarged, substantially more complex, and associated with an apparently novel type of secretory cells. In some clades of the Philanthini the MG was even lost and entirely replaced by the PPG. However, several species showed reversals of and exceptions from this trend. Head gland morphology was significantly more diverse among male than among female Philanthinae.

Conclusion: Our results show considerable variation in male head glands including the loss of an entire gland system and the evolution of a novel kind of secretory cells, confirming the prediction that interspecific diversity in head gland morphology is higher in male than in female Philanthini. We discuss possible causes for the remarkable evolutionary changes in males and we conclude that this high diversity has been caused by sexual selection.

Keywords: Philanthinae, Beewolves, Sexual selection, Interspecific variation, Postpharyngeal gland, Mandibular gland, Comparative morphology, Categorical principal components analysis, Ancestral state reconstruction

Background

Ever since Charles Darwin introduced sexual selection as a distinct evolutionary force [1, 2], its importance relative to other evolutionary processes has been debated [3–10]. In particular, the potential of sexual selection as a driving force for speciation has received much attention [11–15]. Generally, sexual selection is assumed to promote rapid evolutionary change and population divergence ([8, 16–22], but see e.g. [3, 15]) due to different mechanisms like the Fisher-Zahavi processes [23–25] and sexual antagonism [26]. However, as outlined by Panhuis et al. [14], observed diversity in a trait presumably under sexual selection may also have been caused by other evolutionary forces like natural selection, genetic drift, or mutation. Hence, one major problem in the study of sexual selection is the assessment of its effect relative to other potential causes of evolutionary change [14, 27].

* Correspondence: Erhard.Strohm@ur.de
Evolutionary Ecology Group, Institute of Zoology, University of Regensburg, Universitätsstr. 31, 93053 Regensburg, Germany

Whereas the evolution of visual and acoustic courtship signals and their structural basis have been studied extensively (e.g. [28, 29]), the glands involved in the production of sex pheromones have received comparatively little attention [30] although chemical communication is probably the oldest and predominant mode of communication in most animal taxa [31]. Here we test the hypothesis that head glands of male digger wasps that are subject to sexual selection show higher interspecific diversity than the same glands in females, where they are under natural selection.

The mandibular glands (MG) and the postpharyngeal glands (PPG) of the solitary digger wasp subfamily Philanthinae (Hymenoptera, Crabronidae) are an excellent model system to study the relative contribution of sexual selection to evolutionary change since these glands occur in both sexes but are subject to different selection regimes in males and females. The Philanthinae consist of eight genera, separated into three tribes [32]: the Cercerini (comprising the three genera (Cerceris + Eucerceris) + Pseudoscolia), the Aphilanthopsini (comprising Clypeadon and Aphilanthops), and the Philanthini, the so-called beewolves (comprising (Philanthus + Trachypus) + Philanthinus, with Trachypus most probably being a subgenus of Philanthus [32, 33]). The members of the subfamily largely share basic life-history characters, in particular with regard to female nesting behavior (e.g. [34–40]) and male reproductive behavior (e.g. [37, 41–47]).

As best documented for the genus Philanthus, males establish small territories in the vicinity of female nesting aggregations (e.g. [37, 43, 44, 47]) and scent-mark their territories with a secretion from their large head glands to attract receptive females (e.g. [37, 48–50]). Scent marking and territoriality is also known from males of some species of the tribe Cercerini [41, 42, 45, 46] and at least two species of the Aphilanthopsini [41, 46]. Earlier publications on Philanthus assumed that the males' marking secretion is produced and stored in the MG (reviewed in [37]). In the European beewolf Philanthus triangulum the marking secretion is in fact most likely synthesized in the gland cells of the MG [51], but the main storage organ is the remarkably enlarged PPG [50, 52]. The MG and the PPG together are considerably larger than the brain. The huge size of the glands and the tremendous amounts of marking secretion that are produced and stored [49, 50] clearly illustrate the importance of these glands for beewolf males. Moreover, there is evidence that females prefer larger males that produce and store larger amounts of pheromone in their glands and apply more secretion to their territories (Strohm et al., unpublished).

In addition to the quantity of the marking secretion, its composition likely plays a decisive role for male attractiveness. In P. triangulum, the composition of the males' marking secretion has presumably been influenced by a female sensory bias [53–55]. Female P. triangulum use (Z)-11-eicosen-1-ol as a kairomone to identify their only prey, honeybee workers (Apis mellifera), and have evolved a high sensitivity for this compound [53]. Males exploit this pre-existing female sensory bias to increase their territories' conspicuousness to females by using (Z)-11-eicosen-1-ol as the major component of their marking secretion [49, 50, 53]. Taken together, these findings imply that both the amount and the composition of the marking pheromone are important determinants of male reproductive success. Consequently, the secretory cells that produce the marking secretion and the gland reservoirs that store it are subject to strong sexual selection.

Female Philanthinae also possess an MG and a PPG [56–59]. Females of this subfamily mass-provision subterranean brood cells with paralyzed insects as food for their progeny (e.g. [34, 35, 37, 60]). Since the larval provisions are prone to fungal infestation (e.g. [61]), at least some species of the Philanthini have evolved an intriguing defense mechanism that involves the PPG. Females literally embalm their prey with the secretion of the PPG [58, 59, 62–64]. This embalming reduces moisture on the prey's cuticle and hence delays fungal growth [61, 63, 65]. Since all Philanthini appear to face similar challenges regarding fungal infestation of larval provisions, their PPGs can be expected to be subject to similar natural selection pressures. Even though nothing is known about the function of the female MG, it is most likely also subject to natural, rather than sexual selection. The morphology of the PPG and MG has been shown to be rather uniform among female Philanthini [59].

Based on the hypothesis that sexual selection causes greater interspecific diversity than natural selection (e.g. [8, 16–18, 20]), we predict that the morphology of head glands varies more among male than among female Philanthini. Other evolutionary processes like genetic drift and mutations should affect the glands of both sexes in the same way. Since detailed morphological studies on male head glands were only available for two species of the subfamily Philanthinae, P. triangulum (MG: [51], PPG: [52]) and Cerceris rybyensis (MG [56]), we conducted a comparative analysis of the PPG and MG of male Philanthinae. Using histological methods and 3D-reconstructions, we investigated males of 30 species of Philanthini, covering all major phylogenetic lineages. Moreover, we included three species of the closely related tribes Cercerini and Aphilanthopsini. Based on 14 morphological characters, comprising incidence, location, size, shape and structure of gland reservoirs, as well as histological characteristics of associated secretory cells, we performed a multivariate statistical analysis of PPGs and MGs to assess the pattern of interspecific variation in gland morphology. In order to reveal possible phylogenetic trends, we mapped gland morphology on a recent molecular phylogeny of the Philanthinae [33]. To explore the evolutionary origin and fate of important characters, we conducted ancestral state reconstruction analyses [66]. We discuss the interspecific variation in male head gland morphology and assess the role of sexual selection in the evolution of these glands in male Philanthinae. Using the variation of female head glands [59] as a reference under natural selection, we test whether head gland morphology shows higher diversity in males.

Methods
Study material
Overall, males of 33 species and one subspecies from five genera, representing the three tribes of the crabronid subfamily Philanthinae were examined (Table 1). We refer to the phylogeny and phylogeography of the Philanthinae according to Kaltenpoth et al. [33]. Designation of zoogeographic regions follows Holt et al. [67]. Our main focus was on the tribe Philanthini,

Table 1 Species included in the comparative morphological study of head glands of male Philanthinae

Tribe	ID	Species	N	Country	3D
Cercerini	1	Cerceris quinquefasciata	2	Germany	yes
	2	Cerceris rybyensis	2	Germany	yes
Aphilanthopsini	3	Clypeadon laticinctus	5	USA	yes
Philanthini	4	Philanthinus quattuordecimpunctatus	3	Turkey	yes
	5	Philanthus cf. basalis	1	India	no
	6	Philanthus pulcherrimus	1	India	yes
	7	Philanthus spec (India)	1	India	yes
	8	Philanthus venustus	2	Turkey	yes
	9	Philanthus capensis	1	South Africa	yes
	10	Philanthus coronatus	2	Germany	yes
	11	Philanthus fuscipennis	1	South Africa	yes
	12	Philanthus histrio	2	South Africa	no
	13	Philanthus loefflingi	3	South Africa	yes
	14	Philanthus melanderi	1	South Africa	yes
	15	Philanthus rugosus	3	South Africa	yes
	16	Philanthus triangulum triangulum	3	Germany	no
	17	Philanthus triangulum diadema	3	South Africa	yes
	18	Philanthus albopilosus	2	USA	yes
	19	Philanthus barbiger	3	USA	yes
	20	Philanthus bicinctus	2	USA	yes
	21	Philanthus crotoniphilus	2	USA	yes
	22	Philanthus gibbosus	3	USA	no
	23	Philanthus gloriosus	2	USA	yes
	24	Philanthus multimaculatus	2	USA	yes
	25	Philanthus occidentalis	2	USA	no
	26	Philanthus pacificus	1	USA	yes
	27	Philanthus parkeri	1	USA	yes
	28	Philanthus politus	2	USA	yes
	29	Philanthus psyche	1	USA	yes
	30	Philanthus pulcher	1	USA	yes
	31	Philanthus ventilabris	1	USA	yes
	32	Trachypus elongatus	2	Brazil	yes
	33	Trachypus flavidus	2	Brazil	no
	34	Trachypus patagonensis	1	Brazil	no

Tribe: phylogenetic affiliation of the species, ID: identification number of the species, Species: Species name, N: number of specimens examined, Country: collection site of the species, 3D: 3D-reconstruction for this species conducted (yes) or not (no).

the so-called beewolves. The Philanthini can be grouped into five clades, largely coinciding with their geographic distribution [33] and we investigated representatives of all of these clades (Table 1): One species of the basal genus *Philanthinus*, two species of a small clade of Palearctic, Indian, and Afrotropical species of the genus *Philanthus*, forming the sister group to all other *Philanthus*, ten species of a clade comprising all other Palearctic, Indian, and Afrotropical *Philanthus*, 14 Nearctic *Philanthus* species, and three species of the Neotropical subgenus *Trachypus*. The total number of described species is four for *Philanthinus*, 136 for *Philanthus* and 31 for *Trachypus* [68, 69]. Moreover, we included three species of the two other tribes of the Philanthinae, namely one Nearctic *Clypeadon* species (tribe Aphilanthopsini, 13 described species) and two Palearctic *Cerceris* (tribe Cercerini, 905 described species). Each species under study is assigned an ID number (Table 1) that is used throughout the manuscript and Additional files.

Histology

Wasps were caught in the field in their territories or at flowers. They were cold anesthetized, decapitated and heads were fixed either in formalin-ethanol-acetic acid, alcoholic Bouin, or, in four cases, 100% ethanol [70]. After fixation, heads were rinsed, dehydrated in a graded ethanol series and propylene oxide, and embedded in Epon 812 (Polysciences Europe GmbH, Eppelheim, Germany). To facilitate the infiltration of the embedding medium into large heads, lateral parts of both compound eyes were cut off after fixation. Continuous series of sagittal semithin sections (4 µm) were cut with a microtome (Reichert Ultracut; Leica Microsystems AG, Wetzlar, Germany) equipped with a diamond knife and a large trough, mounted on microscope slides, and stained with toluidine blue [70]. The resulting series of histological sections were investigated by light microscopy (bright field, differential interference contrast, and phase contrast; Zeiss Axiophot 2; Carl Zeiss Microscopy GmbH, Oberkochen, Germany; Leica DMLS, Leica GmbH, Wetzlar, Germany).

Designation of glands was done according to the site of their openings. Reservoirs opening near the base of the mandibles were regarded as MGs and reservoirs opening to the pharynx just proximal to the hypopharyngeal plate were regarded as PPGs. Secretory cells associated with the gland reservoirs were classified according to Noirot and Quennedey [71] whenever possible; such cells will be referred to as 'NQ-class cells'. In addition, we detected presumably secretory cells not matching the classification of Noirot and Quennedey [71]. We include these cells as morphological characters in our analysis (see Morphological characters) but will provide extensive histological and ultrastructural details elsewhere.

All species also possessed a hypopharyngeal gland. We did not include this gland in our analysis, because several aspects contradict a role in territory marking: (1) the gland seems to be involved in nutrition and digestion [72–74], (2) it does not have a reservoir, and (3) using gas chromatography and mass spectrometry, we did not find volatile components in this gland (Strohm et al., unpublished).

3D-reconstruction

To visualize the overall morphology of head glands and to facilitate comparison among species, 3D-reconstructions of the head glands were generated for 27 of the 34 investigated taxa (Table 1). For two *Trachypus* and five *Philanthus* species no complete series of sections were available (Table 1); however, also for these species the available histological sections were sufficient to allow for the determination of most gland characters (see Morphological characters). Due to deficient quality of a part of the sections, reconstruction was only possible for one side of the head for *Philanthus capensis* (ID 9), *Philanthus gloriosus* (ID 23), and *Philanthus multimaculatus* (ID 24). For 3D-reconstruction, continuous series of semithin sections of one individual per species (on average 560 sections per head; 14,980 sections in total) were photographed using a digital microscope camera (Olympus DP20; Olympus, Hamburg, Germany) attached to a light microscope (Zeiss Axiophot 2) using 2.5× or 5× PlanNeofluar objectives. The digital images were automatically aligned to each other using the software TrakEM2 [75] for the image processing software Fiji [76]; all alignments were checked and manually corrected if necessary. The outer margin of the epithelium surrounding the reservoirs of the MG and the PPG as well as the pharynx were then marked as 3D-objects in TrakEM2 by manually outlining them in each picture of a series. For *Philanthus rugosus* (ID 15), additionally secretory cells of the MG and the PPG as well as the brain and the ocelli were marked. Finally, 3D-reconstructions were calculated and visualized using Fiji's 3D-viewer plug-in [77].

Statistical analysis of gland morphology
Morphological characters

Based on an extensive examination of both semithin histological sections and 3D-reconstructions, we defined 14 morphological characters of the PPG and MG for a comparative statistical analysis of the head glands of male Philanthinae. These characters comprise information on the incidence, relative size, structure and overall shape of the glands, their location within the head capsule, as well as the type and arrangement of associated gland cells. Character states were categorized and numerically coded for statistical analysis. Due to partial deficiencies in the histological sections not all character states could be determined for all species. Detailed descriptions of the characters and character states are given in section 1 of the Additional file 1. In brief, the defined characters were: (1) 'Overall structure of the PPG', (2) 'Size of the PPG relative to the head capsule', (3) 'Modifications of PPG morphology', (4) 'Branching of the PPG', (5) 'Numbers of openings of the lower part of the PPG to the pharynx', (6) 'Structure of the inner walls of the PPG', (7) 'Type of gland cells associated with the PPG', (8) 'Presence of the MG', (9) 'Overall structure of the MG', (10) 'Size of the MG relative to the head capsule', (11) 'Location of the MG in the head capsule', (12) 'Branching of the MG', (13) 'Structure of the inner walls of the MG', (14) 'Type of gland cells associated with the MG'. While the volume of a gland may vary due to differences in filling status, the longitudinal extension within the head capsule that we used as a measure of gland size is only slightly affected. If several specimens were available for a species, these had very similar morphology and did not differ with regard to the character states.

Data matrix for statistical analysis
The pronounced variation among species (see Results) required the differentiation of many character states. Since only a limited number of species could be analyzed there was only a low number of cases for some character states (see Additional file 1: section 1 and Table S1). Therefore, in addition to a dataset comprising all differences observed among species ('full dataset', Table S1), we created a second dataset, in which we pooled character states wherever reasonable ('combined dataset', Additional file 1: Table S2) and that we used for statistical analyses.

Categorical principal components analysis
To reveal patterns of character distribution among species, a categorical principal component analysis (CATPCA) was conducted using the program 'CATPCA' [78] implemented in the SPSS Categories module (SPSS version 21.0, IBM; Chicago, IL, USA). Two species were excluded from the CATPCA: For the Neotropical *Trachypus patagonensis* (ID 34) the insufficient quality of the single available series of histological sections only allowed to obtain reliable data on MG but not on PPG morphology (Table S1). Moreover, males of the Nearctic *Philanthus albopilosus* (ID 18) lacked well-developed head glands (see Results). Hence, the large difference of *P. albopilosus* to the other philanthine species would have unnecessarily lowered the quality parameters of the CATPCA solution. More details on the implementation of the CATPCA are given in section 2.1 of the Additional file 1.

To test whether there was an opposing trend between MG and PPG with regard to their size and complexity, we conducted phylogenetic generalized least squares regressions based on the molecular phylogeny of Kaltenpoth et al. [33]. As with the CATPCA, *P. albopilosus* was

excluded from this analysis, as well as the Nearctic *Philanthus gibbosus* (ID 22), for which the size of the PPG reservoir could not be assessed (Additional file 1: Tables S1 and S2). Moreover, since the molecular phylogeny comprised only one unidentified *Cerceris* species [33], we included only *C. rybyensis* (ID 2; omitting *Cerceris quinquefasciata*, ID 4). We used the package 'ape' [79] in R (Version 3.3.3, [80]) to test for a correlation between MG and PPG size and MG and PPG complexity with correction for phylogenetic relationships (for more details see section 2.2, Additional file 1).

Hierarchical cluster analysis and phylogenetic trends in gland morphology
We tested for phylogenetic trends in gland morphology using a cophylogenetic analysis between a morphology-based dendrogram resulting from a hierarchical cluster analysis (HCA) and a molecular phylogeny [33]. The HCA was based on 13 of the 14 gland characters (see section 2.2, Additional file 1) and was conducted in PAST (Version 2.08b, [81]) with the Bray-Curtis-index as a measure of dissimilarity and 'unweighted pair-group averages' as clustering algorithm; the number of bootstrap replicates was set to 10,000. Cophylogenetic analyses are mostly employed to test for coevolution of parasites and their hosts. Treating the morphology-based dendrogram as 'parasite tree' and the molecular phylogeny of the Philanthinae [33] as 'host tree', the congruence between the two was tested for statistical significance using the software tool Jane 4 [82]. Details on the implementation of the HCA and the cophylogenetic analysis are given in sections 2.3 and 2.4 of the Additional file 1.

Ancestral state reconstructions
Our investigations revealed that two major aspects of the head glands of male Philanthinae, the MG as well as the presumed secretory cells of the PPG, showed a complex phylogenetic distribution including losses and regains (see Results). Based on the molecular phylogeny [33], we conducted ancestral state reconstructions (ASR) [66] for the presence of both the MG (character 8, Additional file 1: Tables S1 and S2) and the secretory cells of the PPG (state 0 vs. all other states of character 7, Tables S1 and S2) using the software tool Mesquite (Version 3.04, [83]). As above, since the molecular phylogeny comprised only one unidentified *Cerceris* species [33], we conducted the ASR with only *C. rybyensis* (ID 2) and *Clypeadon laticinctus* (ID 3) as outgroup species (omitting *Cerceris quinquefasciata*, ID 4). We applied maximum likelihood (ML) approaches using asymmetrical Markov k-state 2 parameter models with the rate of change between the two character states (i.e. absence vs. presence of the MG and the secretory cells of the PPG, respectively) set to 1. Since for both traits the likelihood of

gain vs. loss is not known, we tested different bias ratios for gains vs. losses ranging from 10 (i.e. gains ten times more frequent than losses) to 0.1 (i.e. losses ten times more frequent than gains).

Comparison of morphological diversity in males and females

To formally evaluate the hypothesis that the diversity among males is larger than among females, we compiled an aggregated matrix of gland characters of males of 32 species and females of 28 species (data for females taken from [59], Additional file 2: Table S4). Most characters are shared by both sexes and the respective character states could be simply combined. However, some characters or character states had to be recoded because they were assessed differently in the sexes or the character states were more finely differentiated in females. Based on the aggregated matrix, we conducted a CATPCA as described above to illustrate the distribution of males and females with regard to their gland morphology. To test for a difference in diversity between males and females, we calculated Shannon diversity indices among the character states of the characters that occur in both sexes (four characters that are restricted to either males or females had to be omitted) and compared these values using an exact Wilcoxon matched pair test. For more details see Additional file 2.

Results
General aspects of gland morphology

In all species of Philanthinae under study, males possess either an MG, or a PPG, or both and, with one exception (*P. albopilosus*, ID 18), at least one of these glands occupies a considerable part of the head capsule. Nineteen of the 33 investigated species possess an MG that is located in the front part of the head capsule anterior to the brain and, depending on its size, may extend behind the brain, lateral from or subjacent to the PPG. The MG comprises paired reservoirs opening at the dorsal side of the mandible base and extending laterally and dorsally on both sides of the head capsule, in some cases even reaching behind the brain (Fig. 1, *P. rugosus*). Some species have only a lower MG reservoir opening at the ventral side of the mandible base and extending backwards. A few species possess both parts. MG reservoirs are surrounded by a monolayered epithelium that is moderately thick in most species. However, in some species with only an upper MG, the epithelium is distinctly thinner. The epithelial cells bear an apical cuticular intima that regularly forms a variety of conspicuous structures. Moreover, there is interspecific variation with regard to the types of secretory cells associated with the MG (see below).

All 33 investigated species possess a PPG, clearly identified by its connection to the pharynx anterior to the

brain and posterior to the hypopharyngeal plate (Fig. 1). The PPG also shows considerable interspecific variation. In most species, the main upper part of the PPG basically consists of two pairs of lateral evaginations: one pair extending dorsally and in some species even around the brain (dPPG in Fig. 1) and a second pair located anterior to the brain and extending laterally towards the ventral rims of the compound eyes (aPPG in Fig. 1, for the delineation of the two parts see also Additional file 1: Figure S1 A and B). The anterior part may reach the compound eyes and, in some species, the base of the mandibles. In 14 of the 33 investigated species, there is an additional, smaller, lower part of the PPG consisting of an unpaired ventral evagination of the pharynx (Additional file 1: Figure S1 F). The walls of all parts of the PPG consist of a (partly very thin) monolayered epithelium with an apical cuticular intima. The epithelial cells generally bear hairs or scales that extend into the lumen of the gland.

The reservoirs of both glands may be associated with different types of cells (Fig. 2 and Additional file 1: Figure S2; see also Fig. 1 for the location of the cells). These cells presumably have secretory functions given their close proximity or direct contact to the reservoirs and the abundance of vesicles and nucleoli (Fig. 2). The gland cells of the MG can be differentiated into three types. In some species there are typical NQ-class 3 cells [71], i.e. complexes of a secretory cell and a canal cell, the latter forming conspicuous end apparatus and canals that connect the secretory cell to the lumen of the MG (Fig. 2a). In other species, several NQ-class 3 cells are aggregated in acini (Fig. 2b). The third type comprises secretory cells that are located directly at the wall of the reservoir and bear end apparatus but no canals (Fig. 2c). Though these cells appear to be complexes of two cells, thus resembling NQ-class 3 cells, we assign them to a different character state to account for the lack of visible canals (see also section 1, Additional file 1).

The cells associated with the PPG can occur either as aggregations of mononuclear cells (superficially resembling the acini of the MG) (Fig. 2d) or as multinuclear syncytia (Fig. 2e) (see below and Additional file 1: section 1), both showing clear signs of secretory activity: large nuclei, conspicuous nucleoli and numerous vesicles (black arrowheads in Fig. 2e). However, these cells are clearly not NQ-class 3 cells, since they lack an end apparatus and canals. Moreover, they are not part of the gland epithelium and are, thus, not NQ-class 1 cells either. Remarkably, the PPG reservoir itself is extensively ramified with the thinnest branches reaching into the cell aggregations or syncytia (black arrow and inset in Fig. 2e). In some species the cell aggregations or syncytia are interspersed with conspicuous small rounded cells with barely any cytoplasm (white arrows in Fig. 2d and f).

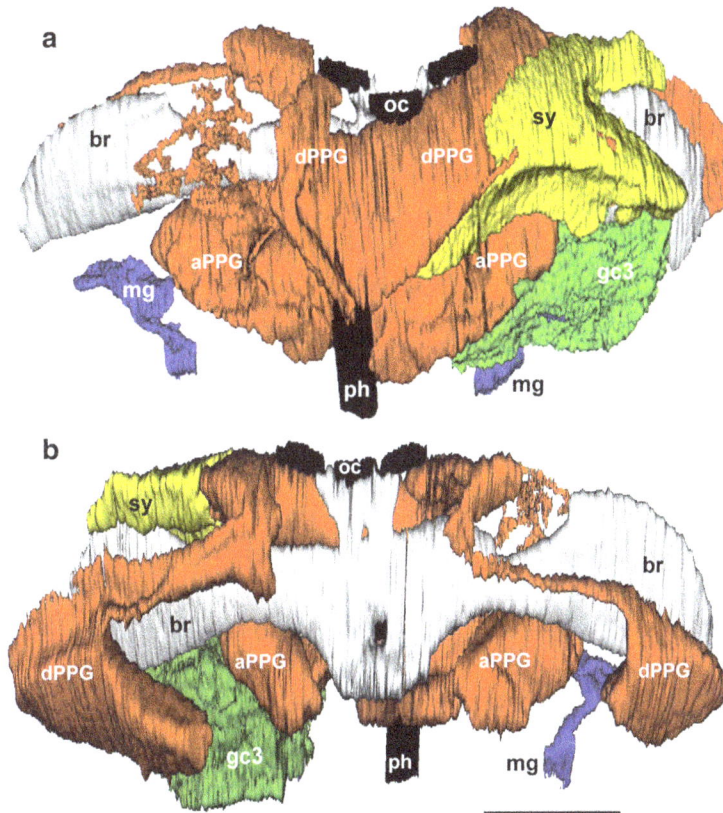

Fig. 1 3D-reconstruction of the internal structures of a male *Philanthus rugosus* head. **a** Anterior view, **b** posterior view. The upper postpharyngeal gland reservoir (PPG; *orange*) originates dorsally from the pharynx (*black*) and basically consists of two pairs of lateral evaginations, one extending dorsally around the brain (*light grey*) (dPPG; see also Additional file 1: Figure S1 A) and one extending laterally anterior to the brain (aPPG; see also Additional file 1: Figure S1 B). The fine branches originating from the dorsal part of the upper PPG (see also Additional file 1: Figure S1 C) are surrounded by syncytia of secretory cells (*yellow*, shown only for the *left side* of the head). The upper mandibular gland reservoirs (MG; *blue*) have their openings at the dorsal mandibular base and extend laterally. The MG is associated with single NQ-class 3 gland cells (*green*, shown only for the *left side* of the head capsule). Abbreviations: aPPG, anterior parts of the upper PPG reservoir; br, brain; dPPG, dorsal parts of the upper PPG reservoir; gc3, single NQ-class 3 gland cells associated with the MG; mg, upper MG reservoir; oc, ocelli; ph, pharynx; sy, syncytia of secretory cells associated with the fine branches of the dorsal part of the upper PPG. Scale bar = 0.5 mm

Pattern of interspecific variation in gland morphology

Both PPG and MG show remarkable interspecific variation with respect to their incidence, size and shape (Fig. 3), as well as the fine structure of the gland reservoirs and the type and arrangement of the associated secretory cells (Fig. 2). Character states for the species under study are given in Additional file 1: Tables S1 and S2. The CATPCA analysis based on 11 morphological characters (Table S2) sorted the species under study into three well defined groups (I-III, see below) and two species largely separated from these groups (Fig. 4). The first two dimensions of the CATPCA together explained 94% (63% and 31%, respectively) of the variance in the dataset and were supported by a total Cronbach's α of 0.99 (maximum value = 1), indicating the high reliability of the detected pattern in the dataset [84]. Size and complexity of MG and PPG strongly contribute to the separation of the groups, and their vectors point in opposite directions. Yet, according to phylogenetic independent regression analyses there was no significant correlation between size ($N = 30$, $r = -0.47$, $p = 0.136$) or complexity ($N = 30$, $r = -0.6$, $p = 0.14$) of MGs and PPGs across species. However, due to the comparatively small set of species in our analysis [85] this result bears some uncertainty.

Group I: Species possessing large MGs but only small PPGs
The first group of species as assigned by the CATPCA (Fig. 4) is characterized by complex and large MGs, but only small and simple PPGs. In all species of this group, the MG reservoir (turquoise in Fig. 3) opens at the dorsal side of the mandible base and is bordered by a rather thin monolayered epithelium in direct contact with gland cells that show the typical end

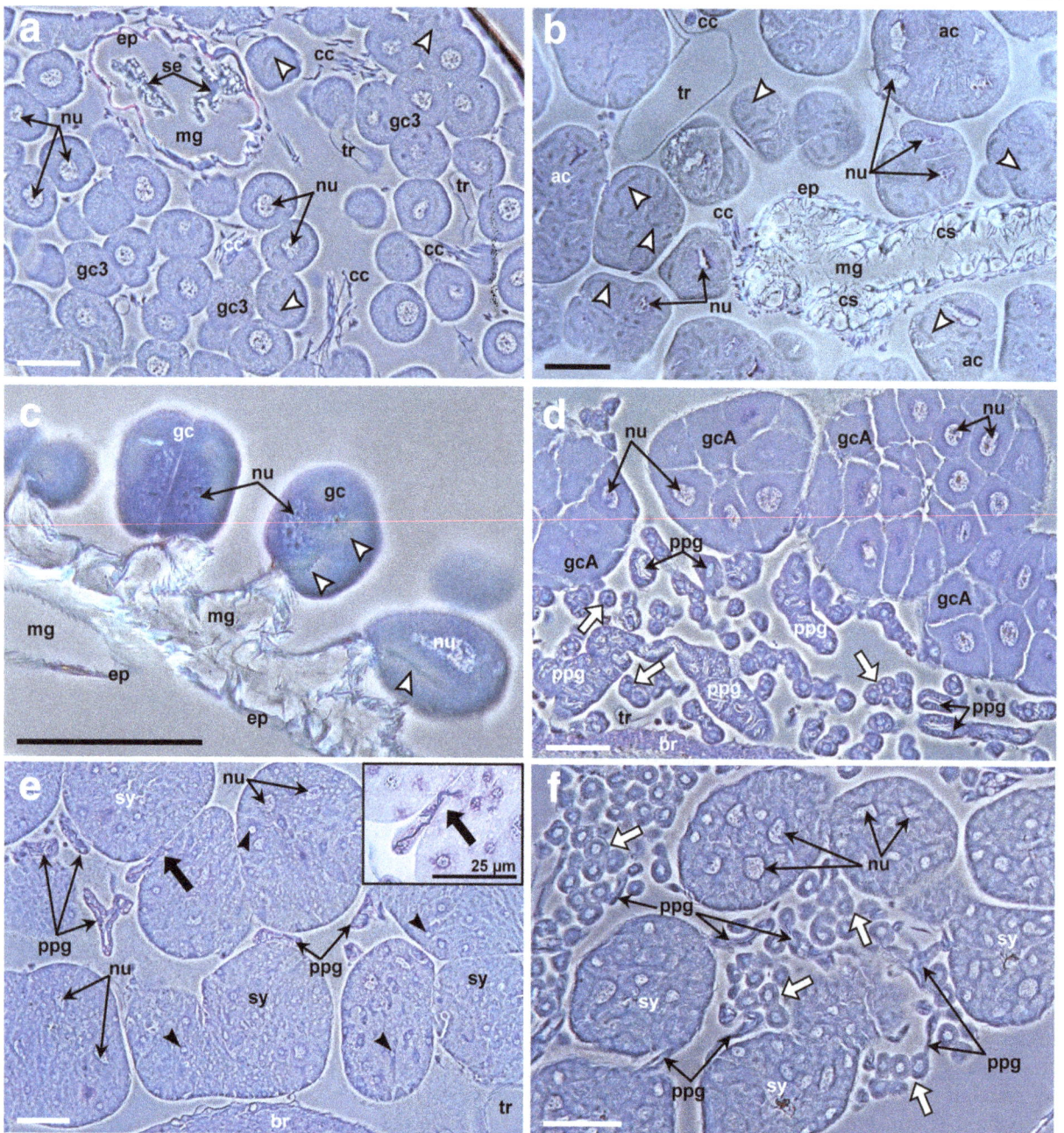

Fig. 2 Semithin sagittal sections through the heads of male Philanthinae. **a** Single NQ-class 3 gland cells, i.e. complexes of a secretory cell and a canal cell, the latter forming a conspicuous end apparatus (*white arrow heads*) and canal that connects the secretory cell to the lumen of the MG (*Philanthus multimaculatus*, ID 24); **b** Acini of NQ-class 3 gland cells with end apparatuses (*white arrow heads*) connected to the MG reservoir by bundles of conducting canals (*Philanthus T. diadema*, ID 17); **c** Single gland cells possessing end apparatus (*white arrow heads*), thus resembling NQ-class 3 cells, but directly associated with the wall of the MG reservoir without canals (*Clypeadon laticinctus*, ID 3); **d** Aggregations of mononuclear secretory cells surrounding the fine branches of the PPG reservoir, interspersed with small rounded cells (*white arrows*) (*Philanthus venustus*, ID 8); **e** Multinuclear syncytia of secretory cells, containing many vesicles (*black arrow heads*), and in close contact to the fine branches of the PPG reservoir (*thick black arrow*; inset: detail of a PPG branch terminating in syncytium) (*Philanthus histrio*, ID 12); **f** Multinuclear syncytia of secretory cells surrounding the fine branches of the PPG reservoir and interspersed with small cells (*white arrows*) (*Philanthus crotoniphilus*, ID 21). Abbreviations: ac, acini of NQ-class 3 cells; br, brain; cc, conducting canal; cs. cuticular spines; ep, epithelium of the MG; gc. secretory cells not resembling NQ-class cells; gcA, aggregations of mononuclear secretory cells; gc3, NQ-class 3 gland cells; mg, lumen of the mandibular gland; nu, nucleus with nucleoli; ppg, fine branches of the postpharyngeal gland; se, secretion within the MG; sy, multinuclear syncytia; tr, tracheole. Scale bars [except inset in (E)] = 50 μm

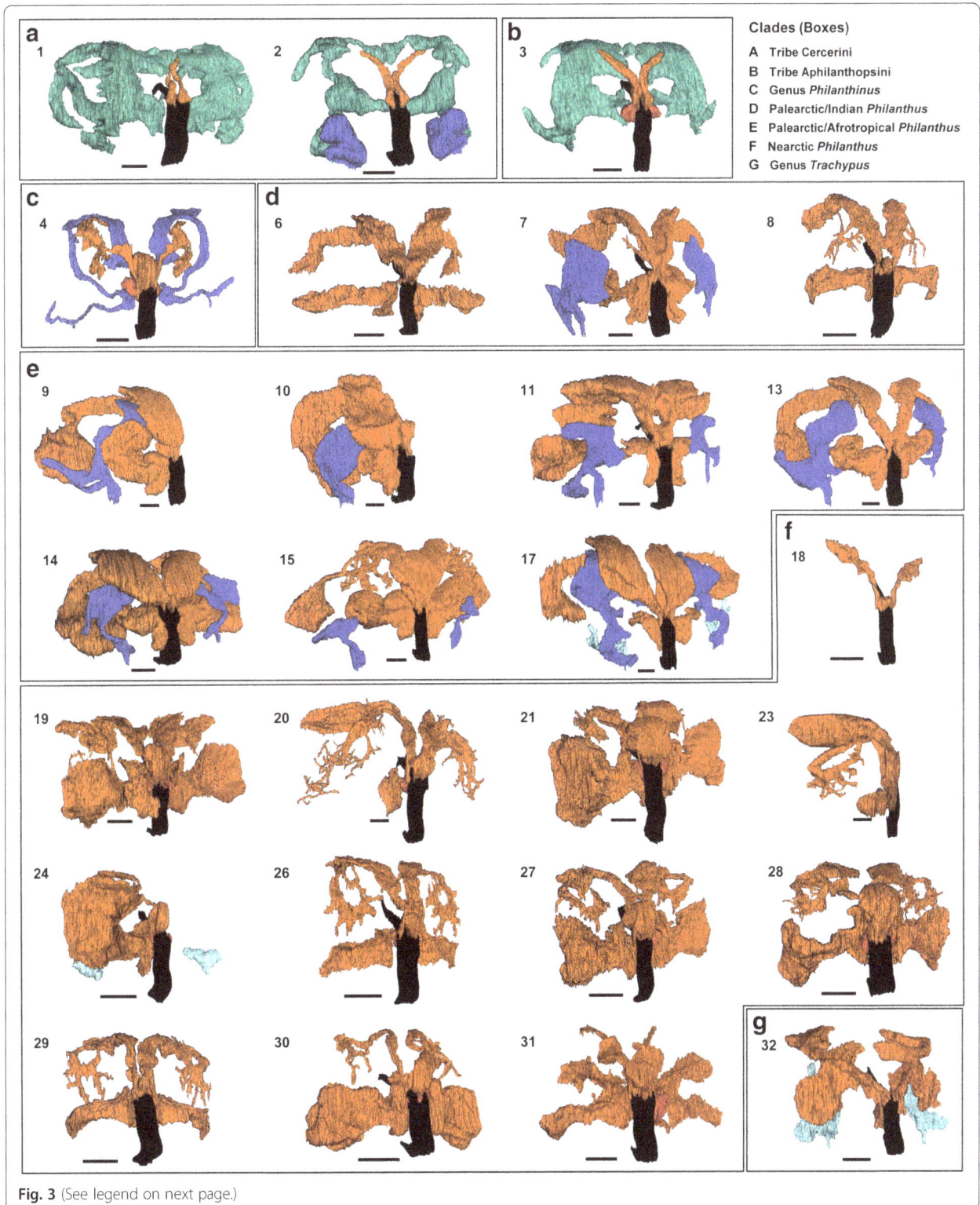

Fig. 3 (See legend on next page.)

(See figure on previous page.)
Fig. 3 3D-reconstructions of the postpharyngeal gland (PPG) and the mandibular gland (MG) of male Philanthinae. Species IDs (corresponding to Table 1): (1) *Cerceris quinquefasciata*, (2) *Cerceris rybyensis*, (3) *Clypeadon laticinctus*, (4) *Philanthinus quattuordecimpunctatus*, (6) *Philanthus pulcherrimus*, (7) *Philanthus spec.* (India), (8) *Philanthus venustus*, (9) *Philanthus capensis*, (10) *Philanthus coronatus*, (11) *Philanthus fuscipennis*, (13) *Philanthus loefflingi*, (14) *Philanthus melanderi*, (15) *Philanthus rugosus*, (17) *Philanthus triangulum diadema*, (18) *Philanthus albopilosus*, (19) *Philanthus barbiger*, (20) *Philanthus bicinctus*, (21) *Philanthus crotoniphilus*, (23) *Philanthus gloriosus*, (24) *Philanthus multimaculatus*, (26) *Philanthus pacificus*, (27) *Philanthus parkeri*, (28) *Philanthus politus*, (29) *Philanthus psyche*, (30) *Philanthus pulcher*, (31) *Philanthus ventilabris*, (32) *Trachypus elongatus*. Boxes a - g indicate phylogeographic classification of species (according to [33], see key in figure). Color code for 3D-structures: *orange*, upper part of the PPG; *red*, lower part of the PPG; *dark blue*, upper part of the MG; *light blue*, lower part of the MG, turquoise, thin-walled MG reservoir of the Cercerini and Aphilanthopsini; *black*, pharynx. Due to limited availability of serial histological sections, for species (9), (10), and (23) only the right side of the paired gland reservoirs could be reconstructed, while for species (24), both reservoirs of the MG, but only the right half of the PPG are depicted; for species (6), (7), (10)-(15), and (31), the fine branches originating from the main PPG reservoir [see e.g. species (8) and (20)] could not be reconstructed based on semithin section due to their very fine structure and high number. Scale bars = 0.25 mm

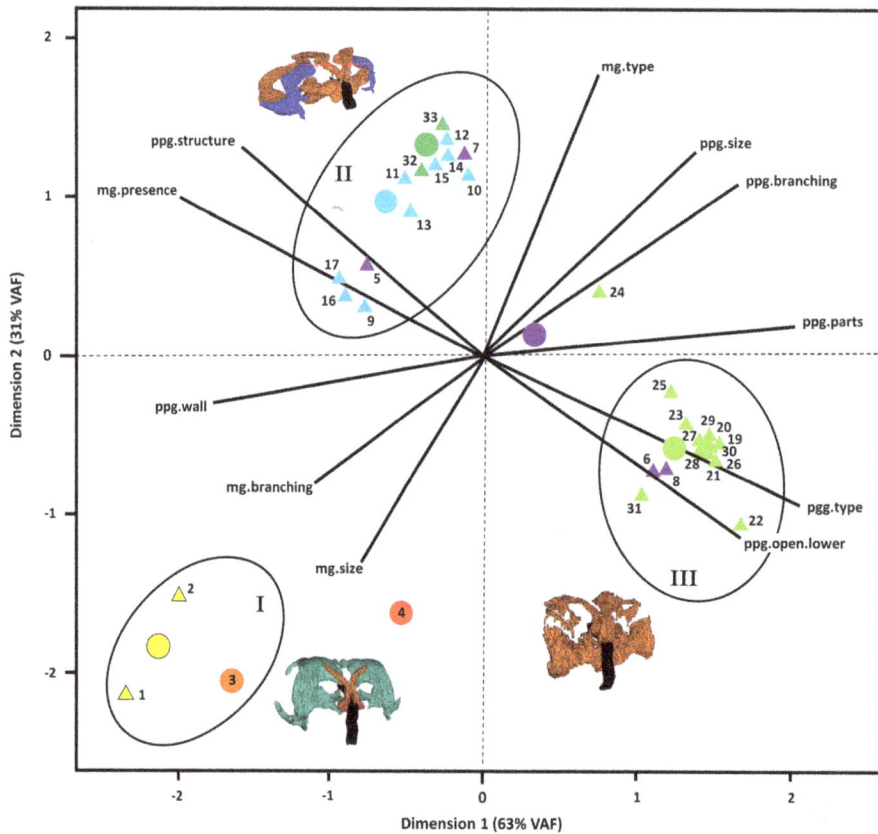

Fig. 4 First two dimensions (VAF: percent of variance accounted for) of the CATPCA of the head gland morphology of male Philanthinae. Based on the morphology of their head glands, the species form three distinct groups (ellipses; exemplary 3D-reconstructions: (I) *Cerceris quinquefasciata*, (II) *Philanthus rugosus*, (III) *Philanthus politus*). Triangles: object scores of single species (IDs correspond to Table 1), *Vectors*: component loadings of morphological characters, *Circles*: Group centroids of the different phylogenetic and phylogeographic clades (according to [33]) included as a supplementary variable. Note that for each of the two genera *Clypeadon* (*orange*) and *Philanthinus* (*red*) only one species was included in the analysis, thus, their object scores are identical to their group centroids. Color code: *yellow*, genus *Cerceris*; *orange*, genus *Clypeadon*; *red*, genus *Philanthinus*; *purple*, Palearctic/Asian *Philanthus*; *blue*, Palearctic/Afrotropical *Philanthus*; *light green*, Nearctic *Philanthus*; *dark green*, genus *Trachypus*. Abbreviations of morphological characters (numbering corresponds to section 2.4.1): ppg.structure, (1) overall structure of the PPG; ppg.size, (2) size of the PPG relative to the head capsule; ppg.parts, (3) modifications of PPG morphology; ppg.branching, (4) branching of the PPG; ppg.open.lower, (5) numbers of openings of the lower part of the PPG to the pharynx; ppg.wall, (6) structure of the inner walls of the PPG; ppg.type, (7) type of gland cells associated with the PPG; mg.presence, (8) presence of the MG; mg.size, (10) size of the MG relative to the head capsule; mg.branching, (12) branching of the MG; mg.type, (14) type of gland cells associated with the MG

apparatus of NQ-class 3 cells, but no canals (Fig. 2c). *Cerceris rybyensis* (ID 2) additionally possesses a second reservoir (dark blue in Fig. 3) with a distinctly thicker, yet likewise monolayered epithelium and being exceptional in having two openings, one dorsally and one ventrally at the mandibular base. This additional reservoir is associated with typical NQ-class 3 cells with end apparatus and canals. The small PPG reservoirs of group I species are not associated with any cells that show signs of secretory activity. Notably, group I solely comprises the three investigated species of the tribes Cercerini and Aphilanthopsini (IDs 1-3).

Group II: Species possessing both well-developed MGs and PPGs

The second group comprises 12 species (including the two subspecies of *P. triangulum*, IDs 16 and 17) (Fig. 4) that possess both large, complex PPGs and mostly medium-sized, yet well-developed MGs with fairly thick epithelia. Most members of this group possess only the upper part of the MG (dark blue in Fig. 3), whereas *Philanthus* cf. *basalis* (ID 5), *P. t. triangulum* (ID 16), and *P. T. diadema* (ID 17) possess both upper and lower parts and *Trachypus elongatus* (ID 32) possesses only the lower part of the MG (light blue in Fig. 3). In nine species of group II, the MG is associated with acini made up of NQ-class 3 cells with canals jointly connecting an acinus with the reservoir (Fig. 2b). Yet, the closely related *Philanthus histrio* (ID 12) and *P. rugosus* (ID 15) as well as the two *Trachypus* species (IDs 32 and 33) possess single NQ-class 3 cells (Fig. 2a).

In eight species of group II, the PPG reservoir is extensively ramified and associated with cells that show clear signs of secretory activity. In seven of these species the cells at the PPG are syncytia (Fig. 2e); only in *Trachypus flavidus* (ID 33) these cells are aggregations of mononuclear cells. The remaining five species of group II, *P.* cf. *basalis* (ID 5), *P. capensis* (ID 9), *P. t. triangulum* (ID 16), *P. T. diadema* (ID 17), and *T. elongatus* (ID 32) possess large, un-ramified more or less tube-shaped PPGs and neither the cells of the PPG epithelium nor surrounding cells show signs of secretory capacity. Group II comprises all but two of the investigated Palearctic, Indian, and Afrotropical species of the genus *Philanthus* (IDs 5, 7 and 9-17), as well as the two Neotropical species *T. elongatus* (ID 32) and *T. flavidus* (ID 33).

Trachypus patagonensis (ID 34) that was not included in the CATPCA (see "Data matrix for statistical analysis") would probably also be placed in this group. Its MG consists of both upper and lower part associated with single NQ-class 3 cells and its PPG is tubular and not associated with secretory cells.

Group III: Species with large, complex PPGs but no MGs

The third group is rather narrowly defined and comprises 14 *Philanthus* species characterized by completely lacking an MG but possessing large and extensively ramified PPGs (Fig. 3) associated with secretory cells. *Philanthus venustus* (ID 8) deviates from the other members of group III in that the secretory cells of its PPG are not syncytia but aggregations of mononuclear cells (Fig. 2 d), similar to *T. flavidus* (ID 33) in group II. Only in species of group III are the syncytia or cell aggregations associated with the PPG branches interspersed with small rounded cells with barely any cytoplasm (white arrows in Fig. 2d and f). Most species of group III have a Nearctic distribution, the exceptions being the Indian *Philanthus pulcherrimus* (ID 6) and the Palearctic *Philanthus venustus* (ID 8).

Divergent species

Two species included in the CATPCA are separated from the three main groups. One is *P. multimaculatus* (ID 24), the only Nearctic species in our dataset whose males have an MG. Like the Neotropical *T. elongatus* (ID 32) it has only the lower part of the MG (Fig. 3). In the CATPCA it is located between its MG-less Nearctic relatives of group III and the Afrotropical, Palearctic and Neotropical species of group II that all possess MGs. The second separated species is *P. quattuordecimpunctatus* (ID 4), whose males have a well-developed tube-shaped MG, associated with cells akin to NQ-class 3 gland cells that, however, lack conducting canals, resembling group I in this respect. The upper part of their PPG extends backwards around the brain, like in the species of group II, and is not associated with any secretory cells. Moreover, the PPG of *P. quattuordecimpunctatus* is unique among all investigated species in that its reservoir consists of an complex network of lamellae (not shown) as opposed to the tubular ramifications of the other species.

Philanthus albopilosus (ID 18; not in CATPCA, see "Categorical principal components analysis" and Discussion) stands out from all other species. Its males not only completely lack an MG, like most of their Nearctic congeners, but also have a largely reduced PPG that consists of only small evaginations of the pharynx (Fig. 3) without any secretory cells, similar to the PPGs of group I.

Phylogenetic trend in gland morphology

As summarized in Fig. 5, the gland morphology of male Philanthinae partly coincided with phylogenetic groups, but there is also considerable diversity within clades and several species deviate from their closest relatives. To test whether there is an overall phylogenetic trend in gland morphology we conducted a HCA (Additional file 1: Figure S3) based on the morphological characters of MG and PPG and compared the resulting dendrogram with the molecular phylogeny of the Philanthinae [33]. The HCA largely corroborated the pattern found in the CATPCA (for details on the clustering of species see

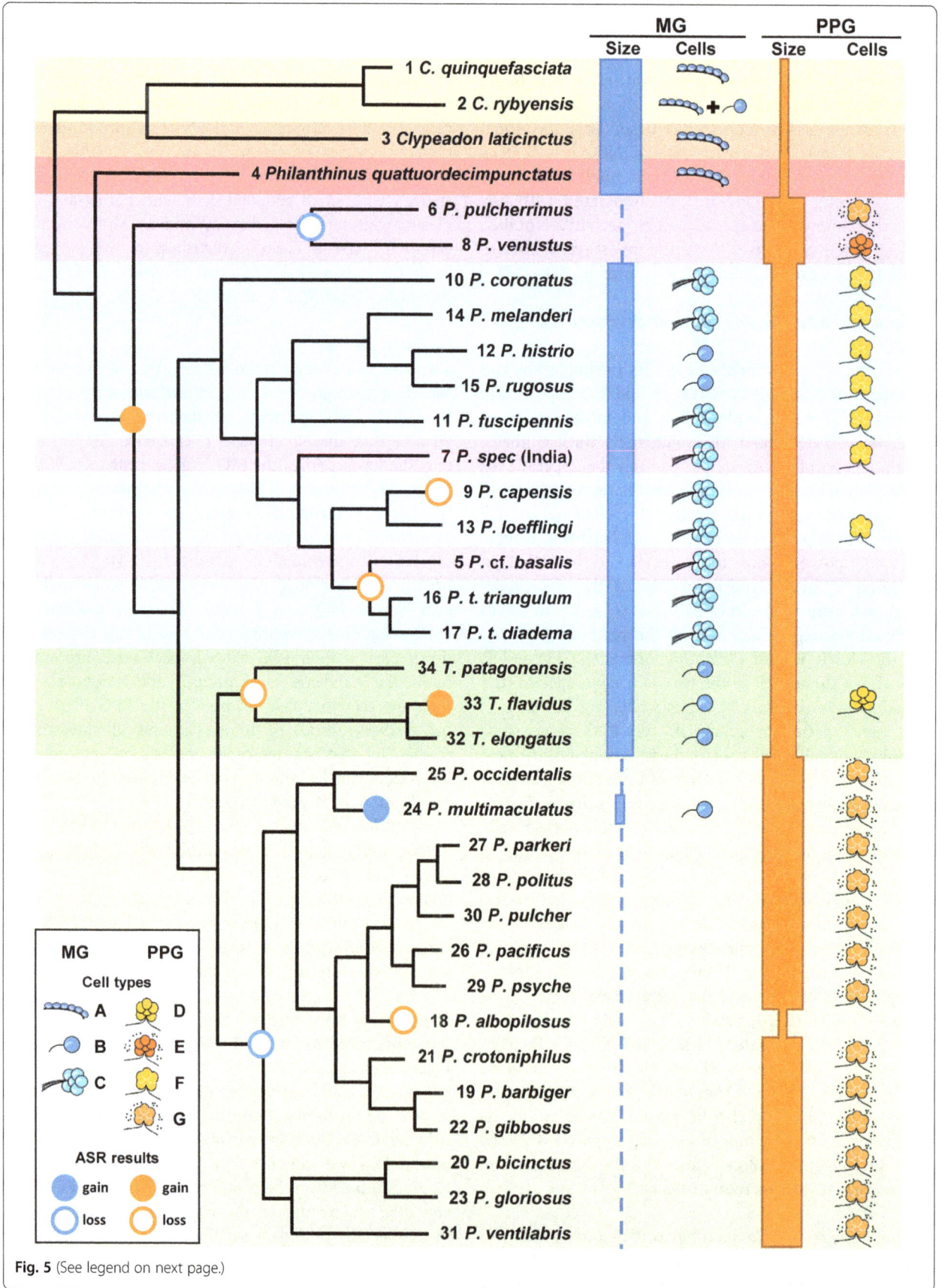

Fig. 5 (See legend on next page.)

Fig. 5 Summary of the phylogenetic trends and deviations in size of the MG and PPG (*bar thickness* indicates relative size, dotted line indicates absence) and type of associated gland cells ('Cells') among male Philanthinae. *Circles at nodes* indicate likely events of gain or loss of the MG or the PPG (symbols see key in figure; for more details see Additional file 1: Figure S4 and S5). Species IDs correspond to Table 1. Color code of phylogeographic clades (according to [33]): *yellow*, genus *Cerceris*; *orange*, genus *Clypeadon*; *red*, genus *Philanthinus*; *purple*, Palearctic/Asian *Philanthus*; *blue*, Palearctic/Afrotropical *Philanthus*; *light green*, Nearctic *Philanthus*; *dark green*, genus *Trachypus*. Pictograms of cell types (labeling see key in figure): (A) single gland cells, showing end apparatuses but directly associated with the wall of the MG reservoir without canal cells (B) single NQ-class 3 gland cells; (C) acini of several NQ-class 3 cells with bundles of conducting canals; (D) aggregations of several gland cells directly associated with very fine branches of the PPG; (E) as in (D), but interspersed with small rounded cells; (F) syncytia of secretory cells directly associated with very fine branches of the PPG; (G) syncytia as in (E), but interspersed with small rounded cells. Dendrogram modified after the molecular phylogeny of Kaltenpoth et al. [33]

Additional file 1). Notably, the HCA dendrogram shows a highly significant congruency with the molecular phylogeny (cophylogenetic analysis, all tested parameter combinations: $p < 0.001$).

Phylogenetic history of the MG
The MG shows a complex phylogenetic pattern of incidence among male Philanthinae (Fig. 5). A maximum likelihood ASR using an unbiased model (bias = 1, Additional file 1: Figure S4) suggests the presence of an MG as the ancestral state of the subfamily Philanthinae as well as of the tribe Philanthini and of the genus *Philanthus* (including *Trachypus*). Accordingly, the MG would have been independently lost twice within the genus *Philanthus*, namely in the last common ancestor of the clade comprising *P. pulcherrimus* (ID 6) and *P. venustus* (ID 8) (ML probability 96%) and in the last common ancestor of the Nearctic *Philanthus* species (ML probability 100%) (Additional file 1: Figure S4). In the Nearctic *P. multimaculatus* (ID 24), however, the MG must have been regained (Additional file 1: Figure S4). This result did not change if losses were assumed to be more frequent than gains (bias <1) and also if gains were assumed to be slightly more likely than losses (up to a bias of 1.5). Varying the bias further in favor of gains (bias ≥2.3), however, led to ambiguous or deviating results for some nodes within the genus *Philanthus* (including *Trachypus*) (Additional file 1: Figure S4). Yet, even with a bias of 10 the analysis indicated the presence of an MG as the ancestral state for both the Philanthinae and the Philanthini (Additional file 1: Figure S4).

Phylogenetic history of the secretory cells of the PPG
The phylogenetic pattern of the presumed secretory cells of the PPG is even more complex (Fig. 5). According to an unbiased maximum likelihood ASR (bias = 1), the secretory cells of the PPG were most likely absent in the last common ancestors of both the Philanthinae and the Philanthini and first occurred in the last common ancestor of *Philanthus* (including *Trachypus*) (ML probability 100%; Fig. 5, Additional file 1: Figure S5). Within *Philanthus/Trachypus*, the secretory cells would then have

been independently lost four times (Fig. 5, Additional file 1: Figure S5), namely in the last common ancestor of the Neotropical *Trachypus* (IDs 32-34) (ML probability 71%), in the Nearctic *P. albopilosus* (ID 18), in the last common ancestor of the clade containing the two subspecies of *P. triangulum* (IDs 16 and 17) and *P.* cf. *basalis* (ID 5) (ML probability 87%) as well as in the related Paleotropical *P. capensis* (ID 9), while the closely related *Philanthus loefflingi* (ID 13) has retained the secretory cells. Hence, one species, *T. flavidus* (ID 33), must have regained the secretory cells of the PPG (Additional file 1: Figure S5). Varying the bias strongly in favor of gains over losses (bias ≥4), resulted in a somewhat different evolutionary scenario in that the secretory cells of the PPG would have been lost in the last common ancestor of both the *P. capensis*-clade and the *P. triangulum*-clade and then regained in *P. loefflingi* (Additional file 1: Figure S5).

Comparison of morphological diversity in males and females
The CATPCA based on the aggregate matrix of character states for males and females reveals a clear distinction between the sexes (Additional file 2: Figure S6). Whereas data points for females are largely clumped, the data points for males are much more scattered and show two main aggregations similar to the CATPCA including only males. Diversity estimates of character states among characters of gland morphology were significantly higher in males (mean ± s.d.: 1.12 ± 0.27) than in females (0.34 ± 0.32; Wilcoxon matched pairs test: $N = 9$ characters, exact $p = 0.004$).

Discussion
There are several comparative phylogenetic studies on secondary sexual traits (e.g. [28, 29, 86–90]), but the present study is, to our knowledge, the first comparative histological study on insect exocrine glands that are under sexual selection. Males of all but one of the investigated Philanthinae bear enormous and elaborate head glands that are considerably larger and more complex than in females of any species of this subfamily [59]. The exaggeration of the male glands emphasizes their significance for mate attraction and the strength of sexual selection acting upon them.

Our comprehensive investigation revealed considerable interspecific variation with numerous species deviating from their close relatives with regard to gland occurrence, size and morphology, as well as the incidence, specific type and arrangement of associated secretory cells. Nevertheless there was a clear phylogenetic trend in gland morphology (summarized in Fig. 5). Ensuing from a plesiomorphic state, two main evolutionary trends emerge: First, the PPG increases in size and complexity and becomes involved in the production and storage of the marking secretion. Second, the MG, in return, decreases in size and is eventually completely lost.

The plesiomorphic state of the Philanthinae
About half of the species under study lacked an MG. To shed light on the plesiomorphic state of the subfamily, we conducted an ancestral state reconstruction. The most likely scenario is the presence of an MG in the predecessor of the Philanthinae, of the Philanthini and of the genus *Philanthus* (including *Trachypus*) with repeated losses in single lineages and one gain (Fig. 5, Additional file 1: Figure S4). This view is corroborated by the fact that such ectal MGs [91–93] as found in male Philanthinae occur in all major lineages of the Aculeata (bees: e.g. [94–96], apoid wasps: [93, 97], vespid wasps: [92], ants: e.g. [98, 99]) and in parasitoid wasps [100, 101]. Moreover, females of all investigated Philanthinae have MGs, albeit small [59], indicating that the genetic information to develop this gland is present throughout the subfamily. Males of all studied species possess a PPG that is probably homologous to the PPGs of ants (Formicidae) [52, 102] and the cockroach wasp *Ampulex compressa* (Ampulicidae) [103]. In the majority of species under study, the PPG is associated with secretory cells. An ASR for the occurrence of these cells revealed that they were probably absent in the last common ancestor of Philanthinae and Philanthini. Accordingly, these cells must have evolved in the last common ancestor of *Philanthus/Trachypus*, but were lost several times within this taxon and regained at least once (Fig. 5, Additional file 1: Figure S5).

The inferred plesiomorphic state of male Philanthinae is represented by the investigated Cercerini and Aphilanthopsini with their large MGs and small PPG reservoirs devoid of secretory cells (Fig. 5). The MGs of these species share a type of gland cells that bear end apparatus but, in contrast to typical NQ-class 3 cells, do not show canals. Such gland cells seem uncommon, but have been described for an ant [104] and some bee species [97]. Notably, there is some variation among the Cercerini in that *C. rybyensis* males have an additional part of the MG with typical NQ-class 3 cells. Our results on *C. rybyensis* are largely consistent with Ågren [56], who, however, did not mention the gland cells with end

apparatus but no canals. The enormous size of the MG reservoir of male Cercerini and Aphilanthopsini and the high number of associated secretory cells suggest that the function of the MG comprises both production and storage of the male marking secretion. In other taxa of Hymenoptera, the MG is known as source of different pheromones like male and female sex pheromones [101, 105–108], the queen pheromone in honeybees (*A. mellifera*) (e.g. [109, 110]) and alarm pheromones in different ants (e.g. [111–113]). The MG can also be the source of defensive secretions in parasitoid wasps [101], bees [114, 115], and ants [116].

The PPGs found in male Cercerini and Aphilanthopsini in the present study largely resemble the PPGs of the respective conspecific females [59]. Moreover, the shape and structure is quite similar to the PPGs of both sexes of the cockroach wasp *A. compressa* [103], a rather basal taxon within the Apoidea [117, 118]. Notably, a PPG had not previously been described for male Cercerini and Aphilanthopsini and currently no information is available on their chemistry. Considering their small size and the lack of secretory cells, we hypothesize that in these tribes the males' PPGs do not play an important role in the production and/or storage of a marking secretion. Instead, as suggested for *A. compressa* [103], the PPG may function as a hydrocarbon reservoir. Until recently, a PPG was only known from ants where it mainly serves to generate the colony odor that is also based on hydrocarbons [119–122] (for a review of other functions of the PPG in ants, see [123]). Such a "social function" of the PPG can be ruled out for the solitary Cercerini and Aphilanthopsini.

The involvement of the PPG
The head glands of male Philanthini differ markedly from the Cercerini and Aphilanthopsini since their MGs are more or less reduced and their PPGs are typically considerably larger and more complex (Fig. 5). Like in most Hymenoptera (e.g. [74, 100, 101, 124–126]), the MGs of male Philanthini are exclusively associated with typical NQ-class 3 cells, either in single units or arranged in acini. As in *P. triangulum* [50–52], the male MG of other Philanthini is presumably also involved in the production of the marking pheromone.

Taking into account its position at the very base of the Philanthini, the genus *Philanthinus* may be expected to represent an intermediate state between the Cercerini and Aphilanthopsini and the Philanthini. In fact, the somewhat smaller MG with typical NQ-class 3 gland cells and the large PPG of *P. quattuordecimpunctatus* (Fig. 3) support this view. However, in contrast to most *Philanthus* species its PPG is not associated with secretory cells (Fig. 5) and *P. quattuordecimpunctatus* stands out from all other Philanthinae with regard to the structural organization of the PPG in lamella-like branches.

In the genus *Philanthus*, males of nearly all studied species possess at least moderately large PPGs (Fig. 5, see also Fig. 3). In most species these PPGs shows extensive ramifications that are closely associated with cells (syncytia or, rarely, cell aggregations) that show clear signs of secretory activity, like large nuclei with several nucleoli and numerous vesicles. Even though these cells do not conform to any previously described type of secretory cell [71, 127, 128], we hypothesize that they synthesize compounds of the marking secretion that are transferred to the PPG reservoir, where they are stored until release during territory marking. How these cells evolved and whether their secretion is transported to the PPG lumen by direct contact as suggested by their close proximity to the PPG ramifications is not known yet. Notably, in species that have lost these secretory cells associated with the PPG (*P. t. triangulum*, *P. t. diadema*, *P.* cf. *basalis*, *P. capensis*, *T. elongatus*, and *T. patagonensis*) the PPG reservoirs consist of voluminous tubes without ramifications that presumably merely store the marking secretion that is produced in the MG [50–52]. Inspection of the mapping of PPG characters on the phylogeny suggests that the secretory cells and the elaboration of the PPG reservoir may have evolved concurrently at the base of the genus *Philanthus* (Fig. 5).

Our results suggest that the PPG contributes to a variable degree to pheromone storage and production in males of most Philanthini. So the question arises why and how its involvement in scent marking came about. Beewolf females have been observed to simply alight in a male's territory and allow mating without additional courtship by males [37, 47]. Therefore, the conspicuousness of the territory, mediated by the composition and amount of marking pheromone, is probably the most important determinant for male reproductive success. Moreover, the spatial proximity of scent marking males in leks, as has been shown for several *Philanthus* species [37, 46, 47], might allow females to directly compare territories and their owners. This results in strong sexual selection on males to maximize both the quantity and quality of the marking secretion.

The original dual role of the MG as site of synthesis and reservoir of the marking secretion (as found in the Cercerini and Aphilanthopsini) might have limited the ability of males to synthesize and store larger amounts of marking secretion or to add novel compounds to the blend. For example, novel classes of compounds might have interfered with the synthesis or storage of the existing components (e.g. due to chemical reactions between acids and alcohols), thus promoting the evolution of novel secretory cells and a separate reservoir. The first evolutionary step towards its prominent role in scent

marking might thus have been a minor participation of the PPG in the storage and production of the marking secretion. Ongoing selection on pheromone quantity and quality would subsequently have enlarged the PPG and augmented its contribution. Whether the involvement of the PPG to pheromone production is accompanied by changes in the chemical composition of marking secretions in the Philanthini, in particular by the addition of novel classes of compounds, should be revealed by a comparative study of the marking secretions among the Philanthini.

An increase in the amount of scent marking secretion would clearly have been an advantage for mate attraction [129–131]. The addition of novel components to a sex pheromone, however, may represent a saltational evolutionary change [30], potentially even hindering mate recognition. Novel compounds might nevertheless be selected for by several not mutually exclusive causes like predation avoidance, male-male competition, and female choice [30, 132, 133]. There is currently no evidence that male scent marking in the Philanthini is effective in repelling predators or in keeping conspecific males at bay. However, different processes related to female choice might explain the evolution of novel pheromone components. First, female sensory biases [134–137] that evolved for prey recognition purposes might influence pheromone composition as in *P. triangulum* [53–55]. Consequently, a shift in the females' prey spectrum might select for changes in the males' marking secretion. Second, Fisher-Zahavi processes [23–25] could cause the addition of novel components. In Fisher's run-away model a female preference might arise accidentally and coevolve with the preferred trait; but this process has rarely been considered for pheromone evolution. Female choice for good or compatible genes could affect the evolution of pheromones [31], in that new components could indicate additional aspects of male quality [138] or improve signal reliability [139]. Finally, since sympatry is widespread among *Philanthus* species (e.g. [37]; G. Herzner, E. Strohm, M. Kaltenpoth, unpublished) the establishment or reinforcement of reproductive isolation between species [30, 140–142] might have selected for novel pheromone components [31].

If the involvement of the PPG enhanced mate attraction in male Philanthini, the question arises why the PPG did not get involved in scent marking (and was not enlarged) in the Cercerini and Aphilanthopsini as well. One possible explanation is that males of these tribes experience weaker sexual selection because, compared to male Philanthini, they have less pronounced territorial behavior and are spatially more dispersed [37, 38, 41, 42, 143]. Different intensities of sexual selection on males could also explain that PPG morphology shows a conspicuously congruent pattern in both sexes among the Philanthinae, with smaller PPGs in the Cercerini and Aphilanthopsini and larger, more complex PPGs in the Philanthini [59]. Owing to correlated evolution between the sexes [144–147], genetic changes underlying the sexually selected elaboration of the PPG in male Philanthini, as documented in this study, could have facilitated an enlargement of the PPG and the evolution of prey embalming in female Philanthini [58, 59, 61–63, 65]. That the PPG has evolved independently in males and females and the observed congruency across tribes is merely accidental seems rather unlikely. Yet, another plausible scenario is that the initial augmentation of the PPG might first have evolved in female Philanthini due to strong natural selection for prey embalming [58, 59, 61–63, 65] and, again assuming correlated evolution between the sexes [144–147], the resulting genetic changes could have facilitated the subsequent enlargement and elaboration of the PPG in male Philanthini by sexual selection. Different natural selection pressures on female Cercerini and Aphilanthopsini [59] may have kept the PPGs of both sexes of these basal tribes comparably small and simple.

The loss of the MG

The loss of the MG in the Nearctic *Philanthus* was surprising, since MGs had previously been reported from males of five of these species [48, 148–150]. We suspect that in these studies the large PPGs were mistaken for MGs, because their conclusions were based on dissections that hardly allow the discrimination of the two glands and at that time PPGs were only known from ants [50, 52].

Notably, in all but one species without MG the PPGs are huge and show extensive ramifications in direct contact with multinuclear syncytia (or aggregations of cells in *P. venustus*) (Fig. 5). Only in species lacking the MG (and in *P. multimaculatus*) the PPG is interspersed with conspicuous small cells (Fig. 5, see also Fig. 2d and f). This might suggest that these cells substitute for some function of the MG. However their small size and little cytoplasm contradict a secretory capacity. Due to the size and complexity of the PPGs and their association with large secretory cell clusters, we conclude that in the species without MGs, the PPG alone is responsible for the synthesis and storage of the marking secretion.

While it appears plausible that the enlargement of the PPG caused a reduction of the MG, its complete loss in several clades of the Philanthini is most puzzling, because it might have been accompanied by the loss of certain components of the marking secretion. Nonadaptive explanations like genetic drift in small populations could hardly explain the disappearance of a whole gland system. According to the above mentioned idea that the involvement of different glands is driven by hybridization avoidance, the loss of components of a sex

pheromone and the respective gland might be possible if the risk of hybridization is lowered. However, since particularly Nearctic species often occur in sympatry (e.g. [37]; G. Herzner, E. Strohm, M. Kaltenpoth, unpublished), a reduced risk of hybridization compared to other clades seems unlikely. An alternative explanation is that a change in female preferences to compounds that can be more efficiently produced in the PPG might make an MG superfluous. Female preferences [134–137] might be altered because of a change in their prey spectrum as explained above. In many Nearctic *Philanthus*, females prey not only on bees but also on wasps, whereas the latter habit seems to be rare in Palearctic and Afrotropical species [37]. Whether such a difference could cause the loss of the MG in males of the Nearctic species cannot be answered yet. Otherwise, there are no conspicuous differences between the Nearctic species and their Palearctic/Afrotropical congeners with regard to scent marking and reproductive behavior [37] that could explain the loss of the MG. Unfortunately, very little is known about the other two species without MG, *P. venustus* and *P. pulcherrimus*.

The loss of a sexual character is becoming increasingly recognized as a common event in the evolution of sexually selected traits and may have different causes [135, 151]. In beewolves, however, the actual trait, scent marking, persists while the source of the secretion is changed. A similar phenomenon has been reported for solitary bees of the genus *Centris*. Depending on the species, males scent mark territories with a secretion from either the MG or tibial glands and the respective other gland is reduced [105, 106, 152].

Taxa deviating from the overall trend

Philanthus albopilosus is the only known species of the genus in which males do not establish and scent mark territories [37]. Therefore, they do not need the respective glands anymore and their PPG has been reduced (Fig. 5). This provides indirect evidence for the role of the PPG in the production and storage of the marking secretion in other male Philanthini. The reduction of a gland following the loss of its function has been reported for fungus-growing ants. In monandrous attine ants, males transfer an antiaphrodisiac from accessory glands during copulation; in polyandrous species, however, males do not mark mated queens and their accessory glands were reduced or completely lost [153].

The regain of the MG in males of the Nearctic *P. multimaculatus* (Fig. 5) is puzzling since there are no conspicuous differences to its Nearctic congeners with regard to their territorial behavior [37]. Also, why in some species (*P. triangulum*, *P.* cf. *basalis*, *P. capensis*) the secretory cells of the PPG were lost while the reservoir became the main storage organ (Fig. 5) cannot be answered yet.

Conclusion

There is substantial evidence that sexually selected traits can undergo rapid evolutionary change, including losses and gains [83, 154–156]. In particular the Fisher-Zahavi processes [23–25] as well as sexual antagonism, like chase-away selection [26] and female sensory biases [135–138] might cause complex phylogenetic patterns in sexually selected characters (e.g. [28, 87–89, 157–159]). Our comparative morphological analyses of male head glands revealed extensive interspecific variation within the Philanthinae, in particular among the Philanthini. While we found clear phylogenetic trends, there are also intriguing deviations and reversals (Fig. 5). The glands of female Philanthini, by contrast, appear virtually uniform with mostly only gradual variation and no loss of a gland system or the addition of novel components like secretory cells [59], probably as a result of stabilizing natural selection. Other evolutionary forces like genetic drift and mutations should affect males and females similarly and can therefore be excluded as causes for the observed higher diversity among males. Taken together our findings support the hypothesis that strong sexual selection acting on male pheromone glands has led to rapid evolutionary changes and to a substantially higher interspecific morphological diversity in males than in females. Taking into account that about 135 of the ca. 170 described species of Philanthini [68] have not been investigated so far, the high diversity observed in this study suggests that there are probably more species with unique and novel gland characteristics yet to be discovered. Further studies on the chemical composition of the marking secretions, male territorial behavior, mate attraction as well as female prey spectrum and mate choice will help to unravel the ecological and evolutionary causes that have given rise to the remarkable diversity and phylogenetic trends in male head gland morphology among the Philanthinae.

Additional files

Additional file 1: Additional methods, additional **Table S1.** showing the data matrix of the analyzed morphological characters, additional **Table S2.** showing the data matrix used for statistical analyses, additional **Table S3.** giving the Eigenvalues of the morphological characters from the categorical principal components analysis, additional **Figure S1.** explaining the morphology of the PPG, additional **Figure S2.** showing sagittal section of the head capsule of *Philanthus rugosus*, additional results of the hierarchical cluster analysis based on PPG and MG morphology, including additional **Figure S3.** showing the dendrogram resulting from the hierarchical cluster analysis, additional **Figure S4.** showing the results of the ancestral state reconstruction of presence vs. absence of the MG, and additional **Figure S5** showing the results of the ancestral state reconstruction of presence vs. absence of secretory cells of the PPG. (PDF 1617 kb)

Additional file 2: Additional methods for coding of morphological characters for the aggregated analysis of male and female head gland morphology, additional **Table S4.** showing the aggregated data matrix used for statistical analyses of male and female gland morphology, additional information on the aggregated categorical principal components analysis, including additional **Table S5.** giving the Eigenvalues of the aggregated morphological characters from the categorical principal components analysis of male and female gland morphology, additional **Figure S6.** showing the plot of the aggregated categorical principal components analysis, and additional methods on the calculation of Shannon diversity indices for the aggregated characters. (PDF 739 kb)

Abbreviations
3D: Three-dimensional; ASR: Ancestral state reconstruction; CATPCA: Categorical principal components analysis; HCA: Hierarchical cluster analysis; ID: Identification number of species; MG: Mandibular gland; ML: Maximum likelihood; NQ-class: Class of secretory cell according to Noirot and Quennedey 1974 (71); PPG: Post pharyngeal gland

Acknowledgements
We thank Margot Schilling for technical assistance and Martin Kaltenpoth, Kerstin Roeser-Mueller, Tobias Engl, Sabrina Köhler, Dirk Koedam, Jon Seger, J. William Stubblefield, Erol Yildirim, and Thomas Schmitt for help in collecting the specimens. Permits were issued by the nature conservation boards of KwaZulu Natal (Permit 4362/2004), Eastern Cape Province (WRO44/04WR, WRO9/04WR, WRO74/06WR, WRO75/06WR, CRO135/11CR, CRO136/11CR, CRO179/10CR, and CRO180/10CR) and Western Cape Province (001-202-00026, 001-506-00001, AAA004-00053-0035, AAA004-00089-0011, AAA004-00683-0035, and 0046-AAA004-00008) of South Africa, and the Brazilian Ministry of the Environment: MMA/SISBIO/22861-1. The study was supported by the Universität Bayern e.V. through a Ph.D. fellowship (K.W.). We are grateful for the comments of anonymous reviewers.

Funding
The study was supported by the Universität Bayern e.V. through a Ph.D. fellowship (K.W.). The funding body played no role in the design of the study, collection, analysis, and interpretation of data and in writing the manuscript.

Authors' contributions
ES, GH, and KW conceived of the study. KW and ES carried out the morphological investigations; KW conducted the 3D-reconstructions and statistical analyses; KW, GH, and ES wrote the manuscript; GH and ES contributed equally to the study. All authors read and approved of the final manuscript.

Competing interests
The authors declare that they have no competing interests.

References
1. Darwin C. On the origin of species by means of natural selection. 1st ed. London: John Murray; 1859.
2. Darwin C. The descent of man and selection in relation to sex. 1st ed. London: John Murray; 1871.
3. Huxley JS. Darwin's theory of sexual selection and the data subsumed by it, in the light of recent research. Am Nat. 1938;72(742):416–33.
4. van Doorn GS, Edelaar P, Weissing FJ. On the origin of species by natural and sexual selection. Science. 2009;326(5960):1704–7.
5. Cornwallis CK, Uller T. Towards an evolutionary ecology of sexual traits. Trends Ecol Evol. 2010;25(3):145–52.
6. Maan ME, Seehausen O. Ecology, sexual selection and speciation. Ecol Lett. 2011;14(6):591–602.
7. Weissing FJ, Edelaar P, van Doorn GS. Adaptive speciation theory: a conceptual review. Behav Ecol Sociobiol. 2011;65(3):461–80.
8. Wagner CE, Harmon LJ, Seehausen O. Ecological opportunity and sexual selection together predict adaptive radiation. Nature. 2012;487(7407):366–70.

9. Safran RJ, Scordato ESC, Symes LB, Rodriguez RL, Mendelson TC. Contributions of natural and sexual selection to the evolution of premating reproductive isolation: a research agenda. Trends Ecol Evol. 2013;28(11):643–50.

10. Scordato ESC, Symes LB, Mendelson TC, Safran RJ. The role of ecology in speciation by sexual selection: a systematic empirical review. J Hered. 2014;105:782–94.

11. West-Eberhard MJ. Sexual selection, social competition, and speciation. Q Rev Biol. 1983;58(2):155–83.

12. Higashi M, Takimoto G, Yamamura N. Sympatric speciation by sexual selection. Nature. 1999;402(6761):523–6.

13. Ritchie MG. Sexual selection and speciation. Annu Rev Ecol Evol Syst. 2007;38:79–102.

14. Panhuis TM, Butlin R, Zuk M, Tregenza T. Sexual selection and speciation. Trends Ecol Evol. 2001;16(7):364–71.

15. van Doorn GS, Dieckmann U, Weissing FJ. Sympatric speciation by sexual selection: a critical reevaluation. Am Nat. 2004;163(5):709–25.

16. Seehausen O, van Alphen JM. Can sympatric speciation by disruptive sexual selection explain rapid evolution of cichlid diversity in Lake Victoria? Ecol Lett. 1999;2(4):262–71.

17. Hosken DJ, Stockley P. Sexual selection and genital evolution. Trends Ecol Evol. 2004;19(2):87–93.

18. Møller AP, Szép T. Rapid evolutionary change in a secondary sexual character linked to climatic change. J Evol Biol. 2005;18(2):481–95.

19. Arnegard ME, McIntyre PB, Harmon LJ, Zelditch ML, Crampton WGR, Davis JK, Sullivan JP, Lavoue S, Hopkins CD. Sexual signal evolution outpaces ecological divergence during electric fish species radiation. Am Nat. 2010;176(3):335–56.

20. Kraaijeveld K, Kraaijeveld-Smit FJL, Maan ME. Sexual selection and speciation: the comparative evidence revisited. Biol Rev. 2011;86(2):367–77.

21. Seddon N, Botero CA, Tobias JA, Dunn PO, MacGregor HEA, Rubenstein DR, Uy JAC, Weir JT, Whittingham LA, Safran RJ. Sexual selection accelerates signal evolution during speciation in birds. Proc R Soc B. 2013;280:1766.

22. Bacquet PMB, Brattström O, Wang HL, Allen CE, Löfstedt C, Brakefield PM, Nieberding CM. Selection on male sex pheromone composition contributes to butterfly reproductive isolation. Proc R Soc B. 2015;282:1804.

23. Prum RO. Phylogenetic tests of alternative intersexual selection mechanisms: trait macroevolution in a polygynous clade (Aves: Pipridae). Am Nat. 1997;149(4):668–92.

24. Kokko H, Brooks R, McNamara JM, Houston AI. The sexual selection continuum. Proc R Soc B. 2002;269(1498):1331–40.

25. Kokko H, Brooks R, Jennions MD, Morley J. The evolution of mate choice and mating biases. Proc R Soc B. 2003;270(1515):653–64.

26. Holland B, Rice WR. Perspective: chase-away sexual selection: antagonistic seduction versus resistance. Evolution. 1998;52(1):1–7.

27. Hosken DJ, House CM. Sexual selection. Curr Biol. 2011;21(2):R62–5.

28. Ord TJ, Martins EP. Tracing the origins of signal diversity in anole lizards: phylogenetic approaches to inferring the evolution of complex behaviour. Anim Behav. 2006;71(6):1411–29.

29. Garamszegi LZ, Eens M, Erritzøe J, Møller AP. Sexually size dimorphic brains and song complexity in passerine birds. Behav Ecol. 2005;16(2):335–45.

30. Symonds MR, Elgar MA. The evolution of pheromone diversity. Trends Ecol Evol. 2008;23(4):220–8.

31. Johansson BG, Jones TM. The role of chemical communication in mate choice. Biol Rev. 2007;82(2):265–89.

32. Alexander BA. A cladistic analysis of the subfamily Philanthinae (Hymenoptera: Sphecidae). Syst Entomol. 1992;17(2):91–108.

33. Kaltenpoth M, Roeser-Mueller K, Koehler S, Peterson A, Nechitaylo TY, Stubblefield JW, Herzner G, Seger J, Strohm E. Partner choice and fidelity stabilize coevolution in a cretaceous-age defensive symbiosis. Proc Natl Acad Sci U S A. 2014;111(17):6359–64.

34. Evans HE. A review of nesting behavior of digger wasps of the genus Aphilanthops, with special attention to the mechanism of prey carriage. Behaviour. 1962;19:239–60.

35. Evans HE. Observations on the nesting behavior of wasps of the tribe Cercerini. J Kansas Entomol Soc. 1971;44(4):500–23.

36. Bohart RM, Menke AS. Sphecid wasps of the world: a generic revision. 1st ed. Ithaca: University of California Press; 1976.

37. Evans HE, O'Neill KM. The natural history and behavior of North American beewolves. 1st ed. Ithaca: Cornell University Press; 1988.

38. Evans HE. Observations on the biology of Cerceris mimica Cresson (Hymenoptera: Sphecidae: Philanthinae). J Kansas Entomol Soc. 2000;73(4):220–4.

39. Polidori C, Boesi R, Isola F, Andrietti F. Provisioning patterns and choice of prey in the digger wasp Cerceris arenaria (Hymenoptera: Crabronidae): the role of prey size. Eur J Entomol. 2005;102(4):801–4.

40. Polidori C, Federici M, Papadia C, Andrietti F. Nest sharing and provisioning activity of females of the digger wasp, Cerceiis rubida (Hymenoptera, Crabronidae). Ital J Zool. 2006;73(1):55–65.

41. Alcock J. Male mating strategies of some philanthine wasps (Hymenoptera: Sphecidae). J Kansas Entomol Soc. 1975;48(4):532–45.

42. Evans HE, O'Neill KM. Male territorial behavior in four species of the tribe Cercerini (Sphecidae: Philanthinae). J New York Entomol Soc. 1985;93(3):1033–40.

43. Strohm E. Allokation elterlicher Investitionen beim Europäischen Bienenwolf Philanthus triangulum Fabricius (Hymenoptera: Sphecidae). Berlin: Verlag Dr, Köster; 1995.

44. Strohm E, Lechner K. Male size does not affect territorial behaviour and life history traits in a sphecid wasp. Anim Behav. 2000;59(1):183–91.

45. Clarke S, Dani F, Jones G, Morgan E, Schmidt J. (Z)-3-hexenyl (R)-3-hydroxybutanoate: a male specific compound in three North American decorator wasps Eucerceris rubripes, E. conata and E. tricolor. J Chem Ecol. 2001;27(7):1437–47.

46. O'Neill KM. Solitary wasps: behavior and natural history. 1st ed. Ithaca: Cornell University Press; 2001.

47. Kroiss J, Lechner K, Strohm E. Male territoriality and mating system in the European beewolf Philanthus triangulum F. (Hymenoptera: Crabronidae): evidence for a "hotspot" lek polygyny. J Ethol. 2010;28(2):295–304.

48. Schmidt JO, Oneill KM, Fales HM, McDaniel CA, Howard RW. Volatiles from mandibular glands of male beewolves (Hymenoptera, Sphecidae, Philanthus) and their possible roles. J Chem Ecol. 1985;11(7):895–901.

49. Schmitt T, Strohm E, Herzner G, Bicchi C, Krammer G, Heckel F, Schreier P. (S)-2,3-dihydrofarnesoic acid, a new component in cephalic glands of male European beewolves Philanthus triangulum. J Chem Ecol. 2003;29(11):2469–79.

50. Kroiss J, Schmitt T, Schreier P, Strohm E, Herzner G. A selfish function of a "social" gland? A postpharyngeal gland functions as a sex pheromone reservoir in males of the solitary wasp Philanthus triangulum. J Chem Ecol. 2006;32(12):2763–76.

51. Goettler W, Strohm E. Mandibular glands of male European beewolves, Philanthus triangulum (Hymenoptera, Crabronidae). Arthropod Struct Dev. 2008;37(5):363–71.

52. Herzner G, Goettler W, Kroiss J, Purea A, Webb AG, Jakob PM, Roessler W, Strohm E. Males of a solitary wasp possess a postpharyngeal gland. Arthropod Struct Dev. 2007;36(2):123–33.

53. Herzner G, Schmitt T, Linsenmair KE, Strohm E. Prey recognition by females of the European beewolf and its potential for a sensory trap. Anim Behav. 2005;70(6):1411–8.

54. Schmitt T, Herzner G, Weckerle B, Schreier P, Strohm E. Volatiles of foraging honeybees Apis mellifera (Hymenoptera: Apidae) and their potential role as semiochemicals. Apidologie. 2007;38(2):164–70.

55. Steiger S, Schmitt T, Schaefer HM. The origin and dynamic evolution of chemical information transfer. Proc R Soc B. 2010;278(1708):970–9.

56. Ågren L. Mandibular gland morphology of Cerceris rybyensis (L) (Hymenoptera: Philanthidae). Zoon. 1977;5(2):91–5.

57. Strohm E, Herzner G, Goettler W. A 'social' gland in a solitary wasp? The postpharyngeal gland of female European beewolves (Hymenoptera, Crabronidae). Arthropod Struct Dev. 2007;36(2):113–22.

58. Herzner G, Kaltenpoth M, Poettinger T, Weiss K, Koedam D, Kroiss J, Strohm E. Morphology, chemistry and function of the postpharyngeal gland in the South American digger wasps Trachypus boharti and Trachypus elongatus. PLoS One. 2013;8:e82780.

59. Weiss K, Strohm E, Kaltenpoth M, Herzner G. Comparative morphology of the postpharyngeal gland in the Philanthinae (Hymenoptera, Crabronidae) and the evolution of an antimicrobial brood protection mechanism. BMC Evol Biol. 2015;15:291.

60. Strohm E, Linsenmair KE. Measurement of parental investment and sex allocation in the European beewolf Philanthus triangulum F. (Hymenoptera: Sphecidae). Behav Ecol Sociobiol. 1999;47(1-2):76–88.

61. Strohm E, Linsenmair KE. Females of the European beewolf preserve their honeybee prey against competing fungi. Ecol Entomol. 2001;26(2):198–203.

62. Herzner G, Schmitt T, Peschke K, Hilpert A, Strohm E. Food wrapping with the postpharyngeal gland secretion by females of the European beewolf *Philanthus triangulum*. J Chem Ecol. 2007;33(4):849–59.

63. Herzner G, Strohm E. Fighting fungi with physics: food wrapping by a solitary wasp prevents water condensation. Curr Biol. 2007;17(2):R46–7.

64. Herzner G, Strohm E. Food wrapping by females of the European Beewolf, *Philanthus triangulum*, retards water loss of larval provisions. Physiol Entomol. 2008;33(2):101–9.

65. Herzner G, Engl T, Strohm E. Cryptic combat against competing microbes is a costly component of parental care in a digger wasp. Anim Behav. 2011; 82(2):321–8.

66. Pagel M. Inferring the historical patterns of biological evolution. Nature. 1999;401(6756):877–84.

67. Holt B, Lessard JP, Borregaard MK, Fritz SA, Araujo MB, Dimitrov D, Fabre PH, Graham CH, Graves GR, Jonsson KA, et al. An update of Wallace's zoogeographic regions of the world. Science. 2013;339(6115):74–8.

68. Pulawski WJ. Number of species. In: Catalog of Sphecidae. San Francisco: California Academy of Sciences; 2016. http://researcharchive.calacademy. org/research/entomology/entomology_resources/hymenoptera/sphecidae/ number_of_species.pdf. Accessed 30 Oct 2016.

69. Pulawski WJ. *Philanthus* - list of species. In: Catalog of Sphecidae. San Francisco: California Academy of Sciences; 2016. http://researcharchive. calacademy.org/research/entomology/entomology_resources/hymenoptera/ sphecidae/genera/Philanthus.pdf. Accessed 30 Oct 2016.

70. Adam H, Czihak G. Arbeitsmethoden der makroskopischen und mikroskopischen Anatomie. 1st ed. Stuttgart: Gustav Fischer Verlag; 1964.

71. Noirot C, Quennedey A. Fine-structure of insect epidermal glands. Annu Rev Entomol. 1974;19:61–80.

72. Cruz-Landim C, Costa R. Structure and function of the hypopharyngeal glands of Hymenoptera: a comparative approach. J Comp Biol. 1998;3(2):151–63.

73. do Amaral JB, Caetano FH. The hypopharyngeal gland of leaf-cutting ants (*Atta sexdens rubropilosa*) (Hymenoptera: Formicidae). Sociobiology. 2005; 46(3):515–24.

74. Billen J, Bauweleers E, Hashim R, Ito F. Survey of the exocrine system in *Protanilla wallacei* (Hymenoptera, Formicidae). Arthropod Struct Dev. 2013; 42(3):173–83.

75. Cardona A, Saalfeld S, Schindelin J, Arganda-Carreras I, Preibisch S, Longair M, Tomancak P, Hartenstein V, Douglas RJ. TrakEM2 software for neural circuit reconstruction. PLoS One. 2012;7:6.

76. Schindelin J, Arganda-Carreras I, Frise E, Kaynig V, Longair M, Pietzsch T, Preibisch S, Rueden C, Saalfeld S, Schmid B, et al. Fiji: an open-source platform for biological-image analysis. Nat Methods. 2012;9(7):676–82.

77. Schmid B, Schindelin J, Cardona A, Longair M, Heisenberg M. A high-level 3D visualization API for java and ImageJ. BMC Bioinformatics. 2010;11:274.

78. Meulman J, Heiser WJ, SPSS. SPSS Categories 13.0. Chicago: SPSS; 2004.

79. Paradis E, Claude J, Strimmer K. APE: analyses of phylogenetics and evolution in R language. Bioinformatics. 2004;20:289–90.

80. R Core Team. R: A language and environment for statistical computing. Vienna: R Foundation for Statistical Computing; 2017. http://www.R-project.org

81. Hammer Ø, Harper DAT, Ryan PD. PAST: paleontological statistics software package for education and data analysis. Paleontol Electron. 2001;4:9.

82. Conow C, Fielder D, Ovadia Y, Libeskind-Hadas R. Jane: a new tool for the cophylogeny reconstruction problem. Algorithms Mol Biol. 2010;5:16.

83. Maddison WP, Maddison DR. Mesquite: A modular system for evolutionary analysis, Version 3.04. 2015. http://mesquiteproject.org. Accessed 13 Jan 2015.

84. Heiser W, Meulman J. Homogeneity analysis: exploring the distribution of variables and their nonlinear relationships. In: Greenacre JB, Blasius J, editors. Correspondence analysis in the social sciences: recent developments and applications. 1st ed. New York: Academic Press; 1994. p. 179–209.

85. Münkemüller T, Lavergne S, Bzeznik B, Dray S, Jombart T, Schiffers K, Thuiller W. How to measure and test phylogenetic signal. Methods Ecol Evol. 2012; 3(4):743–56.

86. Kopp A, True JR. Evolution of male sexual characters in the Oriental *Drosophila melanogaster* species group. Evol Dev. 2002;4(4):278–91.

87. Emlen DJ, Marangelo J, Ball B, Cunningham CW. Diversity in the weapons of sexual selection: horn evolution in the beetle genus *Onthophagus* (Coleoptera: Scarabaeidae). Evolution. 2005;59(5):1060–84.

88. Price JJ, Friedman NR, Omland KE. Song and plumage evolution in the new world orioles (*Icterus*) show similar lability and convergence in patterns. Evolution. 2007;61(4):850–63.

89. Puniamoorthy N, Su KFY, Meier R. Bending for love: losses and gains of sexual dimorphisms are strictly correlated with changes in the mounting position of sepsid flies (Sepsidae: Diptera). BMC Evol Biol. 2008;8:155.

90. Symonds MR, Moussalli A, Elgar MA. The evolution of sex pheromones in an ecologically diverse genus of flies. Biol J Linn Soc. 2009;97(3):594–603.

91. Fortunato A, Turillazzi S, Delfino G. Ectal mandibular gland in *Polistes dominulus* (Christ) (Hymenoptera, Vespidae): ultrastructural modifications over the secretory cycle. J Morphol. 2000;244(1):45–55.

92. Pietrobon TAO, Caetano FH. Ultramorphology and histology of the ectal mandibular gland in *Polistes versicolor* (Olivier) (Hymenoptera: Vespidae). Cytologia. 2003;68(1):89–94.

93. Penagos-Arévalo AC, Billen J, Sarmiento CE. Uncovering head gland diversity in neotropical Polistinae wasps (Hymenoptera, Vespidae): comparative analysis and description of new glands. Arthropod Struct Dev. 2015;44(5):415–25.

94. Cruz-Landim C. Estudo comparativo de algumas glândulas das abelhas (Hymenoptera, Apoidea) e respectivas implicacoes evolutivas. Arqu Zool. 1967;15(3):177–290.

95. Cruz-Landim C, Abdalla FC, Gracioli-Vitti LF. Morphological and functional aspects of volatile-producing glands in bees (Hymenoptera: Apidae). Insect Sci. 2005;12(6):467–80.

96. Galvani GL, Settembrini BP. Comparative morphology of the head glands in species of Protepeolini and Emphorini (Hymenoptera: Apidae). Apidologie. 2013;44(4):367–81.

97. Duffield RM, Shamim M, Wheeler JW, Menke AS. Alkylpyrazines in the mandibular gland secretions of *Ammophila* wasps (Hymenoptera: Sphecidae). Comp Biochem Phys B. 1981;70(2):317–8.

98. Blum MS. Alarm pheromones. Annu Rev Entomol. 1969;14:57–80.

99. do Amaral JB, Machado-Santelli G. Salivary system in leaf-cutting ants (*Atta sexdens rubropilosa* Forel, 1908) castes: a confocal study. Micron. 2008;39(8):1222–7.

100. Zimmermann D, Vilhelmsen L. The sister group of Aculeata (Hymenoptera) - evidence from internal head anatomy, with emphasis on the tentorium. Arthropod Syst Phylog. 2016;74(2):195–218.

101. Stökl J, Herzner G. Morphology and ultrastructure of the allomone and sex-pheromone producing mandibular gland of the parasitoid wasp *Leptopilina heterotoma* (Hymenoptera: Figitidae). Arthropod Struct Dev. 2016;45(4):333–40.

102. Strohm E, Kaltenpoth M, Herzner G. Is the postpharyngeal gland of a solitary digger wasp homologous to ants? Evidence from chemistry and physiology. Insect Soc. 2010;57(3):285–91.

103. Herzner G, Ruther J, Goller S, Schulz S, Goettler W, Strohm E. Structure, chemical composition and putative function of the postpharyngeal gland of the emerald cockroach wasp, *Ampulex compressa* (Hymenoptera, Ampulicidae). Zoology. 2011;114(1):36–45.

104. Billen J, Mandonx T, Hashim R, Ito F. Exocrine glands of the ant *Myrmoteras iriodum*. Entomol Sci. 2015;18(2):167–73.

105. Vinson SB, Williams HJ, Frankie GW, Wheeler JW, Blum MS, Coville RE. Mandibular glands of male *Centris adani* (Hymenoptera, Anthophoridae) - their morphology, chemical constituents, and function in scent marking and territorial behavior. J Chem Ecol. 1982;8(2):319–27.

106. Vinson SB, Williams HJ, Frankie GW, Coville RE. Comparative morphology and chemical contents of male mandibular glands of several *Centris* species (Hymenoptera, Anthophoridae) in Costa Rica. Comp Biochem Phys A. 1984;77(4):685–8.

107. Hefetz A. Function of secretion of mandibular gland of male in territorial behavior of *Xylocopa sulcatipes* (Hymenoptera, Anthophoridae). J Chem Ecol. 1983;9(7):923–31.

108. Ayasse M, Paxton R, Tengö J. Mating behavior and chemical communication in the order Hymenoptera. Annu Rev Entomol. 2001;46(1):31–78.

109. Slessor KN, Kaminski LA, King GGS, Borden JH, Winston ML. Semiochemical basis of the retinue response to queen honey bees. Nature. 1988;332(6162):354–6.

110. Winston ML, Slessor KN. The essence of royalty - honey-bee queen pheromone. Am Sci. 1992;80(4):374–85.

111. Hughes WOH, Howse PE, Vilela EF, Goulson D. The response of grass-cutting ants to natural and synthetic versions of their alarm pheromone. Physiol Entomol. 2001;26(2):165–72.

112. Hughes WOH, Howse PE, Goulson D. Mandibular gland chemistry of grass-cutting ants: species, caste, and colony variation. J Chem Ecol. 2001;27(1):109–24.

113. Lalor PF, Hughes WHO. Alarm behaviour in *Eciton* army ants. Physiol Entomol. 2011;36(1):1–7.

114. Cane JH, Michener CD. Chemistry and function of mandibular gland products of bees of the genus *Exoneura* (Hymenoptera, Anthophoridae). J Chem Ecol. 1983;9:1525–31.

115. Cane JH, Gerdin S, Wife G. Mandibular gland secretions of solitary bees (Hymenoptera, Apoidea) - potential for nest cell disinfection. J Kansas Entomol Soc. 1983;56(2):199–204.

116. Chadha M, Eisner T, Monro A, Meinwald J. Defence mechanisms of arthropods - VII: citronellal and citral in the mandibular gland secretion of the ant *Acanthomyops claviger* (Roger). J Insect Physiol. 1962;8(2):175–9.

117. Melo GAR. Phylogenetic relationships and classification of the major lineages of Apoidea (Hymenoptera), with emphasis on the crabronid wasps. Scientific Papers Nat Hist Mus Univ Kans. 1999;14:1–55.

118. Debevec AH, Cardinal S, Danforth BN. Identifying the sister group to the bees: a molecular phylogeny of Aculeata with an emphasis on the superfamily Apoidea. Zool Scr. 2012;41(5):527–35.

119. Soroker V, Vienne C, Hefetz A, Nowbahari E. The postpharyngeal gland as a "gestalt" organ for nestmate recognition in the ant *Cataglyphis niger*. Naturwissenschaften. 1994;81:510–3.

120. Soroker V, Hefetz A, Cojocaru M, Billen J, Franke S, Francke W. Structural and chemical ontogeny of the postpharyngeal gland in the desert ant *Cataglyphis niger*. Physiol Entomol. 1995;20:323–9.

121. Hefetz A, Errard C, Chambris A, LeNegrate A. Postpharyngeal gland secretion as a modifier of aggressive behavior in the myrmicine ant *Manica rubida*. J Insect Behav. 1996;9:709–17.

122. Lenoir A, Fresneau D, Errard C, Hefetz A. Individuality and colonial identity in ants: the emergence of the social representation concept. In: Information processing in social insects. Basel: Birkhäuser Verlag; 1999. p. 219–37.

123. Eelen D, Borgesen L, Billen J. Functional morphology of the postpharyngeal gland of queens and workers of the ant *Monomorium pharaonis* (L.). Acta Zool. 2006;87:101–11.

124. Cruz-Landim C, Reginato RD. Exocrine glands of *Schwarziana quadripunctata* (Hymenoptera, Apinae, Meliponini). Braz J Biol. 2001;61(3):497–505.

125. Grasso D, Romani R, Castracani C, Visicchio R, Mori A, Isidoro N, Le Moli F. Mandible associated glands in queens of the slave-making ant *Polyergus rufescens* (Hymenoptera, Formicidae). Insect Soc. 2004;51(1):74–80.

126. Boonen S, Eelen D, Børgesen L, Billen J. Functional morphology of the mandibular gland of queens of the ant *Monomorium pharaonis* (L.). Acta Zool. 2013;94(4):373–81.

127. Billen J. Diversity and morphology of exocrine glands in ants. Proceedings XIX Simpósio Mirmecologia, Ouro Preto, Brasil; 2009. p. 17–21.

128. Billen J. Exocrine glands and their key function in the communication system of social insects. Formosan Entomol. 2011;31:75–84.

129. Droney DC, Hock MB. Male sexual signals and female choice in *Drosophila grimshawi* (Diptera: Drosophilidae). J Insect Physiol. 1998;11(1):59–71.

130. Ruther J, Matschke M, Garbe LA, Steiner S. Quantity matters: male sex pheromone signals mate quality in the parasitic wasp *Nasonia vitripennis*. Proc R Soc B. 2009;276(1671):3303–10.

131. Foster SP, Johnson CP. Signal honesty through differential quantity in the female-produced sex pheromone of the moth *Heliothis virescens*. J Chem Ecol. 2011;37(7):717–23.

132. Haynes KF, Yeargan KV. Exploitation of intraspecific communication systems: illicit signalers and receivers. Ann Entomol Soc Am. 1999;92(6):960–70.

133. Raffa KF, Hobson KR, LaFontaine S, Aukema BH. Can chemical communication be cryptic? Adaptations by herbivores to natural enemies exploiting prey semiochemistry. Oecologia. 2007;153(4):1009–19.

134. Morris M. Further examination of female preference for vertical bars in swordtails: preference for 'no bars' in a species without bars. J Fish Biol. 1998;53:56–63.

135. Wiens JJ. Widespread loss of sexually selected traits: how the peacock lost its spots. Trends Ecol Evol. 2001;16(9):517–23.

136. Palmer CA, Watts RA, Gregg RG, McCall MA, Houck LD, Highton R, Arnold SJ. Lineage-specific differences in evolutionary mode in a salamander courtship pheromone. Mol Biol Evol. 2005;22(11):2243–56.

137. Elias DO, Hebets EA, Hoy RR. Female preference for complex/novel signals in a spider. Behav Ecol. 2006;17(5):765–71.

138. Herzner G, Schmitt T, Heckel F, Schreier P, Strohm E. Brothers smell similar: variation in the sex pheromone of male European Beewolves *Philanthus triangulum* F. (Hymenoptera: Crabronidae) and its implications for inbreeding avoidance. Biol J Linn Soc. 2006;89(3):433–42.

139. Mahr K, Evans C, Thonhauser K, Griggio M, Hoi H. Multiple ornaments - multiple signaling functions? The importance of song and UV plumage coloration in female superb fairy-wrens (*Malurus cyaneus*). Front Ecol Evol. 2016;4:43.

140. Smadja C, Butlin RK. On the scent of speciation: the chemosensory system and its role in premating isolation. Heredity. 2009;102(1):77–97.

141. Niehuis O, Buellesbach J, Gibson JD, Pothmann D, Hanner C, Mutti NS, Judson AK, Gadau J, Ruther J, Schmitt T. Behavioural and genetic analyses of *Nasonia* shed light on the evolution of sex pheromones. Nature. 2013; 494(7437):345–8.

142. Weber MG, Mitko L, Eltz T, Ramírez SR. Macroevolution of perfume signalling in orchid bees. Ecol Lett. 2016;19(11):1314–23.

143. Steiner AL. Observations on spacing, aggressive and lekking behavior of digger wasp males of *Eucerceris flavocincta* (Hymenoptera: Sphecidae; Cercerini). J Kansas Entomol Soc. 1997;51(3):492–8.

144. Lande R. Sexual dimorphism, sexual selection, and adaptation in polygenic characters. Evolution. 1980;34(2):292–305.

145. Amundsen T. Why are female birds ornamented? Trends Ecol Evol. 2000; 15(4):149–55.

146. Potti J, Canal D. Heritability and genetic correlation between the sexes in a songbird sexual ornament. Heredity. 2011;106:945–54.

147. Tobias JA, Montgomerie R, Lyon BE. The evolution of female ornaments and weaponry: social selection, sexual selection and ecological competition. Phil Trans R Soc B. 2012;367:2274–93.

148. Gwynne DT. Male territoriality in bumblebee wolf, *Philanthus bicinctus* (Mickel) (Hymenoptera, Sphecidae) - Observations on behavior of individual males. Z Tierpsychol. 1978;47(1):89–103.

149. McDaniel CA, Howard RW, Oneill KM, Schmidt JO. Chemistry of male mandibular gland secretions of *Philanthus basilaris* Cresson and *Philanthus bicinctus* (Mickel) (Hymenoptera, Sphecidae). J Chem Ecol. 1987;13(2):227–35.

150. McDaniel CA, Schmidt JO, Howard RW. Mandibular gland secretions of the male beewolves *Philanthus crabroniformis*, *P. barbatus*, and *P. pulcher* (Hymenoptera, Sphecidae). J Chem Ecol. 1992;18(1):27–37.

151. Porter ML, Crandall KA. Lost along the way: the significance of evolution in reverse. Trends Ecol Evol. 2003;18(10):541–7.

152. Williams HJ, Vinson SB, Frankie GW, Coville RE, Ivie GW. Morphology, chemical contents and possible function of the tibial gland of males of the Costa Rican solitary bees *Centris nitida* and *Centris trigondoides subtarsata* (Hymenoptera, Anthophoridae). J Kansas Entomol Soc. 1984;57(1):50–4.

153. Mikheyev AS. Male accessory gland size and the evolutionary transition from single to multiple mating in the fungus-gardening ants. Insect Sci. 2004;4:37.

154. Meyer A, Morrissey JM, Schartl M. Recurrent origin of a sexually selected trait in *Xiphophorus* fishes inferred from a molecular phylogeny. Nature. 1994;368(6471):539–42.

155. Meyer A. The evolution of sexually selected traits in male swordtail fishes (*Xiphophorus*: Poeciliidae). Heredity. 1997;79:329–37.

156. Kimball RT, Braun EL, Ligon JD, Lucchini V, Randi E. A molecular phylogeny of the peacock-pheasants (Galliformes: *Polyplectron spp.*) indicates loss and reduction of ornamental traits and display behaviours. Biol J Linn Soc. 2001; 73(2):187–98.

157. Pomiankowski A, Iwasa Y. Runaway ornament diversity caused by Fisherian sexual selection. Proc Natl Acad Sci U S A. 1998;95(9):5106–11.

158. Omland KE, Lanyon SM. Reconstructing plumage evolution in orioles (*Icterus*): repeated convergence and reversal in patterns. Evolution. 2000; 54(6):2119–33.

159. Price JJ, Lanyon SM. Patterns of song evolution and sexual selection in the oropendolas and caciques. Behav Ecol. 2004;15(3):485–97.

2

Evolutionary origin of type IV classical cadherins in arthropods

Mizuki Sasaki[1,4], Yasuko Akiyama-Oda[1,2] and Hiroki Oda[1,3*] (iD)

Abstract

Background: Classical cadherins are a metazoan-specific family of homophilic cell-cell adhesion molecules that regulate morphogenesis. Type I and type IV cadherins in this family function at adherens junctions in the major epithelial tissues of vertebrates and insects, respectively, but they have distinct, relatively simple domain organizations that are thought to have evolved by independent reductive changes from an ancestral type III cadherin, which is larger than derived paralogs and has a complicated domain organization. Although both type III and type IV cadherins have been identified in hexapods and branchiopods, the process by which the type IV cadherin evolved is still largely unclear.

Results: Through an analysis of arthropod genome sequences, we found that the only classical cadherin encoded in chelicerate genomes was the type III cadherin and that the two *type III cadherin* genes found in the spider *Parasteatoda tepidariorum* genome exhibited a complex yet ancestral exon-intron organization in arthropods. Genomic and transcriptomic data from branchiopod, copepod, isopod, amphipod, and decapod crustaceans led us to redefine the type IV cadherin category, which we separated into type IVa and type IVb, which displayed a similar domain organization, except type IVb cadherins have a larger number of extracellular cadherin (EC) domains than do type IVa cadherins (nine versus seven). We also showed that *type IVa cadherin* genes occurred in the hexapod, branchiopod, and copepod genomes whereas only *type IVb cadherin* genes were present in malacostracans. Furthermore, comparative characterization of the type IVb cadherins suggested that the presence of two extra EC domains in their N-terminal regions represented primitive characteristics. In addition, we identified an evolutionary loss of two highly conserved cysteine residues among the type IVa cadherins of insects.

Conclusions: We provide a genomic perspective of the evolution of classical cadherins among bilaterians, with a focus on the Arthropoda, and suggest that following the divergence of early arthropods, the precursor of the insect type IV cadherin evolved through stepwise reductive changes from the ancestral type III state. In addition, the complementary distributions of polarized genomic characters related to type IVa/IVb cadherins may have implications for our interpretations of pancrustacean phylogeny.

Keywords: Cadherin, Cell adhesion, Adherens junction, Arthropod, Chelicerate, Crustacean, Insect, Genome, Evolution, Phylogeny

Background

Classical cadherins, a metazoan-specific subfamily of the cadherin superfamily [1–3], are homophilic cell-cell adhesion molecules that play key roles in metazoan morphogenesis [3–7], and as single-pass transmembrane proteins, their ectodomains contain repetitive extracellular cadherin (EC) domains that function to recognize and bind cells that express the same or similar cadherin molecules [8, 9]. The cytoplasmic domains of classical cadherins also bind to catenins [10], through which they interact with the actomyosin network [7] and potentially integrate actomyosin-generated physical forces into tissue-level tension, thereby regulating tissue homeostasis and morphogenesis [11–13].

Genes that encode classical cadherins have been identified in many bilaterian species, as well as in several non-bilaterian metazoans [2, 3, 14–19], and studies in both vertebrate and insect models have firmly

* Correspondence: hoda@brh.co.jp
[1]Laboratory of Evolutionary Cell and Developmental Biology, JT Biohistory Research Hall, 1-1 Murasaki-cho, Takatsuki 569-1125, Osaka, Japan
[3]Department of Biological Sciences, Graduate School of Science, Osaka University, Osaka, Japan
Full list of author information is available at the end of the article

established the role and mechanisms of classical cadherins in animal development [4, 5, 7]. However, despite the conservation of their functions, classical cadherins exhibit remarkable variation in the structure of their ectodomains [3], and members of the classical cadherin family have been categorized as types I, II, III, and IV, or otherwise, based on their phylogenetic grouping and domain organization [1, 3, 20].

Type I and type II cadherins each possess five tandem EC domains, and these cadherin types are common in vertebrates but have not been reported to occur in invertebrates, with the exception of urochordates [14, 21]. Certain subtypes of type I and type II cadherins, including E-cadherin (type I) and cadherin-5 or VE-cadherin (type II), serve as components of adherens junctions in vertebrate epithelial tissues. However, type IV cadherins function as the key adhesion molecules of adherens junctions in insect epithelial tissues and include the *Drosophila melanogaster* E-cadherin, DE-cadherin (Fig. 1a), which is the representative type IV cadherin [22–24]. Type IV cadherins are characterized by their shared domain organization, which includes seven EC domains, followed by the non-chordate classical cadherin (NC), cysteine-rich EGF-like (CE), and laminin-G (LG) domains [25], and they have been identified in insects, non-insect hexapods (e.g., collembolan) and branchiopod crustaceans [15]. Importantly, recent studies have revealed that the structural mechanisms responsible for homophilic binding of type I/II and type IV cadherins are quite different [26, 27]. Moreover, type III cadherins are distributed among a wide range of bilaterian metazoans, including arthropods, echinoderms, and even vertebrates, but they have yet to be identified in non-bilaterian metazoans [2, 15, 20, 28, 29]. The representative type III cadherin is *D. melanogaster* neural cadherin, DN-cadherin (Fig. 1a), the expression and function of which primarily occurs in non-epithelial tissues [30]. In contrast to type I, II, and IV cadherins, type III cadherin molecules contain 14 to 17 EC domains followed by the ectodomain, which includes one NC, three CE (CE1-CE3), and two LG (LG1 and LG2) domains with the following organization: NC-CE1-LG1-CE2-LG2-CE3. In addition, non-categorized/unconventional forms of classical cadherins have also been reported to occur in nematodes, hemichordates, and cephalochordates [15, 31, 32]. Although up to 17 EC domains have been observed in the classical cadherins of bilaterians, 25 or more have been reported in the classical cadherin-encoding genes of non-bilaterian metazoans [2, 17].

The structural variation of the ectodomains of classical cadherins is thought to have resulted from domain losses that occurred at critical points in metazoan or bilaterian evolution [2, 3, 15, 33]. This hypothesis is based on the conclusion that the type III form represents the last common precursor of all bilaterian classical cadherins, a conclusion that is supported by the widespread, albeit scattered, phylogenetic distribution of *type III cadherin* genes among bilaterians, detectable conservation throughout the amino acid sequences of type III cadherins, and the observation that all other forms of cadherins can be recognized as derived states of the type III form [15, 34]. However, it remains unclear whether the various forms of classical cadherins were present in the last common ancestors of the individual phyla, as well as whether the currently recognized derived states, i.e., the type I/II and type IV cadherins, evolved from the ancestral type III state during a distinct event or through progressive evolution. Efforts to answer these questions may contribute to a better understanding of how the structural mechanisms of classical cadherin-mediated adhesion evolved in metazoans and at what points of animal evolution the adhesion mechanisms were changed or modified.

To address these questions, we focused on the phylum Arthropoda, in which growing volumes of genomic and transcriptomic sequence resources are available for a broad range of species. We investigated both genomic and transcriptomic classical cadherin-encoding sequences from a wide range of arthropod and non-arthropod bilaterians, including chelicerates, a myriapod, and several non-branchiopod crustaceans, to determine whether *type IV cadherin* genes evolved from *type III cadherin* genes before, during, or after the early divergence of arthropods, and whether type IV cadherins arose from the type III state abruptly or through an intermediate state (or several intermediate states).

Results

Classical cadherin genes in the chelicerate *P. tepidariorum* genome

In the present study, we first identified a *P. tepidariorum* (common house spider formerly known as *Achaearanea tepidariorum*) type III cadherin-encoding cDNA (Fig. 1a) that was distinct from a copy of At-cadherin cDNA previously reported [15]. Therefore, the previously identified At-cadherin was redesignated Pt1-cadherin, and the newly identified gene product was designated Pt2-cadherin.

RNA sequencing (RNA-seq) of *P. tepidariorum* embryos at stages 5 and 10 demonstrated that the Pt1- and Pt2-cadherin transcripts were expressed in both early and late embryonic stages and that, at both stages, the expression level of the *Pt2-cadherin* gene was greater than that of the *Pt1-cadherin* gene (Additional file 1: Tables S1, S2). The predicted Pt1- and Pt2-cadherins were 2985 and 2961 amino acids long, respectively, and the sequences could be aligned along their entire lengths, exhibiting 66% identity. These sequences could also be aligned with the DN-cadherin sequence; however, the N-terminal regions appeared to have diverged. Using a protein domain search of the PROSITE database

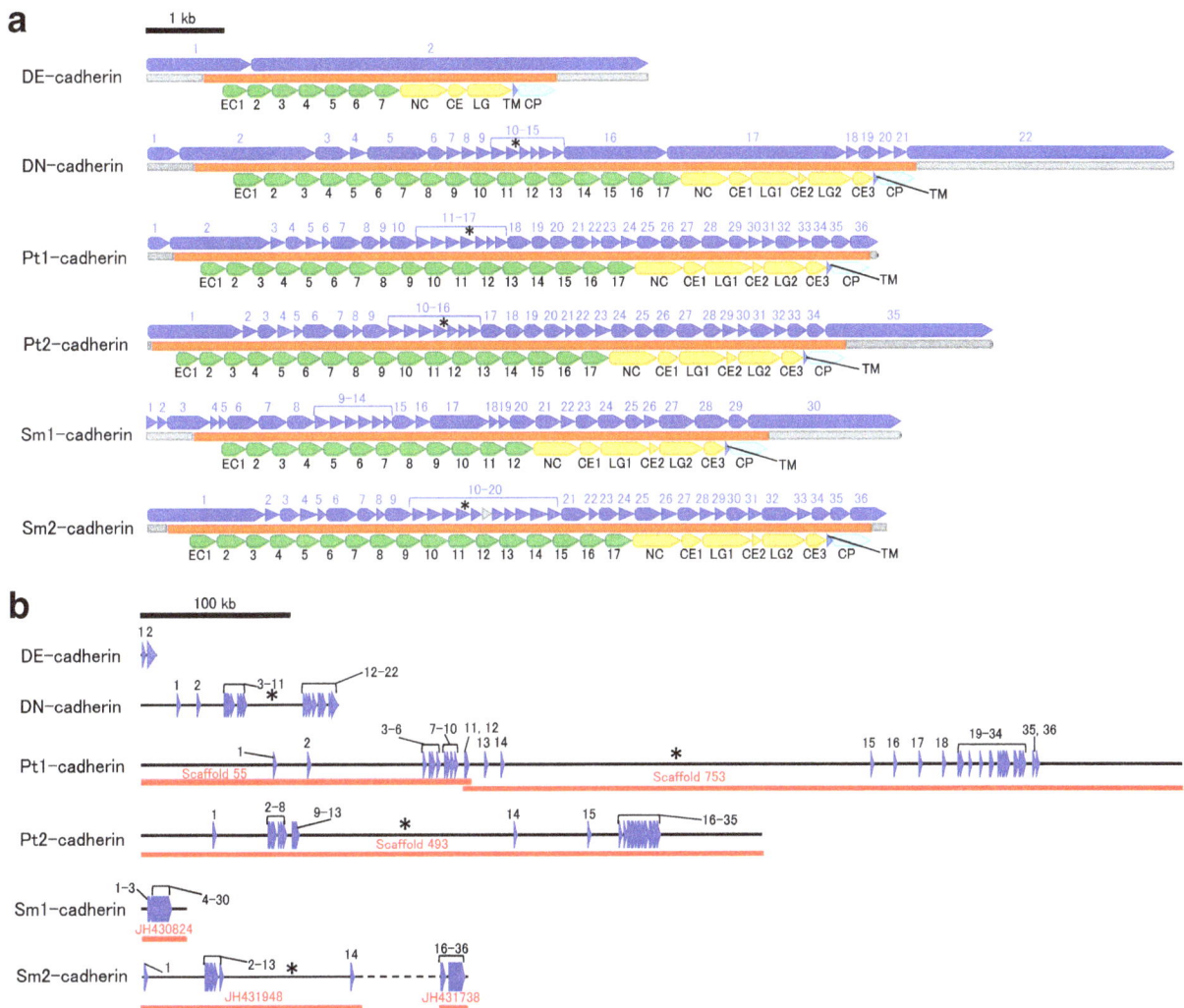

Fig. 1 Genomic and domain organization of classical cadherins in *Parasteatoda tepidariorum* and *Strigamia maritima*. **a**. Schematic representation of exons (upper), transcripts (middle), and domain organization (lower) of Pt1-, Pt2-, Sm1-, and Sm2- cadherins, compared with those of DE- and DN-cadherins. The scale bar indicates 1 Kbp. All exons identified in the genome sequences are depicted in blue, except for the exons depicted in gray for the *Sm2-cadherin* gene, which remained hypothetical because its sequence was not found in the *S. maritima* genome sequence assembly. For each cadherin, the exons are tentatively numbered to facilitate comparison (numbers in blue). The coding region of each transcript is depicted in orange, with the 5'- and 3'-untranslated regions depicted in gray. The domain names are abbreviated as follows: EC, extracellular cadherin domain; NC, non-chordate classical cadherin domain; CE, cysteine-rich EGF-like domain; LG, laminin globular-like domain; TM, transmembrane domain; CP, cytoplasmic domain. **b**. Schematic representation of the genomic organization of the *Pt1-*, *Pt2-*, *Sm1-*, and *Sm2-cadherin* genes, compared with those of the *DN-* and *DE-cadherin* genes. Thin horizontal black lines indicate the genome sequences. The scale bar indicates 100 Kbp. Blue triangles indicate individual exons, which are numbered to facilitate comparison. Red lines indicate scaffold sequences of the *P. tepidariorum* or *S. maritima* genome assemblies. Broken lines indicate missing sequences. Conserved insertions of the largest introns observed in the *Pt1-*, *Pt2-*, *DN-*, and *Sm2-cadherin* genes are indicated by asterisks in both A and B. The genomic sequences annotated for the Pt1- and Pt2-cadherins are available in GenBank (BR001342 and BR001343)

[35], we detected 17 and 16 EC domains in the Pt1- and Pt2-cadherin sequences, respectively, and two LG domains in each. Next, we aligned the sequences of the EC repeats (Additional file 2: Figure S1) and defined the start and end positions of the individual EC domains, which were numbered from 1 to 17 (EC1 to EC17). The more C-terminal regions of the Pt1- and Pt2-cadherins were subdivided into eight domains (NC, CE1, LG1, CE2, LG2,

CE3, TM, and CP; Fig. 1a; Additional file 2; Figure S1). Although some NC domain sequences in classical cadherins have been reported to exhibit weak similarities to typical EC domains [36], the NC domain was not considered an EC domain in this work because of its limited sequence similarity. Practically, the positions of the domains of DN-cadherin and other type III cadherins were defined based on sequence alignment with the Pt1- and Pt2-cadherins.

To investigate the genomic organization of the *Pt1-* and *Pt2-cadherin* genes, we used scaffold sequences of the *P. tepidariorum* isolate Göttingen genome (~1.4 Gbp) (GCA_000365465.1) [37], as well as whole genome shotgun sequencing (WGS) reads of the *P. tepidariorum* isolate Osaka genome (>31× coverage) (Additional file 1: Table S3). The full-length nucleotide sequence of the Pt1-cadherin cDNA was mapped to the ~488 Kbp region of Scaffolds 55 and 753, which could be connected into a continuous sequence (Fig. 1b) and contained at least 35 exons that were separated by introns of various sizes (from 444 bp to more than 240 Kbp). Similarly, the full-length nucleotide sequence of the *Pt2-cadherin* transcript was mapped to the ~298 Kbp region of Scaffold 493 (Fig. 1b), and we found that the *Pt2-cadherin* gene contained at least 36 exons that were separated by introns of various sizes (from 75 bp to more than 140 Kbp). Most, but not all, of the exons in both genes were small (<400 bp), and all of the introns in the protein-coding regions of the *Pt1-* and *Pt2- cadherin* genes were inserted at homologous sites (Additional file 2: Figure S1). In addition, the total lengths of the *Pt1-* and *Pt2-cadherin* genes were much larger than the total length of the *DN-cadherin* gene (Fig. 1a, b); however, the three genes shared at least 13 intron insertion sites, including those for the largest introns (Fig. 1a, b; Additional file 2: Figure S1).

To investigate whether a third *classical cadherin* gene was present in the *P. tepidariorum* genome, we exhaustively searched the genome *P. tepidariorum* isolate Göttingen genome sequence assembly and reads from the *P. tepidariorum* isolate Osaka WGS and RNA-seq. However, there was no sign of a third *classical cadherin* gene in *P. tepidariorum*, which led us to conclude that the *Pt1-* and *Pt2-cadherin* genes are the only *classical cadherin* genes in the species.

Identification of *classical cadherin* genes in other non-hexapod arthropod genomes

To investigate the repertoire of *classical cadherin* genes in other non-hexapod arthropod genomes, we searched the publicly available genome sequence assemblies of four chelicerate species (velvet spider *Stegodyphus mimosarum* [38]; two-spotted mite *Tetranychus urticae* [39]; western predatory mite *Metaseiulus occidentalis* [40]; *Mesobuthus martensii* [41]), a myriapod species (centipede *Strigamia maritima* [42]), and four crustacean species (water flea *Daphnia pulex* [43]; copepod *Eurytemora affinis* [37]; amphipod *Hyalella azteca* [37]; amphipod *Parhyale hawaiensis* [44]) (Table 1). The capability to detect the entire organization of *classical cadherin* genes depended on the quality of the genome sequence assembly and the availability of rich transcriptomic resources. The genome sequence assemblies that were searched comprised scaffolds or contigs with

relatively high N50 values and relatively low proportions of undetermined bases. Although transcript models for classical cadherins were predicted in many of the genome sequence assemblies, we carefully evaluated the organization of all detectable *classical cadherin* genes. RNA-seq reads, if publicly available and necessary, were used to reconstruct the transcript sequence of classical cadherins. In addition to the publicly available sequence resources, we generated RNA-seq reads for both sea slater *Ligia exotica* and freshwater shrimp *Caridina multidentata*, as well as WGS reads with approximately 8× and 13× coverage depths, respectively (Table 1; Additional file 1: Tables S1, S3). These sequence resources were also used to search for *classical cadherin* genes.

Chelicerates
In each of the *S. mimosarum* and *M. martensii* genomes, we detected two *type III cadherin* genes that were closely related to the *Pt1-* and *Pt2-cadherin* genes. In the genomes of both *T. urticae* and *M. occidentalis*, we detected a single *type III cadherin* gene; however, no other *classical cadherin* genes were detected. In addition, we found that all intron insertion sites in the coding regions of all the *type III cadherin* genes of *P. tepidariorum* and *M. martensii* were conserved between them (Additional file 3: Figure S2).

Myriapods
In the *S. maritima* genome sequence assembly, two *classical cadherin* genes were identified (Fig. 1a; Table 1). The predicted products were designated Sm1- and Sm2-cadherin. The Sm2-cadherin was considered a type III cadherin, and its exon-intron structure was similar, but not identical, to that of the *Pt1-* and *Pt2-cadherin* genes, although a small portion of the coding sequence was not mapped to any scaffold (Fig. 1a, b). In contrast, Sm1-cadherin exhibited most of the typical type III cadherin elements, but since it contained only 12 EC domains, it could be classified as neither a type III nor a type IV cadherin. In addition, we also observed that the *Sm1-cadherin* gene contained at least 30 exons that were condensed within a small genomic region (~15 Kbp) (Fig. 1a, b).

Branchiopod crustaceans
In the branchiopod crustacean *Artemia franciscana*, we previously identified both type IV and type III cadherins, i.e., Af1- and Af2-cadherin [15], which were orthologous to DE- and DN-cadherin, respectively, as well as to two predicted products from the *Daphnia pulex* genome [43], hereafter referred to as Dp1- and Dp2-cadherin (Table 1).

Table 1 *Classical cadherin* genes found in publicly available genome sequences of non-hexapod arthropods

Taxon/species	Genome accession	#Scaffold (Gene accession)	Type	Product
Chelicerata, Araneae				
Parasteatoda tepidariorum	GCA_000365465.1	#55/#753 (AB190303)	III	Pt1-cadherin
		#493 (LC110189)	III	Pt2-cadherin
Stegodyphus mimosarum	GCA_000611955.2	#7/#4105/#15,197/#13303[a]	III	(Close to Pt1)
		#10064/#11,847/#1110[a]	III	(Close to Pt2)
Chelicerata, Scorpiones				
Mesobuthus martensii	GCA_000484575.1	#343080[b, c, d]	III	Mma1-cadherin
		#352483[b, c, d]	III	Mma2-cadherin
Chelicerata, Acari				
Tetranychus urticae	GCA_000239435.1	#8 (XP_015784984)	III	
Metaseiulus occidentalis	GCA_000255335.1	#JH621154 (XM_003743492)	III	Mo-cadherin
Myriapoda, Chilopod				
Strigamia maritima	GCA_000239455.1	#JH431948/#JH431738[d]	III	Sm2-cadherin
		#JH430824[d] (SMAR001807)	n.c.	Sm1-cadherin
Crustacea, Branchiopoda				
Daphnia pulex	GCA_000187875.1	#100 (EFX70325)	III	Dp2-cadherin
		#3 (EFX89066)	IVa	Dp1-cadherin
Crustacea, Copepoda				
Eurytemora affinis	GCA_000591075.1	#33[d]	III	Ea2-cadherin
		#103273/#511[d]	IVa	Ea1-cadherin
Crustacea, Isopoda				
Ligia exotica	BDMT010000000	(AB190302)	III	Le2-cadherin
		(LC110190)	IVb	Le1-cadherin
Crustacea, Amphipoda				
Hyalella azteca	GCA_000764305.2	#323 (XM_018161032)	III	Ha2-cadherin
		#236 (XM_018157906)	IVb	Ha1-cadherin
Parhyale hawaiensis	GCA_001587735.1	#25754[c, e] (tra_m.010273)	III	Ph2-cadherin
		#4723 (tra_m.024063)	IVb	Ph1-cadherin
Crustacea, Decapoda				
Caridina multidentata	BDMR010000000	(AB190301)	III	Cm-cadherin

[a]The scaffolds were linked by detected sequences that were very similar to those of Pt1- or Pt2-cadherin
[b]#Contig
[c]Only the scaffold or contig containing exons coding for the CP domain is shown
[d]Sequence details are available in Additional files 11 and 12
[e]#95284 was detected as a partial duplicate. n.c., not categorized

Non-branchiopod crustaceans

In the isopod *L. exotica* and the decapod *C. multidentata*, we previously identified type III cadherins but failed to detect any other forms [15]. In the present study using *L. exotica*, we were able to predict a transcript that encoded a hypothetical classical cadherin that was distinct from the previously identified Le-cadherin (Table 1; Fig. 2a), and the occurrence of the transcript was validated using reverse transcriptase polymerase chain reaction (PCR) amplification and sequencing. Accordingly, the newly identified classical cadherin was designated Le1-cadherin, and the previously identified

Le-cadherin was redesignated Le2-cadherin. Notably, Le1-cadherin was structurally similar to type IV cadherins in that it lacked the CE2, LG2, and CE3 domains that are typical of type III cadherins (Fig. 2a; Additional file 4: Figure S3). However, the protein was distinct from other known type IV cadherins in that it contained two additional EC domains.

Considering that the N-terminal-most four EC domains of type IV DE-cadherin has a folded, globular structure involved in homophilic binding [26, 27], the finding of the unique domain organization of Le1-cadherin raised the question of whether it is functional.

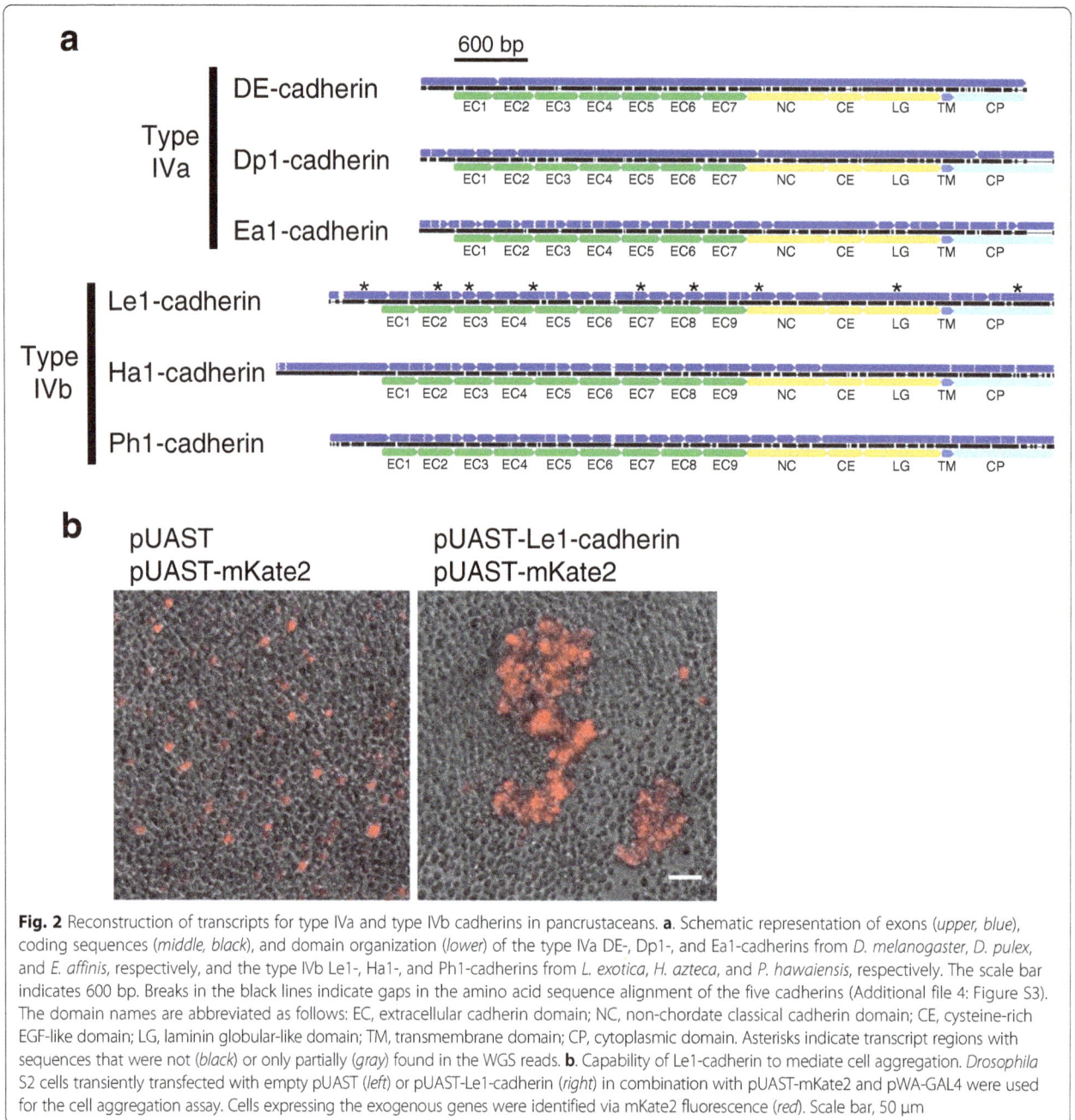

Fig. 2 Reconstruction of transcripts for type IVa and type IVb cadherins in pancrustaceans. **a**. Schematic representation of exons (*upper, blue*), coding sequences (*middle, black*), and domain organization (*lower*) of the type IVa DE-, Dp1-, and Ea1-cadherins from *D. melanogaster*, *D. pulex*, and *E. affinis*, respectively, and the type IVb Le1-, Ha1-, and Ph1-cadherins from *L. exotica*, *H. azteca*, and *P. hawaiensis*, respectively. The scale bar indicates 600 bp. Breaks in the black lines indicate gaps in the amino acid sequence alignment of the five cadherins (Additional file 4: Figure S3). The domain names are abbreviated as follows: EC, extracellular cadherin domain; NC, non-chordate classical cadherin domain; CE, cysteine-rich EGF-like domain; LG, laminin globular-like domain; TM, transmembrane domain; CP, cytoplasmic domain. Asterisks indicate transcript regions with sequences that were not (*black*) or only partially (*gray*) found in the WGS reads. **b**. Capability of Le1-cadherin to mediate cell aggregation. *Drosophila* S2 cells transiently transfected with empty pUAST (*left*) or pUAST-Le1-cadherin (*right*) in combination with pUAST-mKate2 and pWA-GAL4 were used for the cell aggregation assay. Cells expressing the exogenous genes were identified via mKate2 fluorescence (*red*). Scale bar, 50 μm

To investigate this question, we performed cell aggregation assays using *Drosophila* S2 cells transiently transfected with or without a Le1-cadherin expression construct (Fig. 2b). The result indicated that Le1-cadherin was capable of mediating cell-cell adhesion.

A *classical cadherin* gene specifically related to the *Le1-cadherin* gene was also identified in each of the *H. azteca* and *P. hawaiensis* genomes. Both these predicted products had essentially the same domain organization as Le1-cadherin, and they were designated Ha1- and Ph1-cadherin, respectively, (Fig. 2a; Additional file 4: Figure S3). The *Le1-*, *Ha1-* and *Ph1-cadherin* genes

exhibited a relatively complex yet mutually similar exon-intron organization (Fig. 2a). As expected, the amphipod genomes also contained *type III cadherin* genes, and their predicted products were designated Ha2- and Ph2-cadherin (Additional file 3: Figure S2).

Additionally, analysis of the *C. multidentata* WGS reads revealed genomic regions with sequences that were distinct from the previously reported *type III Cm-cadherin* gene and were more closely related to Le1-cadherin than to Le2-cadherin (Additional file 5: Figure S4). However, since the *C. multidentata* RNA-seq data poorly represented sequences specifically related to Le1-cadherin, we were unable to generate a predicted transcript. The de novo assembly of the *C. multidentata* RNA-seq reads, nonetheless, allowed us to detect contigs encoding a part of classical cadherin closely similar but not identical to Cm-cadherin (Additional file 5: Figure S4). These contigs were connected by some raw reads, indicating that the *C. multidentata* genome might have another *classical cadherin* gene, which had retained domain elements characteristic of type III cadherin, rather than type IV cadherin.

In the genome sequence assembly of the copepod *E. affinis*, two hypothetical *classical cadherin* genes were detected, and their predicted products were designated Ea1- and Ea2-cadherin (Table 1). Ea2-cadherin was a type III cadherin, whereas Ea1-cadherin was a type IV cadherin that had essentially the same domain organization as those of DE- and Dp1-cadherin (Fig. 2a; Additional file 4: Figure S3).

Redefinition of the type IV cadherin category

For simplicity and convenience, we redefined the term "type IV cadherin." Irrespective of the total number of EC domains, all the classical cadherins that were characterized by the absence of the CE2, LG2, and CE3 domains were included in the type IV category, and the category was also separated into two subclasses, type IVa and type IVb, based on differences in their EC domains. More specifically, type IVa cadherins were defined as type IV cadherins that contain seven EC domains, whereas type IVb cadherins were defined as type IV cadherins that contain the same seven EC domains as well as two more EC domains (Fig. 2a). The validity of this classification will be further examined below.

Relationships among the domain organizations of type II, III, IVa, and IVb cadherins

To systematically detect the possible homologous regions between the diverse classical cadherins in various arthropods and other bilaterians, we searched for collinear arrangements of similarities between their amino acid sequences using blast-based dot-plot comparisons with the amino acid sequences of the Pt1- and Pt2-cadherins as the reference sequences. Using a sliding window of 120 amino acids, we generated a series of overlapping sequences from the entire amino acid sequences of DN-, Sm1-, Le1-, and DE-cadherins, *Pundamilia nyererei* (teleost fish) Pn-cadherin (vertebrate type III), and *Mus musculus* (mammal) Mm5-cadherin (also known as VE-cadherin; vertebrate type II). The serial sequences were then blasted against each reference (i.e., Pt1- and Pt2-cadherin), and the resulting E-values at the blast-hit positions were plotted to visualize the collinearity and identify evolutionarily conserved regions (Fig. 3).

The resulting dot-plots indicated that DN-cadherin is well conserved with the chelicerate type III cadherins

throughout its length and that Pn-cadherin has 14 EC domains that correspond to the EC4-EC17 domains of Pt1- and Pt2-cadherins (Fig. 3). However, the three Pn-cadherin (vertebrate type III) EC domains that corresponded to the EC1-EC3 domains of arthropod type III cadherins were too divergent to detect. To avoid confusion, however, the EC domains of all the type III cadherins in the present study were numbered based on the detected collinear arrangements with the numbered EC domains of Pt1- and Pt2-cadherin (Additional file 2: Figure S1). The blast-based dot-plot comparisons revealed correspondence between the EC1-EC6 region of the type IVa cadherins and the EC8-EC13 region of the type III cadherins, as well as between the EC7-LG region of the type IVa cadherins and the EC17-LG1 region of the type III cadherins (Fig. 3). However, although the conservation between the EC9-LG region of Le1-cadherin and the EC17-LG1 region of the type III cadherins was evident, the comparisons of Le1-cadherin against the Pt1- and Pt2-cadherins yielded less clear patterns than were observed for some of the other comparisons (Fig. 3). In addition, the five EC domains of the type II Mm5-cadherin yielded weak but specifically detectable collinear plots with the five C-terminal EC domains of Pt2-cadherin, although the pattern was less clear in the comparison of Mm5- and Pt1-cadherins (Fig. 3).

Since the difficulty of aligning the Le1-cadherin sequence might have stemmed from its divergence from the Pt1- and Pt2-cadherin sequences, we also compared the Le1-cadherin sequence to the sequences of the DE-, Dp1-, Sm1-, Cm-, Le2-, DN-, and Pn-cadherins (Fig. 4). These comparisons provided clearer patterns of collinearity and revealed correspondences between the EC3-EC9 region of Le1-cadherin and the EC1-EC7 region of the type IVa cadherins, although the sequence of the Le1-cadherin EC8 domain appeared to diverge from that of the type IVa cadherin EC6 domain. The correspondence between the EC1-EC5 region of Le1-cadherin and the EC6-EC10 region of the type III cadherins and between the EC2-EC5 region of both Le1-cadherin and Sm1-cadherin were also supported (Fig. 4, green boxes), whereas the EC6-EC8 region of Le1-cadherin demonstrated ambiguous affinities to the EC11-EC13, EC12-EC14, and EC14-EC16 regions of the type III cadherins (Fig. 4, blue boxes). Nonetheless, the correspondence between the EC6-EC7 regions of the Le1- and Sm1-cadherins was specifically supported. Similar results were also obtained with the other type IVb cadherins (Additional file 6: Figure S5).

Comparison of the exon-intron organization of type II, III, IVa and IVb cadherins

To assess the conservation of the exon-intron organization among *classical cadherin* genes, we constructed an alignment of the amino acid sequences

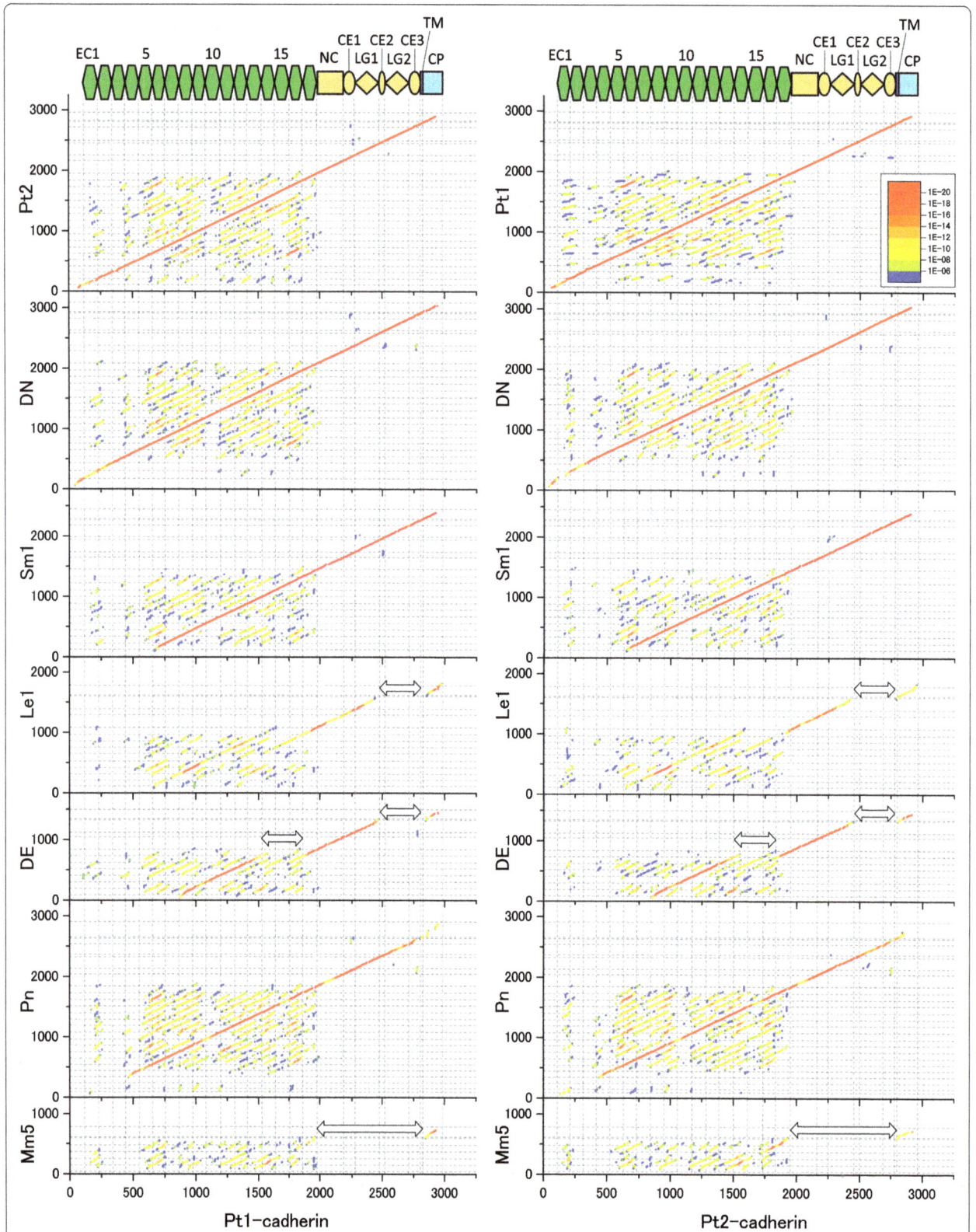

Fig. 3 Blast-based dot-plot comparisons of classical cadherin amino acid sequences. The classical cadherins analyzed are as follows: type III Pt1-cadherin (Pt1, spider); type III Pt2-cadherin (Pt2, spider); type III DN-cadherin (DN, fruit fly); Sm1-cadherin (Sm1, centipede); type IVb Le1-cadherin (Le1, sea slater); type IVa DE-cadherin (DE, fruit fly); type III Pn-cadherin (Pn, fish); type II Mm5-cadherin (Mm5, mouse). Double-headed horizontal arrows indicate gaps detected between regions of collinear similarity

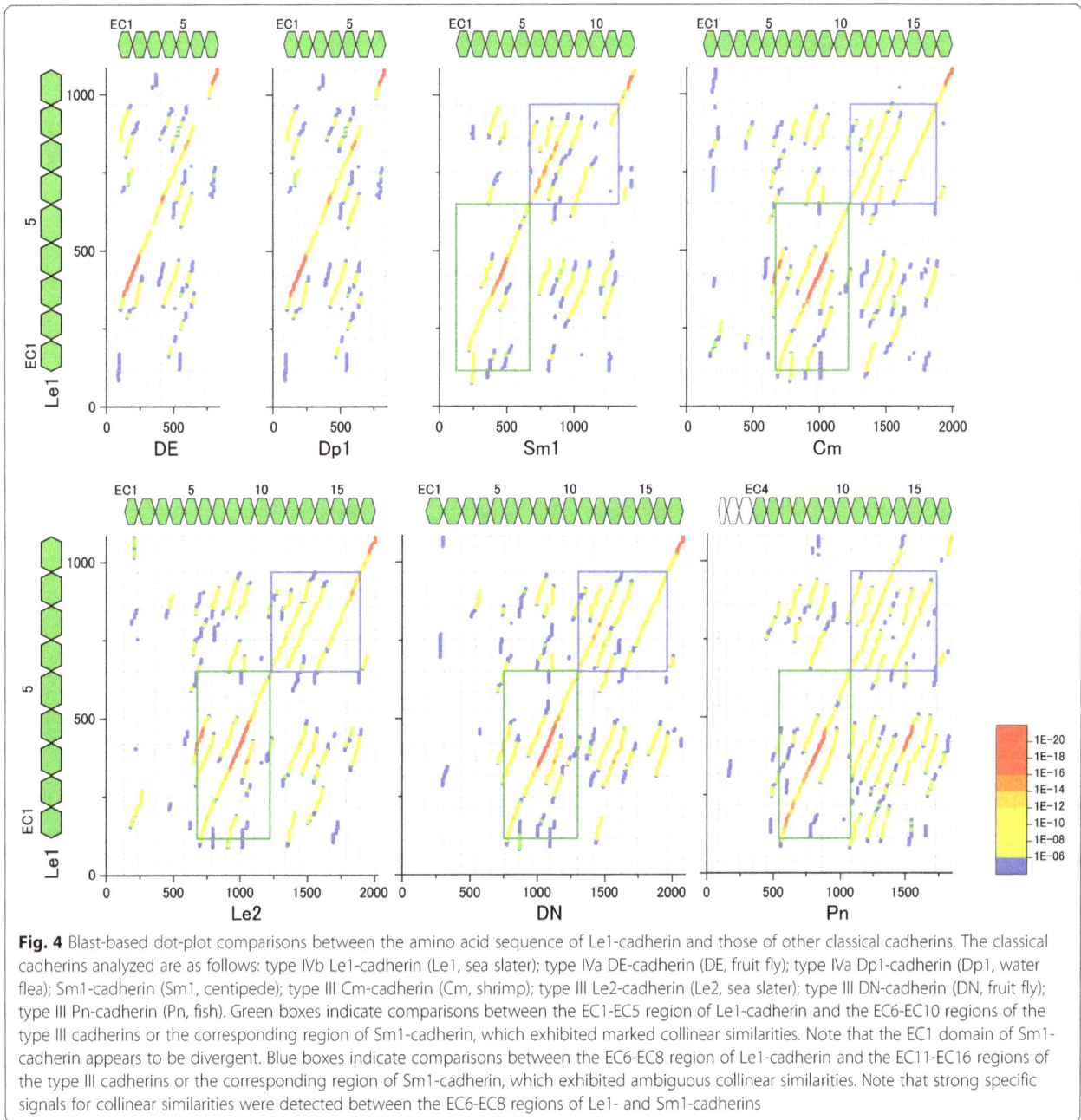

Fig. 4 Blast-based dot-plot comparisons between the amino acid sequence of Le1-cadherin and those of other classical cadherins. The classical cadherins analyzed are as follows: type IVb Le1-cadherin (Le1, sea slater); type IVa DE-cadherin (DE, fruit fly); type IVa Dp1-cadherin (Dp1, water flea); Sm1-cadherin (Sm1, centipede); type III Cm-cadherin (Cm, shrimp); type III Le2-cadherin (Le2, sea slater); type III DN-cadherin (DN, fruit fly); type III Pn-cadherin (Pn, fish). Green boxes indicate comparisons between the EC1-EC5 region of Le1-cadherin and the EC6-EC10 regions of the type III cadherins or the corresponding region of Sm1-cadherin, which exhibited marked collinear similarities. Note that the EC1 domain of Sm1-cadherin appears to be divergent. Blue boxes indicate comparisons between the EC6-EC8 region of Le1-cadherin and the EC11-EC16 regions of the type III cadherins or the corresponding region of Sm1-cadherin, which exhibited ambiguous collinear similarities. Note that strong specific signals for collinear similarities were detected between the EC6-EC8 regions of Le1- and Sm1-cadherins

of 21 bilaterian classical cadherins, including arthropod and non-arthropod type III cadherins, hexapod and branchiopod type IVa cadherins, a non-branchiopod type IVb cadherin, and a vertebrate type II cadherin (Additional file 9: Figure S7). Large gaps were introduced based on the results shown in Figs. 3 and 4, so that the likely homologous regions of the sequences could be aligned. Although many regions of the cadherins, including their N-terminal regions, remained poorly aligned, we only considered unambiguously aligned regions of the sequences during the analysis (Fig. 5; Additional file 9: Figure S7).

As previously mentioned (Fig. 1), there were 34 introns inserted in the respective coding regions of the *Pt1-* and *Pt2-cadherin* genes, and all of the intron insertions were located at identical sites between the two genes. Notably, 33 of the 34 intron insertion sites were shared among four or more of the non-chelicerate bilaterian genes examined (Fig. 5, red, yellow and green lines), and of those 33 intron insertion sites, only six were specific to arthropods (Fig. 5, yellow lines), and two were shared only with non-arthropod genes (Fig. 5, green lines). Despite the differences in domain organization, six of the 10 intron insertions in the *type II cadherin* gene were

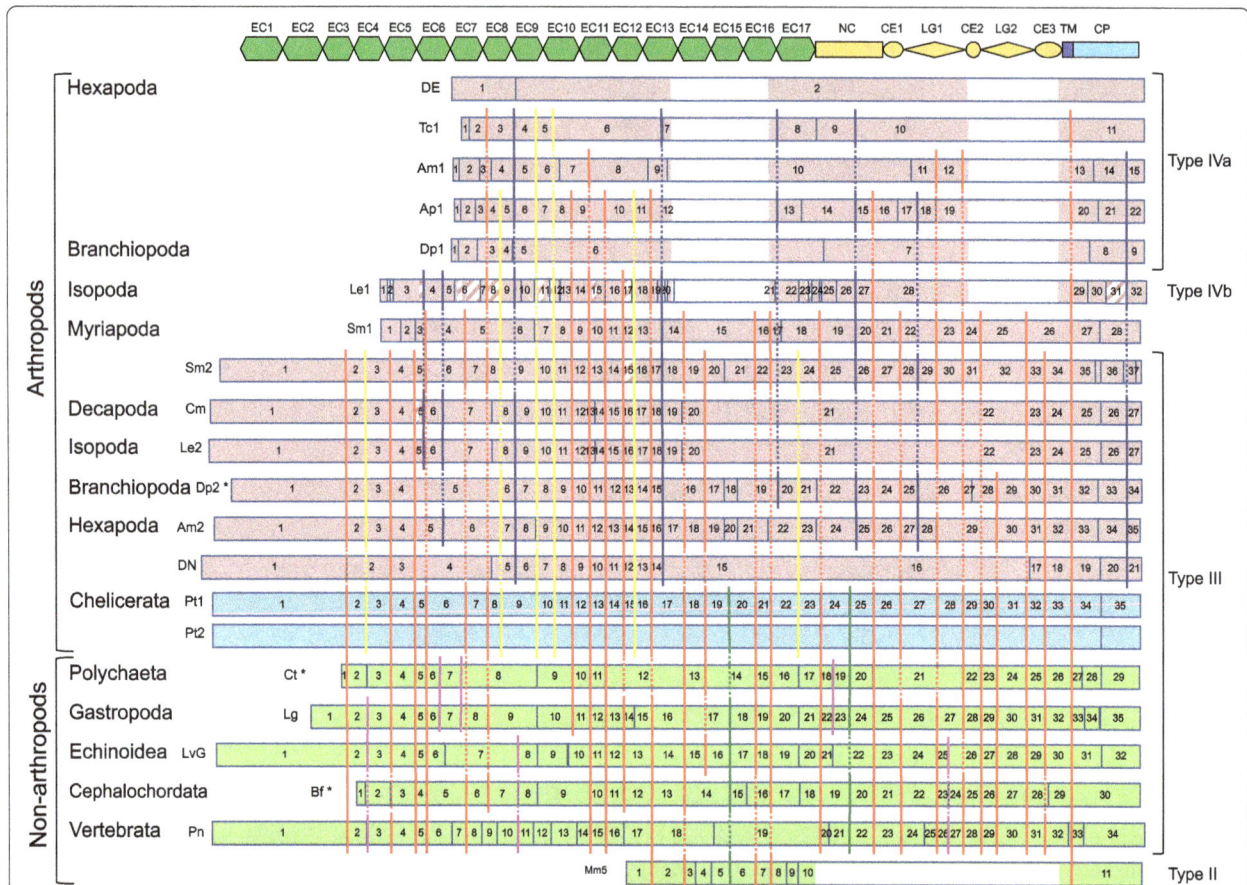

Fig. 5 Schematic representation of conserved intron positions among *classical cadherin* genes. Colored rectangles indicate protein-coding regions of the transcripts of the *Pt1*- and *Pt2-cadherin* (light blue), non-spider arthropod *classical cadherin* (pink), and non-arthropod *classical cadherin* (green) genes. The *classical cadherin* genes shown are as follows: *DE-cadherin* (DE, fruit fly); *Tc1-cadherin* (Tc1, beetle); *Am1-cadherin* (Am1, honey bee); *Ap1-cadherin* (Ap1, aphid); *Dp1-cadherin* (Dp1, water flea); *Le1-cadherin* (Le1, sea slater); *Sm1-cadherin* (Sm1, centipede); *Sm2-cadherin* (Sm2, centipede); *Cm-cadherin* (Cm, shrimp); *Le2-cadherin* (Le2, sea slater); *Dp2-cadherin* (Dp2, water flea); *Am2-cadherin* (Am2, honey bee); *DN-cadherin* (DN, fruit fly); *Pt1-cadherin* (Pt1, spider); *Pt2-cadherin* (Pt2, spider); *Ct-cadherin* (Ct, polychaete); *Lg-cadherin* (Lg, snail); *LvG-cadherin* (LvG, sea urchin); *Bf-cadherin* (Bf, amphioxus); *Pn-cadherin* (Pn, fish); and *Mm5-cadherin* (Mm5, mouse). Regions of the transcripts for which genomic sequences were not available are indicated by slanted stripes. A schematic illustration exhibiting the domain structure of Pt1-cadherin is placed at the top as a positional reference. All identified intron insertion sites are shown, and the exons are tentatively numbered from the 5′-terminal side of each transcript. Conserved intron positions that were identified based on the alignment of the amino acid sequences (Additional file 10: Figure S8) are represented by colored vertical lines. The yellow lines indicate arthropod-specific intron positions, whereas the green lines indicate conserved intron positions between the *Pt1*- and *Pt2-cadherin* genes and some of the non-arthropod bilaterian *classical cadherin* genes but not among any other arthropod genes. The red lines indicate intron positions that are conserved both between the *Pt1*- and *Pt2-cadherin* genes and some of the other arthropod genes and between the spider genes and some of the non-arthropod bilaterian genes. The blue lines indicate conserved intron positions between some of the non-spider arthropod *type III cadherin* genes and some of the *type IV cadherin* genes, and the purple lines show the conservation of intron positions within the non-arthropods

conserved in the *Pt1*- and *Pt2-cadherin* genes, as well as in the gastropod and echinoderm *type III cadherin* genes (Fig. 5, red and green lines). Among the arthropod genes, eight intron insertion sites were conserved between the *type III* and *type IV cadherin* genes but were missing in the spider and non-arthropod *classical cadherin* genes (Fig. 5, blue lines). Taken together, these observations indicated that many of the introns in the *Pt1*- and *Pt2-cadherin* genes were inherited in a complex ancestral state rather than acquired via lineage-specific gains of introns and suggested that at least 33 of the 34 introns in the *Pt1*-

and *Pt2-cadherin* genes predate the earliest divergence of extant arthropod groups.

Notably, comparisons between the exon-intron organizations of type IVb and arthropod type III cadherins in their EC coding regions revealed marked conservation between them despite their divergence at the amino acid sequence level (Fig. 5; Additional file 7: Figure S6), while the ancestral patterns of intron insertions were less conserved in *type IVa cadherin* genes. The EC10-EC13 coding regions of arthropod *type III cadherin* genes contained eight conserved intron insertions, seven of which were conserved in the EC5-EC8 coding region of the *Le1-cadherin* and other *type IVb cadherin* genes (Fig. 5; Additional file 7: Figure S6). Similarly, the EC1 coding region of the *type IVb cadherin* genes had two intron insertion sites conserved in the EC6 coding region of some of the pancrustacean *type III cadherin* genes (i.e., the *Le2-* and *Cm-cadherin* genes). These observations, together with the results of the blast-based dot-plot comparisons, strongly suggested that the EC1-EC7 region of type IVb cadherins and the adjacent EC8 domain were homologous to the EC6-EC12 region of type III cadherins and the adjacent EC13 domain.

Despite the simple exon-intron organization of the *DE-* and *Dp1-cadherin* genes, some other type IVa cadherins shared a considerable number of intron insertions with *type III cadherin* genes (Fig. 5, red, yellow, and blue lines). This finding indicated that the *type IVa cadherin* genes had experienced varying degrees of intron loss, depending on their specific lineage. Conversely, the *Ea1-cadherin* gene contained many additional introns whose positions were not shared with the *type III* or *type IVb cadherin* genes (Fig. 2; Additional file 4: Figure S3). In this case, we concluded that the additional complexity had resulted from lineage-specific gains of introns.

Phylogenetic characterization of type IVa and type IVb cadherins

To validate the proposed classification of the type IVa and type IVb cadherin subtypes in the phylogenetic context, we analyzed the amino acid sequences of type IVb cadherins more extensively. The patterns for type IVb cadherins in the blast-based dot-plot comparisons indicated the divergence of their amino acid sequences. Although the EC1-EC2 region of type IVb cadherins was shown to exhibit the highest affinity to the EC6-EC7 region of type III cadherins among the classical cadherins examined, it might be possible that the N-terminal two EC domains in type IVb cadherins have a unique history. To test this possibility, we blasted the amino acid sequences of the EC1-EC2 region, as well as of the EC3-EC4 region, of Le1-, Ha1- and Ph1-cadherin against the Reference Sequence (RefSeq) protein databases for *D. melanogaster* and *Tribolium castaneum* (Additional file 8: Table S4). In all but one of the blast results, the top hit proteins were type III classical cadherins, and the hit sites were consistent with the detected collinear similarities between the type III and type IVb cadherin EC domains (Fig. 4; Additional file 6: Figure S5). These findings strongly suggested that the EC1-EC2 region of the type IVb cadherins shares a relatively recent common history with the EC6-EC7 region of the type III cadherins, and is compatible with the presence of conserved intron insertions in

the EC1 coding region of the *type IVb cadherin* genes and the EC6 coding region of the *Le2-* and *Cm-cadherin* genes.

Furthermore, we performed phylogenetic analyses of the amino acid sequences of five different extracellular regions (including three EC regions and two non-EC regions) of arthropod type III, type IVa, type IVb and Sm1-cadherins using the maximum likelihood (ML) method (Fig. 6). The results of these analyses revealed the separation of type IVa/IVb cadherins from type III cadherins as well as the separation of type IVb cadherins from type IVa cadherins. The data from all the different regions consistently indicated deep divergence between type IVa and type IVb cadherins, validating the classification of these type IV cadherin subtypes. In addition, the position of Sm1-cadherin was varied among the ML trees, although one of them had support for its association with type IVb cadherin branch.

Conserved cysteine residues in specific subsets of classical cadherins

The alignments of the amino acid sequences of the selected bilaterian classical cadherins (Additional file 9: Figure S7) allowed us to observe that the majority of cysteine residues are conserved in two or more classical cadherins. To determine the phylogenetic range at which each cysteine residue was conserved, we mapped the relative positions of cysteine residues in the amino acid sequences of 28 arthropod and non-arthropod classical cadherins (Fig. 7a). We found that the CE and LG domains demonstrated highly conserved cysteine patterns, and we also found two highly stable cysteine residues in the EC1 domain of type IVa cadherins and the corresponding EC domains of other metazoan classical cadherins, including a *Trichoplax adhaerens* (placozoan) classical cadherin (Additional file 10: Figure S8).

In addition to the presence of stable cysteine residues, other lineage-restricted features were also observed. For example, we identified a short sequence motif (E-S/A-W-C) at the C-terminus as a shared characteristic of both type IVa and type IVb cadherins (with the exception of DE-cadherin; Fig. 7a, b). We also found that EC6 domains of hexapod type IVa cadherins shared a unique pair of cysteine residues (Fig. 7a; Additional file 10: Figure S8) and that the CE domains of insect type IVa cadherins lacked the two other highly conserved cysteine residues (Fig. 7a, c). We also identified six cysteine residues that were specific to the EC5, EC8 and EC9 domains of type IVb cadherins and two cysteine residues that were specific to the EC8 domain of hexapod type IVa cadherins (Fig. 7a; Additional file 4: Figure S3; Additional file 10: Figure S8). The lineage-specific cysteine residues of the type III cadherins included two consecutive cysteine residues in the EC7 domain of

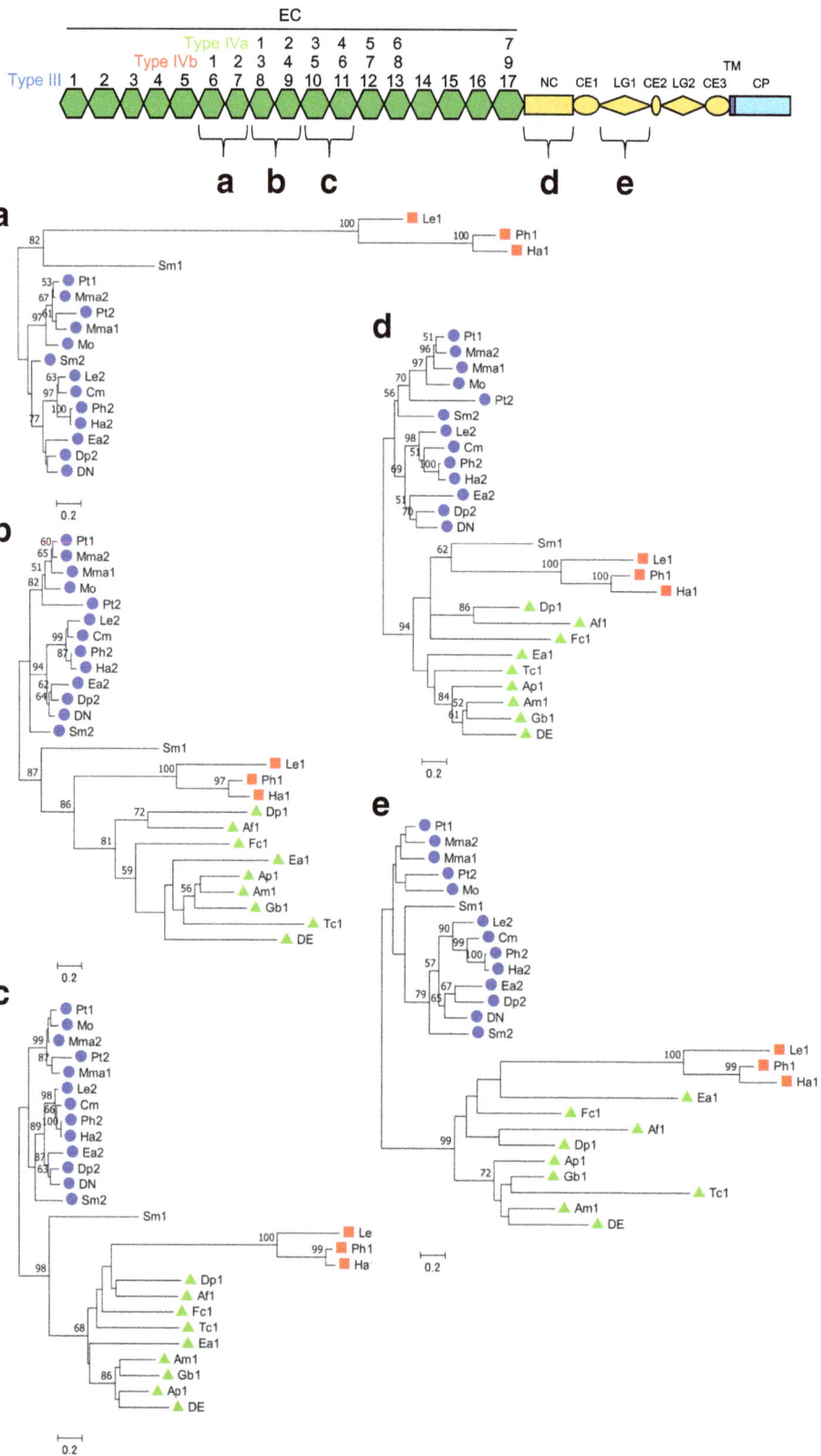

Fig. 6 (See legend on next page.)

(See figure on previous page.)
Fig. 6 Amino acid substitution-based phylogenetic analyses of the amino acid sequences of five different extracellular regions of the arthropod classical cadherins using the ML method. The five analyzed different extracellular regions of the type III, type IVa and type IVb cadherins (**a-e**) are schematically represented at the top. The type III (*blue*), type IVa (*orange*), and type IVb (*red*) cadherins are indicated by blue circles, red squares and green triangles, respectively. Numbers at nodes are bootstrap values based on 100 replicates. Nodes with no numbers have support with lower bootstrap values (<50)

arthropods and annelids/molluscans (Fig. 7a), as well as two other consecutive cysteine residues in the EC14 domain, which was limited to arthropods (Fig. 7a). We noted that two cysteine residues were also conserved in the Ca^{2+}-binding motifs between the EC5 and EC6 domains in the echinoderm and chordate type III cadherins (Fig. 7a; Additional file 10: Figure S8). However, these cysteine residues were also found in a predicted classical cadherin of the non-bilaterian metazoan *T. adhaerens*. These observations indicated that the patterns of cysteine residues among the classical cadherins of metazoans were stable but varied, thus presumably reflecting rare evolutionary changes.

Discussion
Type III cadherin is the ancestral classical cadherin in arthropods as well as in bilaterians
The present study investigated whether type IV or related cadherins were present in all the major arthropod lineages. However, our exhaustive search of multiple chelicerate genomes failed to identify any classical cadherins other than the type III form, which was found in all the arthropods examined. This result corroborates the previous finding that the only *classical cadherin* gene in the genome of the echinoderm *Strongylocentrotus purpuratus* encodes a type III cadherin [16]. It is, therefore, reasonable to assume that, among the various forms of classical cadherin, the type III form is the only one known to have been passed on from the last common ancestor to both the arthropod and echinoderm lineages and that the same form is also the only one known to have been passed on from the earliest arthropods to all the major extant arthropod groups.

Importantly, these hypotheses provide an explanation for the relationships between the various states of classical cadherin domain organization and exon-intron organization observed both within and beyond the Arthropoda. The findings of the present study indicate that the markedly complex exon-intron organization of the *P. tepidariorum type III cadherin* genes is representative of the ancestral state for arthropods (Figs. 1 and 5), and the detection of numerous conserved intron positions between the spider and non-arthropod bilaterian *classical cadherin* genes also indicated that the complex exon-intron organization predates the divergence of the arthropod lineage from other bilaterians. Our finding that some of the *type IV cadherin* genes have retained ancestral states of exon-intron organization

provides genomic evidence for the derivation of type IV cadherin from type III cadherin.

Similarly, the *5-EC cadherin* genes, which are prevalent in vertebrates (i.e., *type I* and *type II cadherin* and *non-classical desmosomal cadherin* genes), also possess a conserved exon-intron organization [45, 46], indicating that they share a common precursor in the lineage that gave rise to vertebrates. The identification of six shared intron positions among arthropod *type III* and vertebrate *type II cadherin* genes (Fig. 5) also indicated a deep link among the different classical cadherins, paving the way toward a comprehensive framework for the divergence of *classical cadherin* genes among bilaterians.

Evolution and divergence of type IV cadherins
An unexpected and important finding of this work was the identification of a novel form of classical cadherin in isopod and amphipod crustaceans that was similar to, but distinct from, the known hexapod and branchiopod type IV cadherins. This finding led us to propose a revision of the type IV cadherin category and to define two subclasses, type IVa and type IVb. This classification was validated based on comparative and phylogenetic analyses of the arthropod classical cadherins.

The identification and characterization of type IVb cadherin provided an opportunity to discuss the transition from the ancestral type III cadherin to the insect type IVa cadherin (Fig. 8a), which is often referred to as E-cadherin. Comparative data presented in this and previous research [15] suggest that the origination of the last common precursor of type IVa and IVb cadherins was associated with a duplication of a preexisting *type III cadherin* gene followed by, or coupled with, the following three changes: the loss of 3 EC domains from the region corresponding to the EC13-EC16 region of type III cadherin (Fig. 8a, Change A), the loss of the region corresponding to the CE2-LG2-CE3 region of type III cadherin (Fig. 8a, Change B), and the gain of the C-terminal motif E-S-W-C (Fig. 8a, Change C).

Our dot-plot and genomic data indicated that both type IVa and type IVb cadherins have tandem EC domains that were derived from the EC8-EC12 region of type III cadherin, and our phylogenetic analyses based on amino acid alignment and substitution supported deep divergence between type IVa and type IVb cadherins. The key issue is what form the last common precursor of type IVa and type IVb cadherins had. There are

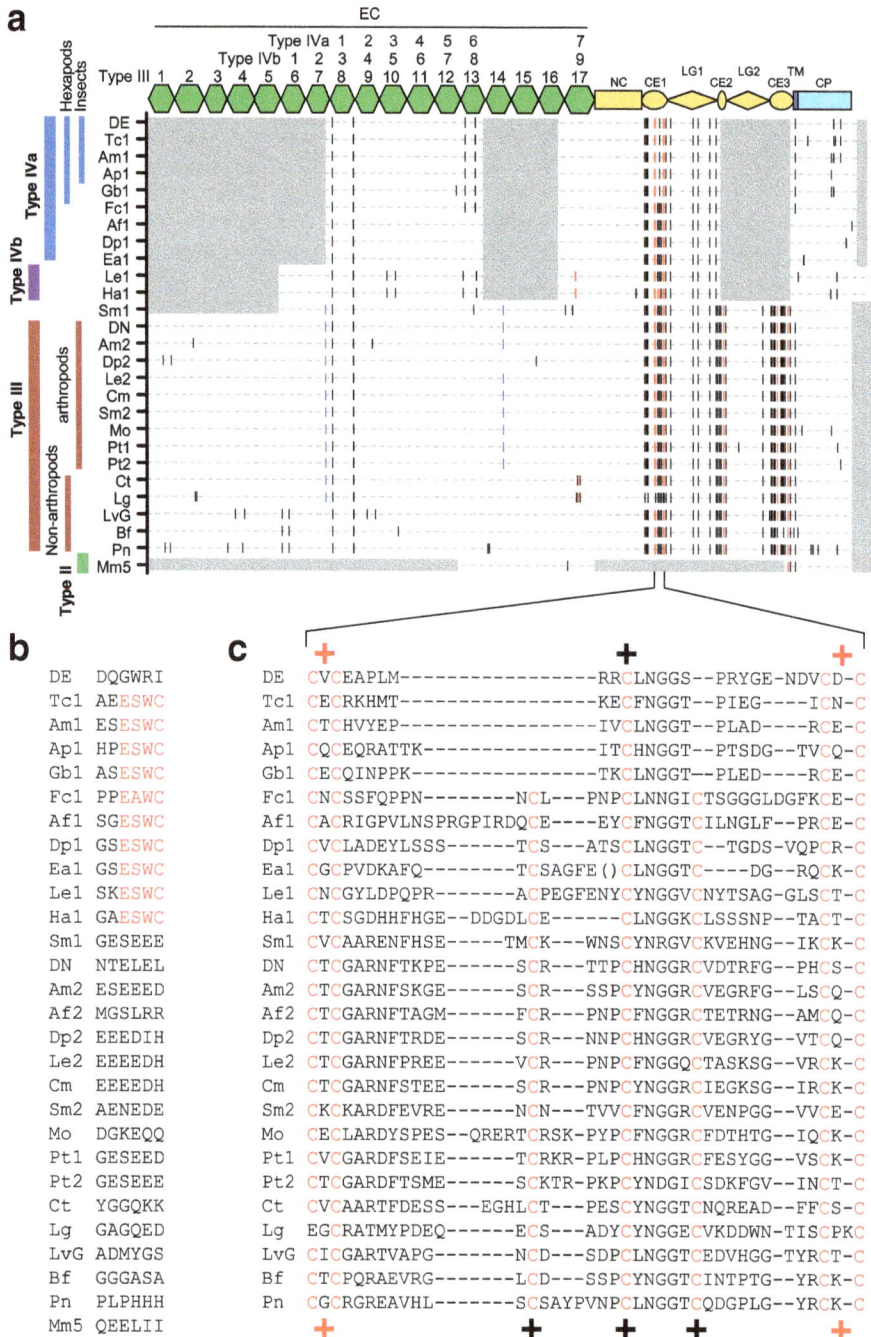

Fig. 7 Comparison of the distribution patterns of cysteine residues among classical cadherins. **a**. Comparative diagram of the distribution patterns of cysteine residue in arthropod and non-arthropod bilaterian classical cadherins. In addition to the 21 classical cadherins shown in Fig. 5, the following seven classical cadherins were used: Gb1-cadherin (Gb1, cricket); Fc1-cadherin (Fc1, springtail); Af1-cadherin (Af1, brine shrimp); Af2-cadherin (Af2, brine shrimp); Ea1-cadherin (Ea1, copepod); Ha1-cadherin (Ha1, amphipod); and Mo-cadherin (Mo, mite). The relative positions of cysteine residues in the classical cadherins are indicated by short vertical bars. The black bar denotes a solitary cysteine residue (–C–), the blue bar denotes two successive cysteine residues (–C-C–), and the red bar denotes two cysteine residues spaced with a single non-cysteine residue (–C-X-C–). The shaded regions indicate that there are no sequences for comparison. **b** Representation of the six C-terminal-most amino acid residues of the various classical cadherins. The short sequence motif, E-S/A-W-C, is shown in red. **c** Alignment of the amino acid sequences in parts of the CE1 or CE domains of the various classical cadherins. All cysteine residues are highlighted in red. Parentheses denote the omission of seven non-cysteine residues. The difference between the cysteine patterns of the insect type IVa cadherins and the other cadherins is indicated by "+" characters

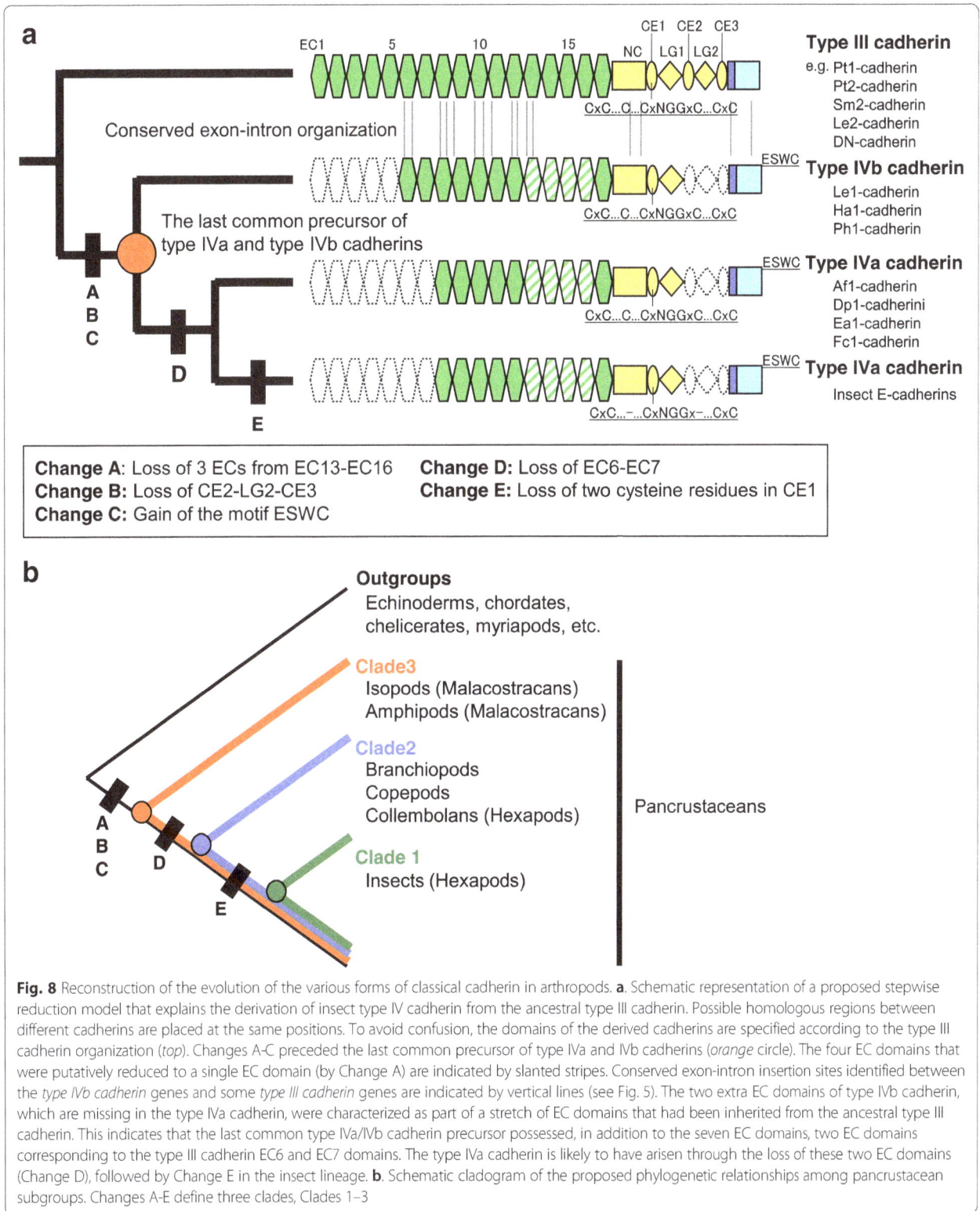

Fig. 8 Reconstruction of the evolution of the various forms of classical cadherin in arthropods. **a**. Schematic representation of a proposed stepwise reduction model that explains the derivation of insect type IV cadherin from the ancestral type III cadherin. Possible homologous regions between different cadherins are placed at the same positions. To avoid confusion, the domains of the derived cadherins are specified according to the type III cadherin organization (*top*). Changes A-C preceded the last common precursor of type IVa and IVb cadherins (*orange* circle). The four EC domains that were putatively reduced to a single EC domain (by Change A) are indicated by slanted stripes. Conserved exon-intron insertion sites identified between the *type IVb cadherin* genes and some *type III cadherin* genes are indicated by vertical lines (see Fig. 5). The two extra EC domains of type IVb cadherin, which are missing in the type IVa cadherin, were characterized as part of a stretch of EC domains that had been inherited from the ancestral type III cadherin. This indicates that the last common type IVa/IVb cadherin precursor possessed, in addition to the seven EC domains, two EC domains corresponding to the type III cadherin EC6 and EC7 domains. The type IVa cadherin is likely to have arisen through the loss of these two EC domains (Change D), followed by Change E in the insect lineage. **b**. Schematic cladogram of the proposed phylogenetic relationships among pancrustacean subgroups. Changes A-E define three clades, Clades 1-3

two lines of evidence for the primitiveness of the type IVb subtype. First, the amino acid sequences of the two extra EC domains (the EC1-EC2 regions) of type IVb cadherins exhibit specific affinities to those of the EC6-EC7 regions of type III cadherins (Fig. 4; Additional file 8: Table S4). Second, the *type IVb cadherin* genes have retained many intron insertions that are shared with the arthropod *type III cadherin* genes (Figs. 5, 8a).

Notably, the conservation of two intron insertion sites in the EC1 coding region of the *type IVb cadherin* genes and the EC6 coding region of the *type III Cm-* and *Le2-cadherin* genes suggest a specific association between the type IVb and type III forms. Altogether, the occurrence of these primitive characters in the *type IVb cadherin* genes is most easily understood if we consider that the two N-terminal-most EC domains and subsequent (at least) five EC domains of the type IVb cadherin are composed of a continuous stretch of EC domains that were inherited from the ancestral type III cadherin (Fig. 8a). This interpretation does not necessarily indicate that the last common precursor of type IVa and type IVb cadherins was a type IVb cadherin but does indicate that it could have at least nine EC domains. If this is the case, the transition from the last common precursor of type IVa and type IVb cadherins to type IVa cadherin should have occurred through the loss of two EC domains corresponding to the EC6-EC7 region of type III cadherin (Fig. 8a, Change D). Duplication of the precursor gene and subsequent differential loss of paralogs might have potentially occurred during the processes that gave rise to the type IVa and type IVb cadherin subtypes in the different crustacean lineages. Despite the primitive characteristics of the type IVb subtype, it could also be hypothesized that type IVb cadherin evolved from type IVa cadherin. Indeed, in the vertebrate lineage, 7-EC non-classical cadherins (e.g., LI-cadherin) are reported to have arisen from a 5-EC cadherin by internal domain duplication [47]. However, our comparative and phylogenetic analyses of type IVa and type IVb cadherins yielded no sign to support such a scenario. As another possibility, gene conversion might have led to a lineage-specific addition of two EC domains to the type IVa cadherin form. However, considering the determined characteristics of the EC1-EC2 coding regions of the *type IVb cadherin* genes, such a genetic event is much less likely to have occurred than the proposed domain loss event (Fig. 8a, Change D).

The predicted Sm1-cadherin could be categorized as neither a type III nor type IV classical cadherin. Because this cadherin, similar to type IVb cadherins, lacks the five EC domains that correspond to the EC1-EC5 region of type III cadherin, we considered whether Sm1-cadherin and the type IVb cadherins share a common precursor. However, we were unable to identify any specific signature that supported such a hypothesis. To fill the possible gaps in reconstruction of the transition processes between type III and type IVa/IVb cadherins, more data from the myriapod group, as well as from the crustacean group, will be required.

The evolutionary conservation of cysteine residues has been previously reported among type I cadherins or among protocadherins, and conserved cysteine residues are apparently involved in intra- or intermolecular disulfide bonds that contribute to the maintenance or stabilization of functional structures [48, 49]. In the present study, our comparative analysis revealed phylogenetically stable, yet varied, patterns of cysteine residues among the metazoan classical cadherins, which might imply transitions of structural mechanisms during classical cadherin evolution. We found that the CE domains of all insect type IVa cadherins specifically lacked two highly conserved cysteine residues. The relationship of the insect type IVa cadherins with other classical cadherins clearly indicated that the simpler pattern of cysteine residues in the CE domains of the insect type IVa cadherins is a derived state that was produced by the evolutionary loss of two cysteine residues (Fig. 8a, Change E). Therefore, we suggest that the transition from the ancestral type III cadherin to the derived insect type IVa cadherin was a multistep process that involved several progressive reductive changes.

Directionality in the evolution of the classical cadherin structure

Accumulated evidence from the arthropods and chordates suggests that the lineage-specific forms of classical cadherin have been shaped by reductive changes from an ancestral type III cadherin. The genomes of certain non-bilaterian metazoans are known to contain genes encoding much larger classical cadherins than type III cadherin [2, 17]. Therefore, it is likely that reductive changes also preceded the establishment of the type III cadherin [2].

In fact, the modification of cadherins by reductive changes from a larger state seems to be a common trend in the evolution of the classical cadherin structure. This directionality could possibly be associated with the cost and efficiency of the mechanical energetic processes by which classical cadherin-based adherens junctions drive morphogenesis. However, it is difficult to imagine that large deletions would instantly enhance the performance of classical cadherin molecules. Nonetheless, if such a deletion mutation did not disrupt the homophilic cell-cell binding properties of the classical cadherin but did alter its binding specificity or strategy, the affected cadherin might have had a chance to evolve independently of the parental cadherin.

Investigating the mechanism by which the classical cadherins were able to evolve via simplification is a typical challenge in experimentation-based evolutionary biology. The multistep transition from the ancestral type III cadherin to the insect type IVa cadherin offers an example with which to investigate the evolvability of type III cadherin. Genetic analysis of a DE-cadherin in which the EC7 and subsequent membrane-proximal extracellular domains had been deleted suggested that the six N-terminal-most EC domains constitute a functional unit that is capable of mediating cell-cell adhesion, whereas the membrane-proximal domains are necessary to ensure the functionality of the DE-cadherin during morphogenesis [26].

Recently, atomic force microscopy imaging revealed that the EC1-EC4 region of DE-cadherin forms a tightly folded, globular structure [27], and the deeply bent conformation of the EC region is likely to be associated with the Ca^{2+}-free linker found between the EC2 and EC3 domains of type IVa cadherin, which is conserved in the corresponding EC domains (EC9 and EC10) of type III cadherin [36]. The EC2-EC4 region of hexapod type IVa cadherins has been demonstrated as the minimal portion capable of mediating exclusive homophilic binding specificity. Intriguingly, three consecutive EC domains of the type III DN-cadherin that correspond to the EC2-EC4 region of type IVa cadherins are able to specifically recognize the DN-cadherin as the binding partner. A series of these findings implies that type IVa cadherin might have inherited part of an as-yet-uncharacterized mechanism of homophilic binding from the type III cadherin. In this context, questions of how the two N-terminal-most ECs of type IVb cadherin contribute to the functioning of the type IVb cadherin and how the large number of EC domains in type III cadherin are utilized to mediate homophilic cell-cell adhesion are key to a better understanding of the stepwise reductive changes involved in the evolution of insect E-cadherin.

Implications for pancrustacean phylogeny

Phylogenetic inferences of deep relationships are often problematic [50–53]. For example, the reconstruction methods used in the majority of modern phylogenetic analyses are based on nucleotide or amino acid substitutions in orthologous genes or proteins; however, the deep branch topologies of the resulting phylogenetic trees can be influenced by the sampling of species, choice of sequence evolution models, and variations in substitution rates among species, among sites, and even at individual sites over time [51]. Therefore, any phylogenetic hypothesis based on such analyses should be verified using alternative methods, one of which is the identification of rare genomic changes [54, 55].

In the present study, we identified stable yet varied character states among the arthropod *classical cadherin* genes, which provided clues to the polarity of cadherin evolution. A thematically similar situation has been described among the phylum Chordata [32, 33]. For example, the 5-EC state of classical cadherin (type I/II cadherin) defines a clade that includes the vertebrates and urochordates, but excludes the cephalochordates, a relationship that is supported by various independent phylogenetic studies [56–58]. In the phylum Arthropoda, the paraphyly of crustaceans has been firmly established, based on analyses of nucleotide and amino acid

substitutions and rare genomic changes, and the crustaceans are now suggested to form a clade with hexapods, termed Pancrustacea, that excludes the myriapods and chelicerates [59–63]. The monophyly of both hexapods and insects has also been strongly supported by recent molecular phylogenies [64–70]. However, the relationships between the hexapods and other pancrustacean subgroups remain a controversial topic [71, 72].

One of the major conflicts regarding the pancrustacean phylogeny concerns the relationships between hexapods, branchiopods, and malacostracans. Neural cladistics and some molecular studies propose a closer relationship between the hexapods and malacostracans than between the hexapods and branchiopods [73–77], whereas various recent studies that have used well-sampled, large-scale sequence data support a closer relationship between the hexapods and branchiopods [64, 67, 68, 78–80], and some phylogenies even suggest that the branchiopods and malacostracans are more closely related to each other than to the hexapods [65].

The complementary distributions of polarized genomic characters related to type IVa/IVb cadherins could be informative for the hexapod-branchiopod-malacostracan relationship. Based on these characters, we propose three successively nested clades, Clades 1–3 (Fig. 8b), in which Clade 1 includes insects but excludes collembolans, branchiopods, copepods, malacostracans, and non-pancrustacean arthropods, Clade 2 includes Clade 1, collembolans, branchiopods, and copepods but excludes malacostracans (such as isopods and amphipods) and non-pancrustacean arthropods, and Clade 3 includes Clade 2 and malacostracans but excludes non-pancrustacean myriapods and chelicerates. This topology indicates that branchiopods have a closer relationship to hexapods than malacostracans.

The matter of which crustacean subgroup is the closest relative of hexapods is also a subject of debate. Branchiopods [64, 66, 67, 81, 82], remipedians [61, 68, 69] and a clade that includes both remipedians and cephalocarids [65] have all been proposed as candidates for the sister group of hexapods. A few studies of ribosomal RNA sequences have also proposed copepods as the sister group of hexapods [78, 83]; however, recent molecular studies have frequently placed the copepods within a clade with the malacostracans [64, 65, 67, 68, 79, 81]. Nevertheless, although the present study did not include several key crustacean subgroups, such as Remipedia and Cephalocarida, our phylogenetic hypothesis could indicate that copepods, as well as branchiopods, should be included in the group of potential candidates for the closest relative of hexapods.

Finally, despite our exhaustive searches of the genome sequences for *classical cadherin* genes in the species, it was difficult to entirely exclude the possibility of additional *classical cadherin* genes due to incompleteness of the genome sequence assemblies. It is potentially possible that the establishment of the type IVa and type IVb cadherin subtypes might have preceded the diversification of the crustacean lineages. For example, if the malacostracan genomes are found to have *type IVa cadherin* genes in addition to the *type IVb cadherin* genes, our phylogenetic proposals should be reconsidered or rejected. Furthermore, because the number and range of the species examined in the present study were limited, our phylogenetic proposals remain highly hypothetical. We believe that the growing availability of arthropod genome sequences [37] will soon facilitate a more comprehensive analysis of type IV cadherin-related character states and help evaluate the conflicting hypotheses for arthropod phylogeny.

Conclusions

In the present study, we provided a genomic perspective of the evolution of classical cadherins among bilaterians, with a focus on the phylum Arthropoda. We demonstrated that the *type III cadherin* genes in the chelicerate *P. tepidariorum* were representative of the ancestral genomic organization of classical cadherins in arthropods, and suggested that the precursor of insect E-cadherin originated through stepwise reductive changes after the earliest divergence of extant arthropod groups. Future studies should investigate the structural mechanisms underlying the multistep transition from the arthropod ancestral type III cadherin to the more recent insect E-cadherin. The varied, polarized, and stable character states of classical cadherins could be widely applicable as indicators of deep phylogenetic relationships, as exemplified in the arthropod and chordate phyla.

Methods
Animals

This work was performed according to the institutional animal care and use committee guidelines (JT Biohistory Research Hall). Laboratory stocks of *Parasteatoda tepidariorum* (formerly *Achaearanea tepidariorum*) were derived from individuals collected at several different sites in Kyoto and Osaka, Japan [84]. Adults of *Ligia exotica* were collected from Kobe, Hyogo, Japan; and adults of *Caridina multidentata* (formerly *Caridina japonica*) were purchased from local suppliers.

Sequencing

For transcriptome sequencing with the Roche GS FLX+ system, total RNA was isolated from *P. tepidariorum* embryos at stages 1, 3, 5, 7, and 9, using the MagExtractor RNA nucleic acid purification kit (Toyobo). Poly (A) + RNA was purified from the total RNA and used to generate a cDNA library (GATC Biotech). First-strand cDNA synthesis was primed with a N6 randomized primer, and adaptors were ligated to the 5′ and 3′ ends of the cDNA, followed by 17 cycles of PCR amplification. The amplified cDNA was normalized using a single cycle of denaturation and re-association, and the re-associated ds-cDNA was separated from the remaining ss-cDNA using a hydroxylapatite column. Subsequently, the ss-cDNA was PCR amplified (7 cycles), and 500–850 bp-long cDNA fragments were eluted from an agarose gel and then sequenced, yielding 842,126 reads with an average length of 426 bp. These raw reads were subjected to adaptor trimming and de novo assembly using the CLC Genomics Workbench Version 7.0.3 (Qiagen) with the following settings: Bubble size, Automatic = 425; Word size, Automatic = 21; Map reads back to contigs, Yes (Mismatch cost, 2; Insertion cost, 3; Deletion cost, 3; Length fraction 0.9; Similarity fraction, 0.9); Update contigs, Yes. Some misassembled contigs were manually corrected. The resultant transcriptome assembly consisted of 23,144 contigs with N50 of 1046 bp (SRA Accession: DRR054577; Sequence Accession: IABY01000000).

For RNA-seq with the Illumina MiSeq system, mRNA was purified from stage-5 and stage-10 *P. tepidariorum* embryos, late stage *L. exotica* embryos, and adult *C. multidentata* muscle and neural tissues, using the QuickPrep Micro mRNA Purification Kit (GE Healthcare). The mRNAs were fragmented using the NEBNext RNase III RNA Fragmentation Module (New England BioLabs) and then used to construct DNA libraries with the NEBNext Ultra Directional RNA Library Prep Kit for Illumina (New England BioLabs) and NEBNext Multiplex Oligos for Illumina (Index Primers Set 1, New England BioLabs). Sequencing reactions were performed using the 150- or 500-cycle formats of the Illumina MiSeq Reagent Kit, and the resulting raw sequence reads were subjected to adaptor trimming and de novo assembly (Additional file 1: Table S1) using the CLC Genomics Workbench with the following settings: Bubble size, Automatic = 50; Word size, Automatic = 24; Perform Scaffolding, Yes; Auto-detect paired distance, Yes; Map reads back to contigs, Yes (Mismatch cost, 2; Insertion cost, 3; Deletion cost, 3; Length fraction 0.9; Similarity fraction, 0.9); Update contigs, Yes. For gene expression analysis, the adaptor-trimmed reads were mapped to the sequences of selected transcripts, using the CLC Genomics Workbench (Mismatch cost, 2; Insertion cost, 3; Deletion cost, 3; Length fraction 0.9; Similarity fraction, 0.9), and counted in order to quantify the expression levels were quantified as reads per kilobase of exon per million total reads.

For genome sequencing with the Illumina MiSeq system, genomic DNA was isolated from late stage *P.*

tepidariorum embryos, late stage *L. exotica* embryos, and adult *C. multidentata* muscle and neural tissues, using the GenomicPrep Cells and Tissue DNA Isolation Kit (GE Healthcare). The isolated DNA was fragmented using an acoustic solubilizer (Covaris), and the fragmented DNA (250–400 bp for *L. exotica* and *C. multidentata* DNA and 250–400, 400–500, 500–600, or 600–800 bp for *P. tepidariorum* DNA) was used to construct DNA libraries with the Truseq DNA Sample Prep Kit (Illumina). Paired-end sequencing of the libraries was performed using the 500- or 600-cycle formats of the Illumina MiSeq Reagent Kit, and the resulting raw sequence reads were subjected to adaptor trimming and de novo assembly (Additional file 1: Table S3) using the CLC Genomics Workbench with the following settings: Bubble size, Automatic = 227 (for *L. exotica*) or 219 (for *C. multidentata*); Word size, 64; Perform Scaffolding, Yes; Auto-detect paired distance; Yes. The approximate coverage depth for the obtained genomic sequences was estimated by mapping the reads to the 1307-, 1226- and 2142-bp regions of the *P. tepidariorum* genome (corresponding to exon 2 of the *Pt1-cadherin* gene and exons 1 and 35 of the *Pt2-cadherin* gene in Fig. 1), the 1010-, 2281-, and 914-bp regions of the *L. exotica* genome (corresponding to exon 28 of the *Le1-cadherin* gene and exons 21 and 22 of the *Le2-cadherin* gene in Fig. 5), or the 1392-, 2275-, and 914-bp regions of the *C. multidentata* genome (corresponding to exons 1, 21, and 22 of the *Cm-cadherin* gene in Fig. 5).

Identification of *classical cadherin* genes

Partial *Pt2-cadherin* sequences were originally found in the *P. tepidariorum* transcriptome that was generated using the Roche GS FLX+ system, and the full-length *Pt2-cadherin* cDNA was isolated from cDNA libraries of *P. tepidariorum* embryos [84] and then sequenced. The *Le1-cadherin* transcript was predicted from the de novo assembly of the *L. exotica* RNA-seq reads. A cDNA fragment that corresponded to the coding region of the *Le1-cadherin* transcript was amplified from oligo-dT primed cDNA of late stage *L. exotica* embryos by PCR using the following primers: 5′-ATAAGAATGCGGCCGCATCGG TGAACAAATCTTCAGGTTCA-3′; 5′-ATAAGAATGC GGCCGCTTAGCACCAAGATTCCTTGCTCTG-3′ (Underlines indicate the *NotI* recognition sites). This product was digested with *NotI* and then cloned into the *NotI* site of pUAST, resulting in pUAST-Le1-cadherin. This cloned cDNA was sequenced to validate the predicted transcript.

We searched for *classical cadherin* sequences in arthropod genomes available from public databases, as well as in genomic and transcriptomic sequences generated from *P. tepidariorum*, *L. exotica* and *C. multidentata* in this study (Table 1; Additional file 1: Tables S1 and S3). For the initial detection of classical cadherin-encoding gene (s), the amino acid sequences of the entire CP domains of Pt1-, Pt2-, DE-, DN-, Le1-, Le2-, Gb1-, Af1-, and Dp1-cadherins were blasted against each genome sequence assembly or the WGS and RNA-seq reads and assemblies (*P. tepidariorum*, *L. exotica* and *C. multidentata*) using the tblastn algorithm with the cutoff E-value of 1×10^{-4}. We also blasted the entire amino acid sequences of the Pt1-, DN-, and Le2-cadherins against the identified scaffolds/contigs to determine whether the detected scaffolds contained all the typical type III cadherin elements (14–17 ECs, NC, CE1, LG1, CE2, LG2, CE3, TM, and CP domains) in the expected order. To exhaustively search for *type IV cadherin* genes, the entire amino acid sequences of DE-, Gb1-, Af1-, Dp1- and Le1-cadherins were blasted against the scaffolds/contigs in which classical cadherin CP domain-related sequences were found.

In cases where only some of the type III or type IV cadherin elements were found in the scaffolds/contigs, we examined the possibility that the remaining elements might be encoded in other scaffolds/contigs. To identify neighboring exons of a gene within the same or different scaffolds/contigs, we also used predicted transcript sequences, RNA-seq reads and transcriptome assemblies that were either publicly available or generated in the present study (Table 1; Additional file 1: Tables S1 and S3; Additional file 11: Table S5). The transcript sequences for the hypothetical Mma1-, Mma2-, Sm1-, Sm2-, Ea1-, and Ea2-cadherins were reconstructed from publicly available RNA-seq reads of *Strigamia maritima* (PRJNA246089), and *Eurytemora affinis* (PRJNA275666) using the CLC Genomics Workbench or Geneious Version 9.0.5 (Biomatters). RNA-seq reads of *Hyalella azteca* (PRJNA277380) [85] were also used to assess the transcript models for Ha1- and Ha2-cadherin. To identify type IV cadherin-related sequences in the *C. multidentata* genome, portions of the amino acid sequence of Le1-cadherin were blasted against the WGS reads.

Sequence analysis and characterization

The sources of genomic, mRNA and protein sequences of classical cadherins used for sequence alignment, dot-plot analysis, exon-intron structure analysis and phylogenetic analysis are listed in Additional file 11: Table S5. The amino acid sequences of classical cadherins were aligned using the ClustalW algorithm with the following settings: Cost matrix, BLOSUM; Gap open cost, 10; Gap extend cost, 0.1. With the exception of the alignments shown in Additional file 3: Figure S2 and Additional file 7: Figure S6, the alignments were followed by manual adjustment, considering the results of the dot-plot analyses (Figs. 3 and 4), as well as other conserved motifs or residues. Exon-intron boundaries of the genes were determined by comparing the transcript and genome assembly sequences, considering the GT-AG mRNA processing rule. Importantly, to determine the exon-intron boundaries of the *L. exotica* and *C. multidentata* genes, the WGS reads were used.

For the dot-plot analysis, a series of overlapping 120-residue amino acid sequences were generated from the amino acid sequences of each cadherin using a sliding window with a step of five amino acids. The individual sequences were blasted against the reference sequences, using the blastp algorithm, and the E-values for the individual hits were plotted using color codes. To map the relative positions of cysteine residues in a given cadherin, the sequence alignment shown in Additional file 9: Figure S7 was used as the positional reference.

For amino acid substitution-based analysis, the amino acid sequences from the multiple regions of the indicated cadherins were individually aligned using the ClustalW algorithm with the same settings as above, and ML analyses of the resultant sequence alignments (without manual adjustment) were performed using *MEGA* version 7.0.25 [86]. Model testing was conducted under the Bayesian Information Criterion, which selected the LG + G model as the best-fit model. ML trees were constructed with the following settings: Substitution model, LG + G; Number of discrete gamma categories, 5; Gaps data treatment, complete deletion; ML heuristic method, Subtree-pruning-regrafting (extensive); Initial tree for ML, NJ/BioNJ.

Cell aggregation assay

The culture of S2 cells, transfection, and cell aggregation assays were conducted as described [25], with some modifications. Briefly, 5×10^6 S2 cells were co-transfected with pUAST-Le1-cadherin, pUAST-mKate2, and pWA-GAL4 (a gift from Yasushi Hiromi, National Institute of Genetics, Japan) at a ratio of 5:5:1, and cells were co-transfected with empty pUAST, pUAST-mKate2, and pWA-GAL4 as a negative control. After 45 h of incubation, the transfected cells were collected and resuspended in 5 ml of culture medium, and then 0.5 ml aliquots of each cell suspension were transferred to individual wells of a 24-well plate and rotated at 150 rpm for 15 min. The cell aggregates formed in the wells were observed and photographed using an Olympus IX71 fluorescence microscope equipped with a UPlanFl 10×/N.A. 0.3 objective, differential interference contrast optics, and a CoolSNAP HQ camera (Roper Scientific).

Additional files

Additional file 1: Table S1. Statistics and accessions of RNA-seq data from *P. tepidariorum*, *L.exotica* and *C. multidentata*. **Table S2.** Expression levels of selected transcripts from *P. tepidariorum* embryos, as indicated by RNA-seq. **Table S3.** Statistics and accessions of WGS data from *P. tepidariorum*, *L.exotica* and *C. multidentata*. (PDF 91 kb)

Additional file 2: Figure S1. Characterization and subdivision of the amino acid sequences of DN-, Pt1-, and Pt2-cadherins. **A.** Alignment of the EC1-EC17 regions of DN-, Pt1-, and Pt2-cadherins (abbreviated as DN, P1 and P2, respectively). The "-"character indicates introduced gaps. Conserved hydrophobic residues (blue), Ca^{2+}-binding motifs or residues (red), and XPXF motif sequences (green) are aligned, all of which represent structural features of EC domains as shown schematically at the top. Thick blue arrows denote the seven β-strands (βA to βG). Each red arrow indicates the inter-EC linker to which the Ca^{2+}-binding motif or residue belongs. No residues are omitted from the alignment, except for three sections where 7–12 residues of the DN-cadherin sequences are placed outside the alignment (parentheses). The N-terminal sequence (Nt) preceding the EC1 domain is also shown for each cadherin. **B.** Alignment of the NC and subsequent domains of the DN-, Pt1-, and Pt2-cadherins. In both A and B, conserved cysteine residues are highlighted in pink, and the residues bordering the start and end of the introns are highlighted with yellow and green. (PDF 362 kb)

Additional file 3: Figure S2. Alignment of the entire amino acid sequences of thirteen type III cadherins in arthropods, and comparison of the exon-intron organizations. The alignment was produced using the ClustalW algorithm without manual adjustment. The classical cadherins shown are Pt1-, Pt2-, Mma1-, Mma2-, Mo-, Sm2-, Cm-, Le2-, Ph2-, Ha2-, Ea2-, Dp2-, and DN-cadherins. The domain organization is indicated above the Pt1-cadherin sequence. Blue lines with breakages indicate exons, and the breaking points indicate intron insertion sites revealed by comparisons with the corresponding genomic sequences. (PDF 5991 kb)

Additional file 4: Figure S3. Characterization of the amino acid sequences of type IVa and type IVb cadherins. **A.** Alignment of the amino acid sequences of all EC domains (EC1-EC7 or EC1-EC9) of the DE-, Dp1-, Ea1-, Le1-, Ha1-, and Ph1-cadherins (abbreviated as DE, D1, E1, L1, H1, and Ph1, respectively). Conserved hydrophobic residues (blue), Ca^{2+}-binding motifs or residues (red), and XPXF motif sequences (green) are aligned. Thick blue arrows denote the seven β-strands (βA to βG), and each red arrow indicates the inter-EC linker to which the Ca^{2+}-binding motif or residue belongs. No residues are omitted from the alignment, except for three instances where 5–7 residues from the Le1- or Ha1-cadherin sequences are placed outside the alignment (parentheses). The N-terminal sequence (Nt) preceding the EC1 domain is also shown for each cadherin. **B.** Alignment of the amino acid sequences of the NC and subsequent domains of the DE-, Dp1-, Ea1-, Le1-, Ha1-, and Ph1-cadherins. In both A and B, the conserved cysteine residues are highlighted in pink, and the residues bordering the start and end of the introns are highlighted with yellow and green. (PDF 379 kb)

Additional file 5: Figure S4. Schematic representation of detected sequences of *C. multidentata* related to classical cadherins. **A.** Nine reconstructed genomic sequences of *C. multidentata* that contain coding sequences closely related to those of Le1-cadherin. The sequences are available under the indicated accession numbers. **B.** Eight transcriptome contigs connected by raw reads. The sequences of these contigs are available under the indicated accession numbers. Contig33642 was modified by an insertion of 5 nucleotide bases (CCGGA) between the nucleotides 349 and 350 based on assessment of raw reads (asterisk). The assembled transcript and protein sequences are available in Additional file 12. Detected domain elements are shown. (PDF 108 kb)

Additional file 6: Figure S5. Blast-based dot-plot comparisons between the amino acid sequence of Ha1- (A) or Ph1- (B) cadherin and those of DE-, Dp1-, Sm1-, Cm-, Le2-, DN-, and Pn-cadherins. Green boxes indicate comparisons between the EC1-EC5 region of Ha1- or Ph1- cadherin and the EC6-EC10 regions of the type III cadherins or the corresponding region of Sm1-cadherin, which exhibited marked collinear similarities. Blue boxes indicate comparisons between the EC6-EC8 region of Ha1- or Ph1-cadherin and the EC11-EC16 regions of the type III cadherins or the corresponding region of Sm1-cadherin, which exhibited ambiguous collinear similarities. (PDF 2623 kb)

Additional file 7: Figure S6. Comparison of the exon-intron organizations of *type IVa*, *type IVb* and *type III cadherin* genes. Alignment of the amino acid sequences of the EC1-EC6 region of type IVa cadherins, the EC1-EC8 region of type IVb cadherins, and the EC6-EC13 region of type III cadherins was produced using the ClustalW algorithm without manual adjustment. The classical cadherins shown are DE-, Dp1-, Ea1-, Le1-, Ha1-, Ph1-, Pt1-, Sm2-, Cm-, Le2-, and DN-cadherins. The EC domains for type IVa, type IVb, and type III cadherin are indicated above the DE-, Le1- and, Pt1-cadherin sequence, respectively. Blue lines with breakages indicate exons, and the breaking points indicate intron insertion sites revealed by comparisons with the corresponding genomic sequences. (PDF 1348 kb)

Additional file 8: Figure S6. Results of blast searches of type IVb cadherin EC domain sequences against the *D. melanogaster, T. castaneum* and *D. pulex* RefSeq protein sequences. (XLSX 11 kb)

Additional file 9: Figure S7. Amino acid alignment of selected classical cadherins from arthropods and non-arthropod bilaterians. The classical cadherins shown are as follows: DE-cadherin (DE, fruit fly); Tc1-cadherin (Tc1, beetle); Am1-cadherin (Am1, honey bee); Ap1-cadherin (Ap1, aphid); Gb1-cadherin (Gb1, cricket); Fc1-cadherin (Fc1, springtail); Af1-cadherin (Af1, brine shrimp); Dp1-cadherin (Dp1, water flea); Ea1-cadherin (Ea1, copepod); Le1-cadherin (Le1, sea slater); Ha1-cadherin (Ha1, amphipod); Sm1-cadherin (Sm1, centipede); DN-cadherin (DN, fruit fly); Am2-cadherin (Am2, honey bee); Dp2-cadherin (Dp2, water flea); Le2-cadherin (Le2, sea slater); Cm-cadherin (Cm, shrimp); Sm2-cadherin (Sm2, centipede); Mo-cadherin (Mo, mite); Pt1-cadherin (Pt1, spider); Pt2-cadherin (Pt2, spider); Ct-cadherin (Ct, polychaete); Lg-cadherin (Lg, snail); LvG-cadherin (LvG, sea urchin); Bf-cadherin (Bf, amphioxus); Pn-cadherin (Pn, fish); Ta-cadherin (Ta, placozoan); and Mm5-cadherin (Mm5, mouse). The amino acid sequence of Pt1-cadherin is duplicated; one of the duplicates is placed at the top as a reference to show the domain subdivisions. The "-" character indicates introduced gaps. All residues of each cadherin sequence are shown, although some parts of the sequences were aligned poorly or not at all. Excluding the reference sequence, the amino acid sequences derived from different exons are distinguished using arbitrary background colors to indicate the exon-exon junctions in the transcripts. (PDF 385 kb)

Additional file 10: Figure S8. Conserved cysteine residues in the EC domains of classical cadherins. Alignments were generated from the EC5-EC6 (A), EC7 (B), EC7-EC8 (C), EC13 (D), EC14 (E) and EC17 (F) regions of type III cadherins and the corresponding regions of other classical cadherins. The cysteine residues are shown in red. The "-" character indicates introduced gaps. The classical cadherins shown are as follows: DE-cadherin (DE, fruit fly); Tc1-cadherin (Tc1, beetle); Am1-cadherin (Am1, honey bee); Ap1-cadherin (Ap1, aphid); Dp1-cadherin (Dp1, water flea); Le1-cadherin (Le1, sea slater); Sm1-cadherin (Sm1, centipede); Sm2-cadherin (Sm2, centipede); Cm-cadherin (Cm, shrimp); Le2-cadherin (Le2, sea slater); Dp2-cadherin (Dp2, water flea); Am2-cadherin (Am2, honey bee); DN-cadherin (DN, fruit fly); Pt1-cadherin (Pt1, spider); Pt2-cadherin (Pt2, spider); Ct-cadherin (Ct, polychaete); Lg-cadherin (Lg, snail); LvG-cadherin (LvG, sea urchin); Bf-cadherin (Bf, amphioxus); Pn-cadherin (Pn, fish); and Mm5-cadherin (Mm5, mouse). (PDF 41 kb)

Additional file 11: Table S5. Sources of genomic, mRNA and protein sequences of classical cadherins used in the present study. (XLSX 14 kb)

Additional file 12: Multi-fasta format file of the nucleotide sequences of reconstructed transcripts for Mma1-, Mma2-, Sm1-, Sm2-, Ea1-, and Ha1-cadherins and the potential third *C. multidentata* classical cadherin, and their predicted protein sequences. (TXT 79 kb)

Abbreviations
CE: Cysteine-rich epidermal growth factor-like domain; CP: Cytoplasmic domain; EC: Extracellular cadherin domain; LG: Laminin globular domain; ML: Maximum likelihood; NC: Non-chordate classical cadherin domain; PCR: Polymerase chain reaction; RNA-seq: RNA sequencing; TM: Transmembrane domain; WGS: Whole genome shotgun sequencing

Acknowledgements
We thank Stephen Richards and the Baylor College of Medicine's Human Genome Sequencing Center i5k pilot project, Alistair McGregor and the *Parasteatoda tepidariorum* Genome Consortium, Carol E. Lee and the *E. affinis* Genome Consortium, and Helen C. Poynton and the *H. azteca* Genome Consortium for allowing us to use their pre-publication genome and transcript data from *P. tepidariorum, E. affinis*, and *H. azteca*; and Ryan Gott and the University of Maryland for allowing us to use their pre-publication RNA-seq data from *H. azteca*. We also thank Yasushi Hiromi for the pWA-GAL4; Akiko Noda for technical assistance; Nakatada Wachi for technical advice; Keiko Nakamura for encouragement; and other members of the JT Biohistory Research Hall for helpful discussions.

Funding
This work was supported in part by Japan Society for the Promotion of Science (JSPS) Grants-in-Aid for Scientific Research (KAKENHI) awards to HO (23,370,095, 15 K07139) and YA (24,870,035, 26,440,130).

Authors' contributions
MS, YA, and HO conceived the project. HO and YA obtained sequence reads from *P. tepidariorum*, and MS obtained sequence reads from *L. exotica* and *C. multidentata*. YA processed all the sequence reads for depositing in public databases. MS and HO performed the experiments and data analyses and prepared all the Figures and Tables. MS and HO wrote the manuscript and incorporated input from YA. All authors read and approved the final manuscript.

Competing interests
The authors declare that they have no competing interests.

Author details
[1]Laboratory of Evolutionary Cell and Developmental Biology, JT Biohistory Research Hall, 1-1 Murasaki-cho, Takatsuki 569-1125, Osaka, Japan. [2]Department of Microbiology and Infection Control, Osaka Medical College, Takatsuki, Osaka, Japan. [3]Department of Biological Sciences, Graduate School of Science, Osaka University, Osaka, Japan. [4]Current address: Department of Parasitology, Asahikawa Medical University, 2-1-1-1 Midorigaoka-higashi, Asahikawa 078-8510, Hokkaido, Japan.

References
1. Hulpiau P, van Roy F. Molecular evolution of the cadherin superfamily. Int J Biochem Cell Biol. 2009;41:349–69.
2. Hulpiau P, Van Roy F. New insights into the evolution of metazoan Cadherins. Mol Biol Evol. 2010;28:647–57.
3. Oda H, Takeichi M. Evolution: structural and functional diversity of cadherin at the adherens junction. J Cell Biol. 2011;193:1137–46.
4. Gumbiner BM. Regulation of cadherin-mediated adhesion in morphogenesis. Nat Rev Mol Cell Biol. 2005;6:622–34.
5. Harris TJC, Tepass U. Adherens junctions: from molecules to morphogenesis. Nat Rev Mol Cell Biol. 2010;11:502–14.
6. Guillot C, Lecuit T. Mechanics of epithelial tissue homeostasis and morphogenesis. Science. 2013;340:1185–9.
7. Takeichi M. Dynamic contacts: rearranging adherens junctions to drive epithelial remodelling. Nat Rev Mol Cell Biol. 2014;15:397–410.
8. Nose A, Tsuji K, Takeichi M. Localization of specificity determining sites in cadherin cell adhesion molecules. Cell. 1990;61:147–55.
9. Vendome J, Felsovalyi K, Song H, Yang Z, Jin X, Brasch J, et al. Structural and energetic determinants of adhesive binding specificity in type I cadherins. Proc Natl Acad Sci U S A. 2014;111:E4175–84.
10. Ozawa M, Ringwald M, Kemler R. Uvomorulin-catenin complex formation is regulated by a specific domain in the cytoplasmic region of the cell adhesion molecule. Proc Natl Acad Sci U S A. 1990;87:4246–50.
11. Mammoto T, Ingber DE. Mechanical control of tissue and organ development. Development. 2010;137:1407–20.
12. Lecuit T, Yap AS. E-cadherin junctions as active mechanical integrators in tissue dynamics. Nat Cell Biol. 2015;17:533–9.
13. Heisenberg CP, Bellaïche Y. Forces in tissue morphogenesis and patterning. Cell. 2013;153:948–62.
14. Sasakura Y, Shoguchi E, Takatori N, Wada S, Meinertzhagen IA, Satou Y, et al. A genomewide survey of developmentally relevant genes in *Ciona intestinalis*. X. Genes for cell junctions and extracellular matrix. Dev Genes Evol. 2003;213:303–13.
15. Oda H, Tagawa K, Akiyama-Oda Y. Diversification of epithelial adherens junctions with independent reductive changes in cadherin form: identification of potential molecular synapomorphies among bilaterians. Evol Dev. 2005;7:376–89.

16. Whittaker CA, Bergeron K-FF, Whittle J, Brandhorst BP, Burke RD, Hynes RO. The echinoderm adhesome. Dev Biol. 2006;300:252–66.

17. Chapman JA, Kirkness EF, Simakov O, Hampson SE, Mitros T, Weinmaier T, et al. The dynamic genome of *Hydra*. Nature. 2010;464:592–6.

18. Fahey B, Degnan BM. Origin of animal epithelia: insights from the sponge genome. Evol Dev. 2010;12:601–17.

19. Nichols SA, Roberts BW, Richter DJ, Fairclough SR, King N. Origin of metazoan cadherin diversity and the antiquity of the classical cadherin/β-catenin complex. Proc Natl Acad Sci U S A. 2012;109:13046–51.

20. Tanabe K, Takeichi M, Nakagawa S. Identification of a nonchordate-type classic cadherin in vertebrates: chicken Hz-cadherin is expressed in horizontal cells of the neural retina and contains a nonchordate-specific domain complex. Dev Dyn. 2004;229:899–906.

21. Levi L, Douek J, Osman M, Bosch TC, Rinkevich B. Cloning and characterization of BS-cadherin, a novel cadherin from the colonial urochordate *Botryllus schlosseri*. Gene. 1997;200:117–23.

22. Oda H, Uemura T, Harada Y, Iwai Y, Takeichi M. A *Drosophila* homolog of cadherin associated with armadillo and essential for embryonic cell-cell adhesion. Dev Biol. 1994;165:716–26.

23. Uemura T, Oda H, Kraut R, Hayashi S, Kotaoka Y, Takeichi M. Zygotic *Drosophila* E-cadherin expression is required for processes of dynamic epithelial cell rearrangement in the drosophila embryo. Genes Dev. 1996;10:659–71.

24. Tepass U, Gruszynski-DeFeo E, Haag TA, Omatyar L, Török T, Hartenstein V. *shotgun* encodes *Drosophila* E-cadherin and is preferentially required during cell rearrangement in the neurectoderm and other morphogenetically active epithelia. Genes Dev. 1996;10:672–85.

25. Oda H, Tsukita S. Nonchordate classic cadherins have a structurally and functionally unique domain that is absent from chordate classic cadherins. Dev Biol. 1999;216:406–22.

26. Haruta T, Warrior R, Yonemura S, Oda H. The proximal half of the *Drosophila* E-cadherin extracellular region is dispensable for many cadherin-dependent events but required for ventral furrow formation. Genes Cells. 2010;15:193–208.

27. Nishiguchi S, Yagi A, Sakai N, Oda H. Divergence of structural strategies for homophilic E-cadherin binding among bilaterians. J Cell Sci. 2016;129:3309–19.

28. Iwai Y, Usui T, Hirano S, Steward R, Takeichi M, Uemura T. Axon patterning requires DN-cadherin, a novel neuronal adhesion receptor, in the *Drosophila* embryonic CNS. Neuron. 1997;19:77–89.

29. Miller JR, McClay DR. Characterization of the role of cadherin in regulating cell adhesion during sea urchin development. Dev Biol. 1997;192:323–39.

30. Takeichi M. The cadherin superfamily in neuronal connections and interactions. Nat Rev Neurosci. 2007;8:11–20.

31. Costa M, Raich W, Agbunag C, Leung B, Hardin J, Priess JR. A putative catenin-cadherin system mediates morphogenesis of the *Caenorhabditis elegans* embryo. J Cell Biol. 1998;141:297–308.

32. Oda H, Akiyama-Oda Y, Zhang S. Two classic cadherin-related molecules with no cadherin extracellular repeats in the cephalochordate amphioxus: distinct adhesive specificities and possible involvement in the development of multicell-layered structures. J Cell Sci. 2004;117:2757–67.

33. Oda H, Wada H, Tagawa K, Akiyama-Oda Y, Satoh N, Humphreys T, et al. A novel amphioxus cadherin that localizes to epithelial adherens junctions has an unusual domain organization with implications for chordate phylogeny. Evol Dev. 2002;4:426–34.

34. Oda H. Evolution of the cadherin-catenin complex. Subcell Biochem. 2012;60:9–35.

35. Sigrist CJ, de Castro E, Cerutti L, Cuche BA, Hulo N, Bridge A, et al. New and continuing developments at PROSITE. Nucleic Acids Res. 2013;41:D344–7.

36. Jin X, Walker MA, Felsővályi K, Vendome J, Bahna F, Mannepalli S, et al. Crystal structures of *Drosophila* N-cadherin ectodomain regions reveal a widely used class of Ca²⁺-free interdomain linkers. Proc Natl Acad Sci USA. 2012;109:E127–34.

37. i5K Consortium. The i5K initiative: advancing arthropod genomics for knowledge, human health, agriculture, and the environment. J Hered. 2013; 104:595–600.

38. Sanggaard KW, Bechsgaard JS, Fang X, Duan J, Dyrlund TF, Gupta V, et al. Spider genomes provide insight into composition and evolution of venom and silk. Nat Commun. 2014;5:3765.

39. Grbić M, Van Leeuwen T, Clark RM, Rombauts S, Rouzé P, Grbić V, et al. The genome of *Tetranychus urticae* reveals herbivorous pest adaptations. Nature. 2011;479:487–92.

40. Hoy MA, Waterhouse RM, Wu K, Estep AS, Ioannidis P, Palmer WJ, et al. Genome sequencing of the phytoseiid predatory mite *Metaseiulus occidentalis* reveals completely atomized Hox genes and superdynamic intron evolution. Genome Biol Evol. 2016;8:1762–75.

41. Cao Z, Yu Y, Wu Y, Hao P, Di Z, He Y, et al. The genome of *Mesobuthus martensii* reveals a unique adaptation model of arthropods. Nat Commun. 2013;4:2602.

42. Chipman AD, Ferrier DE, Brena C, Qu J, Hughes DS, Schröder R, et al. The first myriapod genome sequence reveals conservative arthropod gene content and genome organisation in the centipede *Strigamia maritima*. PLoS Biol. 2014;12:e1002005.

43. Colbourne JK, Pfrender ME, Gilbert D, Thomas WK, Tucker A, Oakley TH, et al. The ecoresponsive genome of *Daphnia pulex*. Science. 2011; 331:555–61.

44. Kao D, Lai AG, Stamataki E, Rosic S, Konstantinides N, Jarvis E, et al. The genome of the crustacean *Parhyale hawaiensis*, a model for animal development, regeneration, immunity and lignocellulose digestion. elife. 2016;5

45. Greenwood MD, Marsden MD, Cowley CM, Sahota VK, Buxton RS. Exon-intron organization of the human type 2 desmocollin gene (DSC2): desmocollin gene structure is closer to "classical" cadherins than to desmogleins. Genomics. 1997;44:330–5.

46. Huber P, Dalmon J, Engiles J, Breviario F, Gory S, Siracusa LD, et al. Genomic structure and chromosomal mapping of the mouse VE-cadherin gene (Cdh5). Genomics. 1996;32:21–8.

47. Jung R, Wendeler MW, Danevad M, Himmelbauer H, Gessner R. Phylogenetic origin of LI-cadherin revealed by protein and gene structure analysis. Cell Mol Life Sci. 2004;61:1157–66.

48. Ozawa M, Hoschützky H, Herrenknecht K, Kemler R. A possible new adhesive site in the cell-adhesion molecule uvomorulin. Mech Dev. 1990;33:49–56.

49. Chen X, Molino C, Liu L, Gumbiner BM. Structural elements necessary for oligomerization, trafficking, and cell sorting function of paraxial protocadherin. J Biol Chem. 2007;282:32128–37.

50. Rokas A, Krüger D, Carroll SB. Animal evolution and the molecular signature of radiations compressed in time. Science. 2005;310:1933–8.

51. Brinkmann H, Philippe H. Animal phylogeny and large-scale sequencing: progress and pitfalls. J Syst Evol. 2008;46:274–86.

52. Salichos L, Rokas A. Inferring ancient divergences requires genes with strong phylogenetic signals. Nature. 2013;497:327–31.

53. Pisani D, Pett W, Dohrmann M, Feuda R, Rota-Stabelli O, Philippe H, et al. Genomic data do not support comb jellies as the sister group to all other animals. Proc Natl Acad Sci U S A. 2015;112:15402–7.

54. Rokas H. Rare genomic changes as a tool for phylogenetics. Trends Ecol Evol. 2000;15:454–9.

55. Delsuc F, Brinkmann H, Philippe H. Phylogenomics and the reconstruction of the tree of life. Nat Rev Genet. 2005;6:361–75.

56. Delsuc F, Brinkmann H, Chourrout D, Philippe H. Tunicates and not cephalochordates are the closest living relatives of vertebrates. Nature. 2006;439:965–8.

57. Bourlat SJ, Juliusdottir T, Lowe CJ, Freeman R, Aronowicz J, Kirschner M, et al. Deuterostome phylogeny reveals monophyletic chordates and the new phylum Xenoturbellida. Nature. 2006;444:85–8.

58. Delsuc F, Tsagkogeorga G, Lartillot N, Philippe H. Additional molecular support for the new chordate phylogeny. Genesis. 2008;46:592–604.

59. Friedrich M, Tautz D. Ribosomal DNA phylogeny of the major extant arthropod classes and the evolution of myriapods. Nature. 1995;376:165–7.

60. Boore JL, Lavrov DV, Brown WM. Gene translocation links insects and crustaceans. Nature. 1998;392:667–8.

61. Shultz JW, Regier JC. Phylogenetic analysis of arthropods using two nuclear protein-encoding genes supports a crustacean + hexapod clade. Proc Biol Sci. 2000;267:1011–9.

62. Dohle W. Are the insects terrestrial crustaceans? A discussion of some new facts and arguments and the proposal of the proper name 'Pancrustacea' for the monophyletic unit Crustacea+Hexapoda. Ann Soc Entomol. 2001;37:85–103.

63. Giribet G, Edgecombe GD, Wheeler WC. Arthropod phylogeny based on eight molecular loci and morphology. Nature. 2001;413:157–61.

64. Meusemann K, von Reumont BM, Simon S, Roeding F, Strauss S, Kück P, et al. A phylogenomic approach to resolve the arthropod tree of life. Mol Biol Evol. 2010;27:2451–64.

65. Regier JC, Shultz JW, Zwick A, Hussey A, Ball B, Wetzer R, et al. Arthropod relationships revealed by phylogenomic analysis of nuclear protein-coding sequences. Nature. 2010;463:1079–83.

66. Timmermans MJ, Roelofs D, Mariën J, van Straalen NM. Revealing pancrustacean relationships: phylogenetic analysis of ribosomal protein genes places Collembola (springtails) in a monophyletic Hexapoda and reinforces the discrepancy between mitochondrial and nuclear DNA markers. BMC Evol Biol. 2008;8:83.

67. Andrew DR. A new view of insect-crustacean relationships II. Inferences from expressed sequence tags and comparisons with neural cladistics. Arthropod Struct Dev. 2011;40:289–302.

68. von Reumont BM, Jenner RA, Wills MA, Dell'ampio E, Pass G, Ebersberger I, et al. Pancrustacean phylogeny in the light of new phylogenomic data: support for Remipedia as the possible sister group of Hexapoda. Mol Biol Evol. 2012;29:1031–45.

69. Oakley TH, Wolfe JM, Lindgren AR, Zaharoff AK. Phylotranscriptomics to bring the understudied into the fold: monophyletic ostracoda, fossil placement, and pancrustacean phylogeny. Mol Biol Evol. 2013;30:215–33.

70. Sasaki G, Ishiwata K, Machida R, Miyata T, Su ZH. Molecular phylogenetic analyses support the monophyly of Hexapoda and suggest the paraphyly of Entognatha. BMC Evol Biol. 2013;13:236.

71. Jenner RA. Higher-level crustacean phylogeny: consensus and conflicting hypotheses. Arthropod Struct Dev. 2010;39:143–53.

72. Giribet G, Edgecombe GD. The Arthropoda: A phylogenetic framework. In Arthropod Biology and Evolution-Molecules, Development, Morphology. Edited by Boxshall G, Fusco G, Minelli A. Berlin: Springer-Verlag; 2013:17–40.

73. Harzsch S. The phylogenetic significance of crustacean optic neuropils and chiasmata: a re-examination. J Comp Neurol. 2002;453:10–21.

74. Fanenbruck M, Harzsch S, Wägele JW. The brain of the Remipedia (Crustacea) and an alternative hypothesis on their phylogenetic relationships. Proc Natl Acad Sci U S A. 2004;101:3868–73.

75. Strausfeld NJ, Andrew DR. A new view of insect-crustacean relationships I. Inferences from neural cladistics and comparative neuroanatomy. Arthropod Struct Dev. 2011;40:276–88.

76. Hwang UW, Friedrich M, Tautz D, Park CJ, Kim W. Mitochondrial protein phylogeny joins myriapods with chelicerates. Nature. 2001;413:154–7.

77. Lim JT, Hwang UW. The complete mitochondrial genome of Pollicipes Mitella (Crustacea, Maxillopoda, Cirripedia): non-monophylies of maxillopoda and crustacea. Mol Cells. 2006;22:314–22.

78. Mallatt J, Giribet G. Further use of nearly complete 28S and 18S rRNA genes to classify Ecdysozoa: 37 more arthropods and a kinorhynch. Mol Phylogenet Evol. 2006;40:772–94.

79. Roeding F, Borner J, Kube M, Klages S, Reinhardt R, Burmester T. A 454 sequencing approach for large scale phylogenomic analysis of the common emperor scorpion (Pandinus imperator). Mol Phylogenet Evol. 2009;53:826–34.

80. Rota-Stabelli O, Campbell L, Brinkmann H, Edgecombe GD, Longhorn SJ, Peterson KJ, et al. A congruent solution to arthropod phylogeny: phylogenomics, microRNAs and morphology support monophyletic Mandibulata. Proc Biol Sci. 2011;278:298–306.

81. Regier JC, Shultz JW, Kambic RE. Pancrustacean phylogeny: hexapods are terrestrial crustaceans and maxillopods are not monophyletic. Proc Biol Sci. 2005;272:395–401.

82. Glenner H, Thomsen PF, Hebsgaard MB, Sørensen MV, Willerslev E. Evolution. The origin of insects. Science. 2006;314:1883–4.

83. von Reumont BM, Meusemann K, Szucsich NU, Dell'Ampio E, Gowri-Shankar V, Bartel D, et al. Can comprehensive background knowledge be incorporated into substitution models to improve phylogenetic analyses? A case study on major arthropod relationships. BMC Evol Biol. 2009;9:119.

84. Akiyama-Oda Y, Oda H. Early patterning of the spider embryo: a cluster of mesenchymal cells at the cumulus produces Dpp signals received by germ disc epithelial cells. Development. 2003;130:1735–47.

85. Gott R: Development of gene expression-based biomarkers of exposure to metals and pesticides in the freshwater amphipod Hyalella azteca, 2016, Doctoral dissertation at University of Maryland, Maryland, doi:10.13016/M2PN3J.

86. Kumar S, Stecher G, Tamura K. MEGA7: molecular evolutionary genetics analysis version 7.0 for bigger datasets. Mol Biol Evol. 2016;33:1870–4.

3

Further study of Late Devonian seed plant *Cosmosperma polyloba*: its reconstruction and evolutionary significance

Le Liu[1,2], Deming Wang[2*] (iD), Meicen Meng[3] and Jinzhuang Xue[2]

Abstract

Background: The earliest seed plants in the Late Devonian (Famennian) are abundant and well known. However, most of them lack information regarding the frond system and reconstruction. *Cosmosperma polyloba* represents the first Devonian ovule in China and East Asia, and its cupules, isolated synangiate pollen organs and pinnules have been studied in the preceding years.

Results: New fossils of *Cosmosperma* were obtained from the type locality, i.e. the Leigutai Member of the Wutong Formation in Fanwan Village, Changxing County, Zhejiang Province, South China. The collection illustrates stems and fronds extensively covered in prickles, as well as fertile portions including uniovulate cupules and anisotomous branches bearing synangiate pollen organs. The stems are unbranched and bear fronds helically. Fronds are dimorphic, displaying bifurcate and trifurcate types, with the latter possibly connected to fertile rachises terminated by pollen organs. Tertiary and quaternary rachises possessing pinnules are arranged alternately (pinnately). The cupule is uniovulate and the ovule has four linear integumentary lobes fused in basal 1/3. The striations on the stems and rachises may indicate a *Sparganum*-type cortex.

Conclusions: *Cosmosperma* further demonstrates diversification of frond branching patterns in the earliest seed plants. The less-fused cupule and integument of this plant are considered primitive among Devonian spermatophytes with uniovulate cupules. We tentatively reconstructed *Cosmosperma* with an upright, semi-self-supporting habit, and the prickles along stems and frond rachises were interpreted as characteristics facilitating supporting rather than defensive structures.

Keywords: *Cosmosperma polyloba*, Frond, Ovule, Pollen organ, Seed plant, Late Devonian, Wutong Formation, South China

Backgrounds

Many ovules have been reported from the Upper Devonian (Famennian) of Europe, North America and China, and they indicate the first major radiation of seed plants or spermatophytes [1–4]. Pollen organs also add to our knowledge about these earliest spermatophytes, although they are usually detached from the ovules or fronds [3, 5–10]. Despite the abundance of fertile structures (>20 genera of ovules and pollen organs) in the Late Devonian, the frond morphology and overall architecture is only known for a few seed plant taxa.

South China was an isolated crustal plate with great plant diversity in the Devonian [11–13]. However, seed plant were only recently found in the Late Devonian of this plate, displaying cupulate ovules, pollen organs and stem anatomy [3, 4, 8–10, 13]. These findings suggest that China is an important area for understanding the early evolution of seed plants. Among them, *Cosmosperma polyloba* represents the first Devonian ovules known from China and East Asia that are associated with pollen organs and pinnules [3], while the details of the ovules are unclear due to poor preservation. Based on new specimens from the type locality, we now emend the diagnoses of *Cosmosperma*, compare its frond morphology to related taxa and provide further information regarding its overall

* Correspondence: dmwang@pku.edu.cn
[2]Key Laboratory of Orogenic Belts and Crustal Evolution, Department of Geology, Peking University, Beijing 100871, China
Full list of author information is available at the end of the article

architecture. The entire plant is reconstructed and its evolutionary significance is discussed.

Material and Methods

Over 100 new specimens of *Cosmosperma polyloba* were obtained from the Wutong (Wutung) Formation in a quarry near Fanwan Village, Hongqiao Town, Changxing County, Zhejiang Province, China. The information regarding the locality and stratigraphy has been provided in recent studies [3, 14, 15]. At the Fanwan section, the Wutong Formation is divided into the Guanshan Member, with quartz sandstone and conglomerate, and the overlying Leigutai Member, with interbedded quartz sandstone and mudstone. The fossil plant occurs at the 13th bed of the Wutong Formation (in the Leigutai Member), i.e. the same bed from which former specimens of *Cosmosperma* and strobili of lycopsid *Changxingia* sp. were collected [3, 15]. The LC (*Knoxisporites literatus-Reticulatisporites cancellatus*) spore assemblage suggests that the upper part of the Leigutai Member is of the latest Famennian age [16].

In siltstone with tiny crystals of quartz and white micas, the plant is preserved as dark-brown compressions and impressions, displaying great contrast to the yellowish matrix. Steel needles were applied to expose the plant morphology and a digital camera and a stereoscope were used for photographs. All the specimens are housed at the Department of Geology, Peking University, Beijing, China.

Systematics

Division Spermatophyta sensu Rothwell and Serbet 1994

Class Lagenospermopsida sensu Cleal 1994

Order and Family Incertae sedis

Genus *Cosmosperma* Wang et al. 2014 emend.

Emended diagnosis: (emended and additional generic characters are in brackets).

[Seed plant with unbranched stems bearing dimorphic fronds, dichotomized fertile rachises terminated by synangiate pollen organs, and cupulate ovules. Fronds with a swollen pulvinus-shaped base. Majority of fronds bifurcate, with primary rachis dichotomizing into two secondary rachises. The other fronds trifurcate, with primary rachis ended by two subopposite secondary rachises and one median rachis. Tertiary rachises and ultimate pinnae (with quaternary rachis) borne alternately and pinnately.] Nonlaminate pinnules planate, highly dissected and alternately arranged on [quaternary rachis]. Pollen organs synangiate, with each terminating a stalk and consisting of [four] to eight elongate microsporangia that are basally fused and distally free. Uniovulate cupules with [up to approximately 16 tips]; cupule [tips] free for a length of half to two thirds that of cupules. [Ovule connected to cupule by a short stalk. Four linear integumentary lobes fused

in the basal 1/3. Tiny conical prickles occurring on stems, four orders of frond rachises, cupules and fertile rachises.]

Type species *Cosmosperma polyloba* Wang et al. 2014 emend.

Holotype: PKUB13401a, b ([3], original Fig. 2b).

Specimens examined herein: PKUB13501-PKUB13517 (Figs. 1, 2, 4, 5, 6 and 7).

Repository: Department of Geology, Peking University, Beijing, China.

Locality & horizon: Fanwan Village, Hongqiao Town, Changxing County, Zhejiang Province, China; Leigutai Member of Wutong Formation, Upper Devonian (Famennian).

Emended diagnosis: (Emended and additional specific characters are in brackets).

As for generic diagnosis. [Stems up to 25.9 cm long and 2.2 cm wide, with internodes 0.6–6.2 cm long. Fronds departing at 40–70°. Primary rachises 10.3–21.2 cm long and 3.0–12 mm wide. The secondary rachises are up to 14.3 cm long and 1.8–3.6 mm wide. Median rachises of trifurcate fronds up to 10.9 cm long and ca. 4.0 mm wide. Tertiary rachises up to 10.9 cm long and 1.7–2.9 mm wide. First tertiary rachises occurring on outside of frond. Ultimate pinna] up to 54 mm long and 28 mm wide, with [quaternary rachis] about 0.7 mm wide; pinnules [6.0]–13.3 mm long and [3.0]–13.0 mm wide, borne at angles of 70–90°, and consisting of one terminal unit and four alternately arranged lateral units. Pinnule units 4.2–7.2 mm long and 2.8–8.3 mm wide, equally dichotomous for one to three times. [Fertile rachises dichotomizing 3–6 times at 50–120°, with intervals between adjacent bifurcating points 1.4–19.3 mm long and 0.3–1.2 mm wide.] Pollen organs borne in pairs, 2.2–[2.5] mm long and [2.0]–2.9 mm wide, with stalks 1.0 mm long and 0.2–0.3 mm wide; microsporangia 2.3 mm long and [0.3]–0.7 mm wide, and distally tapered. Cupules [5.3]–8.8 mm long and [3.0]–9.0 mm wide; pedicels 1.0 mm long and 0.4 mm wide; ovules 3.7–[4.7] mm long and 1.6–[2.2] mm wide; [ovule stalk ca. 0.2 mm long and ca. 0.5 mm wide; integumentary lobes ca. 3.8 mm long and ca. 0.5 mm wide. Prickles on stems and proximal parts of fronds, ca. 0.3 mm long and ca. 0.5 mm wide at the base; those on cupules and distal fertile rachises, ca. 0.2 mm long and ca. 0.3 mm wide at the base].

Results

General morphology

Plant organs of *Cosmosperma polyloba* described here include stems (Figs. 1 and 2), dimorphic fronds (Figs. 3, 4, 5 and 6a-c), isolated cupulate ovules (Fig. 6d-h), and synangiate pollen organs terminating anisotomous fertile rachises (Fig. 7). One stem, some fronds and fertile rachises with pollen organs are represented in interpretive line-drawings (Additional files 1, 2 and 3: Figures S1–S3).

Fig. 1 Stems of *Cosmosperma polyloba*. **a** Stem with primary rachises. PKUB13501a. Scale bar = 20 mm. **b** Combined figure of two counterparts of specimen shown in Fig. 1a, exhibiting stem with primary rachises. *Arrow* indicating portion enlarged in Fig. 1c. PKUB13501b (part in *dashed box*), PKUB13501c. Scale bar = 20 mm. **c** Enlargement of Fig. 1 (**b**, *arrow*), showing conical prickles (*arrows*) and parallel vertical striations on the stem. Scale bar = 2 mm. **d** Stem with the widest primary rachis. PKUB13502. Scale bar = 20 mm. **e** Stem and helically arranged primary rachises with swollen bases. *Arrows* indicating portions enlarged in Fig. 1**f**, **g**. PKUB13503. Scale bar = 20 mm. **f** Enlargement of Fig. 1 (**e**, *left arrow*), showing conical prickles (*arrows*). Scale bar = 2 mm. **g** Enlargement of Fig. 1 (**e**, *right arrow*), showing parallel vertical striations (*white arrow*) and conical prickles preserved as pit-like impressions (*black arrows*). Scale bar = 2 mm. **h** Combined figure of part and counterpart of one specimen, showing a stem with primary rachises arranged in irregular helix. PKUB13504a, PKUB13504b (*dashed box*). Scale bar = 20 mm

Fig. 2 Stems of *Cosmosperma polyloba*. **a-c** Stems illustrating helically arranged primary rachises with swollen bases. *Arrow* in Fig. 2**b** indicating portion enlarged in Fig. 2**e**. PKUB13505-PKUB13507. Scale bars = 20 mm. **d** Stem with one dichotomous frond rachis. PKUB13508. Scale bar = 20 mm. **e** Enlargement of Fig. 2 (**b**, *arrow*), showing conical prickles (*arrow*) and parallel vertical striations. Scale bar = 5 mm

Fronds are arranged in irregular helices on the stem (Figs. 1a, b, e, h and 2a–c, Additional file 1: Figure S1a) and consist of up to four orders of rachises (Figs. 4a, 5a–c and 6a–c, Additional files 1 and 2: Figures S1b–d, S2a) and pinnate pinnules (Figs. 4c and 5e, Additional file 2: Figure S2b–e). Morphological descriptors for fronds are illustrated in Fig. 3. Tiny conical prickles of different size are present on the stems (Fig. 1c, arrows, f, arrows, g, black arrows, 2e, arrow), frond rachises (Fig. 4b, arrows and Fig. 5d, arrows), cupules (Fig. 6d, arrow) and fertile

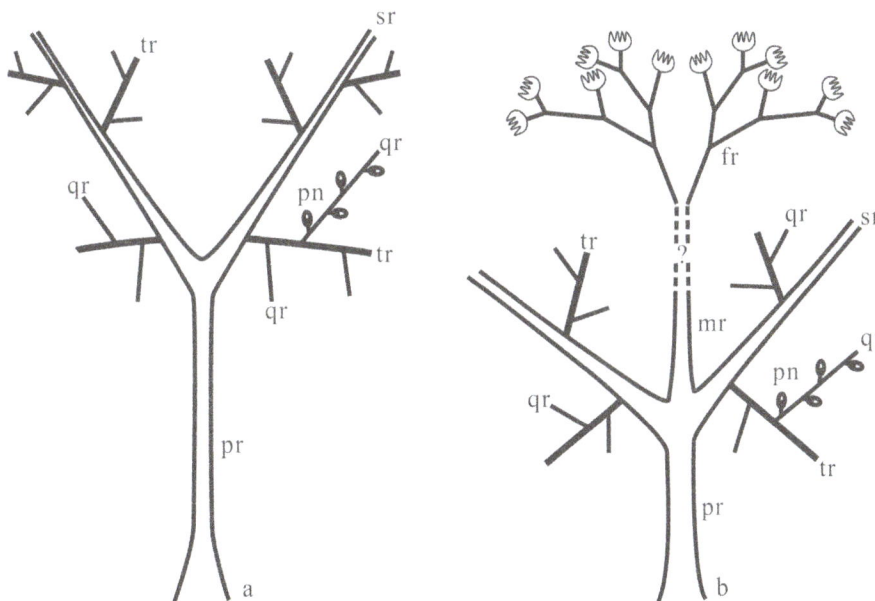

Fig. 3 Interpretative diagram showing architecture of bifurcate (**a**) and trifurcate (**b**) fronds of *Cosmosperma polyloba*. The connection between trifurcate frond and fertile rachises bearing synangia (*dashed lines with a question mark*) is speculative. Abbreviations: pr, primary rachis; sr, secondary rachis; mr, median rachis; tr, tertiary rachis; qr., quaternary rachis; fr, fertile rachis; pn, pinnule

Fig. 4 Bifurcate fronds of *Cosmosperma polyloba*. **a** Combined figure of part and counterpart of one specimen, exhibiting frond with primary rachis ending in a dichotomy (*arrow* 1, enlarged in Fig. 4**b**), two secondary rachises, alternately (pinnately) arranged tertiary rachises and an attached ultimate pinna (quaternary rachis and pinnules; *arrow* 2, enlarged in Fig. 4**c**). A cupulate ovule (*arrow* 3, enlarged in Fig. 6**d**) is associated with the frond. PKUB13509a, PKUB13509b (above the *dashed line*). Scale bar = 20 mm. **b** Enlargement of Fig. 4(**a**, *arrow* 1), showing the dichotomy of primary rachis. Note parallel vertical striations and conical prickles (*arrows*) along the primary rachis and basal part of secondary rachises. Scale bar = 5 mm. **c** Enlargement of Fig. 4 (**a**, *arrow* 2), showing one ultimate pinna with highly dissected, alternate and planate pinnules. *Arrows* indicating conical prickles along quaternary rachis. Scale bar = 5 mm. **d** Longest primary rachis with base but without distal portion preserved. PKUB13508. Scale bar = 20 mm

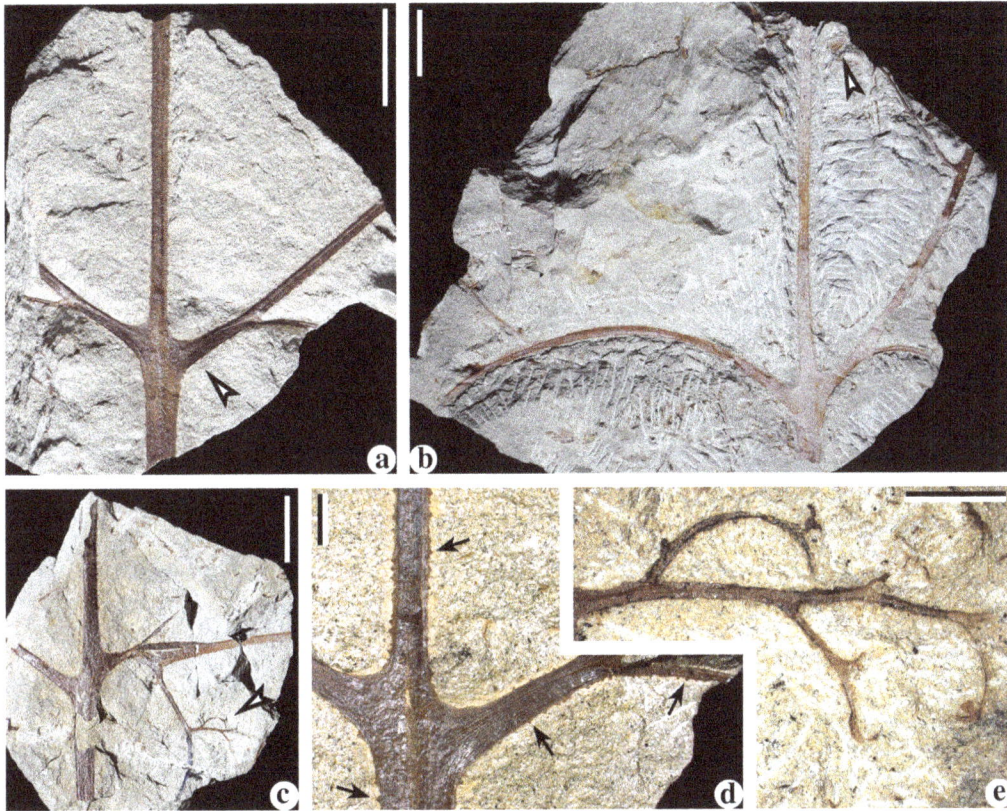

Fig. 5 Trifurcate fronds of *Cosmosperma polyloba*. **a** Frond consisting of a primary rachis, a median and two subopposite secondary rachises and two tertiary rachises. *Arrow* indicating portion enlarged in Fig. 5(**d**). PKUB13510a. Scale bar = 20 mm. **b** Frond consisting of a primary rachis, a median and two subopposite secondary rachises and alternately borne tertiary rachises. One cupulate ovule (*arrow*, enlarged in Fig. 6**e**) associated with the frond. PKUB13511. Scale bar = 20 mm. **c** Frond consisting of a primary rachis, a median and two subopposite secondary rachises, a tertiary rachis and an ultimate pinna. *Arrow* indicating portion enlarged in Fig. 5e. PKUB13512. Scale bar = 20 mm. **d** Enlargement of Fig. 5 (**a**, *arrow*), showing attachment of a median and two secondary rachises. Parallel vertical striations distributed along primary rachis, basal parts of median and secondary rachises. *Arrows* indicating conical prickles. Scale bar = 5 mm. **e** Enlargement of Fig. 5 (**c**, *arrow*), showing an ultimate pinna with poorly preserved pinnules. Scale bar = 5 mm

rachises bearing pollen organs (Fig. 7a, arrows, c, arrows). Narrow and parallel striations occur vertically on the surface of the stems (Fig. 1c, g, white arrow; Fig. 2e), primary rachises and basal part of secondary rachises (Figs. 4b and 5d), implying the *Sparganum*-type cortex.

Stems

Stems are 0.7–2.2 cm wide (Figs. 1 and 2) and up to 25.9 cm long (Fig. 1b). No evidence indicates that the stems are branched. The large stems (Fig. 1) suggest basal or mature parts of the plant, while the slender ones (Fig. 2) may represent the upper or immature portions. The prickles are ca. 0.3 mm long and ca. 0.5 mm wide at the base (Fig. 1c, arrows, f, arrows and Fig. 2e, arrow), and they sometimes leave pit-like impressions on the stem surface (Fig. 1, g, black arrows).

Fronds

Along the stem, the internodal length between the attachments of two adjacent fronds ranges from 0.6–6.2 cm (Fig. 1a, b, e, h; Fig. 2a–c). The fronds depart at 40–70° (Fig. 1a, b, d, e; Fig. 2a–d), and their bases are swollen and pulvinus shaped (Figs. 1e and 2b–d). Fronds exhibit two types of division (Fig. 3), i.e. the bifurcate type (Figs. 3a and 4) and the trifurcate type (Figs. 3b and 5), which can be distinguished by the primary rachises. Most fronds are bifurcate, showing primary rachises that extend a long distance and then dichotomize at 45–70° into two secondary rachises (Figs. 1h, 2d, 4a and 6a). The trifurcate fronds possess a primary rachis that ends in two subopposite secondary rachises departing at 60–90° and a median rachis (Fig. 1b, dashed box; Fig. 5a–d). The total length of fronds is up to 24.2 cm (Fig. 4a). Primary rachises are 10.3–21.2 cm long (Figs. 2d and 4d), and are usually 3.0–4.0 mm wide, but can be up to 12 mm wide

Fig. 6 Fronds and cupulate ovules of *Cosmosperma polyloba*. **a** Frond consisting of distal part of bifurcated primary rachis, two secondary rachises, and one tertiary rachis bearing 11 ultimate pinnae (*arrows*). PKUB13513. Scale bar = 20 mm. **b** Combined figure of part and counterpart of one specimen, showing vegetative frond with secondary rachis at possible distal portion and three alternate (pinnate) tertiary rachises bearing ultimate pinnae. *Arrow* indicating portion redrawn in Additional file 2: Figure S2(d). PKUB13514a, PKUB13514b (*dashed polygon*). Scale bar = 2 cm. **c** Frond segment including secondary rachis, alternately (pinnately) borne tertiary rachises with ultimate pinnae. *Arrow* indicating portions redrawn in Additional file 2: Figure S2(b, c). PKUB13515. Scale bar = 20 mm. **d** Enlargement of cupulate ovule in Fig. 4(a, *arrow* 3), showing prickles (*arrow*) on the outer surface. Scale bar = 2 mm. **e** Enlargement of cupulate ovule in Fig. 5(b, *arrow*), showing compression of ovule with a stalk connecting the cupule (*lower arrow*, enlarged in Fig. 6f) and four integumentary lobes (*upper four arrows*). Scale bar = 2 mm. **f** Enlargement of Fig. 6 (**e**, *lower arrow*), showing lower part of ovule with a short stalk. Scale bar = 0.5 mm. **g** Ovule in Fig. 6e, remnant after removal of integumentary lobes by dégagement. *Arrow* indicating portion enlarged in Fig. 6**h**. Scale bar = 2 mm. **h** Enlargement of Fig. 6 (**g**, *arrow*), showing cupule tips (*white star*) and remnant of ovule (*black star*). Scale bar = 0.5 mm

(Fig. 1d). The secondary rachises are up to 14.3 cm long and 1.8–3.6 mm wide. The median rachises of trifurcate fronds are up to 10.9 cm long and ca. 4.0 mm wide, demonstrating parallel vertical striations and tiny conical prickles (Fig. 5d). Tertiary rachises are borne alternately (Figs. 4a, 5b and 6b, c, Additional file 1: Figure S1b–d), and up to 10.9 cm long and 1.7–2.9 mm wide. Two proximal tertiary rachises are produced toward the outside of the frond, at the same distance from the base of the secondary rachis (Figs. 4a and 5a). Ultimate pinnae are mainly alternately (i.e., pinnately) arranged (Figs. 4a, 5c and 6a–c, Additional files 1 and 2: Figures S1b–d, S2a), but occasionally folded to one side (Fig. 6a) due to

preservation. The quaternary rachises (ultimate pinna rachises) are up to 4.1 cm long and 0.7 mm wide (Figs. 4c, 5e and 6b, c, Additional file 2: Figure S2b–e). The amount of quaternary rachises on a single tertiary rachis is up to 11 (Fig. 6a, Additional file 2: Figure S2a). The prickles on frond rachises are ca. 0.3 mm long and ca. 0.5 mm wide at the base (Fig. 4b, c, arrows and Fig. 5d). Highly dissected but planate pinnules are alternately arranged along the quaternary rachis, and are 6.0–13.0 mm long and 3.0–10.0 mm wide (Figs. 4c, 5e and 6b, c, Additional file 2: Figure S2, b–e). Each pinnule exhibits an "axis", with several alternately-borne lateral units and one terminal unit. These units dichotomize into several

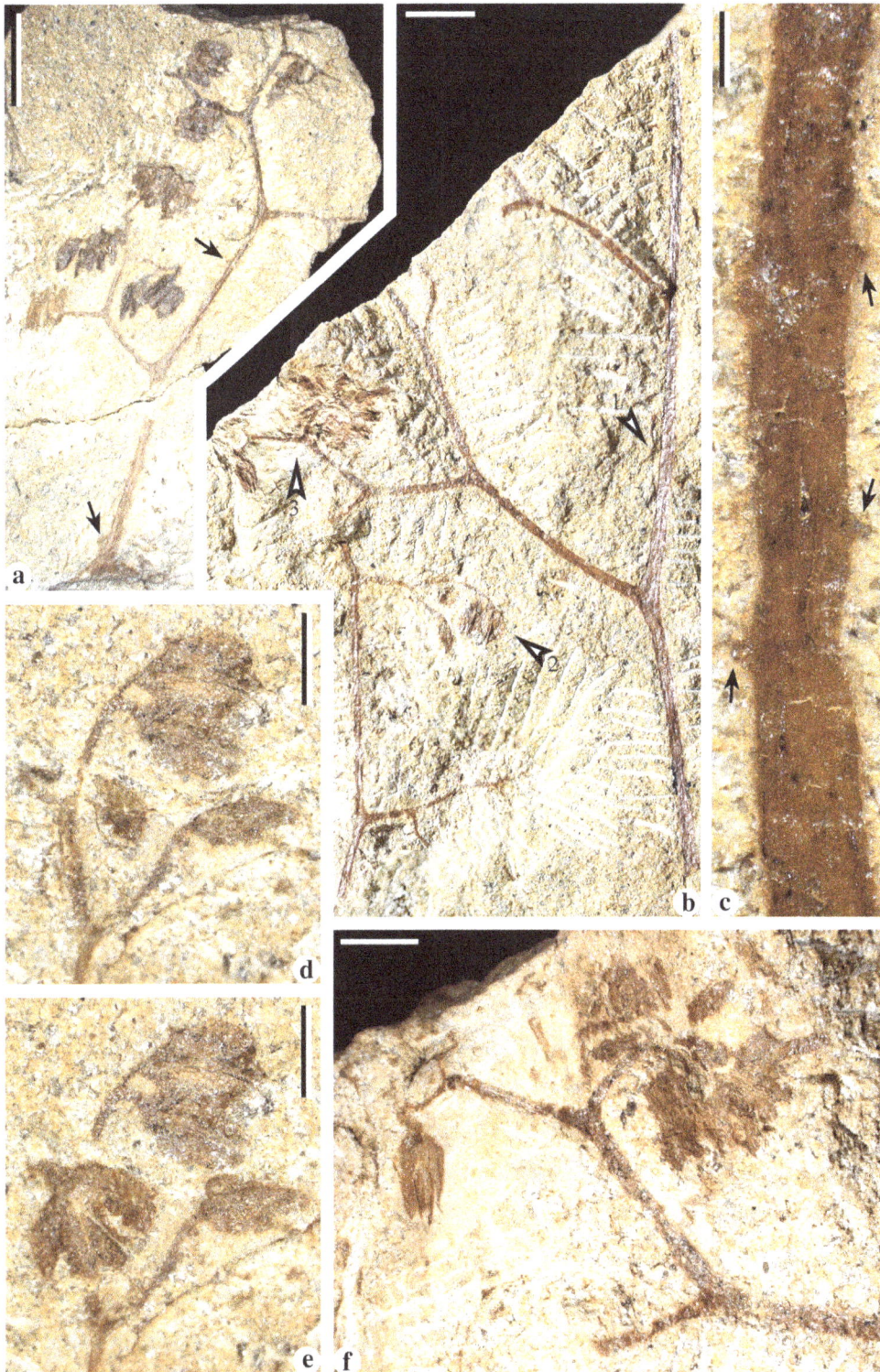

Fig. 7 Synangiate pollen organs on fertile rachises of *Cosmosperma polyloba*. **a** Anisotomous fertile rachises with terminal pollen organs and prickles (*black arrows*). PKUB13516. Scale bar = 5 mm. **b** Two anisotomous fertile rachises with terminal pollen organs. Fertile rachises (*arrow 1*, enlarged in Fig. 7**c** displaying sparse prickles; pollen organ (*arrow 2, 3*, enlarged in Fig. 7**d-f**) born in pairs or singly. PKUB13517. Scale bar = 5 mm. **c** Enlargement of Fig. 7 (**b**, *arrow 1*), showing one fertile rachis bearing conical prickles (*black arrows*). Scale bar = 0.5 mm. **d, e** Enlargement of Fig. 7 (**b**, *arrow 2*). Two stages of dégagement showing pollen organs. Scale bar = 2 mm. **f** Enlargement of Fig. 7 (**b**, *arrow 3*). Pollen organs terminating bifurcated fertile rachises. Scale bar = 2 mm

slender segments (Figs. 4c and 6b, c, Additional file 2: Figure S2b–e). The axis and the segments are ca. 0.5 mm wide.

Cupulate ovules

Cupules are isolated, 5.3–7.7 mm long and 3.0–5.1 mm at the maximum width (Fig. 6d, e). The cupules display minute conical prickles on the outer surface (Fig. 6d, arrow) that are ca. 0.2 mm long and ca. 0.3 mm wide at the base. Each cupule possesses segments with multiple tips that are about half of the total cupule length and are ca. 0.5 mm wide. One specimen illustrates that the cupule is uniovulate (Fig. 6e). The upper part of the ovule (Fig. 6g, arrow) is dégagé to expose several cupule tips (Fig. 6h, white star), which are beneath the ovule remnant (Fig. 6h, black star). Before the dégagement, this ovule was 4.7 mm long and 2.2 mm wide, and connected to the cupule by a short stalk ca. 0.2 mm long and 0.5 mm wide (Fig. 6e, lower arrow, f). Four integumentary lobes are linear and straight (Fig. 6e, black arrows), ca. 3.8 mm long and ca. 0.5 mm wide, and fused to each other in the basal 1/3 of the ovule.

Fertile rachises with terminal pollen organs

The fertile rachises are anisotomous and terminate in pollen organs (Fig. 7a, b, d–f; Additional file 3: Figure S3). These rachises dichotomize 3–6 times and at an angle of 50–120°, with the intervals between two adjacent bifurcating points being 1.4–19.3 mm long and 0.3–1.2 mm wide. Both length and width of the intervals reduce acropetally. Conical prickles are sparse on the branches and ca. 0.2 mm long and 0.3 mm wide at the base (Fig. 7a, c, Additional file 3: Figure S3a). Pollen organs, ca. 2.5 mm long and 2.0 mm wide, are born mainly in pairs, but sometimes singly or incompletely preserved (Fig. 7d–f; Additional file 3: Figure S3). Individual pollen organs are synangiate with basally fused microsporangia. Each synangium consists of 4–8 elongate microsporangia, which are ca. 2.3 mm long and 0.3–0.4 mm wide.

Discussion

Reconstruction of Cosmosperma

Based on the specimens described above, Cosmosperma is characterized by unbranched stem with two types of fronds attached in irregular helices, alternately arranged tertiary and quaternary rachises, uniovulate cupules and synangiate pollen organs terminating anisotomous fertile rachises. We tentatively reconstructed Cosmosperma as shown in Fig. 8, and it is thus one of the best morphologically understood Late Devonian seed plants in the world.

Comparisons with other Devonian seed plants

The cupules and synangiate pollen organs of Cosmosperma have been compared with those of related seed plants [3], and this comparison section primarily focuses on frond morphology. Vegetative fronds have been reported in the Late Devonian seed plants, i.e., Elkinsia from USA [5], Laceya from Ireland [17, 18], Kongshania [8], Yiduxylon [13] and Telangiopsis [10] from China. Among them, Elkinsia, Kongshania and Telangiopsis are also known for fertile rachises with terminal pollen organs. Some selected morphological traits of these plants are listed and compared in Table 1. All of these taxa except Kongshania display bipartite fronds, while Elkinsia exhibits repeatedly bifurcated frond rachises exclusively. Yiduxylon, Telangiopsis and Cosmosperma have highly dissected, planated pinnules, differing from the laminate pinnules of Elkinsia and Kongshania. The fertile rachises bearing pollen organs are anisotomously divided in Cosmosperma, which enables them to be distinguished from the isotomously divided ones in Elkinsia, Kongshania and Telangiopsis. Prickles are extensively distributed on Cosmosperma, but are confined to the stems of Telangiopsis and absent from other coeval seed plants.

Variations in fronds among early seed plants

Early seed plants are characterized by bipartite fronds with dichotomized primary rachises [19, 20], while diversified frond structures are evidenced in the Late Devonian taxa, such as variable dimensions of fronds, different branching manners and flexible locations of ultimate pinnae (Table 1). It has been shown that great morphological disparities have occurred among the Late Devonian spermatophytes. Lyginopterid seed plants in the Early/Late Carboniferous are thought to possess fronds with dichotomized/pinnate branching patterns, respectively [19]. Since Elkinsia is characterized by repeatedly dichotomized fronds [5], while Laceya [17], Yiduxylon [13], Telangiopsis [10] and Cosmosperma show pinnate fronds, it seems that both branching patterns have arisen in the Late Devonian spermatophytes.

The fertile fronds with terminal pollen organs often exhibit cruciate dividing patterns in the Late Devonian seed plants (e.g., Telangium schweitzeri [6] and Elkinsia [5]). Among the Early Carboniferous spermatophytes, the fertile fronds with terminal pollen organs containing trilete prepollen are divided into three types: Rhacopteris/Triphyllopteris-type, Diplopteridium-type and Rhodea-type [21]. The Diplopteridium-type illustrates a trifurcate frond rachis producing a median fertile rachis that is short and dichotomous [21–23]. The trifurcate fronds of Cosmosperma display a unique architecture among coeval seed plants. Such fronds, if connected to the fertile rachises bearing terminal pollen organs (Fig. 8), would greatly resemble the Diplopteridium-type fertile frond. In this case, Cosmosperma exemplifies the diversification of fertile fronds among Late Devonian seed plants,

Fig. 8 Reconstruction of *Cosmosperma polyloba*. The plant is considered to possess an upright, probably semi-self-supporting habit, with adjacent individuals entangled by their bushy, prickle-bearing fronds. Dimorphic fronds are helically arranged along stem, with bifurcate fronds in the majority, and scattered trifurcate fronds displaying median rachises; the connections between trifurcate fronds and fertile parts are speculative

and suggests that some Carboniferous fertile frond types may be traced back to an earlier time.

Different dividing patterns of the fertile and vegetative fronds were present in Carboniferous spermatophytes [22, 23], which is also supported by the anatomical evidence [24, 25]. Both *Elkinsia* [5] and *Cosmosperma* indicate that the dimorphic fronds have occurred in the Late Devonian.

Implications from the ovule of *Cosmosperma*

Nearly all Late Devonian seed plants have cupulate ovules (ovules enclosed in cupules) [2], and the cupules are uniovulate or multiovulate [4]. The uniovulate cupules were considered to be derived from the multiovulate ones [26, 27]. The uniovulate cupule has been proposed [3] and is now confirmed in *Cosmosperma*. Other Devonian seed plants with uniovulate cupules include *Dorinnotheca* [27], *Latisemenia* [4], *Condrusia* [28] and *Pseudosporognites* [2]. Their traits are listed in Table 2. The cupule or integument of the early ovules is considered archaic with numerous, terete and little fused segments or lobes [27, 29, 30]. In this case, *Cosmosperma* appears primitive among the ovules with uniovulate cupules.

Table 1 Comparison of morphological traits among Late Devonian seed plants

Taxon	Frond rachis arrangement	Length of primary rachis prior to bifurcation (cm)	Location of ultimate pinnae	Pinnules	Fertile rachises bearing pollen organs	Prickles
Elkinsia [5]	Equally and repeatedly bifurcated	0-13[a]	on secondary and higher orders of rachises	Laminate	Isotomously and cruciately dichotomized	absent
Laceya [17, 18]	Pinnate with bifurcated primary rachis	up to 15.5	on both primary and secondary rachises	—	—	absent
Kongshania [8]	Pinnate	—	on tertiary rachises	Laminate	Isotomously dichotomized	absent
Yiduxylon [13]	Pinnate with bifurcated primary rachis	ca. 3	on tertiary rachises	Planate and highly dissected	—	absent
Telangiopsis [10]	Pinnate with bifurcated primary rachis	0[a]	on secondary rachises	Planate and highly dissected	Isotomously dichotomized	on stems
Cosmosperma	Pinnate with bifurcated/trifurcated primary rachis	10.3-21.2	on tertiary rachises	Planate and highly dissected	Anisotomously dichotomized	on stems, frond rachises and cupule surfaces

[a]: 0 stand for the basally bifurcated primary rachises

One of the most obvious functions of cupules and integuments is protection for the ovule [1], and a more entire (large and/or widely fused) integument may provide additional protection against water loss [4, 30]. The cupules of *Cosmosperma* enclose the ovule, while those of *Dorinnotheca*, *Pseudosporognites* and *Latisemenia* are recurved or short to extensively expose the ovule. On the other hand, the integrity of the integument is the lowest in *Cosmosperma*, moderate in *Dorinnotheca* and *Pseudosporognites*, and the greatest in *Latisemenia*. Therefore, the protection is largely provided by the cupule in *Cosmosperma*, and by the integument in the other three plants. The evolutionarily primitive status of *Cosmosperma* suggests that the protective function of uniovulate cupules may be replaced by the increasingly developed integument.

Function of prickles and probable growth habit of *Cosmosperma*

The acute outgrowths of epidermis or both epidermis and cortex, without vascular tissues, are usually named prickles, while the sharp-pointed vascularized protuberances modified from axes and leaves are separately called thorns and spines [31, 32]. Commonly, the thorns and spines are only distributed along the axes and, owing to their internal vascular tissues, cannot be easily removed. However, in

Cosmosperma, the tiny conical structures occur on stems, vegetative and fertile rachises and even cupules. They also present a highly variable density corresponding to loss in the transport and/or burial process. Therefore, we tentatively assign such structures to prickles.

The prickles are not common in the Late Devonian spermatophytes, but they have been reported in some later Paleozoic seed plants, including the Early Carboniferous *Medullosa steinii* and Late Permian gigantopterid *Aculeovinea yunguiensis* [33, 34]. It has been suggested that prickles on the cupule surface of *Cosmosperma* may serve as protection [3]. On the other hand, arthropod herbivory was recorded in some Late Devonian myriapods and apterygote hexapods [35], while the major plant defensive adaptations to such herbivory are considered chemical [36]. However, the terrestrial vertebrate herbivory did not occur until the Permian [34]. Since prickles are considered to provide mechanical attachments in other younger Paleozoic seed plants [33, 34], it is plausible that the prickles on the axes and leaves of the Late Devonian seed plants may largely function as supporting structures rather than defense structures against the herbivores.

Previous studies have suggested that the seed plants assigned to the Lyginopteridales are vines/lianas possessing stems generally less than 20 mm wide, and those to

Table 2 Comparison of Late Devonian seeds with uniovulate cupules

Taxon	Number of cupule segments	Structure of cupule segments	Number of cupule tips	Number of integumentary lobes	Shape and fusion of integumentary lobes
Dorinnotheca [27]	8	distally dissected	>40	4	Triangular, basally fused
Condrusia [28]	2	flattened and broad	2	—	—
Pseudosporogonites [2]	1	short, fused and collar/trumpet shaped	—	3-4	Flattened, 1/3 fused
Latisemenia [4]	5	broad and cuneate	5	4	Flattened, 1/2-2/3 fused
Cosmosperma	2?	Distal 1/2-2/3 dissected	up to 16	4	Linear, basal 1/3 fused

the Calamopityales are upright with stems usually over 20 mm wide [13, 25]. Other evidence that supports lyginopterids as vines/lianas includes stems bearing long internodes, the presence of adventitious roots, large fronds with swollen frond bases, wide angle of frond attachment and *Dictyoxylon*-type outer cortex [13, 37, 38]. *Cosmosperma* possesses relatively large fronds with pulvinus-shaped bases, which resemble those of lyginopterids. The extensively born prickles of *Cosmosperma* also remind us of the glands on *Lagenostoma* and *Lyginodendron* [39]. However, in *Cosmosperma*, the width of the stems reaches 22 mm, the internodes are relatively short, the adventitious root is absent, the fronds depart at 40–70° and the cortex is most likely *Sparganum*-type. These traits enable *Cosmosperma* to be tentatively reconstructed as an upright, probably semi-self-supporting plant (Fig. 8), which may support each other by entangled bushy fronds rather than scrambling or climbing. The hypothesis is supported by the preservation that many slabs exhibit pure and dense communities of *Cosmosperma*, without any other arborescent plants. The prickles may help anchor fronds of adjacent individuals. However, the anatomical information is needed to test the suggested growth habits of this plant.

Conclusions

We further studied the seed plant *Cosmosperma polyloba* from the Upper Devonian of South China, and its stems, fronds, cupulate ovules and fertile rachises bearing pollen organs are now known in detail. Based on the morphological evidence mentioned above, we tentatively reconstructed the whole plant with an upright, semi-self-supporting habit. The prickles on stems and rachises may facilitate supporting. The fronds of *Cosmosperma* show bifurcated or trifurcated primary rachises, which further add to the diversity and demonstrate dimorphism of the early spermatophyte fronds. The less-fused cupules and integuments suggest that *Cosmosperma* is primitive among Late Devonian seed plants with uniovulate cupules.

Additional files

Additional file 1: Figure S1. Interpretative line drawings showing branching pattern of *Cosmosperma polyloba*. Abbreviations: st, stem; pr, primary rachis; sr, secondary rachis; tr, tertiary rachis. (a) Stem, primary and secondary rachises and basal part of a tertiary rachis in Fig. 1h. (b) Bifurcate primary rachis, two secondary rachises, and a tertiary rachis bearing ultimate pinnae and conical prickles in Fig. 6c. (c) Secondary rachis with alternate tertiary rachises, ultimate pinnae and conical prickles in Fig. 6b. (d) Bifurcated primary rachis, two secondary rachises and alternate tertiary rachises with ultimate pinnae and conical prickles in Fig. 4a. (TIFF 2282 kb)

Additional file 2: Figure S2. Interpretative line drawings showing frond and ultimate pinnae of *Cosmosperma polyloba*. Abbreviations same as in Figure S1. (a) Bifurcate primary rachis, two secondary rachises, and one tertiary rachis with 11 ultimate pinnae in Fig. 6a. (b-e) Ultimate pinnae in

Fig. 6(c, left arrow), Fig. 6(c, right arrow), Fig. 6(b, arrow) and Fig. 4(c), respectively. Highly dissected and planate pinnules alternately arranged along the quaternary rachis. (TIFF 2025 kb)

Additional file 3: Figure S3. Interpretative line drawing showing synangiate pollen organs on fertile axes of *Cosmosperma polyloba*. (a) Anisotomous fertile rachises with terminal pollen organs in Fig. 7a. Conical prickles sparsely located along the fertile rachises sparsely. (b, c) Two stages of dégagement on pollen organs in Fig. 7d, e, respectively. (TIFF 1395 kb)

Acknowledgements
We thank D. L. Qi (Anhui Geological Survey, Hefei) and T. Liu (Peking University, Beijing) for assistance in the fieldwork, C. C. Labandeira (Smithsonian Institution, Washington, D. C.) and H. Fang (Capital Normal University, Beijing) for suggestions. This study is supported by China Postdoctoral Science Foundation (No. 2016 M600146) and the National Natural Science Foundation of China (No. 41672007).

Authors' contributions
LL and DMW collected the fossils. LL conducted the experiments, prepared the Figures, and wrote the manuscript. All authors discussed the results, read and approved the final manuscript.

Competing interests
The authors declare that they have no competing interests.

Author details
[1]College of Geoscience and Surveying Engineering, China University of Mining and Technology (Beijing), Beijing 100083, China. [2]Key Laboratory of Orogenic Belts and Crustal Evolution, Department of Geology, Peking University, Beijing 100871, China. [3]Science Press, China Science Publishing & Media Ltd., Beijing 100717, China.

References
1. Taylor TN, Taylor EL, Krings M. Paleobotany: the biology and evolution of fossil plants. Burlington: Academic Press; 2009.
2. Prestianni C, Hilton J, Cressler W. Were all Devonian seeds Cupulate? A reinvestigation of *Pseudosporogonites hallei, Xenotheca bertrandii*, and *Aglosperma* spp. Int J Plant Sci. 2013;174(5):832–51.
3. Wang D-M, Liu L, Meng M-C, Xue J-Z, Liu T, Guo Y. *Cosmosperma polyloba* gen. Et sp. nov., a seed plant from the upper Devonian of South China. Naturwissenschaften. 2014;101(8):615–22.
4. Wang D-M, Basinger JF, Huang P, Liu L, Xue J-Z, Meng M-C, Zhang Y-Y, Deng Z-Z. *Latisemenia longshania*, gen. et sp. nov., a new Late Devonian seed plant from China. Proc R Soc B. 2015;282(1817):20151613.
5. Serbet R, Rothwell GW. Characterizing the most primitive seed ferns. I. A reconstruction of *Elkinsia polymorpha*. Int J Plant Sci. 1992;153:602–21.
6. Matten LC, Fine T. *Telangium schweitzeri* sp. nov.: a gymnosperm pollen organ from the upper Devonian of Ireland. Palaeontogr Abt B. 1994;232:15–33.
7. Hilton J. A Late Devonian plant assemblage from the Avon gorge, west England: taxonomic, phylogenetic and stratigraphic implications. Bot J Linn Soc. 1999;129(1):1–54.
8. Wang Y. *Kongshania* gen. Nov. a new plant from the Wutung formation (upper Devonian) of Jiangning County, Jiangsu, China. Acta Palaeontol Sin. 2000;39(SUPP):42–56.
9. Wang D-M, Liu L, Guo Y, Xue J-Z, Meng M-C. A Late Devonian fertile organ with seed plant affinities from China. Sci Rep. 2015;5:10736.
10. Wang D-M, Meng M-C, Guo Y. Pollen organ *Telangiopsis* sp. of Late Devonian seed plant and associated vegetative frond. PLoS One. 2016;11(1): e0147984.
11. Hao S-G, Xue J-Z. The early Devonian Posongchong flora of Yunnan: a contribution to an understanding of the evolution and early diversification of vascular plants. Beijing: Science Press; 2013.

12. Hao S-G, Xue J-Z. Earliest record of megaphylls and leafy structures, and their initial diversification. Chin Sci Bull. 2013;58(23):2784–93.

13. Wang D-M, Liu L. A new Late Devonian genus with seed plant affinities. BMC Evol Biol. 2015;15(1):28.

14. Wang D-M, Meng M-C, Xue J-Z, Basinger JF, Guo Y, Liu L. *Changxingia longifolia* gen. Et sp. nov., a new lycopsid from the Late Devonian of Zhejiang Province, South China. Rev Palaeobot Palynol. 2014;203:35–47.

15. Wang D-M, Qin M, Meng M-C, Liu L, Ferguson DK. New insights into the heterosporous lycopsid *Changxingia* from the upper Devonian Wutong formation of Zhejiang Province, China. Plant Syst Evol. 2017;303:11–21.

16. Ouyang S. Succession of Late Palaeozoic Palynological assemblages in Jiangsu. J Stratigr. 2000;3:230–5.

17. Klavins SD, Matten LC. Reconstruction of the frond of *Laceya hibernica*, a Lyginopterid pteridosperm from the uppermost Devonian of Ireland. Rev Palaeobot Palynol. 1996;93(1):253–68.

18. May BI, Matten LC. A probable pteridosperm from the uppermost Devonian near Ballyheigue, co. Kerry, Ireland. Bot J Linn Soc. 1983;86(1–2):103–23.

19. Taylor TN, Millay MA. Morphologic variability of Pennsylvanian lyginopterid seed ferns. Rev Palaeobot Palynol. 1981;32(1):27–62.

20. Galtier J. The origins and early evolution of the Megaphyllous leaf. Int J Plant Sci. 2010;171(6):641–61.

21. Meyer-Berthaud B. First gymnosperm fructifications with trilete prepollen. Palaeontogr Abt B. 1989;211:87–112.

22. Long AG. The resemblance between the lower Carboniferous cupules of *Hydrasperma* Cf. *tenuis* long and *Sphenopteris bifida* Lindley and Hutton. Trans R Soc Edin. 1979;70:129–37.

23. Rowe N. New observations on the lower Carboniferous pteridosperm *Diplopteridium* Walton and an associated synangiate organ. Bot J Linn Soc. 1988;97(2):125–58.

24. Long AG. —Calathopteris heterophylla gen. Et sp. nov., a lower Carboniferous pteridosperm bearing two kinds of petioles. Trans R Soc Edin. 1976;69:327–36.

25. Galtier J. Morphology and phylogenetic relationships of early pteridosperms. In: Beck CB, editor. Origin and evolution of gymnosperms. New York: Columbia Univ. Press; 1988. p. 135–76.

26. Rothwell GW, Scott AC. Stamnostoma oliveri, a gymnosperm with systems of ovulate cupules from the lower Carboniferous (Dinantian) floras at Oxroad Bay, East Lothian, Scotland. Rev Palaeobot Palynol. 1992;72(3):273–84.

27. Fairon-Demaret M. *Dorinnotheca streelii* Fairon-Demaret, gen. Et sp. nov., a new early seed plant from the upper Famennian of Belgium. Rev Palaeobot Palynol. 1996;93(1):217–33.

28. Prestianni C, Gerrienne P. Lectotypification of the Famennian pre-ovule *Condrusia rumex* Stockmans, 1948. Rev Palaeobot Palynol. 2006;142:161–4.

29. Andrews HN. Early seed plants. Science. 1963;142:925–31.

30. Rothwell GW, Scheckler SE. Biology of ancestral Gymnosperns. In: Beck CB, editor. Origin and evolution of gymnosperms. New York: Columbia Univ. Press; 1988. p. 85–134.

31. Stern KR, Jansky S, Bidlack JE. Introductory plant biology. New York: McGraw-Hill; 2003.

32. Payne WW. A glossary of plant hair terminology. Brittonia. 1978;30(2):239–55.

33. Li H-Q, Taylor DW. *Aculeovinea yunguiensis* gen. Et sp. nov. (Gigantopteridales), a new taxon of gigantopterid stem from the upper Permian of Guizhou Province, China. Int J Plant Sci. 1998;159(6):1023–33.

34. Dunn MT, Krings M, Mapes G, Rothwell GW, Mapes RH, Keqin S. *Medullosa steinii* sp. nov., a seed fern vine from the upper Mississippian. Rev Palaeobot Palynol. 2003;124(3):307–24.

35. Labandeira CC. The four phases of plant-arthropod associations in deep time. Geol Acta. 2006;4:409–38.

36. Shear WA, Kukalová-Peck J. The ecology of Paleozoic terrestrial arthropods: the fossil evidence. Can J Zool. 1990;68:1807–34.

37. Tomescu AMF, Rothwell GW, Mapes G. *Lyginopteris royalii* sp. nov. from the upper Mississippian of North America. Rev Palaeobot Palynol. 2001;116(3):159–73.

38. Dunn MT, Rothwell GW, Mapes G. On Paleozoic plants from marine strata: *Trivena arkansana* (Lyginopteridaceae) gen. Et sp. nov., a lyginopterid from the Fayetteville formation (middle Chesterian/upper Mississippian) of Arkansas, USA. Amer J Bot. 2003;90(8):1239–52.

39. Oliver FW, Scott DH. On the structure of the palaeozoic seed *Lagenostoma lomaxi*, with a statement of the evidence upon which it is referred to *Lyginodendron*. Phil Trans R Soc Lond B. 1905;197:193–247.

4

Evolution of the vertebrate insulin receptor substrate (*Irs*) gene family

Ahmad Al-Salam[1] and David M. Irwin[1,2]* (iD)

Abstract

Background: Insulin receptor substrate (Irs) proteins are essential for insulin signaling as they allow downstream effectors to dock with, and be activated by, the insulin receptor. A family of four Irs proteins have been identified in mice, however the gene for one of these, *IRS3*, has been pseudogenized in humans. While it is known that the *Irs* gene family originated in vertebrates, it is not known when it originated and which members are most closely related to each other. A better understanding of the evolution of *Irs* genes and proteins should provide insight into the regulation of metabolism by insulin.

Results: Multiple genes for Irs proteins were identified in a wide variety of vertebrate species. Phylogenetic and genomic neighborhood analyses indicate that this gene family originated very early in vertebrae evolution. Most *Irs* genes were duplicated and retained in fish after the fish-specific genome duplication. *Irs* genes have been lost of various lineages, including *Irs3* in primates and birds and *Irs1* in most fish. Irs3 and Irs4 experienced an episode of more rapid protein sequence evolution on the ancestral mammalian lineage. Comparisons of the conservation of the proteins sequences among *Irs* paralogs show that domains involved in binding to the plasma membrane and insulin receptors are most strongly conserved, while divergence has occurred in sequences involved in interacting with downstream effector proteins.

Conclusions: The *Irs* gene family originated very early in vertebrate evolution, likely through genome duplications, and in parallel with duplications of other components of the insulin signaling pathway, including insulin and the insulin receptor. While the N-terminal sequences of these proteins are conserved among the paralogs, changes in the C-terminal sequences likely allowed changes in biological function.

Keywords: Insulin receptor substrate, Gene duplication, Protein evolution, Episodic evolution, Phylogeny, Vertebrate, Pseudogene

Background

The intracellular actions of insulin are initiated by the binding of the hormone insulin to its specific cell surface receptor, the insulin receptor [1, 2]. The insulin receptor is a heterotetrameric protein consisting of two extracellular alpha subunits and two transmembrane beta subunits that are connected by disulfide bridges [3, 4]. The binding of insulin to the extracellular alpha subunits of the receptor induces a conformational change that activates the intracellular tyrosine kinase domain found in the beta subunits [5, 6]. Once the tyrosine kinase activity is triggered,

the insulin receptor autophosphorylates key tyrosine residues (Tyr-1158, Tyr-1162, and Tyr1163, in the human sequence) in the intracellular portion of the beta subunit [7]. Phosphorylation of these sites then allows interactions with docking proteins, which are also subsequently tyrosine phosphorylated by the insulin receptor tyrosine kinase activity [8], and downstream signaling via SH-2 domain-containing proteins to yield physiological responses [2]. Insulin can initiate several different signaling pathways that regulate metabolic responses, cell survival, growth, and differentiation [1, 2, 9].

Docking proteins are key molecules as they allow the aggregation of components of signaling cascades [7]. The first insulin receptor docking protein identified in mammalian cells was Insulin receptor substrate (Irs1) [10], with three additional docking proteins (Irs2, Irs3,

* Correspondence: david.irwin@utoronto.ca
[1]Department of Laboratory Medicine and Pathobiology, Faculty of Medicine, University of Toronto, 1 King's College Circle, Toronto, ON M5S 1A8, Canada
[2]Banting and Best Diabetes Centre, University of Toronto, Toronto, ON, Canada

and Irs4) subsequently characterized and found to share similarity in their sequences with Irs1 [11–13]. The four characterized members of the Irs protein family share similar protein architectures, with fairly well conserved N-terminal pleckstrin homology (PH) and phosphotyrosine binding (PTB) domains located near their N-termini and having relatively long C-terminal extensions [14–18]. The C-terminal extensions, which show lower levels of similarity than the N-terminal region, contain multiple tyrosine phosphorylation motifs (as well as serine/threonine phosphorylation motifs) that interact with multiple signaling proteins [14–18]. The PH and PTB domains aid in targeting Irs proteins to the plasma membrane and insulin receptor, respectively [19, 20], while differences in the tyrosine phosphorylation motifs in the C-terminal sequences of the Irs proteins allow interactions with distinct downstream signaling pathways [15, 18]. Only three of the four Irs proteins found in the mouse are functional in humans, as the IRS3 gene sequence has been pseudogenized [21]. Intriguingly, Irs3, at only 494 amino acids in length, is less than half the size of the other three characterized Irs proteins, which are about 1200–1300 amino acids in length [10–13]. Compared to the other Irs proteins, Irs3 has a shorter C-terminal domain but retains similar-sized PH and PTB domains [12, 18]. Additional proteins containing both the PH and PTB domains have been identified (i.e., Dok4 and Dok5) that interact with the insulin receptor, however these proteins lack C-terminal extension with multiple phosphotyrosine motifs [22]. While Irs proteins were initially identified due to their interaction with the insulin receptor, they also interact, as docking proteins, with receptors for other growth factors, such as the insulin growth factor 1 receptor (IGF1R) and the insulin-related receptor (IRR), that also contain intracellular tyrosine domains [23, 24].

Irs proteins exert their unique functions through a combination of tissue-specific expression and differential binding of downstream signaling proteins [14–18, 25]. Irs1 is found in many classical targets of insulin action and is important for insulin sensitivity and embryonic and postnatal body growth [26]. Irs2 is found in an overlapping set of tissues with Irs1, however appears to have a more important role in mediating the neuronal effects of insulin [27] and the growth and survival of pancreatic beta-cells [28]. On the other hand, the function of Irs4 has been difficult to identify as genetic knockouts of this gene have little physiological effect [29]. However, when these knockouts are combined with a brain-specific Irs2 knockout, unique changes in energy regulation and glucose homeostasis are observed [30]. Irs3 is not essential for growth or glucose metabolism [31] and its expression is restricted to white adipocyte tissue in mice [12, 32] (and is absent in humans [21]), suggesting a possible, but non-essential, role for this protein in adipose tissue in rodents. In contrast to other Irs

proteins, the PH domain of Irs3 has an additional role in targeting Irs3 to the nucleus, in addition to the plasma membrane, a localization necessary for Irs3 induced glucose uptake [33]. Loss of the Irs3 gene on the human lineage indicates that the function of this gene is not essential in some mammals, and raises questions about the necessity of multiple Irs proteins.

A single Irs-like protein, named Chico, has been found in Drosophila melanogaster that also interacts with the Drosophila insulin receptor [34]. Like the mammalian Irs proteins, Chico is a large protein of about 1000 amino acids in length that contains PH and PTB domains near its N-termini and multiple phosphotyrosine motifs in its C-terminal region [34]. Only a few studies have examined the origin and evolution of the vertebrate Irs gene family, where it has been concluded that these genes diverged on the vertebrate lineage but these studies have reached differing conclusions concerning the relationships among the 4 Irs proteins [17, 35–37]. A number of questions remain to be answered. While it appears that the Irs genes duplicated and diverged from each other on the vertebrate lineage, before the mouse-human divergence, how early in vertebrate evolution this occurred is currently unknown. Did the duplications occur very early in vertebrate evolution in parallel with the duplications of other members of the insulin signaling pathway such as insulin [38] and the insulin receptor [39, 40]? Irs3 was lost on the human (primate) lineage [21]. Was this loss a unique event, or has this gene been lost on other lineages? Have other Irs genes been lost on other vertebrate lineages? Which gene(s) are best conserved (i.e., potentially most essential), both in terms of retention in genomes and in conservation of their sequences within vertebrates? Why is the Irs3 protein sequence much shorter than for other Irs proteins? When did the protein become smaller? Here we show that the Irs genes duplicated very early in vertebrate evolution, likely at a similar time as the origin of the insulin and insulin receptor gene families [38–40] and as a consequence of the two rounds of genome duplications that occurred in the vertebrate ancestor [41, 42]. Our analyses also show that the Irs3 has been lost on multiple independent lineages, and that the genes for other Irs proteins, including Irs1 and Irs2, have occasionally been lost. The length of the Irs3 protein was reduced on the early tetrapod lineage, after divergence for fish, and was followed by a period of rapid sequence evolution in an early mammalian ancestor. Intriguingly, Irs4 also experienced an episode of rapid evolution, in parallel with Irs3, early in mammalian evolution.

Results
Number of insulin receptor substrate (Irs) genes in vertebrate genomes

To estimate the number of insulin receptor substrate (Irs) genes in the genomes of diverse vertebrate species, we conducted *BLAST* searches [43] of 64 diverse vertebrate genomes in the *Ensembl* database [44]. As the *Ensembl* database does not include a species repressing the class Chondrichthyes (cartilaginous fish), we also searched the Elephant shark genome [45], thus the genomes of a total of 65 vertebrate species, representing all vertebrate classes, were examined. Genes were named (see Additional file 1: Table S1) based on orthology-paralogy relationships derived from sequence similarity as well as our phylogenetic and genomic location analyses described below. The numbers of species searched and the numbers of each type of *Irs* gene found in the different groups of vertebrates is summarized in Table 1. Sequences belonging to all four types of *Irs* genes were found in diverse representatives of mammals, reptiles, amphibians, lobe-finned fish, and bony fish (Table 1 and Additional file 1: Table S1). Within the bony fish, a single copy of each *Irs* gene was identified in the spotted gar (Additional file 1: Table S1), a species that diverged before the fish-specific genome duplication [46]. Among the remaining species of bony fish, all of which experienced the fish-specific genome duplication, most had two copies of the *Irs2*, *Irs3*, and *Irs4*-like genes, but only a few had *Irs1* genes (Table 1 and Additional file 1: Table S1). Of the fish species that are descendants of the fish-specific genome duplication, only two (zebrafish and cavefish/Mexican tetra) had an *Irs1*-like gene, and both of these species had only a single copy of this gene, in contrast to the duplicate copies of the other *Irs* genes found in these (and other) fish species (Additional file 1: Table S1). No *Irs3* genes were found in birds or the single representative of cartilaginous fish, although *Irs1*, *Irs2*, and *Irs4* were identified in both groups (Table 1 and Additional file 1: Table S1). A single genomic sequence encoding an incomplete *Irs*-like coding region was found in the lamprey (jawless fish), which showed some affinity to *Irs2* sequences, but its orthology could not be confidently assessed (Table 1 and Additional file 1: Table S1).

Many of the *Irs* genes identified in our searches of the *Ensembl* database were incomplete (i.e., did not predict complete open reading frames). Some of the incomplete coding sequences contained unsequenced gaps (Ns) in the genome assemblies, while others could have been due to sequencing errors or pseudogenization. To complement the sequences identified from the *Ensembl* database, a *BLAST* search [42] was conducted of the *NCBI* database [47] to identify *Irs* coding sequences (Table 1 and Additional file 2: Table S2). Searches of the *NCBI* database identified a larger number (167) of vertebrate species with *Irs* coding sequences than the *Ensembl* database, but many of these are from species do not contain near complete genome sequences (e.g., *Xenopus laevis*), thus the full complement of *Irs* genes in these species might not have been found. A second limitation of our *NCBI* searches was that we only identified *Irs*-like sequences that had been annotated as coding sequences (i.e., if the gene was not annotated or was a pseudogene it would not be found) (see Additional file 1: Table S1 and Additional file 2: Table S2). The total number of vertebrate species with identified *Irs*-like genes was 172 (59 in both *Ensembl* and *NCBI*, 1 in both the Elephant Shark Genome project and *NCBI*, 107 only in *NCBI*, and 5 only in *Ensembl*, see Table 1 and Additional file 1: Table S1 and Additional file 2: Table S2). The distribution of the *Irs*-like gene paralogs among vertebrate classes identified in the *NCBI* database was similar to that seen with the *Ensembl* database (Table 1 and Additional file 2: Table S2).

Phylogeny of vertebrate insulin receptor substrate (*Irs*) genes

To better establish the orthology-paralogy relationships among the identified *Irs* genes, and determine when duplications of the *Irs* genes occurred, phylogenetic relationships of the sequences were established using

Table 1 Numbers of *Irs*-like genes found in diverse vertebrates in the genome and coding sequence databases

	Species[a]	Irs1[b]	Irs2[b]	Irs3[b]	Irs4[b]
Mammals	43 | 81	42 | 81 (73)	39 | 73 (28)	31 | 38 (29)	42 | 78 (51)
Birds	5 | 54	5 | 51 (4)	5 | 32 (5)	0 | 0 (0)	5 | 54 (4)
Reptiles	2 | 5	2 | 5 (4)	2 | 5 (1)	2 | 5 (4)	2 | 5 (4)
Amphibians	1 | 2	1 | 2 (1)	1 | 1 (1)	1 | 1 (1)	1 | 2 (2)
Lobe-finned fish	1 | 1	1 | 1 (1)	1 | 1 (1)	1 | 1 (1)	1 | 1 (0)
Bony fish	11 | 25	3 | 5 (5)	21 | 44 (43)	21 | 49 (46)	21 | 47 (29)
Cartilaginous fish	1 | 1	1 | 1 (1)	1 | 1 (1)	0 | 0 (0)	1 | 1 (1)
Jawless fish	1 | 0	0 | 0 (0)	1 | 0 (0)	0 | 0 (0)	0 | 0 (0)
Total	65 | 167	55 | 146 (89)	71 | 157 (80)	56 | 94 (81)	73 | 188 (91)

[a]Number of species with identified genes (or searched if no genes were found): Number of species with genome sequences searched | Number of species searched only for coding sequences

[b]Number of unique genes or coding sequences found for each gene: Number of genomic sequence | Number of coding sequences (Number of full-length or near full-length sequences)

maximum likelihood [48, 49] and Bayesian approaches [50, 51]. A total of 341 full-length, or near-full length (those missing only short portions of sequence at the N- or C-termini of their predicted proteins), *Irs*-like coding sequences from 172 vertebrate species (including 89 *Irs1*, 80 *Irs2*, 81 *Irs3*, and 91 *Irs4* sequences (Table 1 and Additional file 1: Table S1, Additional file 2: Table S2 and Additional file 3: Figure S1) were used in this analysis. Maximum likelihood phylogenetic analysis of putative *Irs* orthologs yielded topologies consistent with the expected species topologies (Additional file 4: Figure S2, Additional file 5: Figure S3, Additional file 6: Figure S4 and Additional file 7: Figure S5; similar results were obtained using Bayesian methods, results not shown), suggesting that the analyzed genes were orthologous.

The relationship among the *Irs* paralogs was established using these full-length *Irs* sequences and rooted using *Irs*-like sequences from three non-vertebrate outgroup species (see Additional file 2: Table S2). Both the Maximum liklihood (Fig. 1) and Bayesian (Additional file 8: Figure S6) analyses demonstrated that each of the four *Irs* ortholog groups form strongly supported monophyletic clades that diverged from each other prior to the divergence of jawed and jawless vertebrates. Both analyses displayed the same relationships among the paralogs, with *Irs2* and *Irs4* being most closely related, then both grouping with *Irs1*, and *Irs3* genes being the most distantly related group (Fig. 1 and Additional file 8: Figure S6). Like our finding of distinct Irs paralogs in most classes of vertebrates (see Table 1), these results indicate that the *Irs* gene family originated early in vertebrate evolution. An intriguing feature of both analyses was that the mammalian *Irs3* and *Irs4* sequences have longer ancestral mammalian lineages, suggesting episodes of more rapid sequence evolution for these genes in the early mammal.

Origin of vertebrate insulin receptor substrate (*Irs*) genes
Our phylogenetic analysis indicates that the *Irs* gene family originated early in vertebrate evolution; however, due to the absence of full-length gene sequences from the lamprey, we were unable to show whether any of the duplications preceded the earliest divergence within this group. Genome duplications occurred prior to the divergence of jawed and jawless vertebrates and explain the presence of multiple gene families in vertebrate genomes [41, 42, 52]. With genome duplications, paralogous genome segments are created where different chromosome share sets of paralogous genes [41, 52]. To determine whether the *Irs* genes originated through genome duplications we examined the genomic neighborhoods surrounding the four mouse *Irs* genes. As shown in Fig. 2a, the mouse *Irs1*, *Irs2*, and *Irs4* genes are each found adjacent to a pair of collagen type IV genes (*Col4a4* and *Col4a23*, *Col4a1* and *Col4a2*, and *Col4a5* and *Col4a6*,

respectively) on different chromosomes. The same arrangement was found for the human *IRS1*, *IRS2*, and *IRS4* genes (results not shown). The *Irs3* gene, on the other hand, is not located near any collagen gene (Fig. 2a). Whether this difference seen in the genomic neighborhood for *Irs3* reflects changes in genomic organization after genomic duplications, or whether *Irs3* originated via a different mechanism cannot be determined at this time. However, these results do suggest that the *Irs1*, *Irs2*, and *Irs4* genes originated via genome duplications in an early vertebrate, and as *Irs3* diverged earlier from the other *Irs* genes, this supports origin of this gene family at or before the genome duplications on the early vertebrate lineage.

Duplication of *Irs* genes in Bony fish
Duplicate copies of *Irs2*, *Irs3*, and *Irs4* were found in most species of bony fish examined (Table 1 and Additional file 1: Table S1 and Additional file 2: Table S2). Bony fish experienced an additional genome duplication not shared by other vertebrates [53, 54], thus duplicated *Irs* genes would be expected. Duplicated *Irs* genes were not found in the genome of the spotted gar, a species that diverged from other bony fish prior to the fish-specific genome duplication [46]. Phylogenetic analysis of the *Irs2*, *Irs3*, and *Irs4* sequences (Additional file 5: Figure S3, Additional file 6: Figure S4 and Additional file 7: Figure S5) demonstrated that the duplications of these genes occurred early in bony fish evolution consistent with the fish-specific genome duplication. When the genomic neighborhoods surrounding the zebrafish *Irs* genes were examined, only one of the fish duplicates (*Irs1*, *Irs2b*, *Irs3b*, and *Irs4b*) was located in a genomic neighborhood orthologous to those seen in mice (Fig. 2b), while the second paralogous gene (*Irs2a*, *Irs3a*, and *Irs4a*) resided in genomic regions with no similarity in gene composition to the genomic region found in mice.

Loss of the *Irs3* Gene on the primate lineage
While mice have 4 *Irs* genes, only 3 functional *Irs* genes are found in humans, as *Irs3* contains mutations that introduce a stop codon and delete part of the coding sequence [21]. Genomic sequences similar to *Irs3* were identified in a number of primate genomes in the *Ensembl* database; however, intact coding sequences could only be predicted for the Tree shrew and the Mouse lemur (Additional file 1: Table S1). Similarly, searches of the *NCBI* database for coding sequences similar to *Irs3* only identified potentially functional *Irs3* coding sequences in three primate species, the Mouse lemur, Coquerel's sifaka, and Sunda flying lemur (Additional file 2: Table S2). Complete coding sequences could be predicted for the Mouse lemur and Coquerel's sifaka but the sequences from the other two primates contained

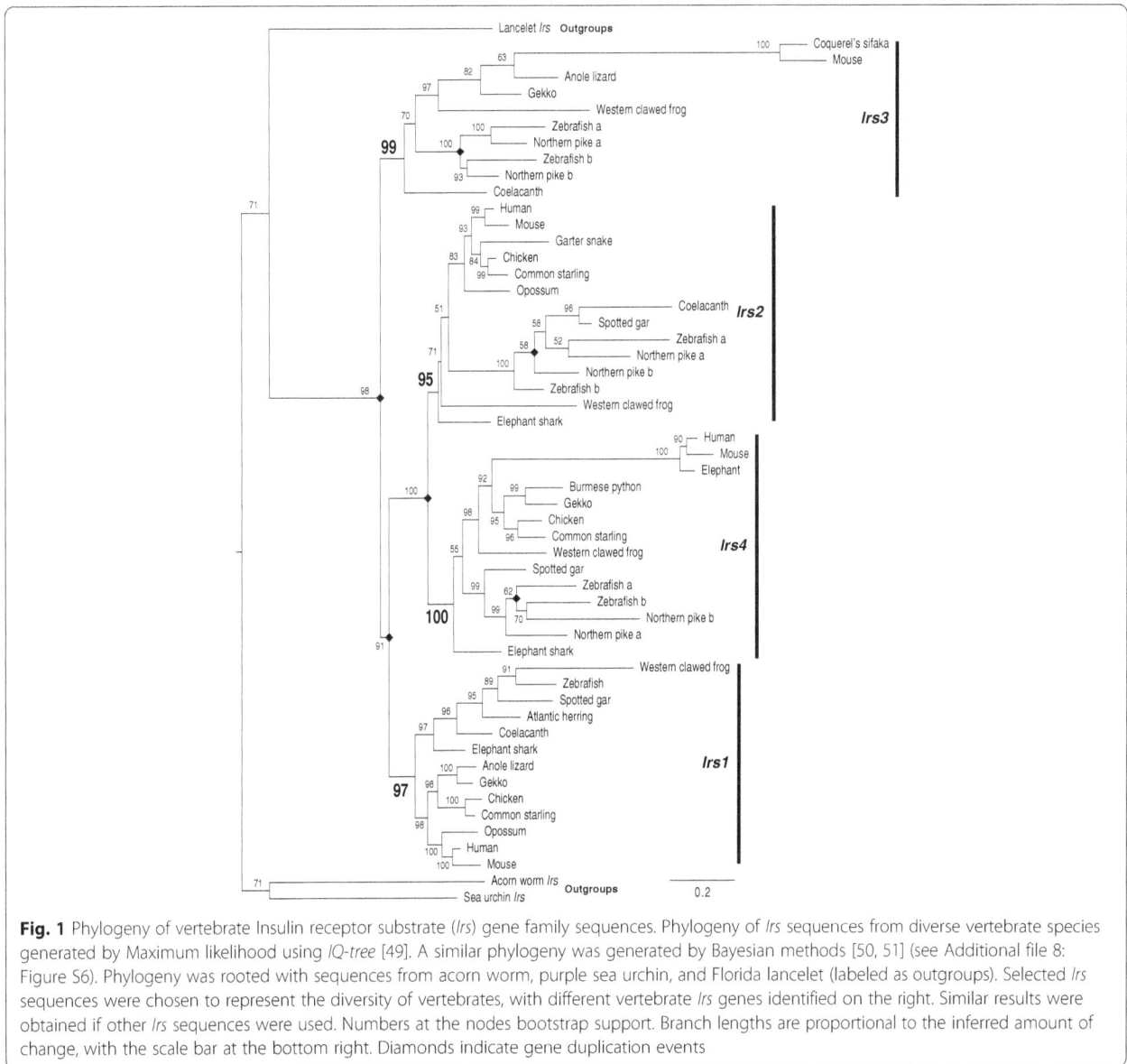

Fig. 1 Phylogeny of vertebrate Insulin receptor substrate (*Irs*) gene family sequences. Phylogeny of *Irs* sequences from diverse vertebrate species generated by Maximum likelihood using *IQ-tree* [49]. A similar phylogeny was generated by Bayesian methods [50, 51] (see Additional file 8: Figure S6). Phylogeny was rooted with sequences from acorn worm, purple sea urchin, and Florida lancelet (labeled as outgroups). Selected *Irs* sequences were chosen to represent the diversity of vertebrates, with different vertebrate *Irs* genes identified on the right. Similar results were obtained if other *Irs* sequences were used. Numbers at the nodes bootstrap support. Branch lengths are proportional to the inferred amount of change, with the scale bar at the bottom right. Diamonds indicate gene duplication events

unsequenced gaps. Importantly, all four of these species with potentially intact *Irs3* gene sequences are early branching lineages within primates [55]. Alignment of the *Irs3* genomic sequences from diverse primates (see Additional file 9: Figure S7) using *MultiPipMaker* [56, 57] demonstrated that the sequences were not well conserved as a large number of frameshift mutations were identified along with large deletions, including those previously identified in the human *IRS3* pseudogene sequence [21]. These results suggest that *Irs3* was inactivated early in primate evolution, but after divergence of the Mouse lemur and Coquerel's sifaka. When *MultiPipMaker* alignments were generated using the human sequence as the master sequence (results not shown), an *Alu* repetitive element that disrupts the human *IRS3* coding region [21] was found to be shared by *Irs3* sequences from primates that lack an intact coding

sequence, suggesting that the insertion of this element into the gene occurred at about the same time as the pseudogenization of the gene.

Loss of the *Irs3* Gene in birds

In addition to the absence of *Irs1* in most bony fish and *Irs3* in most primates, another notable group of animals that lack a specific *Irs* gene is birds, where no *Irs3* coding or gene sequences were identified (Table 1 and Additional file 1: Table S1 and Additional file 2: Table S2). In contrast to primates, where genomic sequences similar to *Irs3* were found containing mutations that disrupt the coding sequences (see above), genomic sequences similar to *Irs3* were not found in any of the bird genomes examined (Additional file 1: Tables S1). To exclude the possibility that the avian *Irs3* sequences had rapidly evolved, and

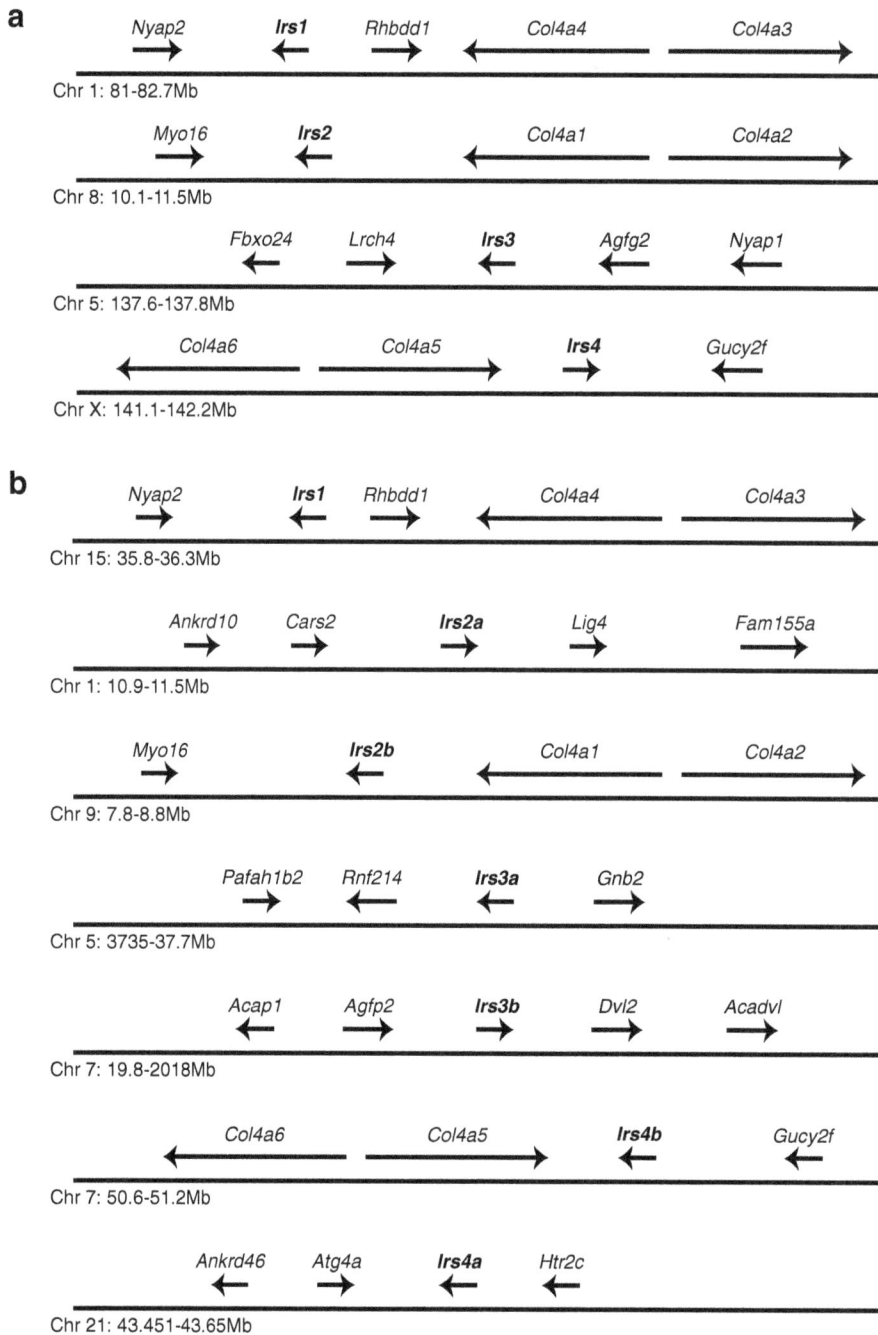

Fig. 2 Genomic organization of genes near *Irs* genes in the mouse and zebrafish genomes. The relative organization and orientation of genes near insulin receptor substrate (*Irs*) genes in (**a**) mouse and (**b**) zebrafish. Chromosomes and genomic locations are from Ensembl [44] (see Additional file 1: Table S1). *Irs* genes are labeled in red. Gene sizes and distances between genes are not to scale. Arrowheads indicate direction of transcription. Gene symbols are: *Irs1–4*, insulin receptors substrates 1–4; *Col4a1–6*, collagen, type IV, alpha1–6; *Rhbdd1*, rhomboid domain containing 1; *Nyap2*, Neuronal tyrosine-phophorylated phosphoinositide 3-kinase adaptor 2; *Myo16*, Myosin XVI; *Fbxo24*, F-box protein 24; *Lrch4*, Leucine-rich repeats and calponin homology (CH) domain containing 4; *Agfp2*, ArfGAP with FG repeats 2; *Nyap1*, Neuronal tyrosine-phosphorylated phosphoinositide 3-kinase adaptor 1; *Gucy2f*, Guanylate cyclase 2f; *Ankrd10*, Ankyrin repeat domain 10; *Ankrd46*, Ankyrin repeat domain 46; *Cars2*, Cysteinyl-tRNA synthetase 2; *Lig4*, Ligase IV, DNA, ATP-dependent; *Fam155a*, Family with sequence similarity 155, member A; *Pafah1b2*, Platelet-activating factor acetylhydrolase, isoform 1b, subunit 2; *Rnf214*, Ring finger protein 214; *Gnb2*, Guanine nucleotide binding protein (G protein), beta 2; *Acap1*, ArfGAP with coiled-coil, ankyrin repeat and PH domains 1; *Dvl2*, Dishevelled segment polarity protein 2; *Acadvl*, Acyl-Coenzyme A dehydrogenase, very long chain; *Atg4a*, Autophagy related 4A, cysteine peptidase; *Htr2c*, 5-hydroxytryptamine (serotonin) receptor 2C

thus were not detectable in the *BLAST* searches [38], we attempted to use genomic neighborhoods to identify these genes. However, searches for the genes that flank the mammalian *Irs3* gene (i.e., *Lrch4* and *Agfg2*, see Fig. 2) also failed to find orthologs of these genes (results not shown). These results suggest that the *Irs3* genomic region, including adjacent genes, had been deleted from the genomes of birds.

Episodic evolution of vertebrate insulin receptor substrate (Irs) genes

Visual inspection of the phylogenies generated from the *Irs* coding sequences, using both single gene (Additional file 4: Figure S2, Additional file 5: Figure S3, Additional file 6: Figure S4 and Additional file 7: Figure S5) and gene family (Fig. 1 and Additional file 8: Figure S6) phylogenies, suggested accelerated evolution on the mammalian ancestral lineages for *Irs3* and *Irs4*. Branch lengths displayed in our phylogenetic analysis are proportional to the number of inferred nucleotide substitutions. For both *Irs3* and *Irs4*, mammals have accumulated more changes than sequences from the other vertebrate classes, suggesting that these genes experienced accelerated evolution early in mammalian evolution. To determine whether the longer branches are due to increased numbers of amino acid substitutions in the Irs3 and Irs4 protein sequences we conducted relative rate tests [58] with protein sequences encoded by *Irs* genes from four different mammalian species (if available) and 6 non-mammalian species (Additional file 10: Table S3). For all relative rate comparisons, the mammalian Irs3 and Irs4 protein sequences accumulated significantly higher numbers of amino acid substitutions compared to protein sequences from a diverse array of non-mammalian species. In contrast, only a small number of the comparisons with Irs1 displayed significantly higher numbers of amino acid substitution on the mammalian lineage, with none being significantly higher on the mammalian lineage for Irs2, although there were a few cases of significantly higher numbers on the non-mammalian lineage for this protein (Additional file 10: Table S3). These results show that the proteins encoded by *Irs3* and *Irs4*, but not *Irs1* or *Irs2*, have accumulated increased numbers of amino acid substitutions on the mammalian lineage.

Changes in the lengths of vertebrate insulin receptor substrate (Irs) proteins

We then examined whether changes in the rate of amino acid sequence evolution resulted in changes in the structure of the Irs proteins. Previously, it had been reported that mouse Irs3 is much shorter than any other mouse Irs proteins, or Irs1, Irs2, and Irs4 proteins from other species [10–13]. To determine whether this was a general feature of Irs3 proteins or was specific to a subgroup of species

we calculated the lengths of Irs proteins from species representing diverse groups of vertebrates (Table 2). Lengths of Irs proteins from species not listed in this table were generally similar to their most closely related representative shown in the table (results not shown). Most Irs protein sequences have a length of about 1000–1300 amino acids, except Irs3 from tetrapods (amphibians, reptiles, and mammals, Table 2). Irs3 proteins from zebrafish and coelacanth (and other fish) have length similar to those of the other Irs proteins. These observations suggest that the length of Irs3 progressively shortened from a full-length sequence of 1000–1300 amino acids, which was retained in fish, to one of ~800 residues that is found today in amphibians (*Xenopus*), to ~600 residues and found in reptiles (garter snake), to ~500 residues found in mammals (mouse) (Table 2). Most of the reduction in Irs3 protein length occurred on the lineages leading to tetrapods (ancestor of amphibians and mammals) and amniotes (ancestor of reptiles and mammals), and not on the lineage leading to mammals. This suggests that the reduction in the length of Irs3 is not associated with the accelerated protein sequence evolution observed for this sequence in the early mammalian lineage. Irs4, which also experienced increased rates of amino acid sequence evolution on the lineage leading to mammals, does not show any major changes in protein length among vertebrate classes, nor do Irs1 or Irs2 (Table 2).

Conservation of Irs protein sequences

Since each of the Irs proteins have differing roles in insulin signaling [14–18, 25], we examined whether these roles generated differences in the constraints acting across the Irs protein sequences. To avoid lineage-specific effects, we only examined Irs protein sequences from species where full-length sequences for all four *Irs* genes had been identified (see Additional file 1: Table S1 and Additional file 2: Tables S2). As such, the four genes would then have existed in parallel in the same genomes for their entire evolutionary history and therefore have likely experienced similar evolutionary pressures at the genomic level. A total of 10 species, 9 mammals and one amphibian (mouse, rat, golden hamster, prairie vole, prairie deer mouse, Coquerel's sifaka, mouse lemur, Mouflon sheep, killer whale, and *Xenopus tropicalis*), were found to have complete coding sequences for all 4 Irs proteins (Additional file 1: Table S1 and Additional file 2: Tables S2). If we included sequences from zebrafish, where the single *Irs1* sequence was used and one of the two paralogs for *Irs2*, *Irs3*, and *Irs4* were selected we obtained similar results for the following analyses (results not shown). Conservation of sequences was assessed using Jenson-Shannon Divergence (JS) scores [59] (Additional file 11: Table S4) and was plotted for each Irs protein and the complete set of four Irs proteins in Fig. 3. Irs1 has the

Table 2 Lengths of Irs proteins from representative vertebrate speceis

Protein	Human	Mouse	Snake[a]	Chicken	Xenopus	Coelacanth	Gar[a]	Zebrafish a	Zebrafish b	Shark[b]
Irs1	1242	1231	1186	1178	1091	1076	>1085[c]	1099	NA[b]	1099
Irs2	1338	1321	1105	1148	1006	1069	1069	1032	1062	>1082
Irs3	NA[b]	495	662	NA[b]	809	1034	NA[b]	1181	1245	NA[b]
Irs4	1257	1216	1191	1164	1077	NA[b]	1120	1158	1051	1200

[a]Snake is garter snake; Gar is spotted gar; Shark is elephant shark
[b]NA, not applicable, gene was not found, incomplete, or absent
[c]Spotted gar Irs1 is missing part of its C-terminus; Elephant shark Irs2 is missing part of its N-terminus

highest average conservations score (0.75) followed by Irs2 (0.73), Irs4 (0.67), and Irs3 (0.45). These results suggest that the episodes of more rapid sequence evolution seen on the early mammalian lineages for Irs3 and Irs4 (see above) resulted in an acceleration relative to all Irs proteins, and that the non-mammalian Irs3 and Irs4 sequences might be evolving at rates similar to those for Irs1 and Irs2. As expected, the PH and PTB domains show strong conservation in all 4 Irs proteins, with the protein sequences between and flanking these domains generally showing lower levels of conservation (Fig. 3 and Additional file 11: Table S4). Strong conservation of the PH and PTB domains might be expected as all Irs proteins interact with plasma membranes and insulin receptors, the functions of these domains [19, 20]. Many locations in the C-terminal extensions of the 4 Irs proteins also display high levels of conservation, however it appears that Irs3, and to some extent Irs4, have lower levels than the other two proteins. In comparisons of all 4 Irs proteins, 9 short regions show high levels of conservation, as indicated by having JS score in the top 10% (Fig. 3e). This suggests that only limited parts of the C-terminal region have functions that are conserved across all *Irs* gene family members, while regions that are not conserved across all family members, but conserved within orthologs might have ortholog-specific functions.

Tyrosine phosphorylation of Irs protein sequences
Phosphorylation of tyrosine residues in Irs proteins, especially those in the C-terminal extension, is important for signaling [2, 8, 9]. Potential tyrosine phosphorylation sites were predicted [60] for the Irs protein sequences from the 10 vertebrates that had complete sequences for all 4 family members (Table 3). Approximately 60% of the tyrosine residues in any Irs protein sequence were predicted to be phosphorylation sites. As expected, being the shortest Irs protein, Irs3 has least number of tyrosine residues (average 17.2 residues per sequence) and putative tyrosine phosphorylation sites (average of 10.1), compared to the other Irs proteins (Irs1: 33.8 and 20.4, Irs2: 36.6 and 20.1, Irs4: 31.2 and 15.8, tyrosine and putative tyrosine phosphorylation sites, respectively). Irs3 also showed the lowest conservation of tyrosine residues

(10/17.2 = 58%) and putative tyrosine phosphorylation sites (5/10.1 = 50%) in protein alignments. Lower levels of conservation were also seen for Irs4 (tyrosine residues: 17/ 31.2 = 54%, and tyrosine phosphorylation sites: 9/15.8 = 57%). In contrast, conservation of both tyrosine residues (Irs1: 27/33.8 = 80%, Irs2: 25/36.6 = 68%) and putative tyrosine phosphorylation sites (Irs1: 16/20.4 = 78%, and Irs2: 13/20.1 = 65%) were higher for Irs1 and Irs2 (see Table 3 and Additional file 12: Figure S8). Only 4 tyrosine residues were conserved across all Irs sequences in all 10 species (see Additional file 12: Figure S8), with two of these being predicted tyrosine phosphorylation sites for all sequences. Of these four conserved residues, the two sites that are not conserved putative tyrosine phosphorylation sites are located in the PH domain, while the two putative tyrosine phosphorylation sites that are conserved are in the C-terminal extension (residues 608 and 628, 649 and 671, 350 and 361, and 672 and 689 in mouse Irs1, Irs2, Irs3, and Irs4, respectively) (see Additional file 12: Figure S8). The two putative tyrosine phosphorylation sites located in the C-terminal extension are located in regions that have strong conservation among the four Irs protein sequences (Additional file 12: Figure S8).

Discussion
Origin of the *Irs* gene family
While multiple *Irs*-like genes have been previously characterized in several mammalian species [10–13, 17, 36], only a few non-mammalian *Irs*-like genes have been identified, which limited the ability to resolve when this gene family originated and how the different genes are related to each other [17, 35–37]. Here, our searches have identified a large number of *Irs*-like genes from a diverse array of vertebrate classes, which should allow better estimation of the time when this gene family originated and how the different genes are related to each other. Searches of vertebrate genomes identified multiple *Irs*-like sequences in the genomes of representative species for all vertebrate classes except Agnatha (Jawless fish) (Table 1 and Additional file 1: Table S1 and Additional file 2: Tables S2). However, given the low coverage of the sea lamprey somatic genome [61] and the loss of DNA in this species due to genomic remodeling in somatic tissue [62], *Irs*-like sequences may have been missed in this jawless

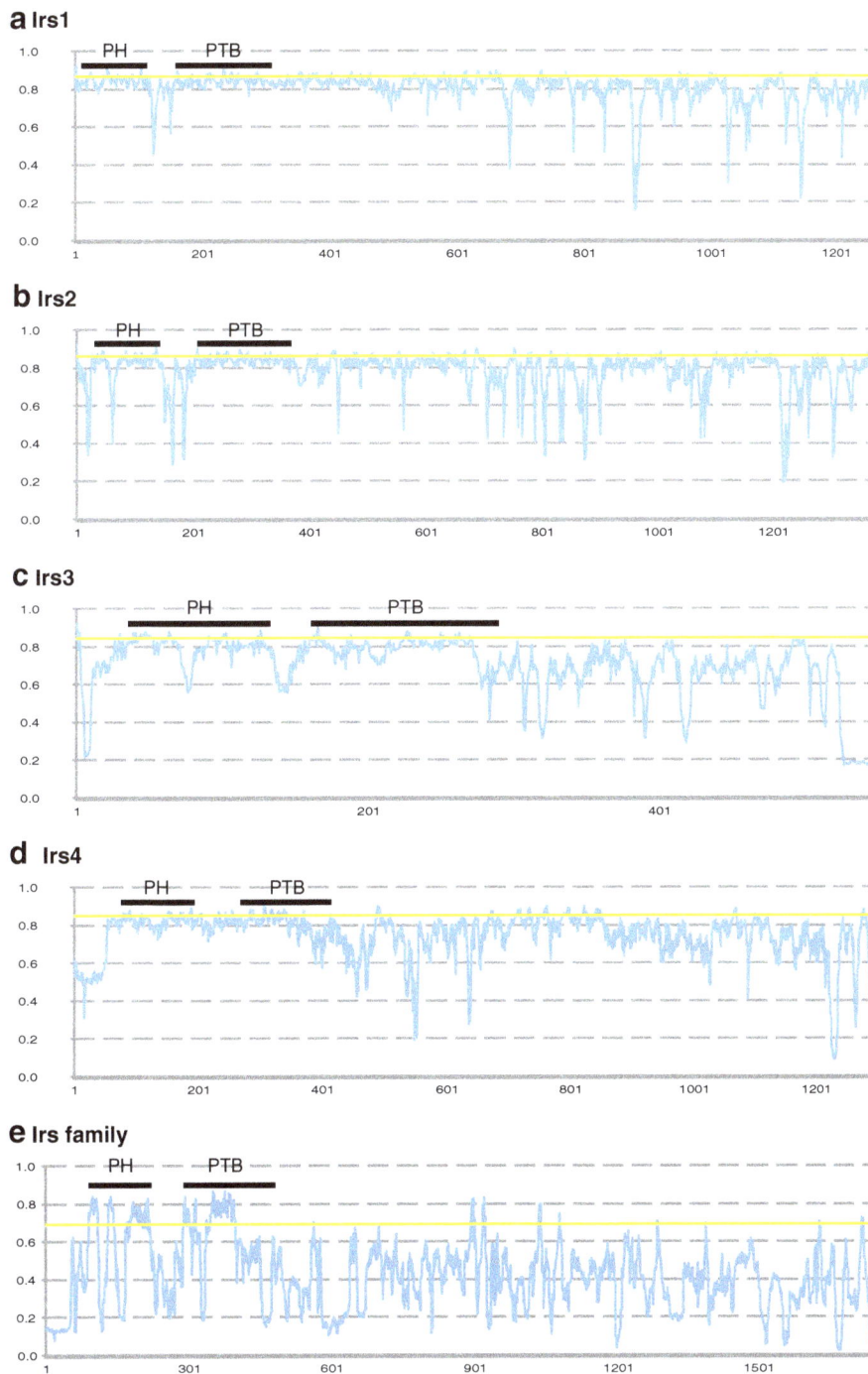

Fig. 3 Conservation of Irs protein sequences. JS divergence scores for aligned Irs protein sequences from 10 vertebrate species. (**a**) Irs1, (**b**) Irs2, (**c**) Irs3, (**d**) Irs4, (**e**) all Irs family members. JS scores are presented in Additional file 11: Table S4. Position in alignment is shown at the bottom of each graph. The locations of the PH and PTB are shown as bars near the top of each graph. JS scores above the *yellow horizontal line* are in the top 10% of JS scores for that alignment

fish. These observations suggest that the *Irs* gene family originated early in vertebrate evolution, and possibly before the earliest divergence of extant vertebrate species.

Phylogenetic analyses of the sequences (Fig. 1 and Additional file 8: Figure S6) strengthened this conclusion, demonstrating that the multiple genes originated early in vertebrate evolution and were not due to parallel duplications on diverse lineages. Analysis of genomic neighborhoods is a powerful tool for identifying orthologs [63], especially in gene families where multiple sequences have

Table 3 Tyrosine phosphoryation of Irs proteins

Species	Irs1		Irs2		Irs3		Irs4	
	Y[a]	pY[b]	Y	pY	Y	pY	Y	pY
Mouse	34	20	36	21	17	10	29	13
Rat	35	22	36	21	18	13	30	17
Golden hamster	34	20	37	20	14	8	31	15
Prairie vole	34	21	37	20	17	11	34	13
Prarie deer mouse	34	20	37	20	18	12	27	15
Coqurel's sifka	34	20	38	21	16	7	33	15
Mouse lemur	34	20	38	20	16	8	31	16
Mouflon sheep	34	19	38	20	14	8	30	15
Killer whale	33	20	38	21	17	8	28	15
Xenopus tropicalis	32	22	31	17	25	16	39	24
Average	33.8	20.4	36.6	20.1	17.2	10.1	31.2	15.8
Conserved[c]	27	16	25	13	10	5	17	9

[a]Number of tyrosine residues in the sequence
[b]Number of tyrosine residues predicted to be phosphorylated
[c]Number of residues conserved across the 10 sequences

similar levels of similarity to a putative ortholog, where only the true othologs share genomic neighborhoods [64]. In this context, we used genomic neighborhoods to confirm the orthology of many of the diverse *Irs* genes found in vertebrates. When the genomic locations of *Irs*-like genes were examined (Fig. 2), three of the 4 *Irs* genes were found to be in genomic neighborhoods that shared similar gene contents. The sharing of paralogus genes among genomic neighborhoods is consistent with these genes originating through genome duplications [65], which suggests that at least 3 of the 4 *Irs* genes originated via the two rounds of genome duplication that occurred in the common ancestral vertebrate lineage [41, 42]. Interestingly, both the insulin [38] and the insulin receptor [39, 40] gene families originated very early in vertebrate evolution, and potentially via the same genome duplications. Irs proteins not only interact with the insulin receptor, but also with other receptors, including the Insulin growth factor I (IGF-1) receptor and the Insulin-related receptor (Irr) [23, 24]. These observations suggest that duplications of the genes for the ligands, receptors, and docking proteins could lead to increased specialization in these signaling pathways, and the possibility to evolve new functions.

Change in number of *Irs* genes
While *Irs* gene originated very early in vertebrate evolution, the number of *Irs* genes is found to vary between species. Similar variations in the numbers of genes within gene families involved in insulin signaling in vertebrates have previously been reported [38, 66, 67]. Early studies demonstrated that the *Irs3* gene was lost on the human lineage [21], and our analysis indicates that it

was possibly inactivated by the insertion of a repetitive DNA element early in primate evolution (results not shown). *Irs3* genes were also lost on the lineage leading to birds. A number of genes involved in insulin-regulated metabolism have been lost in the chicken [68], some of which have been shown to be missing in wide variety of birds (e.g., Resistin [64]), suggesting that the loss of Irs3 might have been part of an adaptation by birds to their new locomotive style. Teleost fish experienced a genome duplication [53], however rapid loss of many of the duplicates occurred [54]. Here we found duplicated copies of *Irs2*, *Irs3*, and *Irs4* in most teleost fish genomes, but most of these species have lost both copies of *Irs1* (see Table 1 and Additional file 1: Table S1 and Additional file 2: Tables S2). The presence of multiple *Irs* genes, and the overlap in the functions of the Irs proteins [14–18] suggests a degree of redundancy among these genes allowing species to adapt to the loss of one (or more) of these genes.

Evolution of Irs proteins
Duplication of genes should allow the specialization of distinct proteins to unique biological roles [69, 70], thus duplication of the *Irs* genes might have allowed the evolution of novel regulatory roles for the insulin signaling pathway. While all Irs proteins are involved in insulin signaling, they each appear to have unique, but to some extent overlapping, biological roles [14–18]. Changes in the numbers of *Irs* genes also shows that the genes have retained a degree of redundancy and have not completely sub-functionalized since their origin. Despite the overlap in function, differences in evolutionary patters can be seen among the *Irs* genes. Irs3 and Irs4 both experienced episodes of more rapid protein sequence evolution on the common ancestral lineage leading to mammals (Additional file 10: Table S3), which suggests either a temporary relaxation of evolutionary constraints on these sequences on this lineage or that the rapid evolution was driven by positive selection. Both patterns of evolution could have resulted in changed biological functions for these proteins, and might explain why Irs3 and Irs4 might have functions that are less essential than Irs1 or Irs2. *Irs3* is non-essential as loss of this gene in humans is tolerated [21], and our data shows that a number of primates, birds and potentially other vertebrates can survive without this gene. Knockout of *Irs4* has little physiological effect [29], while *Irs1* or *Irs2* knockout mice have much more pronounced physiological defects [26, 27, 71, 72].

Further evidence for the diversification of the function of the Irs proteins is derived from the conservation plots. When each Irs protein is individually examined, areas of strong sequence conservation are seen across the entire protein sequence, although to a lower extent for Irs3, which might be due to the rapid evolution on the early mammalian lineage (Fig. 3a-d). However, when conservation is examined across the family of Irs proteins (Fig. 3e), most of the conservation is concentrated in the regions encoding the PH and PTB domains, sequences that are important for localizing these proteins to the plasma membrane [19] and insulin receptors [20], respectively. The plasma membrane localization, and insulin receptor interactions of these proteins have been conserved, but the C-terminal extension, which allow interaction with downstream signaling partners [15, 18], show greater levels of divergence to account for changes in downstream functions. However, there are a few areas of the C-terminal extension that are strongly conserved among all Irs, including two putative tyrosine phosphorylation sites that have been shown to be important in Irs1 and Irs2 for interactions with phosphatidylinositol 3-kinase (PI3K) [73–76], a key downstream signaling protein of insulin receptors [76]. Thus, interaction with PI3K appears to be conserved among all Irs proteins, but changes in interactions with other signaling proteins might explain the differences in biological function of the different Irs proteins.

Conclusions

Here we have shown that the Irs gene family originated early in vertebrate evolution, with at least three of the genes likely generated during the two rounds of genome duplication that occurred in the vertebrate ancestor. Most groups of vertebrates have retained all 4 Irs genes, although some groups have lost genes, including primates and birds that have lost Irs3 and most fish that have lost Irs1. Duplication of Irs genes is only seen in fish that have experienced the fish-specific genome duplication, leading to duplicated Irs2, Irs3, and Irs4 genes. This suggests that while there are redundancies in the function of Irs gene, thus can tolerate the loss of a gene, gain of Irs genes is likely harmful, except when other genes in the insulin signaling pathway are duplicated. This conclusion is agreement with the finding of an increased number of retained duplicated genes involved in signal transduction pathways found in fish after the fish-specific genome duplications [77]. The protein sequences of Irs1 and Irs2 are strongly conserved across vertebrates while Irs3 and Irs4 show lower levels of conservation. In addition to lower sequence conservation, the length of Irs3 progressively shorted along the lineage leading to mammals. Comparisons among the paralogous Irs sequences shows that most of the sequence is well conserved within a paralog, but only the PH and TTB domains, those responsible for binding to plasma membranes and the insulin receptor, are conserved between paralogs. Only a few regions within the C-terminal extensions of these proteins are conserved among Irs paralogs,

suggesting that divergence in these sequences has allowed divergence in function.

Methods
Database searches
Molecular sequence databases maintained by *Ensembl* [44] and the National Center for Biotechnology Information (*NCBI*) [47] were searched in January 2016 for insulin receptor substrate (*Irs1, Irs2, Irs3,* and *Irs4*)-like coding sequences. We initially searched the databases using the *tBLASTn* algorithm [43] using previously characterized mouse Irs1, Irs2, Irs3, and Irs4 protein sequences as queries. Putative Irs-like protein sequences identified were then used in subsequent tBLASTn searches. We also investigated the elephant shark (the sole representative of cartilaginous fish with a near-complete genome sequence) genome generated by the *Elephant Shark Genome Project* [45, 78]. All sequences that had E-scores below 0.01 were examined. Sequences identified by *BLAST* were used in reciprocal *BLASTx* searches of the mouse proteomes to ensure that their best matches were Irs-like sequences.

To examine genomic neighborhoods near *Irs*-like genes genomic comparisons were conducted using *PipMaker* and *MultiPipMaker* [56, 57]. Genes neighboring the *Irs*-like genes were identified from the genome assemblies in *Ensembl* [44] and the *Elephant Shark Genome Project* [78]. The organization of genes adjacent to the *Irs*-like genes was used to determine whether the genes of interest reside in conserved genomic neighborhoods.

Phylogenetic analysis
Phylogenies of vertebrate *Irs*-like gene coding sequences were generated using full-length, or near full-length (i.e., missing a short part of their N- or C-termini), *Irs1, Irs2, Irs3,* and *Irs4* coding sequences from diverse vertebrate and outgroups (see Additional file 1: Table S1 and Additional file 2: Tables S2) and outgroups. *Irs*-like coding sequences were aligned using *MAFFT* [79] as implemented at the *Guidance* web site [80, 81], using default parameters. Similar results were obtained if *Clustal Omega* [82] was used as the alignment program. DNA sequence alignments were based on codons to retain protein alignments. The reliability of the alignments was examined using *Guidance* [80, 81] and trimmed alignments using sites that had values above the default cutoff of 0.93 were generated.

Phylogenetic trees of the sequences were generated using Bayesian methods with *MrBayes* 3.2 [50, 51, 83], maximum likelihood with *IQ-tree* [49, 84], and neighbor-joining distance approaches with *MEGA6.06* [85]. Bayesian trees were generated from coding sequences with *MrBayes* 3.2 using parameters selected by *ModelFinder* [86], whose results are presented in Additional file 13: Figure S9. *MrBayes* was run for 2,000,000 generations with four simultaneous

Metropolis-coupled Monte Carlo Markov chains sampled every 100 generations. The average standard deviation of split frequencies dropped to less than 0.02 for all analyses. The first 25% of the trees were discarded as burn-in with the remaining samples used to generate the consensus trees. Trace files generated by *MrBayes* were examined by *Tracer* [87] to verify if they had converged. Maximum likelihood trees, constructed with 1000 replications by the ultrafast approximation [88], were generated with *IQ-tree* [49] on the *IQ-tree* webserver [84] using parameters for the substitution model suggested by *ModelFinder* [86]. The maximum likelihood search was initiated from a tree generated by *BIONJ* and the best tree was identified after heuristic searches using the nearest neighbor interchange (NNI) algorithm. *MEGA6.06* [85] was used to construct bootstrapped (1000 replications) neighbor-joining distance trees, using either Maximum Composite Likelihood distances for the DNA sequences or JTT distances for the proteins sequences. Similar results were obtained, but with lower confidence (bootstrap or posterior probabilities) intervals if alternative outgroups were used (results not shown).

With respect to orthology-paralogy issues, choice of outgroup, alignment method (*MAFFT* [79] or *Clustal* [82]), or the use of full-length or trimmed (based on *Guidance* scores [81]) alignments had little influence on the key findings of these analyses. Methods that relied on shorter sequences (i.e., trimmed alignments or protein sequences) or simpler models of sequence evolution (i.e., neighbor-joining or parsimony) tended to yield weaker support for the earlier diverging lineages, but none of our analyses were in significant conflict with the key inferences of the phylogeny presented in Fig. 2 or Additional file 11: Figure S8.

Analysis of protein sequence conservation
Conservation of proteins sequences was assessed using Jenson-Shannon (JS) divergence scores [62] on the *JS Divergence* web server [89], using a window size of 3 and the BLOSUM62 matrix as background. Putative tyrosine phosphorylation sites in the protein sequences were predicted using *NetPhos* [63, 90].

Additional files

Additional file 1: Table S1. This file is in Excel format. Genomic locations of *Irs*-like genes in sequenced vertebrate genomes identified from the *Ensembl* database. (XLSX 40 kb)

Additional file 2: Table S2. This file is in Excel format. Accession numbers of *Irs*-like coding sequences from vertebrates and outgroup species identified from the NCBI database. (XLSX 218 kb)

Additional file 3: Figure S1. This file is in Word format. Coding sequences for full-length and near full-length Irs genes from diverse vertebrates and outgroups. (PDF 200 kb)

Additional file 4: Figure S2. This file is in PDF format. Phylogeny of vertebrate *Irs1* sequences. (PDF 172 kb)

Additional file 5: Figure S3. This file is in PDF format. Phylogeny of vertebrate *Irs2* sequences. (XLSX 58 kb)

Additional file 6: Figure S4. This file is in PDF format. Phylogeny of vertebrate *Irs3* sequences. (XLSX 53 kb)

Additional file 7: Figure S5. This file is in PDF format. Phylogeny of vertebrate *Irs4* sequences. (DOCX 627 kb)

Additional file 8: Figure S6. This file is in PDF format. Phylogeny of the vertebrate *Irs* gene family rooted with non-vertebrate *Irs*-like genes. (PDF 525 kb)

Additional file 9: Figure S7. This file is in PDF format. Alignment of primate *Irs3* genomic sequences. (PDF 518 kb)

Additional file 10: Table S3. This file is in Excel format. Relative rates of evolution of *Irs* genes in mammals and non-mammals. (PDF 518 kb)

Additional file 11: Table S4. This file is in Excel format. JS Divergence Scores from alignments of Irs proteins from 10 vertebrates. (PDF 499 kb)

Additional file 12: Figure S8. This file is in PDF format. Alignment of Irs protein sequences with conserved residues highlighted. (PDF 450 kb)

Additional file 13: Figure S9. This file is in PDF format. *ModelFinder* results for the coding sequences used in the phylogenetic analyses. (PDF 224 kb)

Acknowledgements
This work has been supported by a grant from the Canadian Institutes of Health Research CCI-109605 (to DMI). The funding body did not have any role in the design, analysis, or interpretation of data or in the writing of the manuscript and the decision to submit the manuscript for publication.

Authors' contributions
AA and DMI designed the research and outlined the manuscript, obtained and analyzed the data, and drafted the manuscript. The authors have read, edited, and approved the final manuscript.

Competing interests
The authors declare that they have no competing interests.

References
1. Pirola L, Johnston AM, Van Obberghen E. Modulation of insulin action. Diabetologia. 2004;47:170–84.
2. Myers MG Jr, White MF. Insulin signal transduction and the IRS proteins. Annu Rev Pharmacol Toxicol. 1996;36:615–58.
3. De Meyts P, Whittaker J. Structural biology of insulin and IGF1 receptors: implications for drug design. Nat Rev Drug Discov. 2002;1:769–83.
4. Hubbard SR. The insulin receptor: both a prototypical and atypical receptor tyrosine kinase. Cold Spring Harb Perspect Biol. 2013;5:a008946.
5. Ward CW, Lawrence MC. Ligand-induced activation of the insulin receptor: a multi-step process involving structural changes in both the ligand and the receptor. BioEssays. 2009;31:422–34.
6. Du Y, Wei T. Inputs and outputs of insulin receptor. Protein Cell. 2014;5:203–13.
7. Wei L, Hubbard SR, Hendrickson WA, Ellis L. Expression, characterization, and crystallization of the catalytic core of the human insulin receptor protein-tyrosine kinase domain. J Biol Chem. 1995;270:8122–30.
8. Brummer T, Schmitz-Peiffer C, Daly RJ. Docking proteins. FEBS J. 2010;277:4356–69.
9. Jensen M, De Meyts P. Molecular mechanisms of differential intracellular signaling from the insulin receptor. Vitam Horm. 2009;80:51–75.
10. Sun XJ, Rothenberg P, Kahn CR, Backer JM, Araki E, Wilden PA, et al. Structure of the insulin receptor substrate IRS-1 defines a unique signal transduction protein. Nature. 1991;352:73–7.
11. Sun XJ, Wang LM, Zhang Y, Yenush L, Myers MG Jr, Glasheen E, et al. Role of IRS-2 in insulin and cytokine signalling. Nature. 1995;377:173–7.
12. Lavan BE, Lane WS, Lienhard GE. The 60-kDa phosphotyrosine protein in insulin-treated adipocytes is a new member of the insulin receptor substrate family. J Biol Chem. 1997;272:11439–43.

13. Lavan BE, Fantin VR, Chang ET, Lane WS, Keller SR, Lienhard GE. A novel 160-kDa phosphotyrosine protein in insulin-treated embryonic kidney cells is a new member of the insulin receptor substrate family. J Biol Chem. 1997;272:21403–7.

14. White MF. The IRS-signalling system: a network of docking proteins that mediate insulin action. Mol Cell Biochem. 1998;182:3–11.

15. Giovannone B, Scaldaferri ML, Federici M, Porzio O, Lauro D, Fusco A, et al. Insulin receptor substrate (IRS) transduction system: distinct and overlapping signaling potential. Diabetes Metab Res Rev. 2000;16:434–41.

16. Withers DJ. Insulin receptor substrate proteins and neuroendocrine function. Biochem Soc Trans. 2001;29:525–9.

17. White MF. IRS proteins and the common path to diabetes. Am J Physiol Endocrinol Metab. 2002;283:E413–22.

18. Thirone AC, Huang C, Klip A. Tissue-specific roles of IRS proteins in insulin signaling and glucose transport. Trends Endocrinol Metab. 2006;17:72–8.

19. Jacobs AR, LeRoith D, Taylor J. Insulin receptor substrate-1 pleckstrin homology and phosphotyrosine-binding domains are both involved in plasma membrane targeting. J Biol Chem. 2001;276:40795–802.

20. Wolf G, Trüb T, Ottinger E, Groninga L, Lynch A, White MF, et al. PTB domains of IRS-1 and Shc have distinct but overlapping binding specificities. J Biol Chem. 1995;270:27407–10.

21. Björnholm M, He AR, Attersand A, Lake S, Liu SC, Lienhard GE, et al. Absence of functional insulin receptor substrate-3 (IRS-3) gene in humans. Diabetologia. 2002;45:1697–702.

22. Cai D, Dhe-Paganon S, Melendez PA, Lee J, Shoelson SE. Two new substrates in insulin signaling, IRS5/DOK4 and IRS6/DOK5. J Biol Chem. 2003;278:25323–30.

23. De Meyts P. Insulin and its receptor: structure, function and evolution. BioEssays. 2004;26:1351–62.

24. Marino-Buslje C, Martin-Martinez M, Mizuguchi K, Siddle K, Blundell TL. The insulin receptor: from protein sequence to structure. Biochem Soc Trans. 1999;27:715–26.

25. Lavin DP, White MF, Brazil DP. IRS proteins and diabetic complications. Diabetologia. 2016;59:2280–91.

26. Araki E, Lipes MA, Patti ME, Brüning JC, Haag B 3rd, Johnson RS, et al. Alternative pathway of insulin signalling in mice with targeted disruption of the IRS-1 gene. Nature. 1994;372:186–90.

27. Schubert M, Brazil DP, Burks DJ, Kushner JA, Ye J, Flint CL, et al. Insulin receptor substrate-2 deficiency impairs brain growth and promotes tau phosphorylation. J Neurosci. 2003;23:7084–92.

28. Withers DJ, Burks DJ, Towery HH, Altamuro SL, Flint CL, White MF. Irs-2 coordinates Igf-1 receptor-mediated beta-cell development and peripheral insulin signalling. Nature Genet. 1999;23:32–40.

29. Fantin VR, Wang Q, Lienhard GE, Keller SR. Mice lacking insulin receptor substrate 4 exhibit mild defects in growth, reproduction, and glucose homeostasis. Am J Physiol Endocrinol Metab. 2000;278:E127–33.

30. Sadagurski M, Dong XC, Myers MG Jr, White MF. Irs2 and Irs4 synergize in non-LepRb neurons to control energy balance and glucose homeostasis. Mol Metab. 2013;3:55–63.

31. Liu SC1, Wang Q, Lienhard GE, Keller SR: Insulin receptor substrate 3 is not essential for growth or glucose homeostasis. J Biol Chem 1999, 274:18093–18099.

32. Sciacchitano S, Taylor SI. Cloning, tissue expression, and chromosomal localization of the mouse IRS-3 gene. Endocrinology. 1997;138:4931–40.

33. Maffucci T, Razzini G, Ingrosso A, Chen H, Iacobelli S, Sciacchitano S, et al. Role of pleckstrin homology domain in regulating membrane targeting and metabolic function of insulin receptor substrate 3. Mol Endocrinol. 2003;17:1568–79.

34. Böhni R, Riesgo-Escovar J, Oldham S, Brogiolo W, Stocker H, Andruss BF, et al. Autonomous control of cell and organ size by CHICO, a drosophila homolog of vertebrate IRS1-4. Cell. 1999;97:865–75.

35. Uhlik MT, Temple B, Bencharit S, Kimple AJ, Siderovski DP, Johnson GL. Structural and evolutionary division of phosphotyrosine binding (PTB) domains. J Mol Biol. 2005;345:1–20.

36. Chakraborty C, Agoramoorthy G, Hsu MJ. Exploring the evolutionary relationship of insulin receptor substrate family using computational biology. PLoS One. 2011;6:e16580.

37. McGaugh SE, Bronikowski AM, Kuo CH, Reding DM, Addis EA, Flagel LE, et al. Rapid molecular evolution across amniotes of the IIS/TOR network. Proc Natl Acad Sci U S A. 2015;112:7055–60.

38. Olinski RP, Lundin LG, Hallböök F. Genome duplication-driven evolution of gene families: insights from the formation of the insulin family. Ann N Y Acad Sci. 2005;1040:426–8.

39. Hernández-Sánchez C, Mansilla A, de Pablo F, Zardoya R. Evolution of the insulin receptor family and receptor isoform expression in vertebrates. Mol Biol Evol. 2008;25:1043–53.

40. Rentería ME, Gandhi NS, Vinuesa P, Helmerhorst E, Mancera RL. A comparative structural bioinformatics analysis of the insulin receptor family ectodomain based on phylogenetic information. PLoS One. 2008;3:e3667.

41. Huminiecki L, Heldin CH. 2R and remodeling of vertebrate signal transduction engine. BMC Biol. 2010;8:146.

42. Hokamp K, McLysaght A, Wolfe KH. The 2R hypothesis and the human genome sequence. J Struct Funct Genom. 2003;3:95–110.

43. Altschul SF, Madden TL, Schäffer AA, Zhang J, Zhang Z, Miller W, et al. Gapped BLAST and PSI-BLAST: a new generation of protein database search programs. Nucleic Acids Res. 1997;25:3389–402.

44. Ensembl Genome Browser [http://www.ensembl.org/index.html].

45. Venkatesh B, Lee AP, Ravi V, Maurya AK, Lian MM, Swann JB, et al. Elephant shark genome provides unique insights into gnathostome evolution. Nature. 2014;505:174–9.

46. Braasch I, Gehrke AR, Smith JJ, Kawasaki K, Manousaki T, Pasquier J, Amores A, Desvignes T, Batzel P, Catchen J, Berlin AM, Campbell MS, Barrell D, Martin KJ, Mulley JF, Ravi V, Lee AP, Nakamura T, Chalopin D, Fan S, Wcisel D, Cañestro C, Sydes J, Beaudry FE, Sun Y, Hertel J, Beam MJ, Fasold M, Ishiyama M, Johnson J, Kehr S, Lara M, Letaw JH, Litman GW, Litman RT, Mikami M, Ota T, Saha NR, Williams L, Stadler PF, Wang H, Taylor JS, Fontenot Q, Ferrara A, Searle SM, Aken B, Yandell M, Schneider I, Yoder JA, Volff JN, Meyer A, Amemiya CT, Venkatesh B, Holland PW, Guiguen Y, Bobe J, Shubin NH, Di Palma F, Alföldi J, Lindblad-Toh K, Postlethwait JH: The spotted gar genome illuminates vertebrate evolution and facilitates human-teleost comparisons. *Nat Genet* 2016, 48:427–37.

47. National Center for Biotechnology Information [http://www.ncbi.nlm.nih.gov/].

48. Felsenstein J. Evolutionary trees from DNA sequences: a maximum likelihood approach. J Mol Evol. 1981;17:368–76.

49. Nguyen L-T, Schmidt HA, von Haeseler A, Minh BQ. IQ-TREE: a fast and effective stochastic algorithm for estimating maximum likelihood phylogenies. Mol Biol Evol. 2015;32:268–74.

50. Huelsenbeck JP, Ronquist F, Nielsen R, Bollback JP. Bayesian inference of phylogeny and its impact on evolutionary biology. Science. 2001;294:2310–4.

51. Ronquist F, Teslenko M, van der Mark P, Ayres DL, Darling A, Höhna S, et al. MrBayes 3.2: efficient Bayesian phylogenetic inference and model choice across a large model space. Syst Biol. 2012;61:539–42.

52. Cañestro C, Albalat R, Irimia M, Garcia-Fernàndez J. Impact of gene gains, losses and duplication modes on the origin and diversification of vertebrates. Semin Cell Dev Biol. 2013;24:83–94.

53. Glasauer SM, Neuhauss SC. Whole-genome duplication in teleost fishes and its evolutionary consequences. Mol Gen Genomics. 2014;289:1045–60.

54. Inoue J, Sato Y, Sinclair R, Tsukamoto K, Nishida M. Rapid genome reshaping by multiple-gene loss after whole-genome duplication in teleost fish suggested by mathematical modeling. Proc Natl Acad Sci U S A. 2015;112:14918–23.

55. Perelman P, Johnson WE, Roos C, Seuánez HN, Horvath JE, Moreira MA, et al. A molecular phylogeny of living primates. PLoS Genet. 2011;7:e1001342.

56. Schwartz S, Zhang Z, Frazer KA, Smit A, Riemer C, Bouck J, et al. PipMaker–a web server for aligning two genomic DNA sequences. Genome Res. 2000;10:577–86.

57. Schwartz S, Elnitski L, Li M, Weirauch M, Riemer C, Smit A. NISC comparative sequencing program, green ED, Hardison RC, Miller W: MultiPipMaker and supporting tools: alignments and analysis of multiple genomic DNA sequences. Nucleic Acids Res. 2003;31:3518–24.

58. Tajima F. Simple methods for testing molecular clock hypothesis. Genetics. 1993;135:599–607.

59. Capra JA, Singh M. Predicting functionally important residues from sequence conservation. Bioinformatics. 2007;23:1875–82.

60. Blom N, Gammeltoft S, Brunak S. Sequence- and structure-based prediction of eukaryotic protein phosphorylation sites. J Mol Biol. 1999;294:1351–62.

61. Smith JJ, Kuraku S, Holt C, Sauka-Spengler T, Jiang N, Campbell MS, Yandell MD, Manousaki T, Meyer A, Bloom OE, Morgan JR, Buxbaum JD, Sachidanandam R, Sims C, Garruss AS, Cook M, Krumlauf R, Wiedemann LM, Sower SA, Decatur WA, Hall JA, Amemiya CT, Saha NR, Buckley KM, Rast JP, Das S, Hirano M, McCurley N, Guo P, Rohner N, Tabin CJ, Piccinelli P, Elgar G, Ruffier M, Aken BL, Searle SM, Muffato M, Pignatelli M, Herrero J, Jones M, Brown CT, Chung-Davidson YW, Nanlohy KG, Libants SV, Yeh CY, McCauley DW, Langeland JA, Pancer Z, Fritzsch B, de Jong PJ, Zhu B, Fulton LL, Theising B, Flicek P, Bronner ME, Warren WC, Clifton SW, Wilson RK, Li W: Sequencing of the sea lamprey (*Petromyzon marinus*) genome provides insights into vertebrate evolution. *Nat Genet*. 2013, 45:415–21.

62. Smith JJ, Antonacci F, Eichler EE, Amemiya CT. Programmed loss of millions of base pairs from a vertebrate genome. Proc Natl Acad Sci U S A. 2009;106:11212–7.

63. Kurokawa T, Uji S, Suzuki T. Identification of cDNA coding for a homologue to mammalian leptin from pufferfish, *Takifugu rubripes*. Peptides. 2005;26:745–50.

64. Hu Q, Tan H, Irwin DM. Evolution of the vertebrate Resistin Gene family. PLoS One. 2015;10:e0130188.

65. Kuraku S, Meyer A. Detection and phylogenetic assessment of conserved synteny derived from whole genome duplications. Methods Mol Biol. 2012;855:385–95.

66. Arroyo JI, Hoffmann FG, Opazo JC. Gene turnover and differential retention in the relaxin/insulin-like gene family in primates. Mol Phylogenet Evol. 2012;63:768–76.

67. Hoffmann FG, Opazo JC. Evolution of the relaxin/insulin-like gene family in placental mammals: implications for its early evolution. J Mol Evol. 2011;72:72–9.

68. Daković N, Térézol M, Pitel F, Maillard V, Elis S, Leroux S, et al. The loss of adipokine genes in the chicken genome and implications for insulin metabolism. Mol Biol Evol. 2014;31:2637–46.

69. Massingham T, Davies LJ, Liò P. Analysing gene function after duplication. *Bioessays*. 2001;23:873–6.

70. Freeling M, Scanlon MJ, Fowler JE. Fractionation and subfunctionalization following genome duplications: mechanisms that drive gene content and their consequences. Curr Opin Genet Dev. 2015;35:110–8.

71. Tamemoto H, Kadowaki T, Tobe K, Yagi T, Sakura H, Hayakawa T, Terauchi Y, Ueki K, Kaburagi Y, Satoh S, Sekihara H, Yoshioka S, Horikoshi H, Furuta Y, Ikawa Y, Kasuga M, Yazaki Y, Aizawa S: Insulin resistance and growth retardation in mice lacking insulin receptor substrate-1. Nature. 1994;372:182–6.

72. Withers DJ, Gutierrez JS, Towery H, Burks DJ, Ren JM, Previs S, et al. Disruption of IRS-2 causes type 2 diabetes in mice. Nature. 1998;391:900–3.

73. Sun XJ, Crimmins DL, Myers MG Jr, Miralpeix M, White MF. Pleiotropic insulin signals are engaged by multisite phosphorylation of IRS-1. Mol Cell Biol. 1993;13:7418–28.

74. Esposito DL, Li Y, Vanni C, Mammarella S, Veschi S, Della Loggia F, et al. A novel T608R missense mutation in insulin receptor substrate-1 identified in a subject with type 2 diabetes impairs metabolic insulin signaling. J Clin Endocrinol Metab. 2003;88:1468–75.

75. Landis J, Shaw LM. Insulin receptor substrate 2-mediated phosphatidylinositol 3-kinase signaling selectively inhibits glycogen synthase kinase 3β to regulate aerobic glycolysis. J Biol Chem. 2014;289:18603–13.

76. Asano T, Fujishiro M, Kushiyama A, Nakatsu Y, Yoneda M, Kamata H, et al. Role of phosphatidylinositol 3-kinase activation on insulin action and its alteration in diabetic conditions. Biol Pharm Bull. 2007;30:1610–6.

77. Sato Y, Hashiguchi Y, Nishida M. Temporal pattern of loss/persistence of duplicate genes involved in signal transduction and metabolic pathways after teleost-specific genome duplication. BMC Evol Biol. 2009;9:127.

78. Elephant Shark Genome Project [http://esharkgenome.imcb.a-star.edu.sg/].

79. Katoh K, Misawa K, Kuma K, Miyata T. MAFFT: a novel method for rapid multiple sequence alignment based on fast Fourier transform. Nucl Acids Res. 2002;30:3059–66.

80. Penn O, Privman E, Ashkenazy H, Landan G, Graur D, Pupko T. GUIDANCE: a web server for assessing alignment confidence scores. Nucl Acids Res. 2010;38:W23–8.

81. Guidance2 Web Server [http://guidance.tau.ac.il/ver2/].

82. Clustal Omega Web Server [http://www.ebi.ac.uk/Tools/msa/clustalo/].

83. MrBayes 3.2.2 Web Site [http://mrbayes.sourceforge.net/].

84. IQ-tree Web Server [http://iqtree.cibiv.univie.ac.at/].

85. Tamura K, Stecher G, Peterson D, Filipski A, Kumar S. MEGA6: molecular evolutionary genetics analysis version 6.0. Mol Biol Evol. 2013;30:2725–9.

86. Kalyaanamoorthy S, Minh BQ, Wong TKF, von Haeseler A, Jermiin LS. ModelFinder: fast model selection for accurate phylogenetic estimates. Nat Methods. 2017;2017 in press

87. Tracer v1.6 Web Site [http://tree.bio.ed.ac.uk/software/tracer/].

88. Minh BQ, Nguyen MAT, von Haeseler A. Ultrafast approximation for phylogenetic bootstrap. Mol Biol Evol. 2013;30:1188–95.

89. JS Distance Web Server [http://compbio.cs.princeton.edu/conservation/].

90. NetPhos 2.0 Web Server [http://www.cbs.dtu.dk/services/NetPhos/].

5

Evolving mutation rate advances the invasion speed of a sexual species

Marleen M. P. Cobben[1,2*], Oliver Mitesser[2] and Alexander Kubisch[2,3]

Abstract

Background: Many species are shifting their ranges in response to global climate change. Range expansions are known to have profound effects on the genetic composition of populations. The evolution of dispersal during range expansion increases invasion speed, provided that a species can adapt sufficiently fast to novel local conditions. Genetic diversity at the expanding range border is however depleted due to iterated founder effects. The surprising ability of colonizing species to adapt to novel conditions while being subjected to genetic bottlenecks is termed 'the genetic paradox of invasive species'. Mutational processes have been argued to provide an explanation for this paradox. Mutation rates can evolve, under conditions that favor an increased rate of adaptation, by hitchhiking on beneficial mutations through induced linkage disequilibrium. Here we argue that spatial sorting, iterated founder events, and population structure benefit the build-up and maintenance of such linkage disequilibrium. We investigate if the evolution of mutation rates could play a role in explaining the 'genetic paradox of invasive species' for a sexually reproducing species colonizing a landscape of gradually changing conditions.

Results: We use an individual-based model to show the evolutionary increase of mutation rates in sexual populations during range expansion, in coevolution with the dispersal probability. The observed evolution of mutation rate is adaptive and clearly advances invasion speed both through its effect on the evolution of dispersal probability, and the evolution of local adaptation. This also occurs under a variable temperature gradient, and under the assumption of 90% lethal mutations.

Conclusions: In this study we show novel consequences of the particular genetic properties of populations under spatial disequilibrium, i.e. the coevolution of dispersal probability and mutation rate, even in a sexual species and under realistic spatial gradients, resulting in faster invasions. The evolution of mutation rates can therefore be added to the list of possible explanations for the 'genetic paradox of invasive species'. We conclude that range expansions and the evolution of mutation rates are in a positive feedback loop, with possibly far-reaching ecological consequences concerning invasiveness and the adaptability of species to novel environmental conditions.

Keywords: Local adaptation, Spatial sorting, Individual-based model, Evolvability, Dispersal evolution, Metapopulation

Background

Many species are currently expanding their ranges as a response to increasing global temperatures under climate change [1]. Range expansions are known to have profound effects on the genetic composition of populations, regarding both neutral and adaptive genetic diversity [2–5]. Mutations, even deleterious ones, can surf the wave of the range expansion and reach high frequencies in newly established populations [6]. In addition, traits that act to increase species' dispersal capabilities and population growth rates are selected for under range expansions due to spatial sorting [7, 8] and kin competition [9]. This may lead to higher dispersal rates [10, 11], larger dispersal distances [5] and higher effective fertilities [12] at the expanding front of species' ranges due to microevolution. An increasing dispersal rate under range expansion will increase the invasion speed [5], but only if the species is able to adapt sufficiently rapid to novel local conditions (and assuming the absence of strong Allee effects [13]). The depletion of genetic diversity at

* Correspondence: m.cobben@nioo.knaw.nl
[1]Department of Animal Ecology, Netherlands Institute of Ecology
(NIOO-KNAW), PO Box 50, 6700, AB, Wageningen, The Netherlands
[2]Theoretical Evolutionary Ecology Group, Institute for Animal Ecology and
Tropical Biology, University of Würzburg, Emil-Fischerstr. 32, 97074 Würzburg,
Germany
Full list of author information is available at the end of the article

the expanding range border due to iterated founder effects [3, 4, 14] could however be expected to limit the invasion speed as low genetic diversity will lead to lower rates of local adaptation and thereby delayed population establishment [15, 16]. The surprising ability of colonizing species to adapt to novel conditions while being subjected to genetic bottlenecks is termed 'the genetic paradox of invasive species' [17].

Several possible explanations for 'the genetic paradox of invasive species' are reviewed in Stapley et al. [18], particularly highlighting the role of mutational processes as a source of new genetic diversity, and focusing on transposable elements. Another, more classic example of a mutational process is the evolution of mutation rates. Here selection can act on allelic variation in the processes of DNA repair and replication and as such result in increased mutation rates. These cause higher levels of genetic diversity and can thus enable adaptation to changing selection pressures [19–26]. However, the rate at which mutations occur is not a phenotype and thus cannot be selected for. Since selection acts on the mutation that occurs at a gene under selection and thus not on the rate at which such mutations occur, the establishment of a particular mutation rate is restricted to genetic hitchhiking. This is the phenomenon that an allele increases in frequency, because its locus is linked to another locus at which the allele is under positive selection. Usually this link is caused by physical proximity of the two loci, causing the different alleles to be inherited together. In this case, a genetic modifier that increases the mutation rate, increases in frequency by hitchhiking on beneficial mutations, and decreases in frequency by hitchhiking on deleterious mutations [27], inducing linkage disequilibrium (LD). Two loci are said to be in LD when their alleles are more (or less) frequently associated than can be expected under the assumption that they are independently inherited. When the genetic modifier produces a beneficial mutation, this modifier is likely to be passed on to the next generation because the beneficial mutation and the genetic modifier occur in the same individual, causing such LD. Recombination can however easily break up this joint inheritance and thus the linkage disequilibrium. Theoretical studies have thus concluded that in sexual populations, where there is recombination, the effect of beneficial mutations on the evolution of mutation rates is negligible [22] to small under changing environments [28, 29]. However, Johnson [27] has extended these studies by assuming a series of unique beneficial mutations. This increases the strength of linkage disequilibrium between the modifier locus and the mutation locus, due to constant recurring events of genetic hitchhiking. He showed theoretically that under these conditions, beneficial mutations can play a role in determining the evolutionarily stable

mutation rate in sexual populations when environment conditions are constantly changing and costs of lower fidelity replication are low.

A species range consists of, more or less connected, separate populations of individuals, in contrast to the investigated single populations in previous theoretical studies [22, 27–29]. We argue that population structure, and in addition under range expansion, spatial sorting and iterated founder events, benefit the build-up and maintenance of linkage disequilibrium. Therefore, the evolution of mutation rate might be a considerable factor in the invasion ability of a sexually reproducing metapopulation. Here we investigate if the evolution of mutation rates could play a role in explaining the 'genetic paradox of invasive species' [17] for a sexually reproducing species colonizing a landscape with gradually changing conditions. We use a spatially explicit individual-based metapopulation model of a sexual species establishing its range on a spatial gradient to investigate 1) whether mutation rates increase during range expansion, 2) if this is related to the rate of dispersal during range expansion, and 3) if such increase of mutation rates is adaptive.

Methods

We use a spatially explicit individual-based metapopulation model of a sexually reproducing species with discrete generations, which expands its range along a gradient in temperature. We allow the mutation rate to evolve, and investigate its interplay with the evolution of dispersal probability and temperature adaptation during and after range establishment.

Landscape

The simulated landscape consists of 250 columns (x-dimension) of 20 patches each (y-dimension). We assume wrapped borders in y-direction, building a tube. Hence, if an individual leaves the world in y-direction during dispersal, it will reenter the simulated world on the opposite side. However, if it leaves the world in the x-direction, it is lost from the simulation. To answer our research questions the model requires a need for local adaptation during range expansion. Thus every column of patches (x-position) is characterized by its specific mean temperature τ_x. This mean local temperature is used for the determination of the level of local adaptation of individuals. To simulate a large-scale habitat gradient, τ_x changes linearly from $\tau_{x=1} = 0$ to $\tau_{x=250} = 10$ along the x-dimension, i.e. by $\Delta\tau = 0.04$ when moving one step in x-direction.

Population dynamics and survival of offspring

Local populations are composed of individuals, each of which is characterized by several traits: 1) its sex, 2) its dispersal probability determined by the alleles at the dispersal locus l_d, 3) its optimal temperature τ_{opt}, i.e. the temperature under which it survives best, determined by the alleles at its adaptation locus l_a (see below for details), 4) its genetic mutation rate determined by the alleles at the mutator locus l_m (see below under Genetics), and 5) a diploid neutral locus l_n, for the sake of comparing the levels of genetic diversity with other loci.

Local population dynamics follow the time-discrete Beverton–Holt model [30]. Each individual female in patch x, y is therefore assigned a random male from the same habitat patch (males can potentially mate several times) and gives birth to a number of offspring drawn from a Poisson distribution with mean population growth rate λ. The offspring's sex is chosen at random. Density-dependent survival probability s_1 of offspring due to competition is calculated as:

$$s_1 = \frac{1}{1 + \frac{\lambda - 1}{K} \cdot N_{x,y,t}}$$

with K the carrying capacity and $N_{x,y,t}$ the number of individuals in patch x,y at time t. Finally, the surviving offspring experience a further density-independent mortality risk $(1 - s_2)$ that depends on their level of local adaptation, so the matching of their genetically determined optimal temperature (τ_{opt}) to the temperature conditions in patch x, y (τ_x) according to the following equation:

$$s_2 = \exp\left[\frac{-1}{2} \cdot \left[\frac{\tau_{opt} - \tau_x}{\eta}\right]^2\right]$$

where η describes the niche width or 'tolerance' of the species. We performed simulations for the species with a niche width of $\eta = 0.5$, equivalent to a decrease of survival probability of about 0.032 when dispersing one patch away from the optimal habitat. In this approach we assume that density-dependent mortality $(1 - s_1)$ acts before mortality due to maladaptation to local conditions $(1 - s_2)$. In addition, each population has an extinction probability ε per generation. Surviving offspring disperse with probability d that is determined by their dispersal locus (see below). If an individual disperses it dies with probability μ, which is 0.2 throughout the landscape. This mortality accounts for various costs that may be associated with dispersal in real populations, like fertility reduction or predation risk [31]. We assume nearest-neighbor dispersal, i.e. successful dispersers settle randomly in one of the eight surrounding habitat patches.

Genetics

As mentioned above, each individual carries three unlinked, diploid loci coding for its dispersal probability, its optimum temperature (and thus its degree of local adaptation), and its genetic mutation rate, respectively, and an additional neutral locus. The phenotype of an individual is determined by calculating the means of the two corresponding alleles, with no dominance effect involved. Hence, dispersal probability d is given by $d = (l_{d,1} + l_{d,2})/2$ (with $l_{d,1}$ and $l_{d,2}$ giving the two 'values' of the two dispersal alleles), optimal temperature τ_{opt} is calculated as $\tau_{opt} = (l_{a,1} + l_{a,2})/2$ (with $l_{a,1}$ and $l_{a,2}$ giving the 'values' of the two adaptation alleles), and similarly the mutation rate $m = 10^{-exp.}$ (with $exp. = (l_{m,1} + l_{m,2})/2$, and $l_{m,1}$ and $l_{m,2}$ the 'values' of the two mutator alleles). At each of the four loci, newborn individuals inherit alleles, randomly chosen, from the corresponding loci of each of their parents. During transition from one generation to the next an allele at any locus may mutate with the genetically determined probability m given by the value based on the two alleles at the mutator locus l_m as elaborated above. So the mutator locus determines the mutation rate at each of the four loci. Mutations are simulated by adding a random number to the value of the inherited allele. This value is drawn from a Gaussian distribution with mean 0 and standard deviation 0.2. The lethal mutation probability $\Omega = 0.1$ however, so 10 % of the mutations cause immediate death of the individual.

Simulation experiments

Simulations were initialized with a 'native area' (from $x = 1$ to $x = 50$) from where the species was able to colonize the world, while the rest of the world was initially kept free of individuals. Upon initialization, dispersal alleles $(l_{d,i})$ were randomly drawn from the interval $0 < l_{d,i} < 1$, and mutator alleles $l_{m,i}$ were set to 4, which set the initial mutation rate m to 10^{-4}. Populations were initialized with K locally optimally adapted individuals, i.e. adaptation alleles were initialized according to the local temperature τ_x. However, to account for some standing genetic variation we also added to every respective optimal temperature allele a Gaussian random number with mean zero and standard deviation 0.2. Identical copies of these alleles were used to initialize the neutral locus as well, for sake of comparison. We performed 200 replicate simulations, which all covered a time span of 15,000 generations. To establish equilibrium values, the individuals were confined to their native area during the first 10,000 generations. After this burn-in period, the species was allowed to pass the $x = 50$ border and expand its range for the remaining 5000 generations. Table 1 summarizes all relevant model parameters, their meanings and the standard values used for

Table 1 Parameter values

Parameter/ variable	(Initialization) value	Meaning
Individual variables	evolving	
$l_{d,1}, l_{d,2}$	0 to1	alleles coding for the dispersal probability
$l_{a,1}, l_{a,2}$	optimal with std. 0.5	alleles coding for the optimal temperature
$l_{m,1}, l_{m,2}$	4	alleles coding for the mutation rate of the optimal temperature
$l_{n,1}, l_{n,2}$	copy of $l_{a,1}, l_{a,2}$	neutral alleles as control
Simulation parameters:		
K	100	carrying capacity
λ	2	per capita growth rate
ε	0.05	local extinction probability
Ω	0.1	lethal mutation probability
μ	0.2	dispersal mortality
$τ_x$	[0..10]	local temperature
η	0.5	niche width
x_{max}	250	extent of simulated landscape in x-direction
y_{max}	20	extent of simulated landscape in y-direction

these simulations that test whether the evolution of mutation rates actually occurs.

As a follow-up, several other simulation experiments were performed to answer these specific questions:

1. Is the evolution of mutation rates neutral or adaptive? For this, the effect of evolving mutation rates on the speed of colonization was tested by contrast with fixed mutation rates. Therefore, the simulations were repeated with 200 replicates a) with fixed values of the mutation rate m of 10^{-4} and 10^{-5}, combined with evolving dispersal probability, and b) with both values fixed, investigating the combination of $d = 0.2$ and $m = 10^{-4}$, and of $d = 0.2$ and $m = 10^{-5}$.

2. Does the mutation rate evolve in coevolution with the dispersal probability? For this, the dispersal probabilities were fixed to assess the effect on the evolution of mutation rates. The simulations were therefore repeated with 200 replicates for fixed values of dispersal probability, $d = 0.05$, $d = 0.1$ and $d = 0.2$, while allowing the mutation rate to evolve.

3. Can the mutation rate increase if the percentage of deleterious mutations is higher? For this a simulation experiment was performed with 90% lethal mutations, $Ω = 0.9$.

4. How does the evolution of mutation rates depend on directional selection? For this, a landscape was designed in which the temperature varies in the x-

direction, to compare with the initial linearly increasing temperature gradient. We therefore performed an experiment with a variable spatial gradient in temperature. This required the application of a new habitat gradient, where $τ_x$ still changes from $τ_{x=1} = 0$ to $τ_{x = 250} = 10$ along the x-dimension, but at each $τ_x$ we added a random number in the range [−0.5, 0.5].

5. How does the evolution of mutation rates depend on the process of range expansion, i.e. a series of colonizations? For this, we looked into a stable, non-range expanding population to assess the dependency of such repeated colonization, for which we simulated a non-expanding species' range under both temperature gradients, so applying a) a non-variable and b) variable temperature gradient in time (as defined in experiment 4). For these experiments the species was initialized in the whole landscape, so not restricted to the native area, and the temperature was steadily increased in the entire landscape. With this, equal selection pressure for the species was forced, but without a range expansion, by changing the temperature at the same rate as experienced by the marginal populations in the spatial scenarios. Global dispersal was applied here.

Analysis

The individual phenotypes for the three traits were documented in time and space throughout the simulations and averaged per x-position. For the dispersal probability we calculated the arithmetic mean, while the mutation rate was averaged geometrically because the mutator gene codes for the exponent's value. Genetic diversity was calculated as the variance in allelic values at the adaptation locus, the dispersal locus and the neutral locus, per x-position. The analysis of linkage disequilibrium was done using Genepop version 4.5.1 [32] testing for a significant association between diploid genotypes at two loci. We here test every pair of loci, consisting of the locus with the gene coding for mutation rate and all other loci, so pairs of l_a and l_m, l_d and l_m, and l_n and l_m.

Results

The local level of adaptation s_2 is close to one in all simulations, throughout the simulation time and across the entire species' range. This indicates that colonization of new habitat occurs by pre-adapted individuals only. After the burn-in phase, the species' range expands across the landscape (Fig. 1). The landscape is fully colonized after between 1000 and 1500 generations, after which the population density keeps increasing till it reaches 80% of the maximum overall (Fig. 1a). During the range expansion the average dispersal probabilities and mutation rates

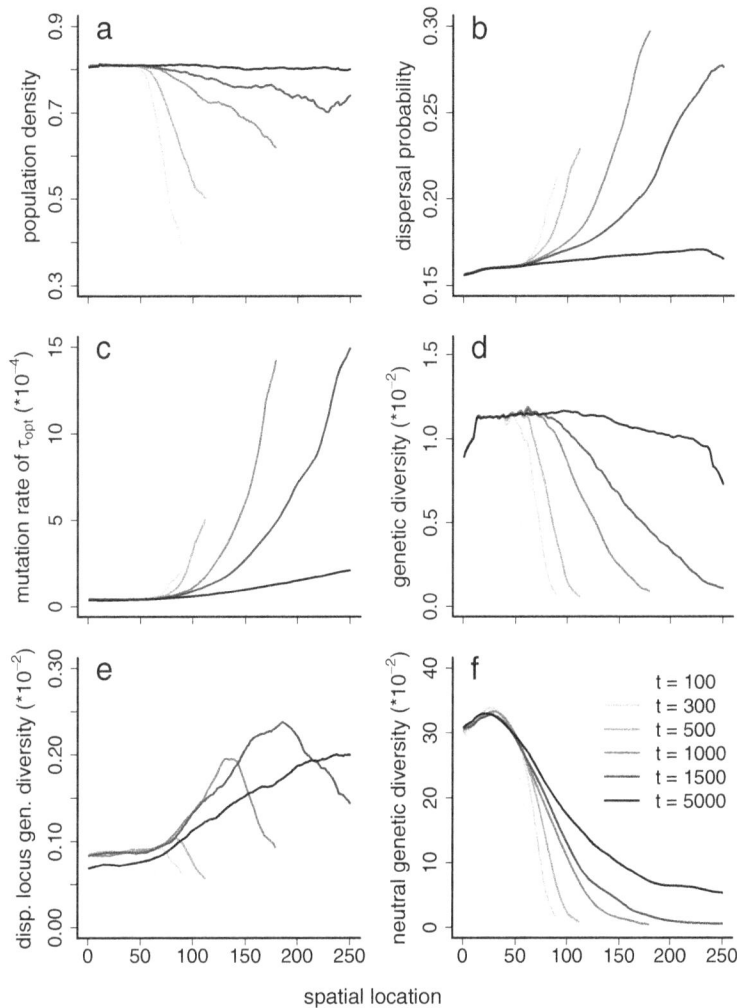

Fig. 1 Base scenario. The average values over 200 simulations during and after range expansion across the gradient (*horizontal axis*) in time (*gray scaling from light to dark, as time proceeds, which is given in a sequence of generations 100, 300, 500, 1000, 1500, 5000*) of **a**. population density, **b**. dispersal probability, **c**. the mutation rate, **d**. genetic diversity at the adaptation locus, **e**. genetic diversity at the dispersal locus, and **f**. neutral genetic diversity, all measured as the variance in allele values. For reasons of clarity, a moving average with a window size of 21 has been applied (each point along the *x*-axis is the average of all points in the range [*x*-10, *x* + 10], data were present in 10-generation intervals)

increase (Fig. 1b/c). This indicates that the individuals carrying alleles for high mutation rates are the first to establish new populations because they allow colonization by carrying these beneficial and novel alleles at the dispersal and temperature loci as well. After the colonization is complete and the range border has stabilized, mutation and dispersal probabilities decrease again (Fig. 1b/c). High dispersal probabilities are indeed only favorable with frequent population extinctions and low dispersal mortality [33]. Once the range border stabilizes, a low dispersal phenotype is more advantageous due to the assumed dispersal mortality. However, these slow dispersers by definition take some time to reach the area (genetic signature of range expansion, [34, 35]), especially when the mutation rate levels off and new dispersal phenotypes only slowly appear locally. The decrease of the mutation rate is

caused by the processes at the adaptation (temperature) locus. Once the maximum level of genetic diversity has been reached here, and population densities are at their equilibrium values, more mutations cause maladaptation. The association with these deleterious mutations causes the modifier of increased mutation rate to decrease in frequency.

Genetic diversity at the different loci shows a typical pattern of range expansion (due to founder effects or spatial sorting) with little genetic diversity at the expanding range margin, which increases with the age of the populations (Fig. 1d-f). The maximum level of genetic diversity differs between the different loci (Fig. 1d-f), reflecting their different functions. At the neutral locus genetic diversity steadily increases, towards a maximum determined by population dynamics. At the gene that determines the individual's optimal temperature the maximum genetic diversity is determined by the number of allele values that allow the individual's survival at that particular local temperature.

To ensure that the observed increase in mutation rate is adaptive and not the result of reduced selection pressure at the range front, we performed the additional simulation experiments with fixed rates (experiments 1a and b). These show that a fixed, lower mutation rate (Fig. 2) causes a lower speed of invasion. This indicates that the observed evolving high mutation rate is indeed not neutral but beneficial, since it allows the faster occurrence of alleles coding for higher dispersal probabilities that establish on the expansion wave and increase the invasion speed. This is the same both for evolving dispersal probabilities (panel A) as for fixed dispersal probability (d = 0.2, panel B), so even if the dispersal probability is fixed to a relatively high value of 0.2, the expansion is slow if the mutation rate is no allowed to evolve. To assess how dispersal rates and mutation rates are connected we look into the results of experiment 2. Under fixed dispersal rates (Fig. 3) we see the evolution of higher mutation rates and faster range expansions with higher dispersal probabilities. This clearly indicates the coevolution of mutation rate and dispersal probability, with a similar advancing effect on the invasion speed due to the production of alleles at the gene for local adaptation to temperature, allowing the faster rate of local adaptation.

The analysis of linkage disequilibrium (Additional file 1: Figure S1) shows that LD is widespread across the species' range, between every combination of loci, and also before and after the range expansion. Particularly at the onset of range expansion (t = 100, with t = 0 is the end of the burn-in period) there is a wide zone of newly colonizing populations at the expanding range margin with no genetic diversity at the temperature (i.e. local adaptation) locus (Additional file 1: Fig. S1a). A high initial population growth rate and the non-random set of invaders in such a newly established population result in a local non-random subset of the available genetic variation at the mutator locus. This causes strong linkage disequilibrium between all pairs of loci. The directional selection caused by the applied temperature gradient and the realistic metapopulation structure contribute to the maintenance and permanent renewal of LD.

The simulation experiment in which 90% of the mutations were lethal (Ω = 0.9, experiment 3) show a clear increase of the dispersal probability and mutation rate compared to the equilibrium conditions in the native area (Additional file 1: Figure S2B/C). Absolute values of both are however much lower than the values observed under Ω = 0.1, with the maximum of the mutation rate a factor 10 lower and the dispersal probability lower by a factor of

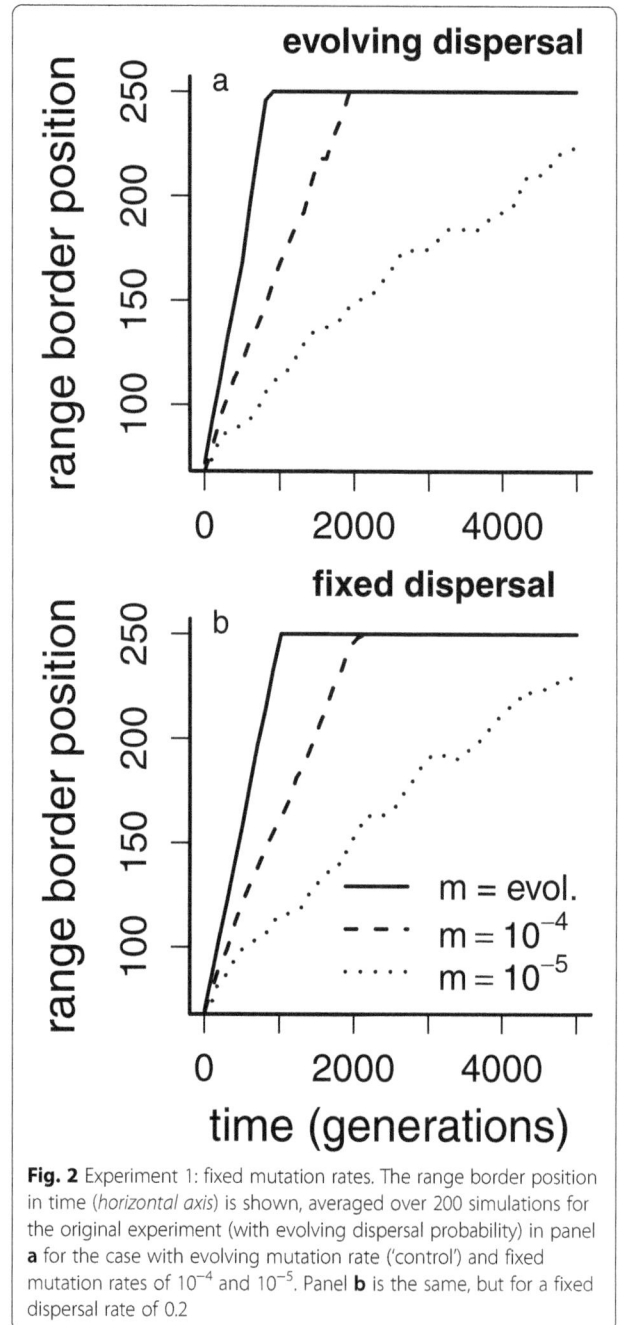

Fig. 2 Experiment 1: fixed mutation rates. The range border position in time (*horizontal axis*) is shown, averaged over 200 simulations for the original experiment (with evolving dispersal probability) in panel **a** for the case with evolving mutation rate ('control') and fixed mutation rates of 10^{-4} and 10^{-5}. Panel **b** is the same, but for a fixed dispersal rate of 0.2

almost 2. This has important consequences for 1) the invasion speed and population growth (Additional file 1: Figure S1A), with 5000 generations not being sufficient to colonize the whole landscape, and for 2) the pattern of genetic diversity at the dispersal locus (Additional file 1: Figure S1E), which is qualitatively different from the Ω = 0.1 simulations (Fig. 1e). The patterns in genetic diversity at the locus for temperature adaptation (Additional file 1: Figure S1D) and the neutral locus (Additional file 1: Figure S1F) only show quantitative differences with those in the base scenario of Ω = 0.1 in Fig. 1.

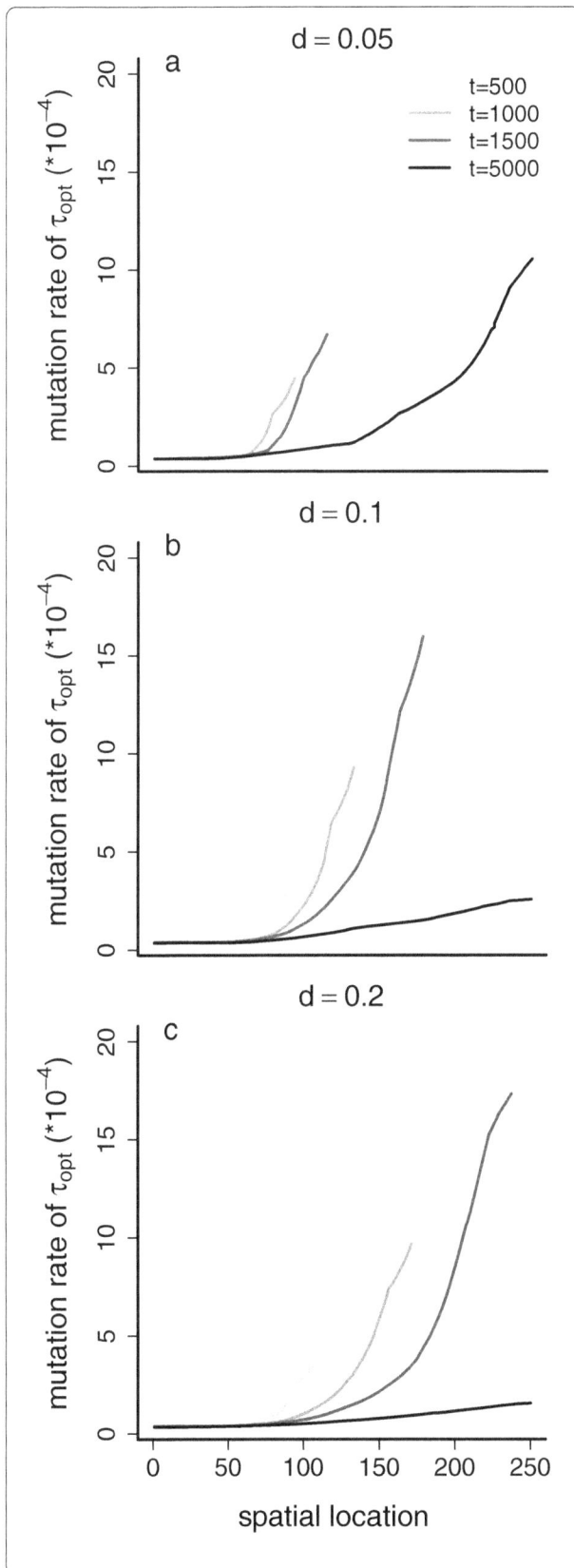

Fig. 3 Experiment 2: fixed dispersal probabilities. The average values of the mutation rate during and after range expansion across the gradient (*horizontal axis*) is shown in time (*gray scaling from light to dark, as time proceeds, which is given in a sequence of generations 500, 1000, 1500, 5000*) under **a**. a fixed dispersal probability of 0.05, **b**. a fixed dispersal probability of 0.1, and **c**. a fixed dispersal probability of 0.2

To determine the dependency of the evolution of high mutation rates on repeated colonizations we performed experiment 5. Here we omitted range expansion, applying a temporal temperature increase only, and added variation to the temperature gradient. For the spatially stable species' range (experiment 5a and b, Additional file 1: Figure S3), a strong directional selection through a steady increase of the temperature induces an increase of dispersal probability (a, panel B) and mutation rate in time (a, panel C). Under variable temperature increase however (Additional file 1: Figure S3b) the mutation rate could not evolve (C) and the dispersal probability hardly changed (B), because the required linkage disequilibrium cannot build up. Overall the population densities in the stable species range are lower (panels A in Additional file 1: Figure S3a/b, ~50% compared to ~80% in the expanding range). Under variable temperature increase, population growth is additionally affected by the non-directionally changing selection pressure, resulting in variable population densities in time (Additional file 1: Fig. S3b, panel A). Under the same scenario of variably increasing temperature, but returning to the expanding population (experiment 4, Additional file 1: Figure S4), the mutation rate contrastingly did increase during colonization (Additional file 1: Figure S4C), as in all other range expanding scenarios. This indicates that a scenario of range expansion can replace a strong directional selection [27] in the build-up of linkage disequilibrium that is required for the evolution of high mutation rates. The other panels of Additional file 1: Figure S4 show qualitatively the same patterns as under the original scenario of steadily increasing temperature across the landscape (compare with Fig. 1). The important difference to notice however is in the absolute level of genetic diversity at the adaptation (temperature) locus (Additional file 1: Figure S4D), which is much higher under variable temperature increase, because more temperature values are now required to maintain locally adapted. Comparing Fig. 1d and Additional file 1: Figure S4D additionally shows that mutation rates above a certain threshold do not add more genetic diversity at the neutral and local adaptation loci. This is in contrast to the dispersal locus, where a higher mutation rate causes a higher level of genetic diversity.

Discussion

In this study we investigate whether the evolution of mutation rates under the range expansion of a sexual species that needs to adapt to novel local temperature conditions can be an explanation for the 'genetic paradox of invasive species' [17], i.e. the ability of colonizing species to adapt to novel conditions while being subjected to genetic bottlenecks. We observe an increase of the mutation rate, which leads to a faster evolution of dispersal probability, faster adaptation to novel local temperature conditions, and thus faster range expansion. This also occurs when we apply variance to the mean temperature gradient in space, and even when we assume that 90% of the mutations are lethal. A simple analytical approach (Additional file 2) shows the increased fitness of individuals with higher mutation rates under large environmental changes, indicating that a stronger change in the environmental conditions should favor higher mutation rates in order to maximize the populations' fitness expectations. The simulation experiments with fixed dispersal probabilities (so non-evolving, Fig. 3) clearly indicate that when dispersal probability is allowed to evolve (in Fig. 1), it coevolves with the mutation rate: the mutation rate only reaches really high levels when the dispersal rate is high as well. So a high mutation rate allows the dispersal rate to evolve to high values, but the high dispersal rate then requires faster local adaptation, indirectly selecting for even higher mutation rates. The genetic modifier that increases the mutation rate can increase in frequency by hitchhiking on beneficial mutations, despite the independent inheritance of the three traits in sexual populations. This result is particularly interesting as selection for optimum mutation rate is mostly associated with asexual populations [20–22, 26]. Indeed, selection only operates on the dispersal and temperature loci, favoring mutations that increase the speed of range expansion. In sexual populations, strong linkage is required for the (advantageous) alleles at the these loci and the (high) mutation rate allele at the mutator locus to be inherited together, and as such to lead to indirect selection at the mutator locus [26, 27, 36]. In our study, however, all loci are genetically unlinked, but linkage disequilibrium (LD) between all pairs of the different loci is widespread, with LD build-up aided by spatial sorting, and being maintained and renewed under population structure. Hitchhiking on deleterious mutations causes a decrease in the frequency of the genetic modifier that increases mutation rates at a later stage, when the colonization process is completed. In a stable species' range, with the metapopulation not subjected to a range expansion but to a temporal increase of temperature, the mutation rate evolves as well, but not when variance is added to the mean temperature increase.

We have modeled the mutation rate as the probability that an inherited allele mutates. Since these mutation rates are caused by genes that are involved in processes of DNA repair and reproduction [23], high mutation rates will however likely affect the individual itself and not only its offspring as we modelled here. This effect is the result of mutations occurring when DNA is copied during the division of cells other than only the reproduction cells. Such mutations might then cause defects or tumors. Modeling mutation rates that negatively affect the individual's fitness and not only its offspring will likely affect our results, because high mutation rates are then more disadvantageous [20, 21]. On the other hand, there are large differences in mutation rates between (parts of) genomes [37] and DNA repair is not restricted to a single pathway [38]. In addition, if a high mutation rate only affects an individual's survival after reproduction, then high mutation rates might evolve despite their negative effects for the individual after it has completed reproduction.

Our results can be affected by the used genetic architecture, where linkage between traits [39, 40], polygeny, and the magnitude of mutations can be of importance in range dynamics [41–43]. The used mutation model of adding values to the inherited values result in mutations that are at most mildly deleterious at the adaptation locus, while an assumption of random mutations would invoke a stronger selection pressure [44]. However, our results are based on the assumption that 10 % of the mutations are lethal and we still see a significant increase of the mutation rate when we assume 90% lethal mutations (Additional file 1: Fig. S2). This is in contrast to what was found in an experimental study of sexual populations of yeast [45], which has however not taken a spatial perspective.

We observe that mutation rate can increase in combination with the increased dispersal probability and spatial variation, as experienced under range expansion. High dispersal rates, resulting in the immigration of many individuals, are expected to maintain a high local level of genetic variation [46], from which one would expect high levels of dispersal to be accompanied by a low local mutation rate. At the margin, however, relatedness amongst individuals increases at an advancing range front [9], reducing both local genetic diversity and the diversity of immigrants. Under these conditions an increase in the mutation rate evolves, which allows faster adaptation to the spatial variation in local temperature, causing a faster range expansion across the spatial gradient.

Holt et al. [46] investigated niche evolution at species' range margins and found that local evolution is hampered when source populations of immigrating individuals are at low density, as a result of the stochastic processes in such populations [47–49]. Next to population density, the mutation rate also affects the probability of niche evolution in their study [46]: higher dispersal is limiting local evolution in the sink population under a higher mutation rate, because of the increased numbers of maladapted individuals from the source. Holt et al. [46] did, however, not allow the joint evolution of mutation rate and dispersal rate, but instead used fixed rates. As a result, the dispersal rate does not decrease after colonization, while the conditions in the sink population make its persistence dependent on the constant influx of (maladapted) individuals, both in contrast to the model presented here.

In our study we investigate the evolution of mutation rates. Dealing with novel environmental conditions or increased evolvability is however not restricted to mutation rates, but can be modeled in different ways, e.g. an increased magnitude of the phenotypic effect of mutations [50], an epigenetic effect, the evolution of modularity [51], degeneracy [52], or the evolution of generalism or plasticity [53]. Kubisch et al. [9] showed that when dispersal is a means of adaptation, by tracking suitable conditions during periods of change, genetic adaptation does not occur. Which kind of adaptation to change can be expected under specific ecological and environmental conditions is an interesting field of future investigation.

There is an ever-expanding pool of literature discussing the ecological and evolutionary dynamics of dispersal in the formation of species ranges [54]. While individual-based models have recently greatly extended our theoretical knowledge of interactions and evolution of traits during range expansion, empirical data have been restricted to a few well-known cases [5, 11, 12, 55]. Increasing ecological realism in our models [56] can improve the predictability of theoretical phenomena which can then be tested by data from field studies. So far, increased dispersal has been shown to increase invasion speeds [11, 57], affect the fate of neutral mutations [58], as well as the level of local adaptation [59], and local population dynamics [33], and in addition causes strong patterns of spatial disequilibrium [34, 35].

Conclusions

In this study we show novel consequences of the particular genetic properties of populations under spatial disequilibrium, i.e. the coevolution of dispersal probability and mutation rate, even in a sexual species and under realistic spatial gradients, resulting in faster invasions. The evolution of mutation rates can therefore be added to the list of possible explanations for the 'genetic paradox of invasive species'. We conclude that range expansions and the evolution of mutation rates are in a positive feedback loop, with possibly far-reaching ecological consequences concerning invasiveness and the adaptability of species to novel environmental conditions.

Additional files

Additional file 1: Figures S1-S4. See supplementary file for full legends. **Figure S1.** Linkage disequilibrium. **Figure S2.** Experiment 3: 90% probability of lethal mutations. **Figure S3.** Experiment 5: no range expansion. **Figure S4.** Experiment 4: application of variation to the spatial temperature gradient. Analysis of linkage disequilibrium and extra simulations. (DOCX 671 kb)

Additional file 2: A simple analytical approach to calculate optimal mutation rates. Simple analytical approach showing fitness effects of different mutation rates (PDF 429 kb)

Acknowledgements
Many valuable discussions have contributed to the content of this manuscript.

Funding
This work was supported by the Open Program of the Netherlands Organization of Scientific Research and the Alexander von Humboldt Foundation (NWO 822.01.020 and AvH REF3.3-NLD-1189416-HFST-P for M.C.), and the German Science Foundation (KU 3384/1–1 for A.K.).

Authors' contributions
The idea for this study was a joint initiative by MC and AK, as was the model and experimental design. AK implemented the model and performed model analyses. MC performed the LD analysis and wrote the manuscript. OM provided the analytical approach. All authors read and approved the final manuscript.

Competing interests
The authors declare that they have no competing interests.

Author details
[1]Department of Animal Ecology, Netherlands Institute of Ecology (NIOO-KNAW), PO Box 50, 6700, AB, Wageningen, The Netherlands. [2]Theoretical Evolutionary Ecology Group, Institute for Animal Ecology and Tropical Biology, University of Würzburg, Emil-Fischerstr. 32, 97074 Würzburg, Germany. [3]Institute for Landscape and Plant Ecology, University of Hohenheim, August-von-Hartmann-Str. 3, 70599 Stuttgart, Germany.

References
1. Chen IC, Hill JK, Ohlemueller R, Roy DB, Thomas CD. Rapid range shifts of species associated with high levels of climate warming. Science. 2011; 333(6045):1024–6.
2. Cobben MMP, Verboom J, Opdam PFM, Hoekstra RF, Jochem R, Smulders MJM. Wrong place, wrong time: climate change-induced range shift across fragmented habitat causes maladaptation and decreased population size in a modelled bird species. Glob Change Biol. 2012;18:2419–28.
3. Excoffier L, Foll M, Petit RJ. Genetic consequences of range expansions. Annu Rev Ecol Evol Syst. 2009;40:481–501.
4. Hewitt GM. Some genetic consequences of ice ages, and their role in divergence and speciation. Biol J Linn Soc. 1996;58(3):247–76.

5.   Phillips BL, Brown GP, Webb JK, Shine R. Invasion and the evolution of speed in toads. Nature 2006; 439(7078):803–803.
6.   Klopfstein S, Currat M, Excoffier L. The fate of mutations surfing on the wave of a range expansion. Mol Biol Evol. 2006;23(3):482–90.
7.   Hill JK, Griffiths HM, Thomas CD. Climate change and evolutionary adaptations at species' range margins. Annu Rev Entomol. 2011;56:143–59.
8.   Shine R, Brown GP, Phillips BL. An evolutionary process that assembles phenotypes through space rather than through time. Proc Natl Acad Sci U S A. 2011;108(14):5708–11.
9.   Kubisch A, Fronhofer EA, Poethke HJ, Hovestadt T. Kin competition as a major driving force for invasions. Am Nat. 2013;181(5):700–6.
10. Kubisch A, Hovestadt T, Poethke H-J. On the elasticity of range limits during periods of expansion. Ecology. 2010;91(10):3094–9.
11. Thomas CD, Bodsworth EJ, Wilson RJ, Simmons AD, Davies ZG, Musche M, et al. Ecological and evolutionary processes at expanding range margins. Nature. 2001;411(6837):577–81.
12. Moreau C, Bhérer C, Vézina H, Jomphe M, Labuda D, Excoffier L. Deep human genealogies reveal a selective advantage to be on an expanding wave front. Science. 2011;334(6059):1148–50.
13. Garnier J, Roques L, Hamel F. Success rate of a biological invasion in terms of the spatial distribution of the founding population. Bull Math Biol. 2012; 74:453–73.
14. Cobben MMP, Verboom J, Opdam P, Hoekstra RF, Jochem R, Arens P, et al. Projected climate change causes loss and redistribution of genetic diversity in a model metapopulation of a medium-good disperser. Ecography. 2011; 34:920–32.
15. Baker HG, Stebbins GL. The Genetics of colonizing species, proceedings. London: Academic Press; 1965.
16. Frankham R. Challenges and opportunities of genetic approaches to biological conservation. Biol Conserv. 2010;143(9):1919–27.
17. Frankham R. Resolving the genetic paradox in invasive species. Heredity 2005; 94(4):385–385.
18. Stapley J, Santure AW, Dennis SR. Transposable elements as agents of rapid adaptation may explain the genetic paradox of invasive species. Mol Ecol. 2015;24(9):2241–52.
19. Bedau MA, Packard NH. Evolution of evolvability via adaptation of mutation rates. Biosystems. 2003;69(2):143–62.
20. Kimura M. On the evolutionary adjustment of spontaneous mutation rates. Genet Res. 1967;9(01):23–34.
21. Leigh EG Jr. Natural selection and mutability. Am Nat. 1970;104:301–5.
22. Leigh EG Jr. The evolution of mutation rates. Genetics. 1973;73:1–18.
23. Metzgar D, Wills C. Evidence for the adaptive evolution of mutation rates. Cell. 2000;101(6):581–4.
24. Sniegowski PD, Gerrish PJ, Lenski RE. Evolution of high mutation rates in experimental populations of E. coli. Nature. 1997;387(6634):703–5.
25. Taddei F, Radman M, Maynard-Smith J, Toupance B, Gouyon P-H, Godelle B. Role of mutator alleles in adaptive evolution. Nature. 1997;387(6634):700–2.
26. Sniegowski PD, Gerrish PJ, Johnson T, Shaver A. The evolution of mutation rates: separating causes from consequences. BioEssays. 2000;22(12):1057–66.
27. Johnson T. Beneficial mutations, hitchhiking and the evolution of mutation rates in sexual populations. Genetics. 1999;151(4):1621–31.
28. Gillespie JH. Mutation modification in a random environment. Evolution. 1981:468–76.
29. Ishii K, Matsuda H, Iwasa Y, Sasaki A. Evolutionarily stable mutation rate in a periodically changing environment. Genetics. 1989;121(1):163–74.
30. Beverton R, Holt S: On the dynamics of exploited fish populations.: Chapman and Hall; 1957.
31. Bonte D, Van Dyck H, Bullock JM, Coulon A, Delgado M, Gibbs M, et al. Costs of dispersal. Biol Rev. 2012;87(2):290–312.
32. Rousset F. genepop'007: a complete re-implementation of the genepop software for windows and Linux. Mol Ecol Resour. 2008;8(1):103–6.
33. Ronce O. How does it feel to be like a rolling stone? Ten questions about dispersal evolution. Annu Rev Ecol Evol Syst. 2007:231–53.
34. Cobben MMP, Verboom J, Opdam PFM, Hoekstra RF, Jochem R, Smulders MJM. Spatial sorting and range shifts: consequences for evolutionary potential and genetic signature of a dispersal trait. J Theor Biol. 2015;373:92–9.
35. Phillips BL, Brown GP, Shine R. Life-history evolution in range-shifting populations. Ecology. 2010;91(6):1617–27.
36. Tenaillon O, Le Nagard H, Godelle B, Taddei F. Mutators and sex in bacteria: conflict between adaptive strategies. Proc Natl Acad Sci U S A. 2000;97(19):10465–70.
37. Drake JW, Charlesworth B, Charlesworth D, Crow JF. Rates of spontaneous mutation. Genetics. 1998;148(4):1667–86.
38. Rottenberg S, Jaspers JE, Kersbergen A, van der Burg E, Nygren AO, Zander SA, et al. High sensitivity of BRCA1-deficient mammary tumors to the PARP inhibitor AZD2281 alone and in combination with platinum drugs. Proc Natl Acad Sci U S A. 2008;105(44):17079–84.
39. Blows MW, Hoffmann AA. A reassessment of genetic limits to evolutionary change. Ecology. 2005;86(6):1371–84.
40. Hellmann JJ, Pineda-Krch M. Constraints and reinforcement on adaptation under climate change: selection of genetically correlated traits. Biol Conserv. 2007;137(4):599–609.
41. Gomulkiewicz R, Holt RD, Barfield M, Nuismer SL. Genetics, adaptation, and invasion in harsh environments. Evol Appl. 2010;3(2):97–108.
42. Kawecki TJ. Adaptation to marginal habitats: contrasting influence of the dispersal rate on the fate of alleles with small and large effects. Proc R Soc Biol Sci Ser B. 2000;267(1450):1315–20.
43. Kawecki TJ. Adaptation to marginal habitats. Annu Rev Ecol Evol Syst. 2008; 39:321–42.
44. Sanjuán R, Moya A, Elena SF. The distribution of fitness effects caused by single-nucleotide substitutions in an RNA virus. Proc Natl Acad Sci U S A. 2004;101(22):8396–401.
45. Raynes Y, Gazzara MR, Sniegowski PD. Mutator dynamics in sexual and asexual experimental populations of yeast. BMC Evol Biol. 2011;11(1):158.
46. Holt RD, Barfield M. Theoretical perspectives on the statics and dynamics of species' borders in patchy environments. Am Nat. 2011;178(S1):S6–S25.
47. Bridle JR, Polechová J, Kawata M, Butlin RK. Why is adaptation prevented at ecological margins? New insights from individual-based simulations. Ecol Lett. 2010;13(4):485–94.
48. Pearson GA, Lago-Leston A, Mota C. Frayed at the edges: selective pressure and adaptive response to abiotic stressors are mismatched in low diversity edge populations. J Ecol. 2009;97(3):450–62.
49. Turner JR, Wong H. Why do species have a skin? Investigating mutational constraint with a fundamental population model. Biol J Linn Soc. 2010; 101(1):213–27.
50. Griswold C. Gene flow's effect on the genetic architecture of a local adaptation and its consequences for QTL analyses. Heredity. 2006;96(6):445–53.
51. Kashtan N, Parter M, Dekel E, Mayo AE, Alon U. Extinctions in heterogeneous environments and the evolution of modularity. Evolution. 2009;63(8):1964–75.
52. Whitacre J, Bender A. Degeneracy: a design principle for achieving robustness and evolvability. J Theor Biol. 2010;263(1):143–53.
53. Lee CE, Gelembiuk GW. Evolutionary origins of invasive populations. Evol Appl. 2008;1(3):427–48.
54. Kubisch A, Holt RD, Poethke HJ, Fronhofer EA. Where am I and why? Synthesizing range biology and the eco-evolutionary dynamics of dispersal. Oikos. 2014;123(1):5–22.
55. Fronhofer EA, Altermatt F. Eco-evolutionary feedbacks during experimental range expansions. Nat Commun. 2015;6:6844.
56. Cobben MMP, Verboom J, Opdam PFM, Hoekstra RF, Jochem R, Smulders MJM. Landscape prerequisites for the survival of a modelled metapopulation and its neutral genetic diversity are affected by climate change. Landsc Ecol. 2012;27:227–37.
57. Phillips B, Brown G, Shine R. Evolutionarily accelerated invasions: the rate of dispersal evolves upwards during the range advance of cane toads. J Evol Biol. 2010;23(12):2595–601.
58. Travis J, Münkemüller T, Burton O. Mutation surfing and the evolution of dispersal during range expansions. J Evol Biol. 2010;23(12):2656–67.
59. Kubisch A, Degen T, Hovestadt T, Poethke HJ. Predicting range shifts under global change: the balance between local adaptation and dispersal. Ecography. 2013;36(8):873–82.

6

Plant manipulation through gall formation constrains amino acid transporter evolution in sap-feeding insects

Chaoyang Zhao[1] and Paul D. Nabity[2*] (iD)

Abstract

Background: The herbivore lifestyle leads to encounters with plant toxins and requires mechanisms to overcome suboptimal nutrient availability in plant tissues. Although the evolution of bacterial endosymbiosis alleviated many of these challenges, the ability to manipulate plant nutrient status has evolved in lineages with and without nutritional symbionts. Whether and how these alternative nutrient acquisition strategies interact or constrain insect evolution is unknown. We studied the transcriptomes of galling and free-living aphidomorphs to characterize how amino acid transporter evolution is influenced by the ability to manipulate plant resource availability.

Results: Using a comparative approach we found phylloxerids retain nearly all amino acid transporters as other aphidomorphs, despite loss of nutritional endosymbiosis. Free living species show more transporters than galling species within the same genus, family, or infraorder, indicating plant hosts influence the maintenance and evolution of nutrient transport within herbivores. Transcript profiles also show lineage specificity and suggest some genes may facilitate life without endosymbionts or the galling lifestyle.

Conclusions: The transcript abundance profiles we document across fluid feeding herbivores support plant host constraint on insect amino acid transporter evolution. Given amino acid uptake, transport, and catabolism underlie the success of herbivory as a life history strategy, this suggests that plant host nutrient quality, whether constitutive or induced, alters the selective environment surrounding the evolution and maintenance of endosymbiosis.

Keywords: Herbivore, Endosymbiosis, Phylloxeridae, Effector, Aphid, Sternorrhyncha

Background

To subsist as an herbivore, an organism must overcome substantial barriers in the form of physical or chemical plant defenses and less than optimal nutrient availability. In some instances, the plant defenses interact directly with nutrient availability by decreasing uptake (e.g., plugged sieve tubes) or impeding digestion (e.g., protease inhibitors), although myriad mechanisms have been described for how herbivores adapt to or avoid defenses [1]. In addition to these deterrents, plant tissues typically maintain high carbon to nitrogen ratios, and plant fluids are depleted in many essential amino acids, making it more difficult for herbivores to acquire nitrogen-based nutrients. To overcome these dietary limitations, herbivores evolved partnerships with bacteria that facilitated transitions to new feeding niches, e.g., on phloem or xylem, or otherwise augmented plant palatability by attenuating defenses [2].

Symbioses can fail, however, when symbiont genomes degrade [3] or limit host range (e.g., plant choice, thermal tolerance; [4]). Thus, there is likely selection pressure to either replace symbionts with more efficient ones [5], or to evolve novel feeding strategies to avoid symbiont dependence. Indeed, several hemipteran lineages, including leafhoppers (Membracoidea: Typhlocybinae) and the Phylloxeridae (Sternorrhyncha: Phylloxeroidea), have transitioned to novel plant-feeding strategies and lost their obligate symbiont associations [4]. The transitions in and out of symbioses have left genomic signatures such as reduced genome structure and function for many obligate symbionts [3, 4], although the effects of symbiosis on herbivore genomes with or without symbionts is unknown.

* Correspondence: pauln@ucr.edu
[2]Department of Botany and Plant Sciences, University of California, Riverside, 900 University Avenue, Batchelor Hall room 2140, Riverside, CA 92521, USA
Full list of author information is available at the end of the article

The metabolic coordination in amino acid synthesis and usage between bacteria and host requires amino acid transporters (AATs) that function in transporting amino acids across the insect/symbiont interface, membranes that separate the cytoplasm of symbionts from insect hemolymph [6]. Two types of AATs mediate this transport: the amino acid polyamine organocation transporter superfamily (APC, transporter classification #2.A.3) and the amino acid/auxin permease transporters family (AAAP, TC #2.A.18). Although both groups belong to the APC superfamily, members of the AAAP family have relatively divergent amino acid sequences, varying substrate specificities, and 11 transmembrane domains, compared to other transporters of the APC family [7, 8]. Expression profiles of these two families of AAT genes for several herbivorous species and their bacterial endosymbionts support a role for these transporters in the evolution of nutritional endosymbiosis [9, 10].

A growing body of evidence has demonstrated that insects induce nutrient sinks in plants in the form of galls that abundantly supply minerals, carbohydrates, and free amino acids [11–16]. Given that numerous insect taxa form galls [17], an intriguing question arises: how does the accessibility of gall-enriched nutritive compounds influence the evolution of insect hosts and/or their symbionts? Among the Sternorrhyncha, few lineages secondarily lost endosymbionts concurrent with a shift to parenchyma feeding [4], and some taxa, such as the Phylloxeridae also induce galls. Insects within the Phylloxeridae are considered sister to the aphids (Aphidoidea: Aphididae) and adelgids (Phylloxeroidea: Adelgidae), groups that also retain galling and free-living species [18]. In contrast with aphids and adelgids that harbor symbionts in bacteriocytes, Phylloxeridae species lack stable intracellular symbionts [19, 20]. Further, Phylloxeridae comprises numerous life history strategies, including galling and free-living species that allow a phylogenetically controlled comparison to understand how these strategies arose with respect to their nutrient acquisition and metabolism. As an important grape pest worldwide, the grape phylloxera (*Daktulosphaira vitifoliae*) is capable of making leaf and root galls and its interaction with plant hosts has been the most investigated among the Phylloxeridae. Studies showed that infestation of *D. vitifoliae* reprograms plant metabolism, leading to the accumulation of nutrients such as carbohydrates and free amino acids [21–23]. Recently, *D. vitifoliae* AATs were compared to paralogs in aphids to help pinpoint which transporters underlie the maintenance of nutrient symbiosis between aphids and *Buchnera* with an emerging conclusion that ecological context may contribute to AAT gene copy number and evolution [10].

To expand the understanding of amino acid metabolism associated with herbivorous insects, we compared species that manipulate plant host amino acid content by gall forming to free-living species, and among species with and without stable nutritional endosymbionts. We sequenced the transcriptomes of nine Phylloxeridae species including *D. vitifoliae* and eight from the genus *Phylloxera*. Oak phylloxera (*P. quercus*) has a free-living life history and thus was compared to other galling phylloxerid species regarding AAT evolution whereas two aphid species whose genomes are sequenced were compared to four other galling aphids. Our results indicated that galling insects, in Phylloxeridae and among aphidomorphs, experienced increased constraints on the evolution of AATs likely because of their ability to manipulate plant host metabolism.

Methods
Insect collection
We collected nine known species within the Phylloxeridae for RNA sequencing. Phylloxeridae is a sister family of Adelgidae under the superfamily Phylloxeroidea, which is sister to the Aphidoidea: Aphididae all within the suborder Sternorrhyncha (Fig. 1). These nine species include three that gall stems/petioles (*Phylloxera caryaecaulis, P. subelliptica, P. caryaemagna*), one that feeds across hosts causing crinkles/folds in leaf veins (*P. caryaevenae*), three that form spheres on leaves (*P. caryaefallax, P. foveata, P. foveola*), one freely living (*P. quercus*), and one that galls both roots and leaves of *Vitis* species (*Daktulosphaira vitifoliae*). In contrast to *P. quercus* that lives freely on oak trees (*Quercus spp.*), all the other *Phylloxera* species induce galls on different hickory (*Carya*) species and/or tissues (Table 1). Although the description of *P. foveata* places it on *C. cordiformis* [18] and the individuals collected in this study came from *C. glabra*, we are considering the insects to be the same species for this study because of similarity in the induced phenotype. The Phylloxeridae represents an unresolved taxon [but see 24], where ongoing research is delineating species. All *Carya* originating phylloxerids were collected from the Arnold Arboretum of Harvard University, Boston, Massachusetts. *Phylloxera quercus* was collected from *Quercus sp.* at a horticultural nursery in Bellevue, WA. *Daktulosphaira vitifoliae* was collected from native grapes near Madera Canyon, Arizona. Collected insect samples were stored in RNAlater solution (Qiagen) at room temperature initially, transferred to 4 °C within eight hours for temporal storage (≤ 7 days), and later kept at −80 °C until RNA isolation. Insects at all stages (including eggs) were collected initially but separated to include only juveniles and adults for sequence analysis.

Transcriptome sequencing and assembly
We used 10–20 individuals for RNA extraction per biological replicate, and performed two replicates for *P. caryaefallax, P. subelliptica, P. caryaemagna, P. caryaecaulis, P. quercus* and *D. vitifoliae*, and one replicate for *P. foveata,*

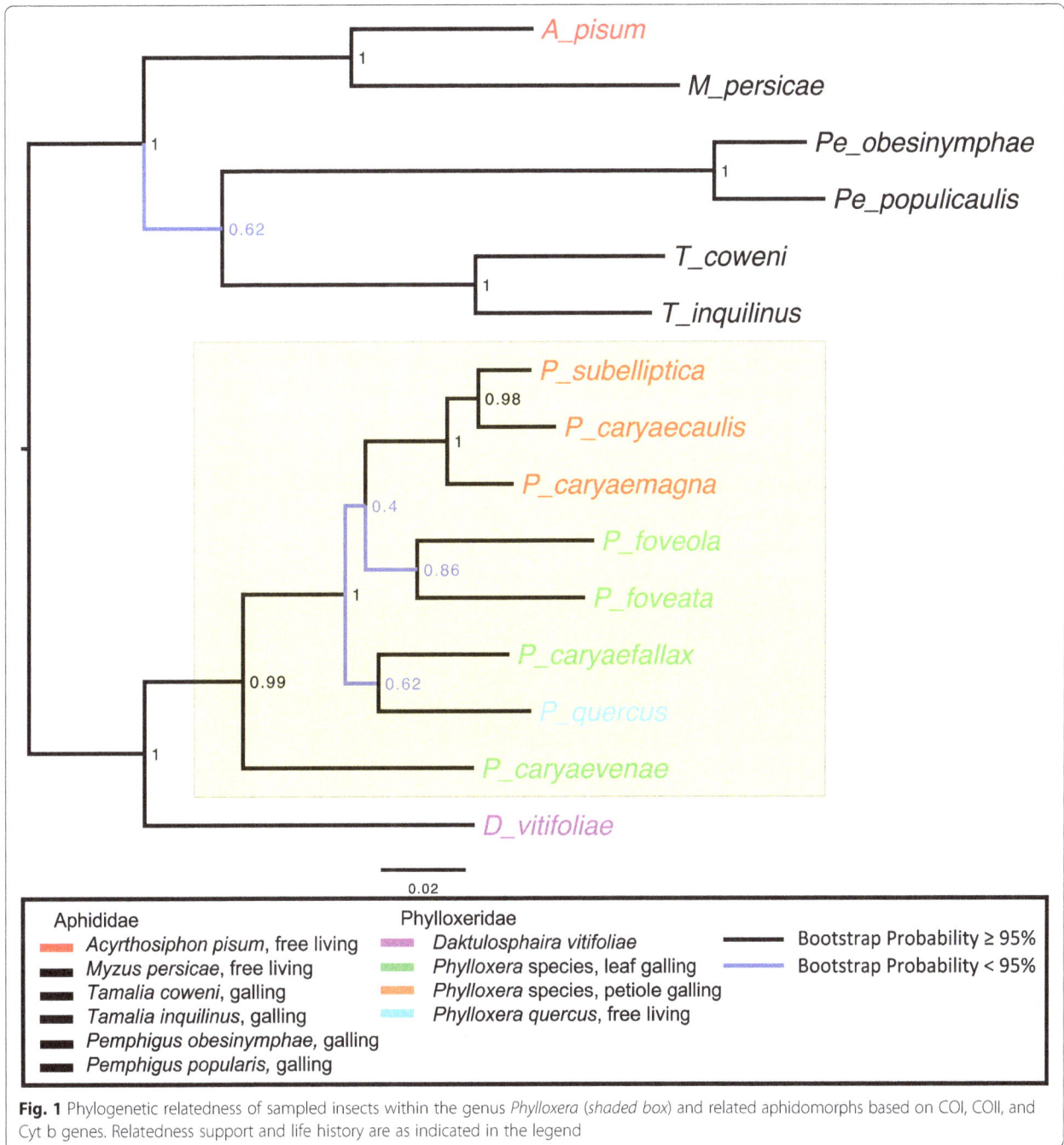

Fig. 1 Phylogenetic relatedness of sampled insects within the genus *Phylloxera* (*shaded box*) and related aphidomorphs based on COI, COII, and Cyt b genes. Relatedness support and life history are as indicated in the legend

P. foveola and *P. caryaevenae*. Whole bodies of all insect specimens were homogenized in RTL buffer (Qiagen) and then processed for total RNA extraction using RNeasy Mini kit (Qiagen) following the protocol provided. RNA integrity was examined using a fragment analyzer and samples of the RNA integrity number (RIN) > 7 were used for sequencing. The mRNA library construction and RNA sequencing were performed at the Genomics Core, Washington State University, Spokane, Washington. Briefly, mRNA molecules were enriched using the oligo-dT beads and libraries were constructed for paired-end sequencing on the Illumina

HiSeq platform. Raw reads were adaptor-trimmed and filtered to a minimum quality score of 30 over 95% of the read. A single transcriptome reference was generated for each taxon by assembling filtered reads in Trinity with default setting (version 2.1.1; [25]) and assembled sequences were subsequently clustered at a minimum identity of 95% using CD-HIT-EST included in the CD-HIT package (version 4.6.1; [26]).

Raw RNA reads of five Aphididae species, including the green peach aphid (*Myzus persicae*) and four plant-gall related species: *Pemphigus obesinymphae*, *Pe. populicaulis*,

Table 1 Number of amino acid transporters (AAT) in sampled insects relative to their life history

Species	Life History	Host	APC	AAT AAAP	Total
Daktulosphaira vitifoliae*	Root & leaf galls	*Vitis*	11	12	23
Phylloxera caryaecaulis	Petiole gall	*Carya glabra*	10	16	26
P. caryaemagna	Petiole gall	*C. cordiformis*	10	15	25
P. subelliptica	Petiole gall	*C. ovata*	10	12	22
P. caryaevenae	Leaf gall: fold	*C. glabra*	8	12	20
P. foveola	Leaf gall: round	*C. glabra*	7	14	21
P. foveata	Leaf gall: round	*C. glabra*	13	17	30
P. caryaefallax	Leaf gall: round	*C. ovata*	9	14	23
P. quercus	Free-living	*Quercus*	19	24	43
Acyrthosiphon pisum*	Free-living	Fabaceae	18	21	39
Myzus persicae*	Free-living	Diverse	16	20(20)	36
Pemphigus obesinymphae	Petiole gall	*Populus*	11(12)	14(14) + 4	29
Pemphigus populicaulis	Petiole gall	*Populus*	12	18	30
Tamalia coweni	Leaf gall	*Arctostaphylos*	12(9)	16(13) + 5	33
Tamalia inquilinus	Inquiline	*Arctostaphylos*	10	17	27
Average	Galling		10	14	24
Average	Free-living		17	21	38
Outgroups: Sternorrhyncha					
Planococcus citri	Free-living	Diverse	10	28	38
Bemisia tabaci	Free-living	Diverse	12	24	36
Bactericera cockerelli	Free-living	Diverse	10	25	28
Outgroup: Non-herbivore					
Drosophila melanogaster	Free-living	NA	10	15	25

All gene counts for insects in this study followed the substitution rate method described in the methods except where genomes were available (*). Gene counts are compared to those from [9] show within parentheses, which used a similar gene coalescing method. Some AAAP genes that are expanded in non-arthropods were not reported in previous studies but were in the current study and are designated using "+". Outgroup Sternorrhyncha herbivores gene counts (from [9]) and non-herbivore *Drosophila melanogaster* are shown

Tamalia coweni and *T. inquilinus*, were downloaded from the NCBI database (BioProject # PRJNA296778 for *M. persicae*, BioProject # PRJNA301746 for the two *Pemphigus* species, and BioProject # PRJNA297665 for the two *Tamalia* species). Unlike *Pe. obesinymphae, Pe. populicaulis,* and *T. coweni* that induce galls in plant tissues, *T. inquilinus* does not induce galls but inhabits galls induced by other galling insects [27]. De novo transcriptome references were generated for these five species using Trinity and CD-HIT-EST as described above. The *M. persicae* and draft *D. vitifoliae* genomes available from BIPAA (http://bipaa.genouest.org/is/) were used to compare results from the de novo assembled transcriptomes to help assess how accurate transcript counts were to the true number of annotated genes; however, only *M. persicae* sequences used in this study were taken from the available genome.

Amino acid transporter annotation
Amino acid transporters in the APC (TC #2.A.3) and AAAP (TC # 2.A.18) families were annotated for all

phylloxerid and downloaded aphid sequences following the previously described methods [9, 10, 28]. All bioinformatics tools used here were run at default setting unless explicitly stated. Briefly, longest open reading frames (ORFs) for all transcripts were predicted and translated into protein sequences using a stand-alone PERL script [29]. The protein sequences were searched against the Pfam domain database (Pfam29.0) for functional domains PF03024 (APC) and PF01490 (AAAP) (evalue <0.001) using the HMMSCAN program included in the HMMER software suite (version 3.1b1, [30]). Transcripts with HMMER APC or AAAP hits were verified subsequently by BLAST searching (evalue <0.001) against the NCBI non-redundant protein database. We excluded transcripts derived from possible plant tissue contaminants or other organisms that co-inhabit within the galls induced by Phylloxeridae species, and those of non-APCs or -AAAPs such as Na-K-Cl cotransporters by retaining only transcripts whose best BLAST hits were hemipteran APC or AAAP members.

Because RNA sequencing and assembling approaches assign unique sequence ID for each splicing variant and truncated transcript that are encoded by same gene loci, the identified amino acid transporter transcripts were subsequently collapsed into putative representative loci following the methods previously described [9, 10]. The genomes of *M. persicae* and *D. vitifoliae* were used to map amino acid transporter transcripts to genome scaffold locations using BLASTN searches. Transcripts mapping to the same location were collapsed into the one encoding the longest ORF, or, when partial- or non-overlapping, merged into a single locus. To recover all possible AATs that are encoded by the genomes but were not identified from *M. persicae* and *D. vitifoliae* de novo transcriptome assemblies, we performed BLAST searches (evalue <0.001) using an APC or AAAP transcript of *M. persicae* and *D. vitifoliae*, respectively, against their own genome databases, and those recovered, if any, were subsequently verified at the NCBI non-redundant protein database as described above. For the remaining species lacking draft genome sequences, we: 1) collapsed transcripts having the same Trinity component number into the one encoding the longest ORF, and 2) collapsed closely related transcripts into the one encoding the longest ORF if they have a pairwise synonymous substitution rate (Ks value) less than 0.25 [9] determined using PAML (version 4.8; [31]) or if two transcripts have less than 50-bp of overlapping region, as performed in [10]. All chosen representative transcripts were translated into the longest protein sequences in Blast2GO Pro [32]. Amino acid transporters encoded by *Acyrthosiphon pisum* and *Drosophila melanogaster* genomes were annotated and previously reported [28].

Phylogenetic analyses

We used DNA sequences of three protein-coding mitochondrial genes, cytochrome c oxidase subunit I (COI), cytochrome c oxidase subunit II (COII) and cytochrome b (CYTB), to resolve the phylogenetic relationship among the nine Phylloxera species and six Aphididae species as described above. COI and COII are widely used to infer insect phylogeny at a variety of hierarchy levels, from closely related species to orders, and CYTB is fast-evolving and thus useful for the phylogenetic analysis of closely-related taxa [33].

The DNA sequences of these three genes were either retrieved from the de novo transcriptomes we assembled or downloaded from the Genbank database (accession # FJ411411.1 for three *A. pisum* genes; accession # NC_029727.1 for three *M. persicae* genes; accession # AM748716.1 for *Pe. obesinymphae* COII). Three gene sequences (COI, COII and CYTB) in each taxon were concatenated to a single one and then aligned using MAFFT (version 7.130) with 'auto' setting [34]. The poorly aligned and divergent regions were eliminated on the Gblocks server with default settings [35]. The best-fit nucleotide substitution model was determined in MEGA6 [36], using GTR + G + I. The maximum likelihood method was then run in MEGA6 to construct the phylogenetic trees by testing 1000 bootstrap replications [36].

Phylogenetic analyses of AATs were performed using putative APC protein sequences and AAAP arthropod expanded clade sequences, respectively. AAAP members are composed of the arthropod and non-arthropod expanded clades, between which the sequences are highly divergent [9, 10, 28]. The arthropod expanded clade was so designated because of its multiple gene duplications in the common ancestor of arthropods in contrast to those AAAPs that fall outside this clade [28]. Two *A. pisum* Na-K-Cl transporters (ACYPI001649 and ACYPI007138) and two human SLC36 proteins (SLC36A1 and SLC36A2), which were previously used as outgroups for the phylogenetic analyses of APC and AAAP members, respectively [9, 10, 28], were used likewise in this study. Sequences were aligned using MAFFT with 'auto' setting and the alignments were trimmed using TRIMAL (version 1.2) based on a gap threshold of 0.25 [37]. We used MEGA6 to determine the best-fit models of protein evolution, which are LG + G + F for APC proteins and LG + G for AAAP proteins. Because the LG model is not available in the phylogenetics program MRBAYES, we chose the WAG + G + F model for APC and WAG + G for AAAP arthropod expanded clade proteins, and ran the analyses using two runs with 4 chains per run in MRBAYES (version 3.2.1) until the standard deviation of split frequencies between runs dropped below 0.05. The first 25% of generations were discarded and the remaining generations were used to build a 50% majority-rule consensus tree. Lastly, we used the same alignment from above to perform a maximum likelihood inference (RAxML-HPC2) on XSEDE in the CIPRES computing environment [38] for comparison and to generate a consensus topology.

To test if phylogenetically dependent gene families differed per life history for transcript counts within a gene family, we used a PGLS model (counts ~ life history) with a Brownian correlation and a phylogenetic tree using the mitochondrial sequences generated above. Unique AAT sequences were counted for each gene family for each insect used in this study (see Fig. 1) and combined with known counts from three free-living Sternorrhyncha [10] to increase sample size prior to assessing for differences, as a conservative approach. The mitochondrial sequences for the additional insects were obtained from NCBI (NC_030055.1; *Bactericera cockerelli*, KU877168.1; *Bemisia tabaci*, KP692637.1 and AY691419.1; *Planococcus citri*). Sequences were aligned, concatenated using Gblocks to identify conserved mitochondrial sequences, and aligned for a final tree output using RAxML on the CIPRES environment, as described above. The PGLS model was run using the R computing environment and the library 'picante' [39].

Results

Our de novo transcriptome assemblies revealed similar contig numbers and quality metrics as other previously assembled aphidomorph transcriptomes, and among species sequenced for this study (Additional file 1: Table S1). Based upon these assemblies, we present the first multi-locus (CO1, CO11, CYTB) phylogeny across multiple species within the Phylloxeridae (Fig. 1). Our phylogeny indicated phylloxerids first colonized leaves and diversified across host species. Then one lineage evolved to feed on petioles and diversified as it recolonized this tissue across *Carya* species. Thus, species feeding on different tissues of the same plant are likely more distantly related than species feeding on the same tissue of different host species. We also noted that *P. quercus* may represent a unique host switching event given it feeds on oak yet is nested among other hickory feeding *Phylloxera*.

In comparison with other Sternorrhyncha insects, nearly all phylloxerids retained fewer AATs (Table 1). The one exception is the free-living *P. quercus* where more AATs are abundant than other aphidomorphs and nearly twice the number than in other phylloxerids. Free-living aphids (*A. pisum* and *M. persicae*) also retained more AATs than related gall-feeding aphids in *Pemphigus* or *Tamalia* genera. For APC transporters, gall-feeding insects showed 7–13 (mean = 10) APC transporters whereas free-living aphidomorphs showed more transporters (15–19, mean = 17). Two clades (yellow boxes; Fig. 2) likely contain phylloxerid-specific duplications, where either two genes were present in the phylloxerid ancestor compared to one for the aphid ancestor (upper yellow box), or two phylloxerids (*P. quercus* and *P. foveata*) show duplications compared to a lack of this in aphids (lower yellow box). All other phylloxerid transcripts either strongly clustered with annotated aphid genes, or likely cluster, as indicated by lower bootstrap values. In a comparison of whether life history depended on the number of AATs, both APC and AAAP families showed galling herbivores retained fewer APC and AAAP than free-living species (F = 32.3, $P < 0.0001$); F = 21.4, $P < 0.0001$, respectively). In comparison with genome data, both *M. persicae* and *D. vitifoliae* transcript counts matched the AAT annotated genes with two exceptions for *M. persicae*: 1) two related transcripts mapped to the same gene in the genome, likely caused by an assembly error, rather than indicating a duplication, 2) and two clades where all aphids except *M. persicae* showed gene copies actually had *M. persicae* genes. Thus, the transcript counts were identical to the genome *D. vitifoliae*, and nearly so for *M. persicae*.

For AAAP transporters, aphidomorphs showed similar abundance profiles, yet have fewer genes than other Sternorrhyncha (see Table 1). However, among aphidomorphs, free-living species showed more AAAP transcripts than galling species, with nearly twice the number in *P. quercus* than other phylloxerids. Among Arthropoda-specific AAAPs, *P. quercus* showed more representative transcripts for seven gene clades whereas free-living aphids differed from galling aphids more variably, with more transcripts in clade two (Additional file 1: Table S2, Fig. 3). Some lineage specificity occurred with *Pemphigus* showing more AAAPs in clade three and *Tamalia* more in clade four. Among non-arthropod genes free-living aphidomorphs show more transcripts than galling aphidomorphs across clades (Additional file 1: Table S2).

Discussion

Amino acid uptake, transport, and catabolism underlie the success of herbivory as a life history strategy [29, 40]. Here we present the first multigene tree for members within the Phylloxeridae; a family with both galling and free-living herbivores. We also present transcript profiles across fluid feeding herbivores that support plant host constraint on insect amino acid transporter evolution. Galling sap-feeding insects show fewer AAT transcripts than free-living species within the same insect families and within the same genus of *Phylloxera*. The ability of galling insects to manipulate plant nutrient content likely altered selection to retain or duplicate the number of functioning AATs within the insect. Previous research suggests some AATs facilitate the evolution of endosymbioses but also that ecological context may interact with nutrient transporter evolution to shape adaptive duplication or loss [10]. Our data advance this idea by highlighting how complex the selective environment is and suggest specialized interactions with plants play a large role in determining the evolution of herbivore genomes, especially when nutrient manipulating strategies are involved.

Previous research on some AATs correlates gene expression and presence with the maintenance of endosymbioses; however, phylloxerids lack stable endosymbionts and still retain many of these same AATs. We found members of the Phylloxeridae family retain at least one copy of each APC found among other aphidomorphs (as in [10]) with the exception of two clades (yellow boxes; Fig. 2) that show duplications. Otherwise, phylloxerids retained at least one APC similar to many other Sternorrhyncha insects and *D. melanogaster* [28]. Interestingly, free-living *P. quercus* often showed multiple AAT copies within clades where galling phylloxerids possessed only one copy (blue asterisks; Fig. 2). Free-living aphids also show a similar pattern compared to galling aphids for many clades (black and red asterisks; Fig. 2). This increase within clades suggests that these paralogs may function generally to support nutrient transport when feeding on host parenchyma, a

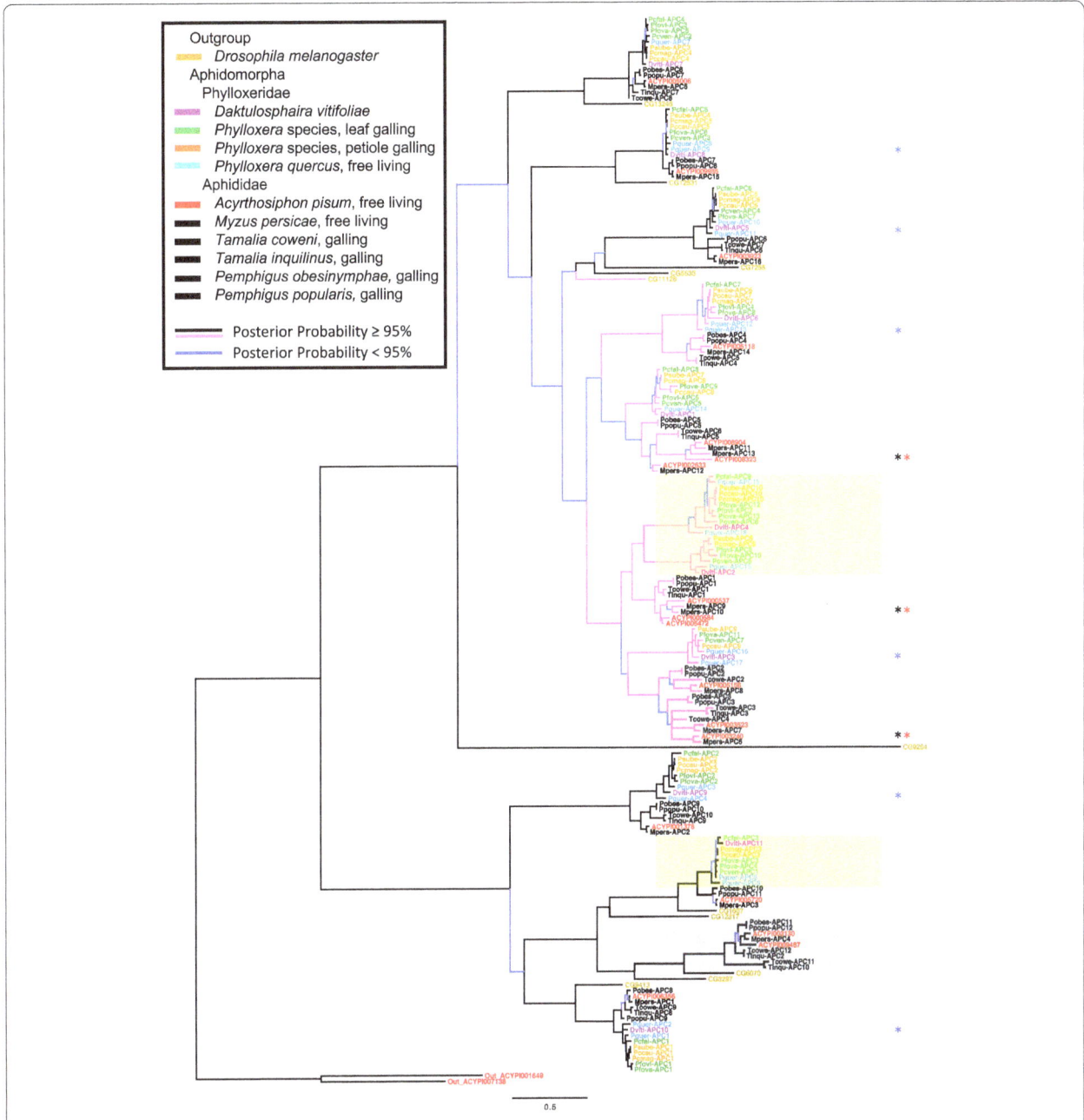

Fig. 2 Phylogeny of APC family transporters for sampled insects, representative aphids, and a nonherbivore outgroup. Taxa are color coded per the key, with pink branches denoting the *slimfast* gene expansion shared among aphidomorphs. Shaded regions indicate likely lineage-specific evolution for free-living aphids (*gray boxes*) and phylloxerids (*yellow boxes*). *Asterisks* indicate clades where paralogs occur in free-living aphidomorphs but are lacking in galling species

tissue where nutrients are lower than when feeding on gall tissue where nutrients can be enriched by and for the galling insect. We hypothesize then that host nutrient manipulation altered the selection environment to maintain certain AATs. In support of this we identified fewer AATs in galling insects than free-living relatives. In some instances, galling phylloxerids did not accumulate specific AAT transcripts; however, lack of accumulation may result from a missing gene or lack of conditions under which

expression occurs. While we recognize the limitations of transcriptome information to resolve this, the use of the *D. vitifoliae* and *M. persicae* genomes suggests all phylloxerid and nearly all aphid genes were accounted for, and that variation in AATs among genera occurs within aphidomorphs. Some *Phylloxera* species show accumulation of AATs absent from the *D. vitifoliae* genome whereas galling *Phylloxera* species show different numbers of AAT transcripts across clades. Similarly, some galling aphids

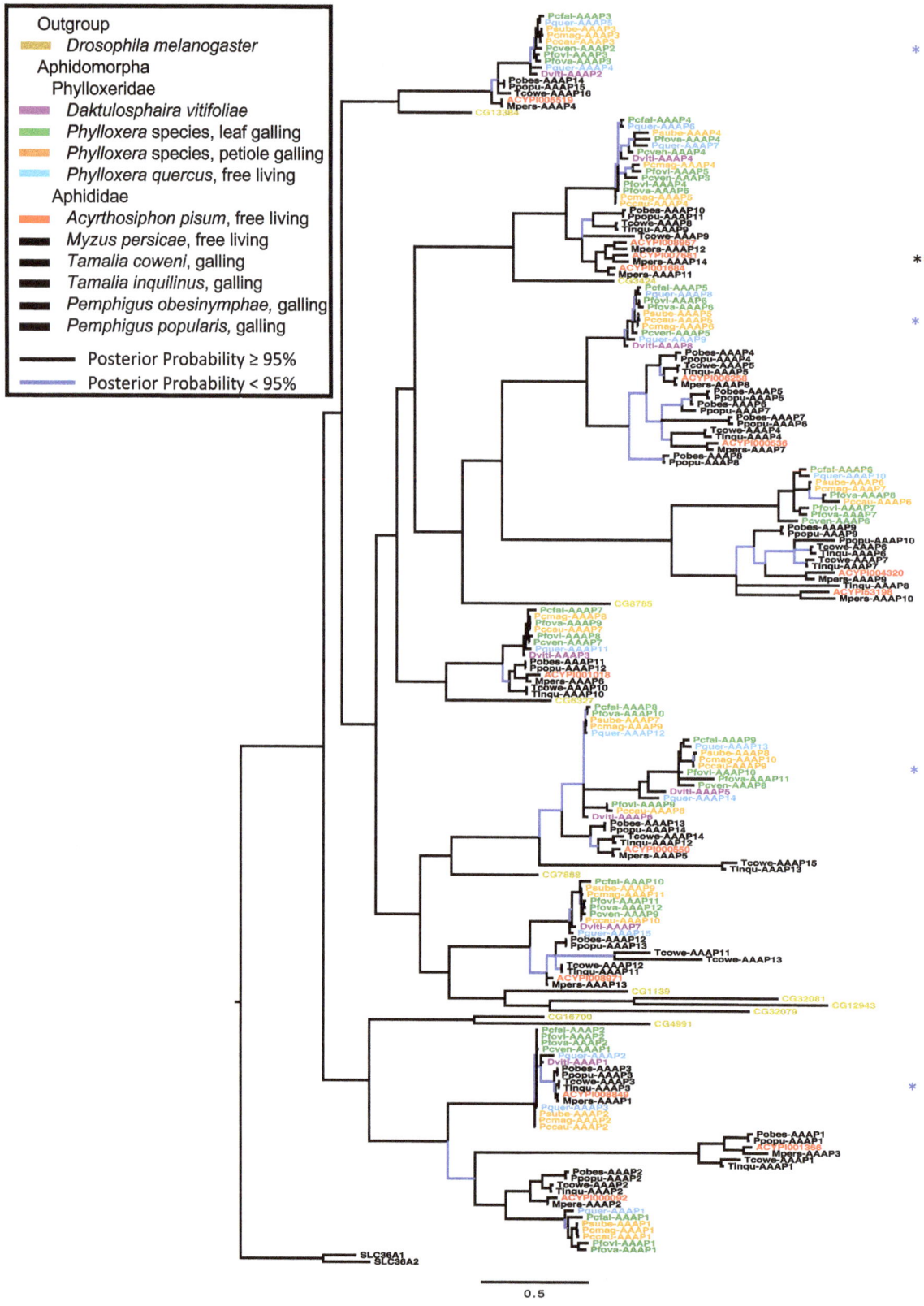

Fig. 3 Phylogeny of AAAP family transporters for sampled insects, representative aphids, and a nonherbivore outgroup. Taxa are color coded per the key, with the outgroup human genes SLC1/2 coded in black. *Asterisks* indicate clades where paralogs occur in free-living aphidomorphs but are lacking in galling species

(e.g., *T. coweni*) also lack transcripts for some clades where free-living aphids retain one if not more transcripts. This provides support that differences in the nutrient environments across plant hosts differentially alter selection to retain certain AATs. Little information exists for comparing extensive metabolite profiles of hosts across galled taxa, but the diversity in morphology, color, and specialized tissues that are induced in plant hosts by galling insects [41] suggests nutrient pools that insects feed upon differ widely. Future studies examining metabolite pools among closely related taxa will help resolve what limitations, if any, are present in induced plant phenotypes, and provide additional tests of the role of host nutrient manipulation in the evolution of insect AATs.

The microbial community plays a fundamental role in animal nutrient acquisition from food, especially for sap-feeding insects where coevolution with endosymbiotic bacteria alleviates low amino acid content provided by phloem diets [42]. Galling or less apparent manipulation of host nutrients (e.g., delayed host senescence by leaf miners; [43]) increases nutrient flux to feeding sites, potentially altering selection on the stability of endosymbiotic relationships. Our data and previous transcriptome profiles support increases in paralogs for two free-living aphids, but no galling aphids share these increases. This pattern suggests lineage specificity [10]; however, until more insects are profiled in a way that controls for phylogeny while spanning the range of plant nutrient manipulation, either host manipulation, lineage specific evolution, or both may alter selection on AAT gene evolution.

Prior transcriptome assessments [9, 10] correlated transcript abundance and presence with maintaining endosymbiosis. By examining more phylloxerids, we increased resolution of *slimfast* gene evolution, providing support for previous data that all aphidomorphs experienced *slimfast* duplication (Fig. 2). Thus, *slimfast* expansions likely occurred in the ancestral aphidomorph. Because the ancestral state of phylloxerids is unresolved without a phylogeny, it is possible that the ancestor lost AATs with the evolution of galling or gained AATs with the transition away from galling. Either scenario would link this clade to nutrient acquisition strategies. Although phylloxerids lack stable nutritional endosymbionts, numerous microbial partners exist within the gut microbiome (PDN unpublished data, [20]). How gut microbial community dynamics modify sap-feeder fitness is less well understood. Profiling partner presence and abundance alongside plant host nutrients may highlight specific roles, if any, for microbes in an insect that lost endosymbionts.

The patterns in AAT transcripts within the Phylloxeridae, and among galling and free-living Aphidomorpha, provide insight into the selection imparted through host manipulation and the evolution (or loss) of endosymbiosis. *Slimfast*

paralogs appear across many phylloxerids, and thus appear not to drive endosymbiosis [10]. Rather the presence of *slimfast* duplications in free-living *Phylloxera* suggest an additional role in nutrient transport when the host cannot be manipulated. Although our data support the *slimfast* expansion among aphidomorphs, we hypothesize that plant host nutrient availability may have facilitated some duplications found in free-living pea and green peach aphids because several paralogs appear absent from galling aphids. Recent cellular localization screens indicated one of these genes (ACYPI008904) increases in expression prior to and after *Buchnera* transmission as the bacteriocyte develops [44]. Because galling aphids lack ACYPI008904 paralogs, selection to duplicate specific nutrient transporters may be relaxed when plant host nutrient status can be manipulated. Notably, the emergence of phylloxerid-specific duplications (yellow shading; Fig. 2) related to these aphid genes, highlights a bacteriocyte-independent role for some *slimfast* orthologs.

Conclusion

Transcriptional profiling of AATs across Sternorrhyncha insects has revealed patterns of co-expression between symbiont and host, and identified gene candidates that may underlie the maintenance and evolution of endosymbiosis. While lineages that contain stable nutritional symbionts provide a model for dissecting how gene expression correlates with symbiosis, insect relatives that lack nutritional endosymbionts provide additional context on how symbioses can evolve or what may lead to losses in symbionts. Here we show that phylloxerids retain many of the AATs of other symbiont-harboring aphidomorphs, and that a pattern is emerging for plant manipulation to constrain AAT evolution. How insects alter host-plant phenotypes, whether chemical or morphological, is largely unknown. Growing evidence suggests that secreted peptides called effectors underlie these induced changes [45], and may target immune function to enable colonization, but also regulate fundamental development pathways that coordinate nutrient transport [46, 47]. Going forward, it is likely that the evolution of plant manipulation interacts with the maintenance of symbiosis to perturb host-symbiont relationships. Understanding genomic patterns in effector families, nutritional symbionts, and host-plant nutrient allocation within a phylogenetic context may provide key insight into the evolution of insect nutrient acquisition, whether through a symbiont, microbe derived effector, or de novo effector evolution to target conserved plant signaling or nutrient mobilization networks. Undoubtedly, the evolution of herbivory is complex, but comparative studies across taxa will continue to provide context on processes, ecological or otherwise, that facilitated and maintain herbivory.

Additional file

Additional file 1: Tables S1. Quality control and assembly statistics for transcriptomes used in analyses. **Tables S2.** Mitochondrial gene sequences used to generate species-level phylogeny of insects assessed in this study. **Tables S3.** AAT gene sequences for APC and AAAP gene families for each species assessed in this study. **Tables S4.** Summary of AAT gene counts per clade for each species assessed in this study. **Tables S5.** Summary of dS values or overlapping states used to merge transcripts for APC genes. **Tables S6.** Summary of dS values or overlapping states used to merge transcripts for AAAP genes. (XLSX 204 kb)

Abbreviations
AAAP: Amino acid/auxin permease transporter; AAT: Amino acid transporter; APC: Amino acid polyamine organocation transporter; COI: Cytochrome oxidase I; COII: Cytochrome oxidase II; CYTB: Cytochrome b

Acknowledgments
We thank L. Fléchon for preliminary data analyses. We also thank the Arnold Arboretum for granting access to their *Carya* collections. We are grateful to the International Aphid Genomic Consortium (IAGC) for allowing us to use the *Myzus persicae* and *Daktulosphaira vitifoliae* genome assembly data housed at AphidBase.

Funding
PDN and CZ were supported on internal funds from Washington State University and the University of California, Riverside.

Authors' contributions
PDN designed the study, CZ and PDN conducted the analyses, CZ and PDN wrote the manuscript. Both authors read and approved the final manuscript.

Competing interests
The authors declare that they have no competing interests.

Author details
[1]Department of Botany and Plant Sciences, University of California, Riverside, Riverside, CA 92521, USA. [2]Department of Botany and Plant Sciences, University of California, Riverside, 900 University Avenue, Batchelor Hall room 2140, Riverside, CA 92521, USA.

References
1. Karban R, Agrawal AA. Herbivore offense. Annu Rev Ecol Systemat. 2002;33: 641–64.
2. Chung SH, Rosa C, Scully ED, Peiffer M, Tooker JF, Hoover K, et al. Herbivore exploits orally secreted bacteria to suppress plant defenses. Proc Natl Acad Sci. 2013;110:15728–33.
3. McCutcheon JP, Moran NA. Extreme genome reduction in symbiotic bacteria. Nature Rev Microbiol. 2012;10:13–26.
4. Bennett GM, Moran NA. Heritable symbiosis: the advantages and perils of an evolutionary rabbit hole. Proc Natl Acad Sci. 2015;112:10169–76.
5. Koga R, Bennett GM, Cryan JR, Moran NA. Evolutionary replacement of obligate symbionts in an ancient and diverse insect lineage. Environ Microbiol. 2013;15:2073–81.
6. Shigenobu S, Wilson AC. Genomic revelations of a mutualism: the pea aphid and its obligate bacterial symbiont. Cell Mol Life Sci. 2011;68(8):1297–309.
7. Jack DL, Paulsen IT, Saier MH. The amino acid/polyamine/organocation (APC) superfamily of transporters specific for amino acids, polyamines and organocations. Microbiol. 2000;146(8):1797–814.
8. Saier MH. Families of transmembrane transporters selective for amino acids and their derivatives. Microbiol. 2000;146(8):1775–95.
9. Duncan RP, Husnik F, Van Leuven JT. Gilbert, D. G., Dávalos LM, McCutcheon JP, Wilson AC. Dynamic recruitment of amino acid transporters to the insect/symbiont interface. Mol Ecol. 2014;23(6):1608–23.
10. Duncan RP, Feng H, Nguyen DM, Wilson AC. Gene family expansions in aphids maintained by endosymbiotic and Nonsymbiotic traits. Genome Biol Evol. 2016;8(3):753–64.
11. Bagatto G, Shorthouse JD. Accumulation of copper and nickel in plant tissues and an insect gall of lowbush blueberry, *Vaccinium angustifolium*, near an ore smelter at Sudbury, Ontario. Canada Canad J Bot. 1991;69:1483–90.
12. Larson KC, Whitham TG. Manipulation of food resources by a gall-forming aphid: the physiology of sink-source interactions. Oecologia. 1991;88(1):15–21.
13. Koyama Y, Yao I, Akimoto SI. Aphid galls accumulate high concentrations of amino acids: a support for the nutrition hypothesis for gall formation. Entomol Exp Appl. 2004;113(1):35–44.
14. Harris MO, Freeman TP, Rohfritsch O, Anderson KG, Payne SA, Moore JA. Virulent hessian fly (Diptera: Cecidomyiidae) larvae induce a nutritive tissue during compatible interactions with wheat. Ann Entomol Soc America. 2006;99(2):305–16.
15. Saltzmann KD, Giovanini MP, Zheng C, Williams CE. Virulent hessian fly larvae manipulate the free amino acid content of host wheat plants. J Chem Ecol. 2008;34(11):1401–10.
16. Marini-Filho OJ, Fernandes GW. (2012). Stem galls drain nutrients and decrease shoot performance in *Diplusodon orbicularis* Lythraceae. Arthropod Plant Interact 2012;6(1):121–128.
17. Shorthouse JD. Rohfritsch O. Biology of insect-induced galls: Oxford University Press; 1992.
18. Blackman RL, Eastop VF. Aphids on the world's crops. England: West Sussex; 2000.
19. Vorwerk S, Martinez-Torres D, Forneck A. *Pantoea agglomerans*-associated bacteria in grape phylloxera (*Daktulosphaira vitifoliae*, Fitch). Agric For Entomol. 2007;9(1):57–64.
20. Medina RF, Nachappa P, Tamborindeguy C. Differences in bacterial diversity of host-associated populations of *Phylloxera notabilis* Pergande (Hemiptera: Phylloxeridae) in pecan and water hickory. J Evol Biol. 2011;24(4):761–71.
21. Kellow AV, Sedgley M, Van Heeswijck R. Interaction between *Vitis vinifera* and grape phylloxera: changes in root tissue during nodosity formation. Ann Bot. 2004;93(5):581–90.
22. Nabity PD, Haus MJ, Berenbaum MR, DeLucia EH. Leaf-galling phylloxera on grapes reprograms host metabolism and morphology. Proc Natl Acad Sci. 2013;110(41):16663–8.
23. Griesser M, Lawo NC, Crespo-Martinez S, Schoedl-Hummel K, Wieczorek K, Gorecka M, et al. Phylloxera (*Daktulosphaira vitifoliae* Fitch) alters the carbohydrate metabolism in root galls to allowing the compatible interaction with grapevine (*Vitis* Ssp.) roots. Plant Sci. 2015;234:38–49.
24. Favret C, Blackman R, Miller GL, Victor B. Catalog of the phylloxerids of the world (Hemiptera, Phylloxeridae). ZooKeys. 2016;629:83–101.
25. Haas BJ. Papanicolaou, Yassour M, Grabherr M, blood PD, Bowden J, Couger mB, Eccles D, li B, Lieber M, et al. Nat Protocol. 2013;8(8):1494–512.
26. Li W, Godzik A. Cd-hit: a fast program for clustering and comparing large sets of protein or nucleotide sequences. Bioinformatics. 2006; 22(13):1658–9.
27. Miller DG, Lawson SP, Rinker DC, Estby H, Abbot P. The origin and genetic differentiation of the socially parasitic aphid *Tamalia inquilinus*. Mol Ecol. 2015;24:5751–66.
28. Price DR, Duncan RP, Shigenobu S, Wilson AC. Genome expansion and differential expression of amino acid transporters at the aphid/*Buchnera* symbiotic interface. Mol Biol Evol. 2011;28(11):3113–26.
29. Min XJ, Butler G, Storms R, Tsang A. OrfPredictor: predicting protein-coding regions in EST-derived sequences. Nucleic Acids Res. 2005;33(suppl 2):W677–80.
30. Finn RD, Clements J, Eddy SR. HMMER web server: interactive sequence similarity searching. Nucleic Acids Res. 2011:gkr367.
31. Yang Z. PAML 4: phylogenetic analysis by maximum likelihood. Mol Biol Evol. 2007;24(8):1586–91.
32. Conesa A, Götz S, García-Goméz JM, Terol J, Talón M, Robles M. Blast2GO: a universal tool for annotation, visualizartion and analysis in functional genomics research. Bioinformatics. 2005;21(18):3674–6.
33. Patwardhan A, Ray S, Roy A. Molecular markers in phylogenetic studies – a review. J Phylogen Evol Biol. 2014;2:131. doi:10.4172/2329-9002.1000131.
34. Katoh K, Misawa K, Kuma K, Miyata T. MAFFT: a novel method for rapid multiple sequence alignment based on fast Fourier transform. Nucleic Acids Res. 2002;30:3059–66.
35. Castresana J. Selection of conserved blocks from multiple alignments for their use in phylogenetic analysis. Mol Biol Evol. 2000;17:540–52.

36. Tamura K, Stecher G. Pet4erson D, Filipiski a, Kumar S. MEGA6: molecular evolutionary genetics analysis version 6.0. Mol Biol Evol. 2013;30(12):2752–9.

37. Capella-Gutiérrez S. Martínez JM. Gabaldon T trimAl: a tool for automated alignment trimming in large-scale phylogenetic analyses Bioinformatics. 2009;25:1972–3.

38. Miller MA, Pfeiffer W, Schwartz T. Creating the CIPRES science gateway for inference of large phylogenetic trees. Proceedings of the Gateway Computing Environment Workshop (GCE) 14 Nov New Orleans LA 1–8. 2010.

39. Kembel SW, Cowan PD, Helmus MR, Cornwell WK, Morlon H, Ackerly DD, et al. Picante: R tools for integrating phylogenies and ecology. Bioinformatics. 2010; 26:1463–4.

40. Haribal M, Jander G. Stable isotope studies reveal pathways for the incorporation of non-essential amino acids in *Acythrosiphon pisum* (pea aphid). J Exp Biol. 2015;218:3797–806.

41. Stone G, Schönrogge K. The adaptive significance of insect gall morphology. Trends Ecol Evol. 2003;18:512–22.

42. Gündüz EA, Douglas AE. Symbiotic bacteria enable insect to use a nutritionally inadequate diet. Proc Roy Soc London B: Biol Sci. 2009; 276(1658):987–91.

43. Kaiser W, Huguet E, Casas J, Commin C, Giron D. Plant green-island phenotype induced by leaf miners is mediated by bacterial symbionts. Proc Roy Soc London B: Biol Sci. 2010;277:2311–9.

44. Lu H, Chang C, Wilson ACC. Amino acid transporters implicated in endocytosis of *Buchnera* during symbiont transmission in the pea aphid. EvoDevo. 2016;7:24.

45. Nabity PD. Insect-induced plant phenotypes: revealing mechanisms through comparative genomics of insects and their hosts. Am J Bot. 2016;103:979–81.

46. Mukhtar MS, Carvunis AR, Dreze M, Epple P, Steinbrenner J, Moore J, Tasan M, Galli M, Hao T, Nishimura MT, Pevzner SJ, Donovan SE, Ghamsari L, Santhanam B, Romero V, Poulin MM, Gebreab F, Gutierrez BJ, Tam S, Monachello D, Boxem M, Harbort CJ, McDonald N, Gai L, Chen H, He Y, European Union Effectoromics Consortium, Vandenhaute J, Roth FP, Hill DE, Ecker JR, Vidal M, Beynon J, Braun P, Dangl JL Independently evolved virulence effectors converge onto hubs in a plant immune system network. Science 2011;333:596–601.

47. Hillmer RA, Tsuda K, Rallapalli G, Asai S, Truman W, Papke MD, Sakakibara H, Jones JDG, Myers CL, Katagiri F. The highly buffered Arabidopsis immune signaling network conceals the functions of its components. PLoS Genet 2017; doi.org/10.1371/journal.pgen.1006639

7

Segmental duplications: evolution and impact among the current Lepidoptera genomes

Qian Zhao[1,2,3,4], Dongna Ma[1,2,3,4], Liette Vasseur[1,2,5] and Minsheng You[1,2,3,4*] (iD)

Abstract

Background: Structural variation among genomes is now viewed to be as important as single nucleoid polymorphisms in influencing the phenotype and evolution of a species. Segmental duplication (SD) is defined as segments of DNA with homologous sequence.

Results: Here, we performed a systematic analysis of segmental duplications (SDs) among five lepidopteran reference genomes (*Plutella xylostella*, *Danaus plexippus*, *Bombyx mori*, *Manduca sexta* and *Heliconius melpomene*) to understand their potential impact on the evolution of these species. We find that the SDs content differed substantially among species, ranging from 1.2% of the genome in *B. mori* to 15.2% in *H. melpomene*. Most SDs formed very high identity (similarity higher than 90%) blocks but had very few large blocks. Comparative analysis showed that most of the SDs arose after the divergence of each linage and we found that *P. xylostella* and *H. melpomene* showed more duplications than other species, suggesting they might be able to tolerate extensive levels of variation in their genomes. Conserved ancestral and species specific SD events were assessed, revealing multiple examples of the gain, loss or maintenance of SDs over time. SDs content analysis showed that most of the genes embedded in SDs regions belonged to species-specific SDs ("Unique" SDs). Functional analysis of these genes suggested their potential roles in the lineage-specific evolution. SDs and flanking regions often contained transposable elements (TEs) and this association suggested some involvement in SDs formation. Further studies on comparison of gene expression level between SDs and non-SDs showed that the expression level of genes embedded in SDs was significantly lower, suggesting that structure changes in the genomes are involved in gene expression differences in species.

Conclusions: The results showed that most of the SDs were "unique SDs", which originated after species formation. Functional analysis suggested that SDs might play different roles in different species. Our results provide a valuable resource beyond the genetic mutation to explore the genome structure for future Lepidoptera research.

Keywords: Segmental duplications, Lepidoptera, Evolution

Background

Segmental duplications (SDs) are DNA fragments with near-identical sequences that are greater than 1Kb [1]. They have been recognized as important mediators of gene and genome evolution, and are considered the origins for gene gain, functional diversification, and gene family expansion [1, 2]. The outcomes of a gene duplication event may lie on lineage-specific selection. In this situation, the new gene copy has the opportunity to acquire novel or modified functions or become nonfunctional [3, 4]. These new copies are often important for the adaption of the species to certain environments [2]. SDs can lead to various types of genome rearrangements [5] and other genome structural changes between and within species [6–8].

Characterization and annotation of SDs are important for understanding the structure and evolution of a

---
* Correspondence: msyou@iae.fjau.edu.cn
[1]State Key Laboratory for Ecological Pest Control of Fujian/Taiwan Crops and College of Life Science, Fujian Agriculture and Forestry University, Fuzhou 350002, China
[2]Institute of Applied Ecology, Fujian Agriculture and Forestry University, Fuzhou 350002, China
Full list of author information is available at the end of the article

genome and have been explored in many organisms' whole genomes [9–15]. Few systematically comparative analyses of SDs however have been performed until now. The most important example is primate genomes, used to understand the pattern and rates of SDs during hominid evolution [6]. Here, we performed the comparative analysis of SDs in the whole genomes of five Lepidoptera insects, diamondback moth (*Plutella xylostella*), Monarch butterfly (*Danaus plexippus*), silkworm (*Bombyx mori*), Carolina sphinx moth (*Manduca sexta*), and postman butterfly (*Heliconius melpomene*), to understand the roles of SDs during the evolution of Lepidoptera. Our analysis revealed that duplication activities varied in terms of number of base pairs or events among these different species. The marked difference of transposable elements (TEs) content in the flanking regions of SDs among these species of Lepidoptera suggested various formation mechanisms of SDs. Our functional analysis of the SDs indicated that gene families embedded in the SDs were different among the five genomes and these gene families may be related to species-specific adaptive evolution.

Methods
Data sources
The five Lepidoptera insect species, *P. xylostella*, *B. mori*, *D. plexippus*, *M. sexta*, and *H. melpomene*, were used to construct the SD map. The genome and predicted transcripts of diamondback moth was downloaded from DBM database (http://www.insect-genome.com/) [16]. The other genomes resources of Lepidoptera insects were downloaded from SilkDB (http://silkworm.genomics.org.cn/) [17], Heliconius Genome Project (http://www.butterflygenome.org/) [18], MonarchBase (http://www.insect-genome.com/) [19] and Carolina sphinx dataset (ftp://ftp.bioinformatics.ksu.edu/pub/Manduca/OGS2/).

Computational analysis of lepidoptera segmental duplications
We used the Whole-Genome Assembly Comparison (WGAC) method to detect the segmental duplications in the five Lepidoptera species. The insect genomes were first masked at 15% divergence level from transposable elements (TEs), high-copy repeats or simple sequence repeats (SSR) using RepeatMasker (Smit and Green http://www.repeatmasker.org/, version 4.0.6). We then used silkworm TE dataset [20] as repeat database to re-run the RepeatMasker to mask as much TEs as possible. All these repeats were deleted from the sequences and the remaining genome sequences were used to perform BLASTN searches against themselves with reduced affine gap extension parameters, which allowed gaps up to 1000 bp and e value ($1e^{-20}$).

After discarding self-alignments, the repeat sequences were reinserted back into these alignments. These seed alignments were subsequently used as queries to search against the unmasked genome using BLASTN, which generated accurate alignment statistics. Considering the high rate of heterozygosity of these Lepidoptera species (except silkworm, which has a long history of domestication and inbreeding) [16, 18, 21], we conservatively lowered the identity threshold to 75% for alignments in order to capture more divergent SDs than under the 90% usual threshold. Selected alignments were those with a length longer than 1 kb and identity higher than 75%.

Gene content and functional annotation
Gene content of segmental duplications was accessed using the GFF files obtained from the dataset above (see data sources). We also assessed whether the molecular function, biological process, and pathway terms were over-represented in SDs using Blast2Go [22]. For each SD, we computed an expected number of genes for different biological processes based on their curated representation in the reference genome. The statistical significance of the functional GO Slim enrichment was evaluated using the Fisher's exact test ($p < 0.05$). This analysis showed the GO terms that were significantly enriched among genes within SDs. Pfam was also used to annotate the function of the genes in the SDs [23].

RNA-seq analysis
We collected the RNA-seq data from published sources to access the gene expression level within and outside SDs regions. These data included different tissues or different developmental stages of diamondback moth [16], silkworm [24] and Carolina sphinx moth [25]. All the reads were mapped back to its genome using TopHat [26]. The expression abundance (RPKM) was calculated using CuffLinks [27]. The expression levels were assessed as $Log_{10}^{(RPKM)}$. Gene expression levels within and outside SDs regions as well as the variables were compared using a T-test with a Bonferroni correction.

Results and discussion
Segmental duplication maps among different Lepidoptera species
Using WGAC, we developed segmental duplication maps for each of the five Lepidoptera species' genomes (Table 1). SD contents greatly varied among the five Lepidoptera species, ranging from 1.2% in *Bombyx mori* to 15.2% in *Heliconius melpomene* (Table 1, Additional file 1: Table S1). SDs with highest identity (≥90%) was the majority (ranging from 80% in *M. sexta* to 93% in *D. plexippus*) (Table 1). Based on our analysis, duplications varied in size from 5.6 Mbp in silkworm to 43 Mbp in *P. xylostella*. *P. xylostella* and *H. melpomene* showed the highest number of duplications (Table 1) suggesting that their genomes could be unstable or capable of tolerating

Table 1 Characterization of the SDs of the five Lepidoptera species

Species	P. xylostella	M. sexta	H. melpomene	D. plexippus	B. mori
Total number of SDs	21,369	11,141	23,942	10,799	3667
Number of SDs with 90% identity	18,064	8892	21,572	10,070	3221
Number of SDs with 80-90% identity	3204	2171	2239	668	416
Number of SDs with 75-80% identity	99	78	127	60	5
Total (Mb)	43	19.1	40.5	23.5	5.6
% of genome	11	5.2	15.2	9.9	1.2
Number of genes	2235	1040	1453	1564	332
% of genes	12.4	6.8	11	10.3	2.3

extensive levels of variation. For example, in human, segmental duplications play an "expanding" role in genomic instability [28].

The analysis of the length of SDs in all five species indicated that the Lepidoptera genomes were significantly poor in large blocks (>4 Kb) (T-test, $P < 0.025$; Fig. 1). This is consistent with SDs data reported in *Drosophila* genome (Fiston-Lavier et al. 2007) and silkworm genome [29, 30]. The number of SDs in Lepidoptera decreased along with the increase in SDs length (Fig. 1a) and this was true for all five species (Fig. 1b). Eichler [31] has suggested that SDs in invertebrates are much smaller in length than in vertebrates. These differences probably reflect some evolutionary constraints imposed by the smaller size of the invertebrate genome [32].

We use RepeatMasker (Smit and Green http://www.repeatmasker.org/, version 4.0.6) to mask the transposable elements (TEs; masked at 15% divergence level), high-copy repeats or simple sequence repeats (SSR). Then silkworm TE dataset [20] was used as repeat database to rerun the RepeatMasker to mask as much TEs as possible. Thus, we used different repeat databases to mask the target genomes. The result showed that almost 22.6% of the silkworm genome was masked while 2.05% - 4.97% of other Lepidoptera genomes were masked (*P. xylostella*: 3.12%, *D. plexippus*: 2.05%, *M. sexta*: 4.97% and *H. melpomene*: 2.25%). Osanai-Futahashi et al. [33] have shown that TEs are enriched in the genome of silkworm and TEs may play important roles during the domestication of silkworm [34]. Thus, the high proportion of SDs in *H. melpomene* may result from some TEs left in the genome.

Comparative analysis of duplication maps among five Lepidoptera species

We further characterized each SD as "unique" or "shared", depending on whether they exist in only one

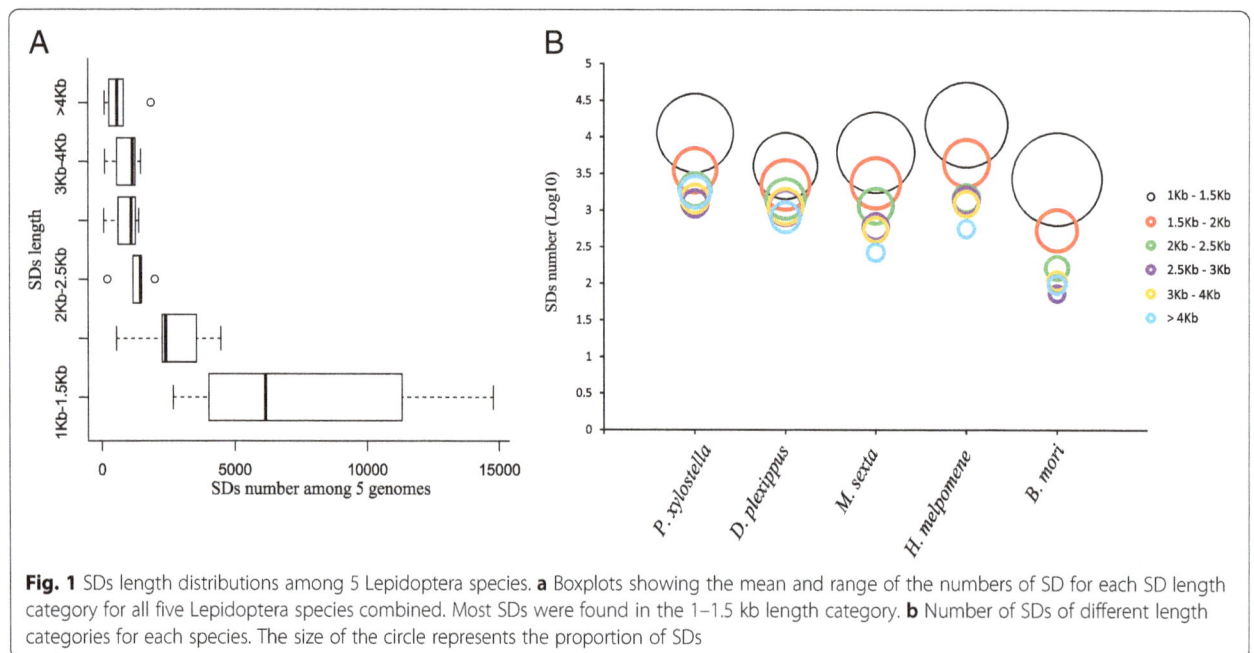

Fig. 1 SDs length distributions among 5 Lepidoptera species. **a** Boxplots showing the mean and range of the numbers of SD for each SD length category for all five Lepidoptera species combined. Most SDs were found in the 1–1.5 kb length category. **b** Number of SDs of different length categories for each species. The size of the circle represents the proportion of SDs

or multiple genomes. The comparative SD maps revealed that most of the segmental duplications were "unique" SDs (Fig. 2). For example, the number of shared SDs among the five Lepidoptera species varied from 83 in *B. mori* (e.g. 83 SDs from *B. mori* shared with other Lepidoptera genomes) to 1817 in *H. melpomene* (e.g. 1817 SDs from *H. melpomene* shared with other Lepidoptera genomes) (Fig. 3a).

Butterflies (*D. plexippus* and *H. melpomene*) shared more SDs with each other than with the other species (Fig. 2) indicating their closer relationship. Silkworm and Carolina sphinx moth also shared more common SDs than with the other species, indicating their close

relationship. These results are consistent with the phylogeny of Lepidoptera published by Regier et al. [35].

Based on the phylogeny of Lepidoptera [35], it was possible to assess the origins of some SDs within specific lineages and ancestral events of SDs. Since segments might have mutated after divergence, we attempted to map duplication events onto the phylogenetic tree using reconciliation method (software like NOTUNG). However, based on the blast search analysis, we found that the "unique" SDs could not find the homologous sequences in other Lepidoptera species. We had two speculations to explain this result: (1) The segments might have mutated after duplications or (2) SDs arose after

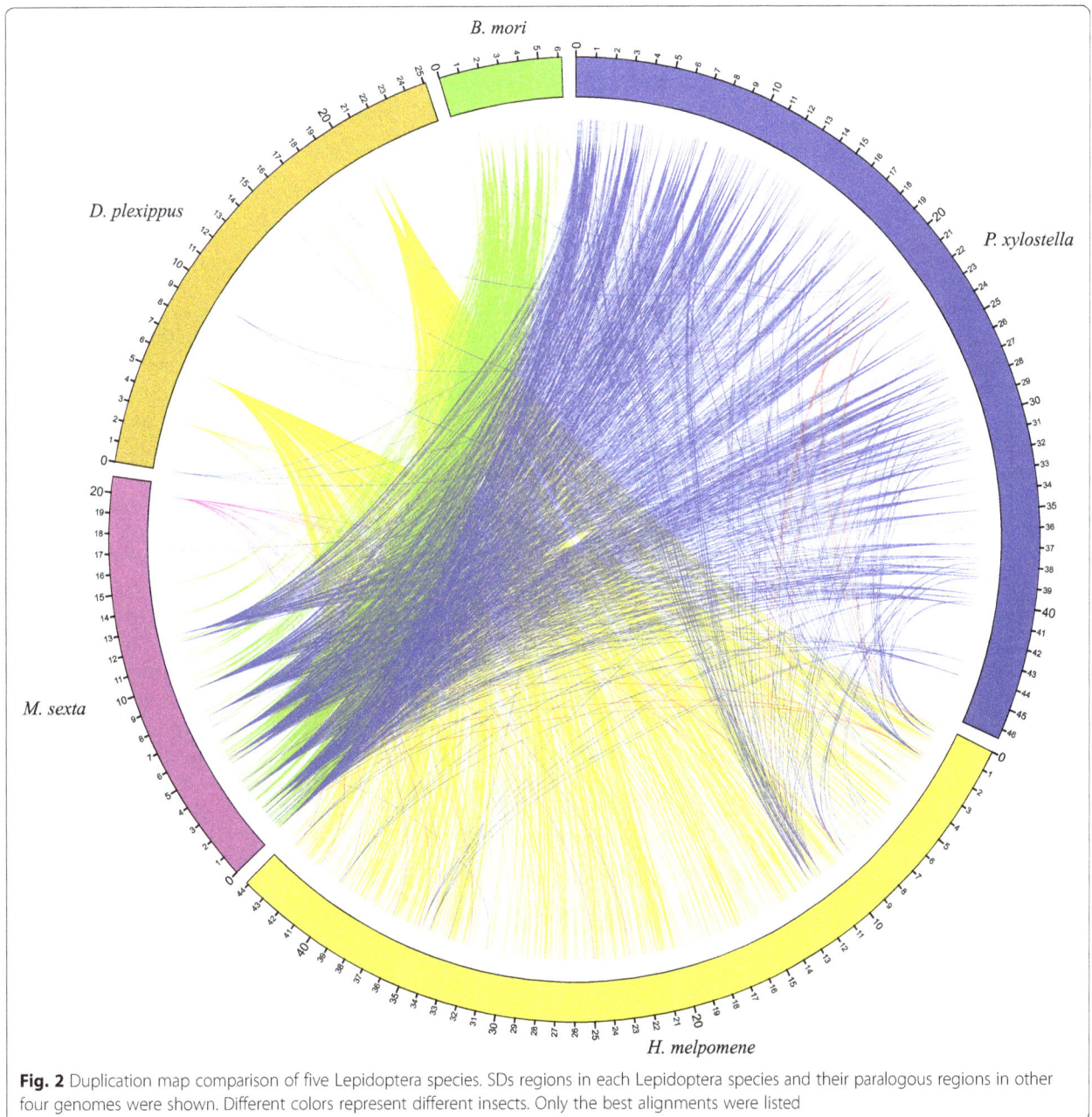

Fig. 2 Duplication map comparison of five Lepidoptera species. SDs regions in each Lepidoptera species and their paralogous regions in other four genomes were shown. Different colors represent different insects. Only the best alignments were listed

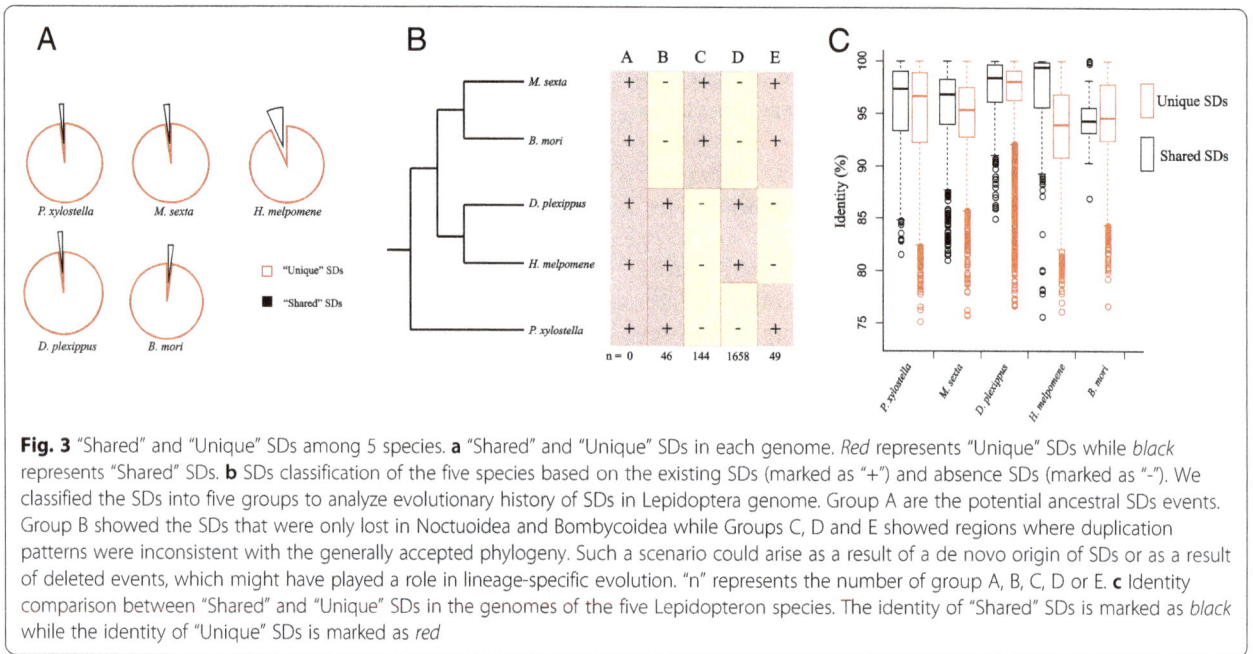

Fig. 3 "Shared" and "Unique" SDs among 5 species. **a** "Shared" and "Unique" SDs in each genome. *Red* represents "Unique" SDs while *black* represents "Shared" SDs. **b** SDs classification of the five species based on the existing SDs (marked as "+") and absence SDs (marked as "-"). We classified the SDs into five groups to analyze evolutionary history of SDs in Lepidoptera genome. Group A are the potential ancestral SDs events. Group B showed the SDs that were only lost in Noctuoidea and Bombycoidea while Groups C, D and E showed regions where duplication patterns were inconsistent with the generally accepted phylogeny. Such a scenario could arise as a result of a de novo origin of SDs or as a result of deleted events, which might have played a role in lineage-specific evolution. "n" represents the number of group A, B, C, D or E. **c** Identity comparison between "Shared" and "Unique" SDs in the genomes of the five Lepidopteron species. The identity of "Shared" SDs is marked as *black* while the identity of "Unique" SDs is marked as *red*

the divergence of each lineage. In the first situation, all the copies should have mutated and evolved rapidly, resulting in sequence variation being too high to find a blast hit in other genomes. If so, it would be difficult to trace the ancestral sequences onto the phylogenetic tree using reconciliation method such as Notung. Thus, we classified the SDs into five groups (Fig. 3b) and focused on analyzing evolutionary history of SDs in Lepidoptera genome. To identify the potential ancestral SDs events, we initially focused on shared duplications among all five species (Group A) but none were identified (Fig. 3b), suggesting that the original SDs might have been lost early during the evolution of Lepidoptera or the origins of the SDs are along with the speciation of the Lepidoptera. The second situation would lead to some SDs that only exist in one or a few genomes.

We then analyzed the SDs that were only lost in Noctuoidea and Bombycoidea (Group B, Fig. 3b). There were 46 cases found in this group. As part of our comparative analyses, we also found some regions where duplication patterns were inconsistent with the generally accepted phylogeny (Groups C, D, and E; Fig. 3b). Such a scenario could arise as a result of a de novo development of SDs or as a result of deleted events, which might play a role in lineage-specific evolution. Groups C and D (Fig. 3b) were more common than the other groups due to their closer relationship with species based on the evolution history (Fig. 3b). The previous study of Marques-Bonet et al. [6] reports that humans share a greater number of SDs with chimpanzees than macaque or orangutan. Only 49 common SDs were lost in *D. plexippus* and *H.*

melpomene (Group E). Since the time of Lepidoptera speciation is relatively long, we cannot test the complete phylogeny of SDs, and a greater number of sequenced Lepidoptera genomes would be necessary to elucidate this aspect.

We speculated that the "shared" SDs between species might represent the "ancestral sequences" as they remained conserved in the genomes during the evolution of Lepidoptera. To test this speculation, we analyzed the alignment identity of "shared SDs" and "unique SDs" and found that "shared SDs" had significantly higher identity comparing with the "unique SDs" ($P < 0.01$, T-test), except for silkworm (Fig. 3c). The results indicated that "shared SDs" might be more conserved than the "unique SDs". Silkworm has diverged from the other species due to its domestication and inbreeding history leading to extremely low level of heterozygosity [36]. We compared the SDs in silkworm with the artificial selected regions that were identified in Xia et al. [36] and found that eight SD regions overlapped with the artificial selected regions, suggesting that these SDs may be related with the silkworm domestication. However, none of these eight regions were "shared SDs", which also indicated that these unique SDs may be involved in the lineage-specific domestication. We also tested the difference of variance between "shared" and "unique" SDs (Fig. 3c) and showed that in *P. xylostella*, *D. plexippus* and *H. melpomene*, the differences were not significant ($p = 0.05755$; $p = 0.5304$ and $p = 0.6278$, respectively). Only *B. mori* and *M. sexta* showed significant differences ($p = 1.218e-05$ and $p = 0.03909$).

Sequence properties of the SDs in the five studied species

The analysis and comparison of the composition of genes in the SDs among the five Lepidoptera species showed that 2235, 1036, 1453, 1564 and 332 putative genes could be identified in *P. xylostella, M. sexta, H. melpomene, D. plexippus* and *B. mori* respectively (Table 1, Additional file 2: Table S2). Most of the segmental duplication intervals identified contained gene duplicates, ranging from 58% in silkworm to 94% in *H. melpomene* (Additional file 3: Table S3). We further characterized the genes as "shared" or "unique" based on whether they were located in the "shared SDs" or "unique SDs". The results showed that most of the genes belonged to the "unique" genes, with only 31, 26, 13, 6, and 3 genes belonging to "shared" genes in *P. xylostella, M. sexta, D. plexippus, H. melpomene,* and *B. mori,* respectively. These results suggested that most of the genes in SDs could play different roles in different species. We hypothesized that these genes might be involved in lineage-specific evolution and particular gene classes might be overrepresented in the SDs.

To test the hypothesis, we used Gene Ontology (GO) to annotate all the genes and showed that each species had different GO enrichments and gene families (Table 2; Additional file 4: Table S4). In *P. xylostella*, 25 proteins were identified such as serine-type endopeptidase activity (GO: 0004252), structural constitute of cuticle (GO: 0042302), and nucleic acid binding (GO: 0003676) (Table 2). Based on previous study of differential expression in response to host-plant on Swedish comma, *Polygonia c-album*, these genes may be related to host-feeding [37, 38]. Thus, we suggested that the genes in SDs of diamondback moth might be related with its host-feeding behavior.

In *M. sexta*, we identified a GO enrichment of prothoracicotrophic hormone activity (GO: 0018445). The prothoracicotropic hormone (PTTH) is well studied in tobacco hornworm (Rountree and Bollenbacher 1986) and in *M. sexta*, it is related to molting and metamorphosis [39, 40]. In *D. plexippus*, we identified the GO enrichment of glucuronosyltransferase activity (GO: 0015020). In silkworm, UDP-glucuronosyltransferase (UGT) plays a role in detoxification processes, such as minimizing the harmful effects of ingested plant allelochemicals [41]. Also, we identified the enrichment of Rho guanyl-nucleotide exchange factor activity (GO: 0005089), which is a modulator in the signaling pathway of Ras/MAPK and Wnt. Previous studies have shown that this activity is associated with neuronal growth cone and planar cell polarity formation [42, 43]. In *B. mori*, consistent with [30], we identified the enrichment of monooxygenase activity (GO: 0004497), which might be associated with detoxification.

To further clarify the functions of SDs in each Lepidoptera species, we annotated the gene functions in the SDs regions using Pfam and although the GO enrichments differed among species, some of the gene families embedded in the SDs were the same for the five species (Additional file 5: Table S5). For example, genes in SDs can be classified into three categories: (1) detoxification, (2) immunity, and (3) environmental signal recognition, which are similar to other mammals and insects [30, 44]. These genes are very important in drug detoxification, defense, and receptor and signal reorganization. The cytochrome P450s (P450s), for example, are important proteins for insect growth and development and have been found to play various functions such as biosynthesis of hormones, and inactivation and metabolism of xenobiotic compounds such as pesticides [45–47]. In this study, P450s were identified in all five species (8, 13, 15, 12, 10 SDs regions in *P. xylostella, B. mori, M. sexta, D. plexippus* and *H. melpomene,* respectively). In *P. xylostella*, Yu et al. [48] report strong expression of 84 functional cytochrome P450 genes, many of them, especially CYP367s, contributing to detoxification or metabolic processing of environmental chemicals.

We also identified the trypsin in the five species (55, 7, 24, 19, 2 SDs regions in *P. xylostella, B. mori, M. sexta, D. plexippus* and *H. melpomene,* respectively), which may be involved in immunity [49]. The glucose-methanol-choline (GMC) oxidoreductases, shown to be involved in developmental and physiological processes, and immunity [50], were also identified in four of the five species (1, 4, 5, 4 SDs regions in *B. mori, M. sexta, D. plexippus* and *H. melpomene,* respectively).

Some species-specific genes in SDs regions were also identified including 13 Lepidopteran-specific Lipoprotein_11 in silkworm. Zhang et al. [51] have shown that this family is involved in various physiological processes such as energy storage, embryonic development and immunity. These SDs might have played a role in the silkworm-specific evolution. Some lineage-specific expansion genes were also embedded in the SDs regions. For example, we identified 167 zinc-finger proteins in the SD regions of *P. xylostella*, which was much more than in any other species (20, 73, 91 and 8 in *B. mori, M. sexta, D. plexippus* and *H. melpomene*). A recent study (data unpubl.) of transcription factors in diamondback moth indicates that zinc-finger proteins may be expanded, also suggesting their potential important functions in the DBM. The zinc-finger has been shown to function in a variety of biological processes, such as DNA-binding, RNA-binding, protein-protein interactions, developmental processes and differentiation [52]. Further studies on expression patterns showed that the expression of some zinc-fingers were significantly different between

Table 2 GO enrichment for some proteins within the SDs regions among the five Lepidoptera species

GO term	p-value	Number of proteins
P. xylostella		
Nucleic acid binding [GO:0003676]	5.78E-06	149
Oxidoreductase activity [GO:0016491]	1.64E-05	16
Oxidation-reduction process [GO:0055114]	0.0005938	38
Serine-type endopeptidase activity [GO:0004252]	0.001488	54
Protein tyrosine phosphatase activity [GO:0004725]	0.003778	12
Protein dephosphorylation [GO:0006470]	0.005278	14
Structural constituent of cuticle [GO:0042302]	0.006409	3
Zinc ion binding [GO:0008270]	0.008232	151
M. sexta		
Prothoracicotrophic hormone activity [GO:0018445]	2.99E-06	10
Growth factor activity [GO:0008083]	0.0026	5
Phosphorylase kinase complex [GO:0005964]	0.003736	3
SWI/SNF complex [GO:0016514]	0.003736	3
Phosphorylase kinase activity [GO:0004689]	0.003736	3
Phosphoprotein phosphatase activity [GO:0004721]	0.00556	6
Neuropeptide signaling pathway [GO:0007218]	0.007097	13
Defense response [GO:0006952]	0.00955	3
H. melpomene		
ATP-dependent peptidase activity [GO:0004176]	0.002622	2
Misfolded or incompletely synthesized protein catabolic process [GO:0006515]	0.002622	2
DNA integration [GO:0015074]	0.008793	2
Inositol-1,4,5-trisphosphate 3-kinase activity [GO:0008440]	0.008793	2
D. plexippus		
Dephosphorylation [GO:0016311]	0.002541	12
RNA-directed DNA polymerase activity [GO:0003964]	0.002617	16
Glucuronosyltransferase activity [GO:0015020]	0.003223	10
Endonuclease activity [GO:0004519]	0.004409	14
Carbohydrate transport [GO:0008643]	0.005052	12
Pyrophosphatase activity [GO:0016462]	0.008074	4
Riboflavin metabolic process [GO:0006771]	0.009158	8
Rho guanyl-nucleotide exchange factor activity [GO:0005089]	0.00975	10
B. mori		
Heme binding [GO:0020037]	1.30E-06	13
Monooxygenase activity [GO:0004497]	2.09E-06	13
Hormone activity [GO:0005179]	0.0001028	6
Electron transport [GO:0006118]	0.0003995	18
Calcium ion binding [GO:0005509]	0.00205	10
Response to oxidative stress [GO:0006979]	0.003103	3
Odorant binding [GO:0005549]	0.003145	6
Oxidoreductase activity [GO:0016491]	0.00358	16

susceptible and resistant strains (data unpublished). However, more researches are needed to illustrate the functions of these zinc-fingers.

Zhao et al. [30] report in silkworm that SDs are characterized by enrichment of DNA transposons and LTR retrotransposons. These observed enrichments in the flanking regions of SDs in silkworm suggest a potential implication in the formation of repeats in SDs. In this study, the TEs composition was analyzed by comparing the sequences near the SDs regions and found that DNA transposons were enriched in SDs regions as well as flanking regions of most species except *H. melpomene* (Table 3). Like in silkworm, DNA transposons and LTR (long terminal repeat) retrotransposons were enriched in the region of SDs and flanking regions in *P. xylostella* (Table 3), suggesting similar potential roles in SD formation. In *M. sexta* and *D. plexippus*, only DNA transposons were found to be enriched (Table 3). In *H. melpomene*, all analyzed TEs, except DNA transposons, were enriched in the SDs and flanking regions (Table 3) with LINEs (long interspersed nuclear elements) being the most abundant. These results suggest that short interspersed nuclear elements (SINEs), LTR and LINEs may also be involved in the formation of SDs in the genome of *H. melpomene*.

Effects of SDs on gene expression

An initial study of lymphoblastoid cell lines in human has shown that CNVs have some effects on gene expression [53]. For example, changes in the number of copies can explain almost 20% of the variation in gene expression [53]. This effect can be the results of gene dosage within SDs or SDs on neighboring genes [53–56]. To assess the effect of SDs on the transcriptomes, we explored the genome-wide expression of three of the Lepidoptera species, *P. xylostella*, *B. mori*, and *M. sexta*, at different developmental stages, different tissues and different strains using RNA-seq data (NCBI website). We found that the gene expression levels embedded within our SDs regions were significantly lower than that of other genes located elsewhere in the genome. This was true for all analyzed available developmental stages or tissues (T-test, $p < 2.20E-16$) (Fig. 4). For example, in *P. xylostella*, we analyzed the expression pattern of genes within and outside the SD regions in different developmental stages. The results showed that the expression values of genes within SDs were significantly lower than the genes outside the SDs regions (T-test, $p < 0.01$, Fig. 4a). We redid the same analysis on the silk gland from different strains of *B. mori* and different tissues of *M. sexta* and found similar expression patterns: genes located in SDs had lower expression values than the genes outside SDs (Fig. 4b and c).

A possible reason for this may be that some regulation mechanisms control the gene expression within SDs.

Table 3 TEs properties of the Lepidoptera genomes, duplications and 2.5 Kb flanking regions

Repeat	Duplication	%	2.5 Kb FR	%	Genome	%	Enrichment in SDs	Enrichment in FR
P. xylostella								
DNA	**72,462**	**0.167**	**55,718**	**0.053**	**180,461**	**0.046**	**3.653**	**1.152**
SINE	3040	0.007	22,679	0.021	258,493	0.066	0.107	0.327
LTR	**18,614**	**0.043**	**10,333**	**0.010**	**23,611**	**0.006**	**7.173**	**1.632**
LINE	249,583	0.576	129,227	0.122	628,430	0.159	3.613	0.767
M. sexta								
DNA	**11,814**	**0.062**	**22,773**	**0.045**	**128,714**	**0.031**	**2.015**	**1.462**
SINE	361	0.002	2714	0.005	48,452	0.011	0.164	0.463
LTR	131	0.001	1537	0.003	12,658	0.003	0.227	1.004
LINE	18,312	0.096	19,895	0.039	156,948	0.037	2.561	1.048
D. plexippus								
DNA	**12,322**	**0.052**	**22,852**	**0.044**	**73,287**	**0.029**	**1.779**	**1.484**
SINE	2783	0.012	4447	0.009	21,783	0.009	1.352	0.972
LTR	555	0.002	1975	0.004	7333	0.003	0.800	1.282
LINE	6026	0.026	4729	0.009	57,808	0.023	1.103	0.389
H. melpomene								
DNA	483	0.001	20,348	0.017	51,129	0.019	0.064	0.938
SINE	**23,660**	**0.058**	**15,872**	**0.014**	**26,036**	**0.009**	**6.169**	**1.436**
LTR	**7735**	**0.019**	**14,801**	**0.013**	**12,676**	**0.004**	**4.143**	**2.751**
LINE	**171,528**	**0.423**	**129,144**	**0.111**	**243,145**	**0.088**	**4.789**	**1.251**

DNA DNA transposons, *SINE* short interspersed nuclear elements, *LTR* long terminal repeat, *LINE* long interspersed nuclear elements
The TEs contents of three regions of the genomes were compared: SDs regions; 2.5 Kb flanking regions (FR) of the SDs and the genome average. Enrichment was defined as the repeat content of duplicated sequences divided by the repeat content of unique sequences. The significance was performed by simulating the repeats in a random sample ($n = 1,00$) of DBM SDs (P-value < 0.05 were in bold)

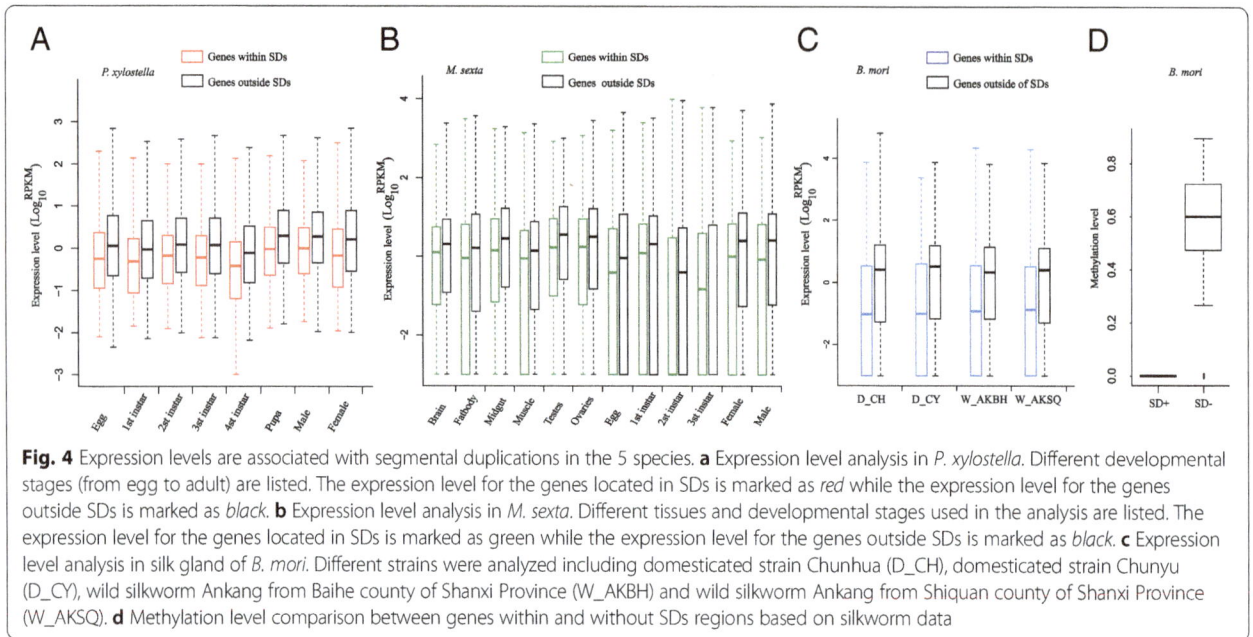

Fig. 4 Expression levels are associated with segmental duplications in the 5 species. **a** Expression level analysis in *P. xylostella*. Different developmental stages (from egg to adult) are listed. The expression level for the genes located in SDs is marked as *red* while the expression level for the genes outside SDs is marked as *black*. **b** Expression level analysis in *M. sexta*. Different tissues and developmental stages used in the analysis are listed. The expression level for the genes located in SDs is marked as green while the expression level for the genes outside SDs is marked as *black*. **c** Expression level analysis in silk gland of *B. mori*. Different strains were analyzed including domesticated strain Chunhua (D_CH), domesticated strain Chunyu (D_CY), wild silkworm Ankang from Baihe county of Shanxi Province (W_AKBH) and wild silkworm Ankang from Shiquan county of Shanxi Province (W_AKSQ). **d** Methylation level comparison between genes within and without SDs regions based on silkworm data

Based on our analysis above, we found that some TEs were enriched in the SDs as well as the SDs' flanking regions (Table 3). In silkworm, methylation levels in TEs regions are extremely low compared to the rest of the genome [57]. Epigenetic regulation in insects can have various effects on biological processes. In silkworm, CG methylation is enriched in gene bodies and is positively correlated with gene expression level, indicating its positive roles in gene transcription [57]. We therefore analyzed the CG methylation level of the genes embedded in the SD regions as well as the genes outside the SDs regions for the five species and did not find any CG methylation in gene bodies of SDs (Fig. 4c). This may explain the low gene expression levels in SDs regions. However, more CG methylation information from other Lepidoptera species may be needed to further validate this conclusion.

Conclusion

Structural variation between genomes is important in phenotype differentiation and genome evolution. Here, we performed a comparative analysis of segmental duplications (SDs) among five lepidopteran reference genomes (*P. xylostella*, *D. plexippus*, *B. mori*, *M. sexta* and *H. melpomene*). We found that the SDs contents greatly varied among the five species. Comparative analyses of SDs showed that most of them arose after the divergence of each lineage. The most closely related species based on the phylogenetic tree also shared more common SDs. Conserved ancestral SDs and species specific SD events were assessed, revealing multiple examples of gain, loss or maintenance of SDs over time. The results indicated that SDs might have undergone loss or gain

during the evolution of the genome. We further analyzed the genes embedded in SDs regions and the result showed that most of the genes were located in the species-specific SDs ("Unique" SDs). Functional analysis of these genes suggested their potential roles in the lineage-specific evolution. Comparison of gene expression between SDs and non-SDs showed that the expression levels of genes embedded in SDs were significantly lower, suggesting that structural changes in the genomes were involved in gene expression differences within each species. Our results suggested that SDs might have been involved in the species-specific evolution. They thus provide a valuable resource beyond the genetic mutation to explore the genome structure for future Lepidoptera research.

Additional files

Additional file 1: Table S1. SDs in the five studied genomes. (XLSX 1727 kb)

Additional file 2: Table S2. Genes embedded in the SDs regions. (XLSX 81 kb)

Additional file 3: Table S3. Gene duplicates located in the SD intervals. (XLSX 68 kb)

Additional file 4: Table S4. GO enrichment for proteins within the SDs regions among the five studied Lepidoptera species. Only *p* < 0.01 were shown. (DOCX 20 kb)

Additional file 5: Table S5. Gene families that were found to be different among the SDs regions from the five studied genomes. (XLSX 10 kb)

Abbreviations
CNV: Copy number variations;; GO: Gene Ontology; SD: Segmental duplications; SSR: Simple sequence repeats; TE: Transposable elements; WGAC: Whole-Genome Assembly Comparison

Acknowledgements

We would like to extend our gratitude to Simon Wade Baxter, for his advice and useful suggestions on this study.

Funding

This work is supported by National Natural Science Foundation of China (No. 31320103922 and No. 31230061), Natural Science Foundation of Fujian Province (No. 2017 J0102). LV is supported by the "111" Program of SAFEA and the Minjiang Scholarships.

Author's contributions

QZ carried out data processing, analysis and wrote the manuscript; MY conceived and designed the study; DM helped to collect the data used in this study and LV helped draft the manuscript. All authors read and approved the final manuscript.

Competing interests

The authors declare that they have no competing interests.

Author details

[1]State Key Laboratory for Ecological Pest Control of Fujian/Taiwan Crops and College of Life Science, Fujian Agriculture and Forestry University, Fuzhou 350002, China. [2]Institute of Applied Ecology, Fujian Agriculture and Forestry University, Fuzhou 350002, China. [3]Fujian-Taiwan Joint Centre for Ecological Control of Crop Pests, Fujian Agriculture and Forestry University, Fuzhou 350002, China. [4]Key Laboratory of Integrated Pest Management for Fujian-Taiwan Crops, Ministry of Agriculture, Fuzhou 350002, China. [5]Department of Biological Sciences, Brock University, 1812 Sir Isaac Brock Way, St. Catharines, ON L2S 3A1, Canada.

References

1. Marques-Bonet T, Girirajan S, Eichler EE. The origins and impact of primate segmental duplications. Trends Genet. 2009;25(10):443–54.
2. Duda TF Jr, Palumbi SR. Molecular genetics of ecological diversification: duplication and rapid evolution of toxin genes of the venomous gastropod Conus. Proc Natl Acad Sci U S A. 1999;96(12):6820–3.
3. Lynch M, Conery JS. The evolutionary fate and consequences of duplicate genes. Science. 2000;290(5494):1151–5.
4. Conant GC, Wolfe KH. Turning a hobby into a job: how duplicated genes find new functions. Nat Rev Genet. 2008;9:938–50.
5. Albano F, Anelli L, Zagaria A, Coccaro N, D'Addabbo P, Liso V, et al. Genomic segmental duplications on the basis of the rearrangement in chronic myeloid leukemia. Oncogene. 2010;29(17):2509–16.
6. Marques-Bonet T, Kidd JM, Ventura M, Graves TA, Cheng Z, Hillier LW, et al. A burst of segmental duplications in the genome of the African great ape ancestor. Nature. 2009;457(7231):877–81.
7. Goidts V, Cooper DN, Armengol L, Schempp W, Conroy J, Estivill X, et al. Complex patterns of copy number variation at sites of segmental duplications: an important category of structural variation in the human genome. Hum Genet. 2006;120(2):270–84.
8. Sharp AJ, Locke DP, McGrath SD, Cheng Z, Bailey JA, Vallente RU, et al. Segmental duplications and copy-number variation in the human genome. Am J Hum Genet. 2005;77(1):78–88.
9. Bailey JA, Gu Z, Clark RA, Reinert K, Samonte RV, Schwartz S, et al. Recent segmental duplications in the human genome. Science. 2002;297(5583): 1003–7.
10. Cheng Z, Ventura M, She X, Khaitovich P, Graves T, Osoegawa K, et al. A genome-wide comparison of recent chimpanzee and human segmental duplications. Nature. 2005;437(7055):88–93.
11. Kim PM, Lam HY, Urban AE, Korbel JO, Affourtit J, Grubert F, et al. Analysis of copy number variants and segmental duplications in the human genome: Evidence for a change in the process of formation in recent evolutionary history. Genome Res. 2008;18(12):1865–74.
12. She X, Jiang Z, Clark RA, Liu G, Cheng Z, Tuzun E, et al. Shotgun sequence assembly and recent segmental duplications within the human genome. Nature. 2004;431(7011):927–30.
13. She X, Cheng Z, Zollner S, Church DM, Eichler EE. Mouse segmental duplication and copy number variation. Nat Genet. 2008;40(7):909–14.
14. Umemori J, Mori A, Ichiyanagi K, Uno T, Koide T. Identification of both copy number variation-type and constant-type core elements in a large segmental duplication region of the mouse genome. BMC Genomics. 2013;14:455.
15. Nicholas TJ, Cheng Z, Ventura M, Mealey K, Eichler EE, Akey JM. The genomic architecture of segmental duplications and associated copy number variants in dogs. Genome Res. 2009;19(3):491–9.
16. You M, Yue Z, He W, Yang X, Yang G, Xie M, et al. A heterozygous moth genome provides insights into herbivory and detoxification. Nat Genet. 2013;45(2):220–5.
17. Duan J, Li R, Cheng D, Fan W, Zha X, Cheng T, et al. SilkDB v2.0: a platform for silkworm (Bombyx mori) genome biology. Nucleic Acids Res. 2009; 38(Database issue):D453–6.
18. Consortium THG. Butterfly genome reveals promiscuous exchange of mimicry adaptations among species. Nature. 2012;487(7405):94–8.
19. Zhan S, Reppert SM. MonarchBase: the monarch butterfly genome database. Nucleic Acids Res. 2012;41(Database issue):D758–63.
20. Xu HE, Zhang HH, Xia T, Han MJ, Shen YH, Zhang Z. BmTEdb: a collective database of transposable elements in the silkworm genome. Database. 2013;2013:bat055.
21. Zhan S, Merlin C, Boore JL, Reppert SM. The monarch butterfly genome yields insights into long-distance migration. Cell. 2011;147(5):1171–85.
22. Gotz S, Garcia-Gomez JM, Terol J, Williams TD, Nagaraj SH, Nueda MJ, et al. High-throughput functional annotation and data mining with the Blast2GO suite. Nucleic Acids Res. 2008;36(10):3420–35.
23. Finn RD, Mistry J, Tate J, Coggill P, Heger A, Pollington JE, et al. The Pfam protein families database. Nucleic Acids Res. 2010;38(Database issue):D211–22.
24. Fang SM, Hu BL, Zhou QZ, Yu QY, Zhang Z. Comparative analysis of the silk gland transcriptomes between the domestic and wild silkworms. BMC Genomics. 2015;16:60.
25. Whittington E, Zhao Q, Borziak K, Walters JR, Dorus S. Characterisation of the Manduca sexta sperm proteome: Genetic novelty underlying sperm composition in Lepidoptera. Insect Biochem Mol Biol. 2015;62:183–93.
26. Trapnell C, Pachter L, Salzberg SL. TopHat: discovering splice junctions with RNA-Seq. Bioinformatics. 2009;25(9):1105–11.
27. Trapnell C, Roberts A, Goff L, Pertea G, Kim D, Kelley DR, et al. Differential gene and transcript expression analysis of RNA-seq experiments with TopHat and Cufflinks. Nat Protoc. 2012;7(3):562–78.
28. Emanuel BS, Shaikh TH. Segmental duplications: an 'expanding' role in genomic instability and disease. Nat Rev Gene. 2001;2(10):791–800.
29. Fiston-Lavier AS, Anxolabehere D, Quesneville H. A model of segmental duplication formation in Drosophila melanogaster. Genome Res. 2007;17(10): 1458–70.
30. Zhao Q, Zhu Z, Kasahara M, Morishita S, Zhang Z. Segmental duplications in the silkworm genome. BMC Genomics. 2013;14:521.
31. Eichler EE. Segmental duplications: what's missing, misassigned, and misassembled–and should we care? Genome Res. 2001;11(5):653–6.
32. Bailey JA, Eichler EE. Primate segmental duplications: crucibles of evolution, diversity and disease. Nat Rev Gene. 2006;7(7):552–64.
33. Osanai-Futahashi M, Suetsugu Y, Mita K, Fujiwara H. Genome-wide screening and characterization of transposable elements and their distribution analysis in the silkworm, Bombyx mori. Insect Biochem Mol Biol. 2008;38(12):1046–57.
34. Sun W, Shen YH, Han MJ, Cao YF, Zhang Z. An adaptive transposable element insertion in the regulatory region of the EO gene in the domesticated silkworm, Bombyx mori. Mol Biol Evol. 2014;31(12):3302–13.
35. Regier JC, Zwick A, Cummings MP, Kawahara AY, Cho S, Weller S, et al. Toward reconstructing the evolution of advanced moths and butterflies (Lepidoptera: Ditrysia): an initial molecular study. BMC Evol Biol. 2009;9:280.
36. Xia Q, Guo Y, Zhang Z, Li D, Xuan Z, Li Z, et al. Complete resequencing of 40 genomes reveals domestication events and genes in silkworm (Bombyx). Science. 2009;326(5951):433–6.
37. de la Paz C-MM, Wheat CW, Vogel H, Soderlind L, Janz N, Nylin S. Mechanisms of macroevolution: polyphagous plasticity in butterfly larvae revealed by RNA-Seq. Mol Ecol. 2013;22(19):4884–95.

38. Hughes J, Vogler AP. Gene expression in the gut of keratin-feeding clothes moths (Tineola) and keratin beetles (Trox) revealed by subtracted cDNA libraries. Insect Biochem Mol Biol. 2006;36(7):584–92.

39. Rountree DB, Bollenbacher WE. The release of the prothoracicotropic hormone in the tobacco hornworm, *Manduca sexta*, is controlled intrinsically by juvenile hormone. J Exp Biol. 1986;120:41–58.

40. Riddiford L, Hiruma K, Zhou X, Nelson C. Insights into the molecular basis of the hormonal control of molting and metamorphosis from *Manduca sexta* and *Drosophila melanogaster*. Insect Biochem Mol Biol. 2003;33(12):1327–38.

41. Luque T, Okano K, O'Reilly DR. Characterization of a novel silkworm (*Bombyx mori*) phenol UDP-glucosyltransferase. Eur J Biochem. 2002;269(3):819–25.

42. Morrison DK. KSR: a MAPK scaffold of the Ras pathway? J Cell Sci. 2001;114:1609–12.

43. Yin A, Pan L, Zhang X, Wang L, Yin Y, Jia S, et al. Transcriptomic study of the red palm weevil Rhynchophorus ferrugineus embryogenesis. Insect Sci. 2015; 22(1):65–82.

44. Liu GE, Ventura M, Cellamare A, Chen L, Cheng Z, Zhu B, et al. Analysis of recent segmental duplications in the bovine genome. BMC Genomics. 2009;10:571.

45. Scott JG. Cytochromes P450 and insecticide resistance. Insect Biochem Mol Biol. 1999;29(9):757–77.

46. Bernhardt R. Cytochromes P450 as versatile biocatalysts. J Biotechnol. 2006; 124(1):128–45.

47. Iga M, Kataoka H. Recent studies on insect hormone metabolic pathways mediated by cytochrome P450 enzymes. Biol Pharm Bull. 2012;35(6):838–43.

48. Yu L, Tang W, He W, Ma X, Vasseur L, Baxter SW, et al. Characterization and expression of the cytochrome P450 gene family in diamondback moth, *Plutella xylostella* (L.). Sci Rep. 2015;5:8952.

49. Kanost M, Gorman M. Phenoloxidases in insect immunity. In: Beckage N.E. (Ed.), Insect Immunol. Oxford: Academic Press; 2008.

50. Sun W, Shen YH, Yang WJ, Cao YF, Xiang ZH, Zhang Z. Expansion of the silkworm GMC oxidoreductase genes is associated with immunity. Insect Biochem Mol Biol. 2012;42(12):935–45.

51. Zhang Y, Dong Z, Liu S, Yang Q, Zhao P, Xia Q. Identification of novel members reveals the structural and functional divergence of lepidopteran-specific Lipoprotein_11 family. Funct Integr Genomic. 2012;12(4):705–15.

52. Munoz-Descalzo S, Terol J, Paricio N. Cabut, a C2H2 zinc finger transcription factor, is required during Drosophila dorsal closure downstream of JNK signaling. Dev Biol. 2005;287(1):168–79.

53. Stranger BE, Forrest MS, Dunning M, Ingle CE, Beazley C, Thorne N, et al. Relative impact of nucleotide and copy number variation on gene expression phenotypes. Science. 2007;315(5813):848–53.

54. Merla G, Howald C, Henrichsen CN, Lyle R, Wyss C, Zabot MT, et al. Submicroscopic deletion in patients with Williams-Beuren syndrome influences expression levels of the nonhemizygous flanking genes. Am J Hum Genet. 2006;79(2):332–41.

55. Henrichsen CN, Vinckenbosch N, Zollner S, Chaignat E, Pradervand S, Schutz F, et al. Segmental copy number variation shapes tissue transcriptomes. Nat Genet. 2009; 41(4):424–9.

56. Blekhman R, Oshlack A, Gilad Y. Segmental duplications contribute to gene expression differences between humans and chimpanzees. Genetics. 2009; 182(2):627–30.

57. Xiang H, Zhu J, Chen Q, Dai F, Li X, Li M, et al. Single base-resolution methylome of the silkworm reveals a sparse epigenomic map. Nat Biotechnol. 2010;28(5):516–20.

8

Using the Neandertal genome to study the evolution of small insertions and deletions in modern humans

Manjusha Chintalapati, Michael Dannemann and Kay Prüfer[*] (ID)

Abstract

Background: Small insertions and deletions occur in humans at a lower rate compared to nucleotide changes, but evolve under more constraint than nucleotide changes. While the evolution of insertions and deletions have been investigated using ape outgroups, the now available genome of a Neandertal can shed light on the evolution of indels in more recent times.

Results: We used the Neandertal genome together with several primate outgroup genomes to differentiate between human insertion/deletion changes that likely occurred before the split from Neandertals and those that likely arose later. Changes that pre-date the split from Neandertals show a smaller proportion of deletions than those that occurred later. The presence of a Neandertal-shared allele in Europeans or Asians but the absence in Africans was used to detect putatively introgressed indels in Europeans and Asians. A larger proportion of these variants reside in intergenic regions compared to other modern human variants, and some variants are linked to SNPs that have been associated with traits in modern humans.

Conclusions: Our results are in agreement with earlier results that suggested that deletions evolve under more constraint than insertions. When considering Neandertal introgressed variants, we find some evidence that negative selection affected these variants more than other variants segregating in modern humans. Among introgressed variants we also identify indels that may influence the phenotype of their carriers. In particular an introgressed deletion associated with a decrease in the time to menarche may constitute an example of a former Neandertal-specific trait contributing to modern human phenotypic diversity.

Keywords: Neandertal, Ancient DNA, Indel evolution

Background

Recent advances in sequencing technology and laboratory methods made it possible to sequence complete genomes from ancient DNA preserved in human remains [1, 2]. High-coverage genome sequences were recently generated from ancient humans, including those from a Neandertal individual [3], a member of a group of close extinct relatives of all present-day humans. The sequence of the Neandertal genome provides a unique resource to study evolution since it can be used to sort sequence changes on the human lineage into those that likely occurred recently (i.e. those that are not shared with the Neandertal) and those that occurred earlier. Of

particular interest are those modern human changes that rose to high frequency or reached fixation since the split from Neandertals, since these changes may underlie phenotypes that were advantageous during the evolution of modern humans. Among the sequence changes reaching fixation are also 4113 insertion/deletion variants [3].

The study of the high-coverage Neandertal genome confirmed that modern humans outside of Africa trace a small percentage of their ancestry back to an admixture event with Neandertals [3]. Although likely of small magnitude, the admixture event occurred sufficiently recent so that a large fraction (around 40%) of the Neandertal genome sequence segregates within present-day humans [4, 5]. However, not all regions in the genome show an equal fraction of Neandertal ancestry, suggesting that a substantial fraction of the introgressed material was lost

* Correspondence: pruefer@eva.mpg.de
Max Planck Institute for Evolutionary Anthropology, 04103 Leipzig, Germany

due to negative selection [4–8], while some specific variants rose to higher frequency likely because they conveyed a selective advantage to the carriers [9–13]. Among the introgressed variants are also larger deletions, some of which are overlapping exons [14].

Although most of the sequence variation among human individuals is due to single nucleotide changes, insertion/deletions (indels), which are approximately one order of magnitude less abundant, have a higher probability to affect function than nucleotide substitutions [15]. However, indels are often excluded in evolutionary studies. This is likely due to the particular challenges of indel genotyping [16–18] and the heterogeneous processes generating indels that lead to a large variation in mutation rates along the genome [19, 20]. For example, deletions were found to evolve, on average, under stronger negative selection on the human lineage than insertions by one study that compared fixed to polymorphic indels [21], while a later study found the opposite signal using the allele frequency spectrum between populations [22]. The cause for this discrepancy may lie in homoplasy, i.e. the independent occurrence of identical changes on several lineages, which can lead to the mis-assignment of the ancestral state and type of the mutation (insertion or deletion) [19].

Here, we use the Neandertal genome [3] together with data of present-day humans from the 1000 Genomes data [23] to identify indels and divide the set of indels further into those that likely occurred after the split from Neandertals, those that arose before the split from Neandertals and likely introgressed indels. We test for different patterns of selection between these sets and compile a list of introgressed and modern-human-fixed indels that may contribute to modern human phenotype.

Results

Indels on the human lineage

To identify insertion and deletion events on the modern human lineage and to alleviate the problem of mis-assignment of the ancestral state, we aligned the human reference genome with seven primate genomes and inferred the derived state on the human lineage by requiring an identical ancestral allele in all seven primate genomes. An insertion on the human lineage is called only when all non-human primates show a deletion compared to the human state, and a human-specific deletion when all primates show an insertion. Our method detected 315,513 indels of 1-5 bp in length in the human reference genome. Of these, most indels (315,412) were covered in the high-coverage Neandertal genome [3].

We used data from the 1000 Genomes project phase 3 [23] to further increase the set of variable indels. Variants marked as copy number variants ("<CN>") exceeded the length of variants considered here and were excluded. A total of 2,982,740 were inferred from 1000 Genomes data

after filtering out sites with more than one derived variant. These indels were assigned an ancestral and derived state by comparison to seven non-human primate genomes, and overlapped with the Neandertal genotypes, resulting in 989,138 indels of length 1-5 bp. Combining indels identified using the human reference and those identified using the 1000 Genomes data, yielded 1,232,285 indels of size 1-5 bps on the human lineage (245,520 appear fixed and 986,765 were segregating in present day populations) (Fig. 1, Additional file 1: Figure S1).

We computed the ratio of deletions to insertions for fixed (1.45) and polymorphic indels (2.06) and found ratios higher than 1, consistent with deletions accumulating approximately twice as fast as insertions [21, 24–26].

Modified McDonald–Kreitman test on the human lineage indels

Previous studies have used a modified version of the McDonald-Kreitman test [19, 21, 27] – comparing the ratio of fixed deletions to fixed insertions to the ratio of polymorphic deletions to polymorphic insertions – to test whether insertions and deletions are affected differently by selection. Under neutrality both the fixed and polymorphic ratios are solely dependent on the rates at which insertions and deletions are generated, i.e. at a roughly 2-fold higher rate for deletions than for insertions. Under this assumption, the ratios of deletions to insertions are not expected to differ significantly from each other when comparing fixed to polymorphic sites. However, a departure from this expectation can emerge if one type of change is selectively favored over the other, and is thus biased towards fixation. Note that such a signal requires only the average selection pressures on insertions and deletions to differ; the majority of both types of changes can still be selectively neutral.

We first applied the modified McDonald Kreitman test to all 1–5 base pair long indels described in the previous section and found a significant difference between the ratio of fixed to the ratio of polymorphic indels ($p < 2.2e\text{-}16$). In order to test whether this signal is driven by a certain length of indels, we repeated the test for each length, separately, and found that the signal persists in all comparisons (Table 1). This result is consistent with the results of Kivkstat and Duret [19] and Sjödin et al. [21] suggesting that deletions are under stronger negative selection than insertions.

It is interesting to note, that the ratio of polymorphic insertions and polymorphic deletions also differs significantly between all lengths (pairwise comparisons between lengths 1-5 bps: p-values < 0.05).

Derived allele frequency of the human lineage indels

The derived allele frequency spectra (AFS) of polymorphic insertions and deletions can be used as an alternative to

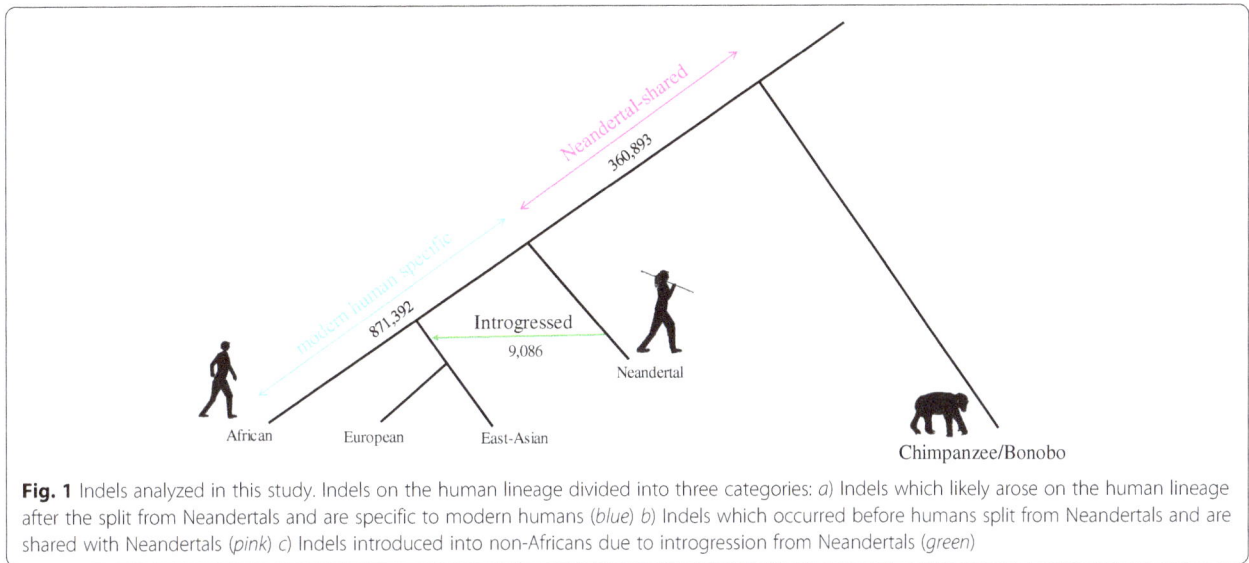

Fig. 1 Indels analyzed in this study. Indels on the human lineage divided into three categories: *a*) Indels which likely arose on the human lineage after the split from Neandertals and are specific to modern humans (*blue*) *b*) Indels which occurred before humans split from Neandertals and are shared with Neandertals (*pink*) *c*) Indels introduced into non-Africans due to introgression from Neandertals (*green*)

test for differences in selection pressure affecting both types of changes [28]. The test is based on the idea that a favorable allele will on average segregate at higher frequency compared to neutral alleles, and neutral alleles will in turn segregate at higher frequencies compared to deleterious alleles [29]. We found that the AFS for deletions differs significantly from the AFS for insertions (two-sided Wilcoxon rank sum test; $p < 2.2e\text{-}16$; Fig. 2), with deletions showing an excess of low-frequency alleles compared to insertions. This signal is detected consistently in all 1000 Genomes populations and for all sizes of indels (1-5 bp) (Additional file 1: Figure S2).

Genomic distribution of the human lineage indels
The previous two tests examined the difference in selection pressure between insertion and deletions by comparing allele frequencies. However, if one type of change is more often deleterious, a difference may also be visible in the fraction of insertions and deletions residing in regions that are more likely functional as compared to regions that are more likely neutral. We tested this hypothesis by annotating indels by their genomic location using the Variant Effect Predictor [30]. As expected,

a major fraction of indels fall in intronic and intergenic regions while a much smaller fraction fall in coding regions. In addition, intergenic regions show a statistically significant higher fraction of deletions than insertions (binomial test; $p = 7.3e\text{-}119$; FDR adjusted $p = 7.8e\text{-}117$) while the opposite is true for intronic regions (p-value $= 3.6e\text{-}59$; FDR adjusted $p = 1.3e\text{-}57$; Fig. 3a). This observation is compatible with the notion that deletions are more constraint than insertions. However, we caution that differences in insertion and deletion frequencies may also be influenced by other factors, such as sequence context [31–33] leading to unequal insertion and deletion mutation rates between classes of genomic regions.

Modern human specific and Neandertal-shared indels
We divided indels into those that were identified in the genomes of the modern human reference and the Neandertal, and those that were only detected in the human reference. A total of 37,443 indels were modern human specific and 265,975 were shared. The frequency of modern human specific indels can be used to calculate a relative divergence of the human reference to the Neandertal genome. We calculate a divergence of 12.3%

Table 1 Fixed and polymorphic indels on the human lineage by length

Category	1 bp	2 bp	3 bp	4 bp	5 bp	Sum: 1–5 bp
Fixed deletions	86,791	26,860	14,802	12,161	4689	145,303
Fixed insertions	66,333	13,589	8022	9406	2867	100,217
Fixed rDI	1.30	1.97	1.845	1.29	1.635	1.449
Polymorphic deletions	344,533	121,548	82,114	84,393	31,607	664,195
Polymorphic insertions	226,712	38,545	21,147	27,180	8986	322,570
Polymorphic rDI	1.519	3.15	3.88	3.10	3.52	2.06

Ratio of deletions to insertions (rDI) is given for polymorphic and fixed indels of different lengths on the human lineage. Fisher's exact tests were applied to the counts of fixed and polymorphic insertions and deletions in each column and yielded p-values $< 2.2e\text{-}16$ in all comparisons

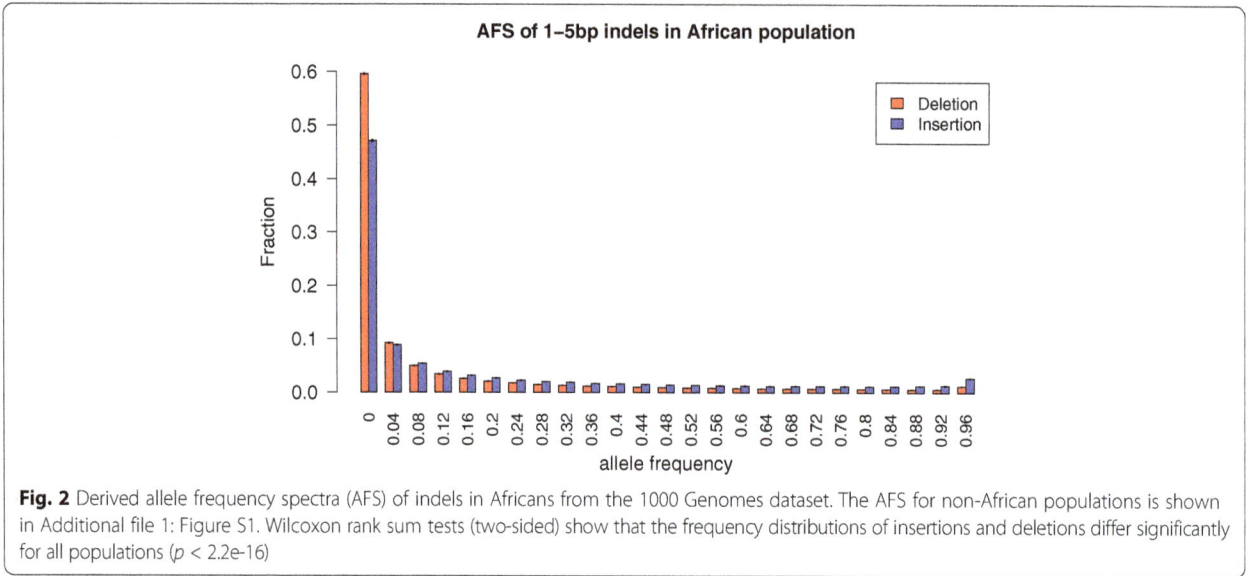

Fig. 2 Derived allele frequency spectra (AFS) of indels in Africans from the 1000 Genomes dataset. The AFS for non-African populations is shown in Additional file 1: Figure S1. Wilcoxon rank sum tests (two-sided) show that the frequency distributions of insertions and deletions differ significantly for all populations ($p < 2.2e-16$)

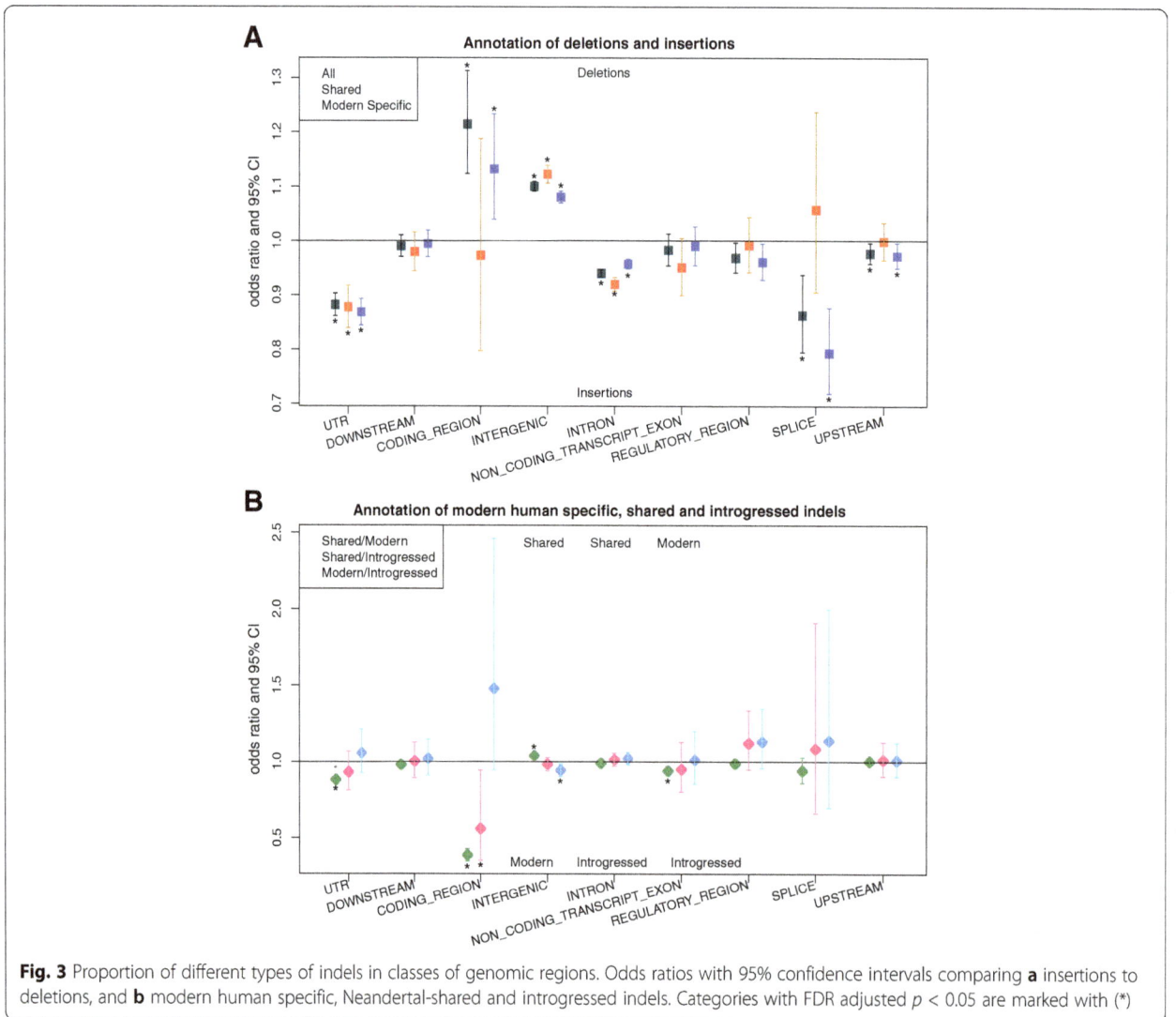

Fig. 3 Proportion of different types of indels in classes of genomic regions. Odds ratios with 95% confidence intervals comparing **a** insertions to deletions, and **b** modern human specific, Neandertal-shared and introgressed indels. Categories with FDR adjusted $p < 0.05$ are marked with (*)

relative to the divergence to the common ancestor with chimpanzee, close to the range of values calculated using nucleotide differences (11.2–11.8%, see SI6a in [3]).

We classified polymorphic indels from the 1000 Genomes Project [23] into those for which the derived variant is shared with the Neandertal and those where the derived variant is only observed in modern humans, and pooled the dataset with human-reference specific indels. As expected by the difference in age, the majority of the 360,893 shared indels were fixed (243,060 fixed and 117,833 polymorphic) while the majority of the 871,392 modern human specific indels were polymorphic (2460 are fixed and 868,932 are polymorphic).

Neandertal-shared indels are expected to be on average older than indels that are specific to modern humans. We use this expectation to test again for differences between the ratios of deletions to insertions of both age-classes, similar to the McDonald-Kreitman test. The ratio of deletions to insertions is significantly lower for shared compared to modern human specific indels (Table 2, Additional file 1: Table S6A) consistent with earlier comparisons between fixed and polymorphic indels. When annotating indels with the class of genomic regions that is most likely to influence phenotype, we find that a significantly higher fraction of Neandertal-shared indels fall in intergenic regions compared to modern human specific indels (Fisher's exact test; $p = 1.77e$-21; False Discovery Rate (FDR) adjusted $p = 9.57e$-21; odds ratio: 0.96) while modern human specific indels fall more often in intronic regions compared to shared indels, although this difference is not significant after multiple testing correction (Fisher's exact test; $p = 0.04$, FDR adjusted $p = 0.08$; odds ratio: 1.009). These signals are consistent with a longer exposure to selection for Neandertal-shared indels as compared to modern human specific indels (Fig. 3b). For both classes, a higher fraction of insertions resides in coding regions compared to deletions and the opposite pattern is observed for intergenic regions (Fig. 3a).

Putatively introgressed indels

A subset of the indel variants segregating in non-African populations trace their ancestry back to Neandertals, through an admixture event between non-Africans and

Neandertals 50–60 thousand years ago [34, 35]. By conditioning on the absence of the derived variant in Africans and the presence of the derived variant in Neandertals and either the East-Asian or European population, we identified 9086 putatively introgressed indels. Of these 6070 are deletions and 3016 insertions with an average allele frequency of 0.027 in Europeans and 0.048 in the East-Asian population (Wilcoxon rank test for European frequencies smaller less than East-Asian frequencies: $p = 1.8e$-35). The difference in allele frequencies between both populations is similar to the one observed for putatively introgressed SNPs (Europeans: 0.026; East-Asians: 0.046; Additional file 1: Figure S4). Following the patterns observed for all indels, we found that a higher fraction of introgressed deletions fall in intergenic regions compared to introgressed insertions (Additional file 1: Figure S3). Our previous results, comparing modern human specific to Neandertal-shared indels, remain significant when putatively introgressed indels are removed (Additional file 1: Tables S6A, 6B).

To gain insight into the selection pressures that acted on introgressed indels, we compared their distribution over classes of genomic regions with those of Neandertal-shared (but without introgressed) and modern human specific indels (Fig. 3b). Interestingly, we find that a slightly smaller proportion of introgressed indels fall in intron regions compared with the other two classes of indels (55.3% versus 55.7% and 55.9% for Neandertal-shared and human specific, respectively), and a slightly larger proportion of introgressed indels fall into intergenic regions (31.5% versus 31.2% and 30.3%) (Additional file 1: Table S5). For Neandertal-shared variants this difference to introgressed indels is not statistically significant (Fisher's exact test, one-sided, $p = 0.23$, odds ratio: 1.016 and $p = 0.26$, odds ratio: 0.985 for intron and intergenic regions, respectively), while modern human specific variants show a significant difference to introgressed variants for intergenic ($p = 0.007$; FDR adjusted $p = 0.02$; odds ratio: 0.945) but not intron regions ($p = 0.13$, odds ratio: 1.024). Coding regions, however, contain a significantly lower proportion of Neandertal-shared variants than introgressed variants (1.2% versus 2.1%, $p = 0.02$; FDR adjusted $p = 0.04$) while the comparison to modern human specific indels shows a non-significant trend in the opposite direction (3.0% versus 2.0%, $p = 0.05$; FDR adjusted $p = 0.10$). These results raise the possibility that introgressed indels have been subjected to stronger negative selection, either before or after the introgression event, compared to modern human specific indels.

Genome wide association studies (GWAS) and Introgressed Indels

To find further evidence for a potential impact of introgressed indels on human phenotypes, we searched for introgressed indels that are in perfect linkage to SNPs

Table 2 Contingency table contrasting modern human specific indels and shared indels

Category	Shared	Modern Human specific
Deletions	205,075	604,423
Insertions	155,818	266,969
Ratio(Deletions/Insertions)	1.316	2.26

The ratios of deletions to insertions are significantly different between the shared and modern human specific classes (Fisher's exact test; $p < 2.2e$-16, odds ratio = 0.58)

that are linked to specific traits by genome wide association studies (Table 3). We found 9 traits (p < 1e-5) related to neurological, immunological, developmental and metabolic phenotypes, among others. Interestingly, one SNP at chromosome 2: 157,096,776 (in perfect linkage disequilibrium (LD) with an indel in chromosome 2: 157,099,707) is associated with menarche [36]. Human carriers of the Neandertal allele showed an earlier menarche compared to non-carriers and the Neandertal allele has a higher prevalence in Europeans (allele frequency = 0.06) compared to Asians (allele frequency = 0.01).

To further corroborate that the menarche associated indel is introgressed, we plotted putatively introgressed variants in the individuals from the 1000 genomes surrounding the location of the indel (Fig. 4). In concordance with the low frequency in present-day Europeans and East-Asians, few individuals showed the homozygous derived state for introgressed variants in the vicinity of the indel. We observe haplotypes of different lengths, two of which encompass an additional introgressed indel upstream. Regions overlapping the indel have also been found to be introgressed in two independent maps of introgressed segments in non-Africans [4, 5].

Considering introgressed variants shared between non-African individuals, we estimate a minimum length of 180,900 bp for the introgressed segment. The recombination rate in this region is 0.23 cM/Mb, which is lower than the genome wide average of ca. 1 cM/Mb [37]. We calculated the probability of a region to retain a length of at least ~180 kb if it was generated by incomplete lineage sorting (see [9, 38]) and found that this scenario is unlikely (p = 0.003).

Gene ontology enrichment

To test whether any group of functionally related genes experienced a shift in constraint from before the split to after the split from Neandertals, we used the Gene Ontology to group and compare the number of shared and modern human specific indels annotated to genes. Two Gene Ontology categories, *ion channel complex* and *transmembrane complex*, showed significant enrichment for modern human specific indels compared to shared indels (Additional file 1: Table S3). This result could be explained by a relaxation of constraint for these genes in modern humans since the split from Neandertals. No significant enrichment was found in the opposite direction, or when comparing introgressed indels to shared indels.

List of potentially disruptive indels

Identifying the molecular basis for modern human specific traits remains a challenge for the study of human evolution. Here we provide a list of candidates that have been fixed in modern humans since the split from Neandertals and that are annotated as a top 1% disruptive change according to the CADD package (Additional file 1: Table S1). Further study is needed to test whether some of these changes play a role in modern human specific traits.

In addition, we provide a list of putatively introgressed indels which have been classified as likely disruptive (Additional file 1: Table S2). Variants with the highest allele frequency differences (measured by F_{ST}) between Europeans and East Asians that also show some evidence for disruptiveness are listed in Additional file 1: Table S4.

Discussion

Small indels are a common type of sequence variation among present-day humans [39]. Here we used several outgroups to divide indels into derived insertions and derived deletions. Each class was further categorized using the Neandertal genome into those derived variants that are shared with Neandertals and those that are only observed in modern humans.

Previous studies have compared allele frequencies and the proportion of fixed to polymorphic insertions and deletions to gain insight into differences in selection pressures affecting each type of change. Some of these studies found that deletions appear to be more deleterious than insertions [21] while others found the opposite

Table 3 Introgressed indels linked to genome-wide association studies candidates

Chr	Indel pos.	SNP pos.	SNP rs ID	P-value	Trait	EAS_AF	EUR_AF	Gene	C-score(indel)	Ref.
1	196,365,712	196,376,474	rs16839886	7.26E-06	Age-related macular degeneration	0.0129	0.0746	KCNT2	8.613	[59]
1	209,987,712	209,988,047	rs10863790	1.00E-14	Cleft lip	0.4286	0.0139	NA	4.657	[60]
1	210,174,981	210,174,417	rs11119388	4.57E-09	Cleft lip	0.4454	0.0089	SYT14	9.739	[60]
14	55,769,446	55,808,151	rs17673930	1.89E-40	Protein biomarker	0.006	0.0805	CHMP4BP1	6.577	[61]
2	157,099,707	157,096,776	rs17188434	1.00E-09	Menarche (age at onset)	0.0099	0.0606	NA	6.499	[36]
3	23,386,162	23,385,942	rs17013049	2.78E-06	Type 2 diabetes	0.1131	0.0239	UBE2E2	6.473	[62]
3	100,671,648	100,647,927	rs13060137	8.96E-08	Suicide attempts in bipolar disorder	0.002	0.1531	RNU6-865P	3.313	[63]
8	20,253,488	20,263,408	rs1016646	9.45E-06	Preeclampsia	0.0923	0.0636	NA	2.605	[59]
9	87,171,753	87,177,586	rs35640669	5.17E-08	Insulin-related traits	0.0546	0.0348	NA	0.207	[64]

EAS AF East Asian allele frequency; *EUR AF* European allele frequency

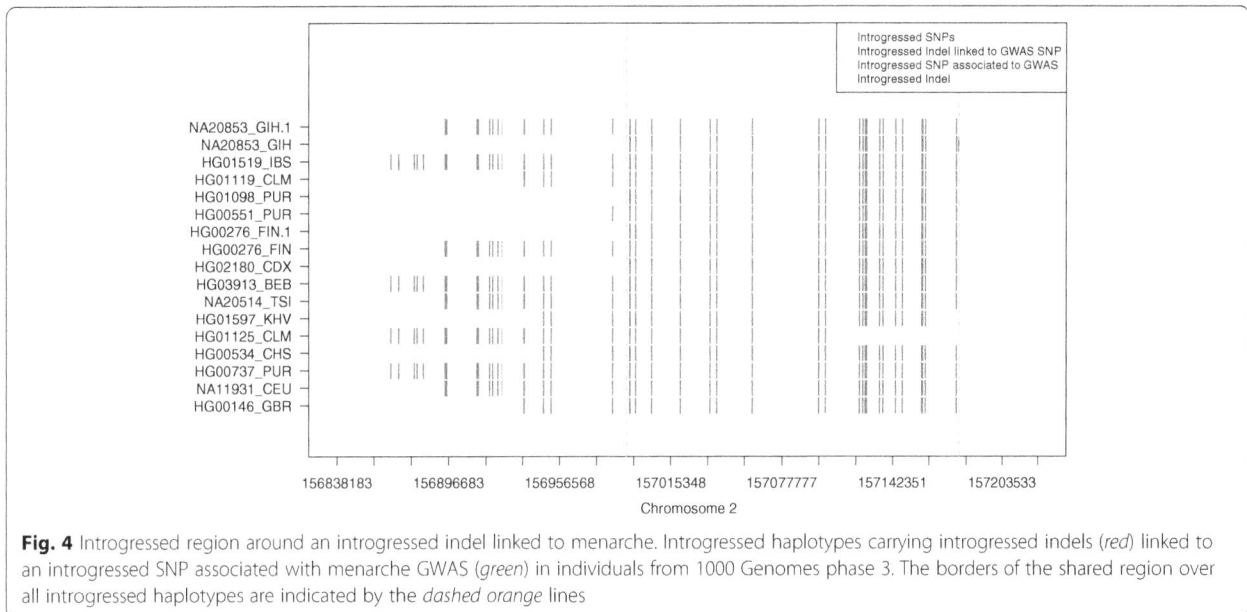

Fig. 4 Introgressed region around an introgressed indel linked to menarche. Introgressed haplotypes carrying introgressed indels (*red*) linked to an introgressed SNP associated with menarche GWAS (*green*) in individuals from 1000 Genomes phase 3. The borders of the shared region over all introgressed haplotypes are indicated by the *dashed orange* lines

[22], a discrepancy that may in parts be explained by homoplasy, i.e. the independent formation of identical indels on several lineages (Additional file 1: Table S7) [19]. Here we used seven primate outgroups to reduce the effect of homoplasy and to confidently call the ancestral state. Comparing allele frequencies, fixed to polymorphic indels, and Neandertal-shared indels to modern human specific, we found that the proportion of deletions is consistently smaller for older time-frames and higher frequencies, suggesting that deletions are on average more deleterious than insertions. Interestingly, this signal is further corroborated by the genomic distribution of insertions and deletions, where we found a higher fraction of insertions in coding regions compared to deletions, which show a higher fraction that fall in intergenic regions. Despite these consistent results, we caution that our strong requirement of several primate outgroups selects for sites that remain stable over millions of years of evolution, and that our results only hold for this subset of indels, which will be biased towards conserved and against repetitive genomic regions. We also caution that insertions and deletions are influenced by other factors than selection [31–33], and that they may form at unequal rates in different functional classes of the genome.

In principle, a Neandertal-shared derived variant could originate through two processes: either the variant came into existence before the Neandertal and modern human populations split, or the variant was contributed to modern humans after the split, through admixture. We make use of previous results that found Neandertal admixture in out-of-African populations to select indels that likely entered through admixture by selecting those Neandertal-shared variants that are only observed in out-of-African

populations. Putatively introgressed indels showed similar differences in the genome-wide distribution of insertions and deletions, with a higher fraction of insertions residing in coding regions and a higher fraction of deletions in intergenic regions. This suggests that introgressed deletions are more deleterious than introgressed insertions.

At least 40% of the introgressing Neandertal genomes can be reconstructed from Neandertal segments segregating in out-of-African populations [4, 5]. However, the distribution of these segments has been found to be non-uniform, with genes and conserved regions of the genome showing an underrepresentation of Neandertal introgression. The patterns of depletion of Neandertal-ancestry near genes have been used to estimate the strength of selection against introgressed segments [7] and simulations suggest that Neandertals may have had a reduction in fitness compared to modern humans [6]. Comparing Neandertal-shared indels, which represent older events and which are mostly fixed, to putatively introgressed indels, we find no evidence for stronger negative selection acting on introgressed variants. However, compared to derived indels on the modern human lineage, Neandertal introgressed variants show some signals that are compatible with more selective constraint, suggesting that selection acted on these variants either before or after introgression.

Some introgressed indels may also convey an advantage to the carrier and there are several examples of variants that have been positively selected after introgression [9, 10, 12, 13]. Among the introgressed indels that were present in both Europeans and East-Asians and that scored highest for affecting phenotype we found a frame shift insertion in PTCHD3 (patched domain-containing protein-3), a gene which has a role in sperm development or sperm function [40] and that has been found to contain a risk-allele for asthma [41]. However, due to the high-frequency in which null-mutations are encountered in present-day humans, the gene has also been suggested to be non-essential in humans [42]. Some introgressed indels were also in perfect linkage with SNPs associated with different traits and diseases in genome-wide association studies. One such indel was linked to a variant associated with a decrease in the time to menarche in humans. The direction of effect for this variant is in line with research suggesting that Neandertals may have reached adulthood earlier than present-day humans [43, 44].

Conclusions

Indels in modern humans contribute not only to genetic variation, but also appear to be subject to stronger selective forces than nucleotide substitutions. Here, we studied the differences between insertions and deletions using the Neandertal genome as an additional outgroup and found signals that suggest that deletions are more often deleterious than insertions. Among the indels segregating in modern humans are those that entered out-of-African populations by admixture with Neandertals. While these introgressed indels show weak signals of negative selection compared to other variants that segregate in modern humans, we find some variants that may contribute to functional variation in present-day humans. Arguably the most interesting variant with phenotype association is an introgressed indel variant associated with a decreased time to menarche, raising the possibility that some of the introgressing Neandertals' life history traits now form part of the modern human variation.

Methods
Primate multiple sequence alignment
Pairwise alignments between the human reference genome (Lander, Linton et al. 2001) (GhRch37/hg19) and six primates (chimpanzee [45] (panTro4), gorilla [46] (gorGor3), orangutan [47] (ponAbe2), gibbon [48] (nomLeu1), rhesus macaque [49] (rheMac3) and marmoset [50] (calJac3)) were downloaded from the UCSC genome browser [51] and converted into MAF format. In addition, the bonobo [52] (panpan1.1) pairwise whole genome alignment to hg19 was prepared in house following the processing applied to genomes for inclusion in the UCSC genome browser. All seven pairwise alignments were joined into one multiple sequence alignment using the reference guided alignment program multiz (Version: roast.v3; Command-line: "roast + E=hg19 '(((((hg19(panTro4,panpan1.1) gorGor3)ponAbe2)nomLeu1)rheMac3)calJac3)' <input_files.sing.maf> <output_file.maf>" , [53]). The resulting file was filtered to retain only those alignment blocks that include sequence from the genomes of all eight species.

Inferring fixed derived and polymorphic indels on the human lineage
Human polymorphic indels were extracted from the 1000 Genomes phase 3 dataset [54]. The indels were further filtered by requiring overlap with the eight species whole genome alignment and requiring all seven non-human reference sequences in this alignment to agree. The ancestral state of polymorphic indels was then called as the non-human state and the alternative labeled as a derived human-specific indel. Further filtering was carried out to remove sites with more than one derived variant and long variants marked as variable in copy number (denoted as <CN> for the derived state in the 1000 Genomes data).

Human-specific derived indels were called fixed if all non-human species showed an identical insertion or deletion difference compared to the human reference sequence and if the position was not listed as polymorphic in the 1000 Genomes data.

Inferring modern human specific indels and putatively introgressed indels using the Neandertal genome
We used the genotype calls of a Neandertal from the Altai Mountains [3] to divide derived human-specific indels into those that are shared with Neandertals and those that are specific to modern humans.

Two percent of the genomes of present day non-Africans show high similarity to the Neandertal genome due to a recent admixture event with Neandertals [3]. To infer putatively introgressed indels we used our set of human polymorphic indels and filtered for variants that are fixed in individuals from sub-Saharan African populations (Luhya, Yoruba, Gambian, Mende and Esan) and show an alternate allele in the Europeans (Utah, Finland, British and Scotland, Iberian, Toscani) or East-Asians (Chinese Dai, Han Chinese, Southern Han Chinese, Japanese, Kinh) that is shared with the Neandertal. We used the same process to infer introgressed SNPs.

Contrasting fixed and polymorphic insertions and deletions
The McDonald–Kreitman test [27] compares the number of polymorphic changes within one species to the number of fixed changes when comparing to another species between two types of sites, neutral and non-neutral. Under neutrality the ratio of non-neutral to neutral changes is expected to be equal when comparing fixed to polymorphic changes. Negative selection is expected to reduce the number of non-neutral changes that reach fixation, while repeated positive selection is expected to increase the number of non-neutral changes due to the rapid fixation of advantageous alleles. Following the approach of Sjödin et al. and Kvikstad and Duret [19, 21], we applied the concept of the McDonald-Kreitman test to indels by

comparing the number of insertions and deletions that are polymorphic to those that are fixed-derived on the human-lineage. *P*-values were calculated using Fisher's exact test as implemented in R [54].

Derived site frequency spectra of polymorphic indels
We used the average allele-frequency for different populations from the 1000 Genomes phase 3 data to tabulate the site frequency spectra. Site frequency spectra were compared by applying a two-sided Wilcoxon rank sum test with continuity correction to the distribution of indel frequencies.

The minor allele frequencies for potentially introgressed indels in the European populations and the East Asian populations from the 1000 Genomes Project phase 3 were tabulated to arrive at an AFS of introgressed indels.

Annotation of indels
Indels were annotated using the variant effect predictor (VEP) [30] version 78 using the option "–most_severe" to limit the output to one annotation per indel. For each annotated region and for each pair of classes of indels, we determined the significance by calculating Fisher's exact test on a 2 × 2 contingency table contrasting the two classes and the counts inside and outside of the annotated region. The combined list of *p*-values from all variance effect predictor tests was FDR adjusted using the p.adjust() function implemented in R.

In addition the Combined Annotation Dependent Depletion (CADD v1.3) tool [55] was used to score the tentative phenotypic impact of indels. CADD annotates each indel with a phred-scaled C-score. A cutoff of 20 on the C-score was applied to generate lists of indels with an increased chance of affecting phenotype.

Genome wide association studies
We used a collection of genome-wide association studies (GWASdb, version: 2015 August, hg19 dbSNP142, [56]) to find potential phenotype associations for introgressed indels. Since indels are typically excluded in the process of GWAS, we sought to detect SNPs that are in perfect LD with introgressed indels in the 1000 Genomes. Indels that showed an identical combination of reference/non-reference genotypes as the GWAS associated SNP in all individuals were considered completely linked. We report phenotype associations for each indel that is in perfect LD with a SNP that has been associated with the corresponding phenotype with a *p*-value of at least 1e-5.

Gene ontology enrichment
Enrichment of indels in specific gene categories was tested using the software package FUNC version 0.4.7

[57]. For this, we selected indels that were assigned to genes based on the VEP annotation and further annotated these indels to gene categories used the Gene Ontology. To account for all the plausible effects, for instance when an indel overlaps more than one gene, we allowed multiple annotations of each indel. Genes were assigned corresponding GO categories using the Ensembl database [version: Ensembl Genes 75 (GRCh37)] [58].

In addition to explanations involving selection, the number of indels in a gene category can vary due to differences in mutation rates or due to a difference in gene-length between categories. In order to avoid these issues, we compared the number of two types of indels per category using the FUNC implementation of the binomial test. The following types of indels were compared:

1. Indels shared with Neandertals to those that are modern human specific
2. Indels that are shared with Neandertals to those that introgressed from Neandertals.

We chose a *p*-value cutoff of less than or equal to 0.05 for the family wise error rate (FWER) to filter for significantly enriched categories.

Abbreviations
AFS: Allele Frequency Spectrum; FDR: False Discovery Rate; Indel: Insertion or Deletion; LD: Linkage Disequilibrium; SNP: Single Nucleotide Polymorphism

Acknowledgements
We thank Stéphane Peyrégne and Steffi Grote for their help in the analysis, Janet Kelso, Martin Petr, Udo Stenzel, Amin Saffari and Rohit Kolora for helpful discussions, and two anonymous reviewers for helpful comments.

Funding
This research was funded by the Max Planck Society.

Authors' contributions
MC carried out analyses. MC, MD and KP wrote the manuscript. KP designed the study. All authors read and approved the final manuscript.

Competing interests
The authors declare that they have no competing interests.

References
1. Der Sarkissian C, Allentoft ME, Avila-Arcos MC, Barnett R, Campos PF, Cappellini E, Ermini L, Fernandez R, da Fonseca R, Ginolhac A, et al. Ancient genomics. Philos Trans R Soc Lond Ser B Biol Sci. 2015; 370(1660):20130387.
2. Kelso J, Prüfer K. Ancient humans and the origin of modern humans. Curr Opin Genet Dev. 2014;29:133–8.
3. Prüfer K, Racimo F, Patterson N, Jay F, Sankararaman S, Sawyer S, Heinze A, Renaud G, Sudmant PH, de Filippo C, et al. The complete genome sequence of a Neanderthal from the Altai Mountains. Nature. 2014;505(7481):43–9.

4.   Sankararaman S, Mallick S, Dannemann M, Prufer K, Kelso J, Paabo S, Patterson N, Reich D. The genomic landscape of Neanderthal ancestry in present-day humans. Nature. 2014;507(7492):354–7.

5.   Vernot B, Akey JM. Resurrecting surviving Neanderthal lineages from modern human genomes. Science. 2014;343(6174):1017–21.

6.   Harris K, Nielsen R. The genetic cost of Neanderthal introgression. Genetics. 2016;203(2):881–91.

7.   Juric I, Aeschbacher S, Coop G. The Strength of Selection Against Neanderthal Introgression. PLoS Genetics. 2016;12(11):e1006340.

8.   Mallick S, Li H, Lipson M, Mathieson I, Gymrek M, Racimo F, Zhao M, Chennagiri N, Nordenfelt S, Tandon A, et al. The Simons genome diversity project: 300 genomes from 142 diverse populations. Nature. 2016;538(7624):201–6.

9.   Dannemann M, Andres AM, Kelso J. Introgression of Neandertal- and Denisovan-like haplotypes contributes to adaptive variation in human toll-like receptors. Am J Hum Genet. 2016;98(1):22–33.

10. Deschamps M, Laval G, Fagny M, Itan Y, Abel L, Casanova JL, Patin E, Quintana-Murci L. Genomic signatures of selective pressures and introgression from archaic Hominins at human innate immunity genes. Am J Hum Genet. 2016;98(1):5–21.

11. Mendez FL, Watkins JC, Hammer MF. A haplotype at STAT2 Introgressed from neanderthals and serves as a candidate of positive selection in Papua New Guinea. Am J Hum Genet. 2012;91(2):265–74.

12. Gittelman RM, Schraiber JG, Vernot B, Mikacenic C, Wurfel MM, Akey JM. Archaic Hominin admixture facilitated adaptation to out-of-Africa environments. Curr Biol. 2016;26(24):3375–82.

13. Racimo F, Gokhman D, Fumagalli M, Ko A, Hansen T, Moltke I, Albrechtsen A, Carmel L, Huerta-Sanchez E, Nielsen R. Archaic adaptive introgression in TBX15/WARS2. Mol Biol Evol. 2017;34(3):509–24.

14. Lin Y-L, Pavlidis P, Karakoc E, Ajay J, Gokcumen O. The evolution and functional impact of human deletion variants shared with archaic Hominin genomes. Mol Biol Evol. 2015;32(4):1008–19.

15. Montgomery SB, Goode DL, Kvikstad E, Albers CA, Zhang ZD, Mu XJ, Ananda G, Howie B, Karczewski KJ, Smith KS, et al. The origin, evolution, and functional impact of short insertion-deletion variants identified in 179 human genomes. Genome Res. 2013;23(5):749–61.

16. Hasan MS, Wu X, Zhang L. Performance evaluation of indel calling tools using real short-read data. Hum Genomics. 2015;9:20.

17. Mullaney JM, Mills RE, Pittard WS, Devine SE. Small insertions and deletions (INDELs) in human genomes. Hum Mol Genet. 2010;19(R2):R131–6.

18. Neuman JA, Isakov O, Shomron N. Analysis of insertion-deletion from deep-sequencing data: software evaluation for optimal detection. Brief Bioinform. 2013;14(1):46–55.

19. Kvikstad EM, Duret L. Strong heterogeneity in mutation rate causes misleading hallmarks of natural selection on indel mutations in the human genome. Mol Biol Evol. 2014;31(1):23–36.

20. Belinky F, Cohen O, Huchon D. Large-scale parsimony analysis of metazoan indels in protein-coding genes. Mol Biol Evol. 2010;27(2):441–51.

21. Sjödin P, Bataillon T, Schierup MH. Insertion and deletion processes in recent human history. PLoS One. 2010;5(1):e8650.

22. Huang S, Li J, Xu A, Huang G, You L. Small insertions are more deleterious than small deletions in human genomes. Hum Mutat. 2013;34(12):1642–9.

23. 1000 Genomes Project Consortium. A global reference for human genetic variation. Nature. 2015;526(7571):68–74.

24. Fan Y, Wang W, Ma G, Liang L, Shi Q, Tao S. Patterns of insertion and deletion in mammalian genomes. Curr Genomics. 2007;8(6):370–8.

25. Matthee CA, Eick G, Willows-Munro S, Montgelard C, Pardini AT, Robinson TJ. Indel evolution of mammalian introns and the utility of non-coding nuclear markers in eutherian phylogenetics. Mol Phylogenet Evol. 2007;42(3):827–37.

26. Ophir R, Graur D. Patterns and rates of indel evolution in processed pseudogenes from humans and murids. Gene. 1997;205(1–2):191–202.

27. McDonald JH, Kreitman M. Adaptive protein evolution at the Adh locus in drosophila. Nature. 1991;351(6328):652–4.

28. Nielsen R, Hellmann I, Hubisz M, Bustamante C, Clark AG. Recent and ongoing selection in the human genome. Nat Rev Genet. 2007;8(11):857–68.

29. Fay JC, Wyckoff GJ, Wu CI. Positive and negative selection on the human genome. Genetics. 2001;158(3):1227–34.

30. McLaren W, Pritchard B, Rios D, Chen Y, Flicek P, Cunningham F. Deriving the consequences of genomic variants with the Ensembl API and SNP Effect Predictor. Bioinformatics. 2010;26(16):2069–70.

31. Kondrashov AS, Rogozin IB. Context of deletions and insertions in human coding sequences. Hum Mutat. 2004;23(2):177–85.

32. Kvikstad EM, Chiaromonte F, Makova KD. Ride the wavelet: a multiscale analysis of genomic contexts flanking small insertions and deletions. Genome Res. 2009;19(7):1153–64.

33. Kvikstad EM, Tyekucheva S, Chiaromonte F, Makova KD. A macaque's-eye view of human insertions and deletions: differences in mechanisms. PLoS Comput Biol. 2007;3(9):1772–82.

34. Lazaridis I, Patterson N, Mittnik A, Renaud G, Mallick S, Kirsanow K, Sudmant PH, Schraiber JG, Castellano S, Lipson M, et al. Ancient human genomes suggest three ancestral populations for present-day Europeans. Nature. 2014;513(7518):409–13.

35. Sankararaman S, Patterson N, Li H, Paabo S, Reich D. The date of interbreeding between Neandertals and modern humans. PLoS Genet. 2012;8(10):e1002947.

36. Elks CE, Perry JR, Sulem P, Chasman DI, Franceschini N, He C, Lunetta KL, Visser JA, Byrne EM, Cousminer DL, et al. Thirty new loci for age at menarche identified by a meta-analysis of genome-wide association studies. Nat Genet. 2010;42(12):1077–85.

37. Hinch AG, Tandon A, Patterson N, Song Y, Rohland N, Palmer CD, Chen GK, Wang K, Buxbaum SG, Akylbekova EL, et al. The landscape of recombination in African Americans. Nature. 2011;476(7359):170–5.

38. Huerta-Sanchez E, Jin X, Asan, Bianba Z, Peter BM, Vinckenbosch N, Liang Y, Yi X, He M, Somel M, et al. altitude adaptation in Tibetans caused by introgression of Denisovan-like DNA. Nature. 2014;512(7513):194–7.

39. Mills RE, Pittard WS, Mullaney JM, Farooq U, Creasy TH, Mahurkar AA, Kemeza DM, Strassler DS, Ponting CP, Webber C, et al. Natural genetic variation caused by small insertions and deletions in the human genome. Genome Res. 2011;21(6):830–9.

40. Fan J, Akabane H, Zheng X, Zhou X, Zhang L, Liu Q, Zhang YL, Yang J, Zhu GZ. Male germ cell-specific expression of a novel patched-domain containing gene Ptchd3. Biochem Biophys Res Commun. 2007;363(3):757–61.

41. White MJ, Risse-Adams O, Goddard P, Contreras MG, Adams J, Hu D, Eng C, Oh SS, Davis A, Meade K, et al. Novel genetic risk factors for asthma in African American children: precision medicine and the SAGE II study. Immunogenetics. 2016;68(6–7):391–400.

42. Ghahramani Seno MM, Kwan BY, Lee-Ng KK, Moessner R, Lionel AC, Marshall CR, Scherer SW. Human PTCHD3 nulls: rare copy number and sequence variants suggest a non-essential gene. BMC Med Genet. 2011;12:45.

43. Ramirez Rozzi FV, Bermudez de Castro JM. surprisingly rapid growth in Neanderthals. Nature. 2004;428(6986):936–9.

44. Smith TM, Tafforeau P, Reid DJ, Pouech J, Lazzari V, Zermeno JP, Guatelli-Steinberg D, Olejniczak AJ, Hoffman A, Radovcic J, et al. Dental evidence for ontogenetic differences between modern humans and Neanderthals. Proc Natl Acad Sci U S A. 2010;107(49):20923–8.

45. The Chimpanzee Sequencing and Analysis Consortium. Initial sequence of the chimpanzee genome and comparison with the human genome. Nature. 2005;437(7055):69–87.

46. Scally A, Dutheil JY, Hillier LW, Jordan GE, Goodhead I, Herrero J, Hobolth A, Lappalainen T, Mailund T, Marques-Bonet T, et al. Insights into hominid evolution from the gorilla genome sequence. Nature. 2012;483(7388):169–75.

47. Locke DP, Hillier LW, Warren WC, Worley KC, Nazareth LV, Muzny DM, Yang SP, Wang Z, Chinwalla AT, Minx P, et al. Comparative and demographic analysis of orang-utan genomes. Nature. 2011;469(7331):529–33.

48. Carbone L, Harris RA, Gnerre S, Veeramah KR, Lorente-Galdos B, Huddleston J, Meyer TJ, Herrero J, Roos C, Aken B, et al. Gibbon genome and the fast karyotype evolution of small apes. Nature. 2014;513(7517):195–201.

49. Gibbs RA, Rogers J, Katze MG, Bumgarner R, Weinstock GM, Mardis ER, Remington KA, Strausberg RL, Venter JC, Wilson RK, et al. Evolutionary and biomedical insights from the rhesus macaque genome. Science. 2007; 316(5822):222–34.

50. The Marmoset Sequencing and Analysis Consortium. The common marmoset genome provides insight into primate biology and evolution. Nat Genet. 2014;46(8):850–7.

51. Speir ML, Zweig AS, Rosenbloom KR, Raney BJ, Paten B, Nejad P, Lee BT, Learned K, Karolchik D, Hinrichs AS, et al. The UCSC Genome Browser database: 2016 update. Nucleic Acids Res. 2016;44(D1):D717–25.

52. Prüfer K, Munch K, Hellmann I, Akagi K, Miller JR, Walenz B, Koren S, Sutton G, Kodira C, Winer R, et al. The bonobo genome compared with the chimpanzee and human genomes. Nature. 2012;486(7404):527–31.

53. Blanchette M, Kent WJ, Riemer C, Elnitski L, Smit AF, Roskin KM, Baertsch R, Rosenbloom K, Clawson H, Green ED, et al. Aligning multiple genomic sequences with the threaded blockset aligner. Genome Res. 2004;14(4):708–15.

54. The 1000 Genomes Project Consortium. A global reference for human genetic variation. Nature. 2015;526(7571):68–74.

55. R Core Team: R: A Language and Environment for Statistical Computing. Vienna: R Foundation for Statistical Computing; 2017.

56. Kircher M, Witten DM, Jain P, O'Roak BJ, Cooper GM, Shendure J. A general framework for estimating the relative pathogenicity of human genetic variants. Nat Genet. 2014;46(3):310–5.

57. Li MJ, Wang P, Liu X, Lim EL, Wang Z, Yeager M, Wong MP, Sham PC, Chanock SJ, Wang J. GWASdb: a database for human genetic variants identified by genome-wide association studies. Nucleic Acids Res. 2012; 40(Database issue):D1047–54.

58. Prüfer K, Muetzel B, Do HH, Weiss G, Khaitovich P, Rahm E, Paabo S, Lachmann M, Enard W. FUNC: a package for detecting significant associations between gene sets and ontological annotations. BMC Bioinformatics. 2007;8:41.

59. Cunningham F, Amode MR, Barrell D, Beal K, Billis K, Brent S, Carvalho-Silva D, Clapham P, Coates G, Fitzgerald S, et al. Ensembl 2015. Nucleic Acids Res. 2015;43(D1):D662–9.

60. Fritsche LG, Chen W, Schu M, Yaspan BL, Yu Y, Thorleifsson G, Zack DJ, Arakawa S, Cipriani V, Ripke S, et al. Seven new loci associated with age-related macular degeneration. Nat Genet. 2013;45(4):433–9. 439e431-432

61. Beaty TH, Murray JC, Marazita ML, Munger RG, Ruczinski I, Hetmanski JB, Liang KY, Wu T, Murray T, Fallin MD, et al. A genome-wide association study of cleft lip with and without cleft palate identifies risk variants near MAFB and ABCA4. Nat Genet. 2010;42(6):525–9.

62. de Boer RA, Verweij N, van Veldhuisen DJ, Westra HJ, Bakker SJ, Gansevoort RT, Muller Kobold AC, van Gilst WH, Franke L, Mateo Leach I, et al. A genome-wide association study of circulating galectin-3. PLoS One. 2012;7(10):e47385.

63. Cho YS, Chen CH, Hu C, Long J, Ong RT, Sim X, Takeuchi F, Wu Y, Go MJ, Yamauchi T, et al. Meta-analysis of genome-wide association studies identifies eight new loci for type 2 diabetes in east Asians. Nat Genet. 2011;44(1):67–72.

64. Perlis RH, Huang J, Purcell S, Fava M, Rush AJ, Sullivan PF, Hamilton SP, McMahon FJ, Schulze TG, Potash JB, et al. Genome-wide association study of suicide attempts in mood disorder patients. Am J Psychiatry. 2010; 167(12):1499–507.

65. Chen G, Bentley A, Adeyemo A, Shriner D, Zhou J, Doumatey A, Huang H, Ramos E, Erdos M, Gerry N, et al. Genome-wide association study identifies novel loci association with fasting insulin and insulin resistance in African Americans. Hum Mol Genet. 2012;21(20):4530–6.

Genetic basis of brain size evolution in cetaceans: insights from adaptive evolution of seven primary microcephaly (*MCPH*) genes

Shixia Xu*, Xiaohui Sun, Xu Niu, Zepeng Zhang, Ran Tian, Wenhua Ren, Kaiya Zhou and Guang Yang* ⓘ

Abstract

Background: Cetacean brain size expansion is an enigmatic event in mammalian evolution, yet its genetic basis remains poorly explored. Here, all exons of the seven primary microcephaly (*MCPH*) genes that play key roles in size regulation during brain development were investigated in representative cetacean lineages.

Results: Sequences of *MCPH2–7* genes were intact in cetaceans but frameshift mutations and stop codons was identified in *MCPH1*. Extensive positive selection was identified in four of six intact MCPH genes: *WDR62*, *CDK5RAP2*, *CEP152*, and *ASPM*. Specially, positive selection at *CDK5RAP2* and *ASPM* were examined along lineages of odontocetes with increased encephalization quotients (EQ) and mysticetes with reduced EQ but at *WDR62* only found along odontocete lineages. Interestingly, a positive association between evolutionary rate (ω) and EQ was identified for *CDK5RAP2* and *ASPM*. Furthermore, we tested the binding affinities between Calmodulin (CaM) and ASPM IQ motif in cetaceans because only CaM combined with IQ, can ASPM perform the function in determining brain size. Preliminary function assay showed binding affinities between CaM and IQ motif of the odontocetes with increased EQ was stronger than for the mysticetes with decreased EQ. In addition, evolution rate of *ASPM* and *CDK5RAP2* were significantly related to mean group size (as one measure of social complexity).

Conclusions: Our study investigated the genetic basis of cetacean brain size evolution. Significant positive selection was examined along lineages with both increased and decreased EQ at *CDK5RAP2* and *ASPM*, which is well matched with cetacean complex brain size evolution. Evolutionary rate of *CDK5RAP2* and *ASPM* were significantly related to EQ, suggesting that these two genes may have contributed to EQ expansion in cetaceans. This suggestion was further indicated by our preliminary function test that *ASPM* might be mainly linked to evolutionary increases in EQ. Most strikingly, our results suggested that cetaceans evolved large brains to manage complex social systems, consisting with the 'social brain hypothesis', as evolutionary rate of *ASPM* and *CDK5RAP2* were significantly related to mean group size.

Keywords: Cetacea, *MCPHs*, Positive selection, Brain size evolution, EQ, Group size

Background

Cetaceans are a group of secondary-adapted marine mammals, the common ancestor of which diverged from terrestrial artiodactyls approximately 53–56 million years ago (Ma) [1]. Cetaceans comprise of one extinct (Archaeoceti) and two extant (Mysticeti and Odontoceti) suborders. Until 40 Ma, archaeocetes were completely aquatic [2]. Extant cetaceans evolved from archaeocetes at about 34 Ma, and distributed nearly all the world's oceans, as well as some freshwater lakes and rivers [1]. During the transition from terrestrial to fully aquatic environments, significant changes affecting sensory systems, locomotion, breathing and feeding took place [3], of which, the large brains of modern cetaceans remains the most enigmatic [4].

* Correspondence: xushixia78@163.com; gyang@njnu.edu.cn
Jiangsu Key Laboratory for Biodiversity and Biotechnology, College of Life Sciences, Nanjing Normal University, 1 Wenyuan Road, Nanjing 210023, China

Fossil and anatomical evidence show that the brain size (or encephalization) of archaeocetes was similar to their ancestor [5]. The brain mass of mysticetes increased, but their body mass increased at a much rapid rate leading to a decrease in relative brain size (as measured by encephalization quotient, i.e. EQ, which accounts for body size) with a mean EQ of 0.21 [6]. Reduced EQ was suggested to be related to the massive biomechanical forces needed to open their mouths when feeding [7]. In contrast, the relative brain size of the odontocetes (mean EQ = 3.10) is higher than that of their ancestors (mean EQ = 2.43) [6]. For example, some species of delphinids have EQs (4–5) significantly larger than nonhuman primates (EQ = 3.3) and are second only to humans (EQ = 7) [4]. What kinds of selective pressures could have led to this rapid increase in brain size among odontocetes? Previous ecological studies have shown that odontocetes have high degrees of encephalization primarily as an adaptation for living in complicated social groups (cooperative actions and fission–fusion societies), as asserted by the 'social brain hypothesis' [6, 8–12]. This hypothesis is supported by the positive correlation between EQ and one measure of social complexity, group size, in many dolphin species [13]. However, the brain size evolution in cetaceans remains poorly tested at the molecular level.

There has been much interest in exploring the genetic basis of adaptive phenotypes using candidate genes or gene families. Primary microcephaly (*MCPH*) genes are thought to play key roles in size regulation during brain development in mammals due to the fact that *MCPH* gene mutations can cause severe defects in the development of cerebral in humans [14, 15]. So far, seven autosomal recessive loci (*MCPH1–7*) have been identified in humans: *MCPH1*, *WDR62*, *CDK5RAP2*, *CEP152*, *ASPM*, *CENPJ* and *STIL* [15]. Specially, there is a 1:1 orthologous relationship of these genes in dolphins and whales. Increasingly, evidence of positive selection was detected at *MCPH* gene, special for *ASPM*, *CDK5RAP2* and *MCPH1* genes, across primate lineages with massive brain size [16, 17]. A recent study suggested that *ASPM* was linked to both evolutionary increases and decreases in brain size in anthropoids with positive selection acting on both lineages [18]. Investigating the genetic basis of brain size evolution in cetaceans was only recently commenced with examination of some exons of the *MCPH1* and *ASPM* genes in cetaceans [19, 20]. Evidence of positive selection was determined on exons 3 and 18 of the *ASPM* gene in odontocetes, especially for species in the superfamily Delphinoidea, which was well matched with the two major events of relative brain size enlargement in cetaceans [20]. However, no significant association was identified between the evolutionary rate of the two *ASPM* exons and brain size phenotypes in cetaceans [21]. For *MCPH1* gene, no compelling evidence of positive

selection and association was examined between *MCPH1* evolution and brain evolution in cetaceans [19]. Different results from these two *MCPH* genes suggest a complex mechanism of brain size evolution in cetaceans. Here, the evolution of seven *MCPH* genes was investigated in representative species of major cetacean lineages. First, we tested whether different selection patterns acted on the seven *MCPH* genes in cetacean lineages and whether positive selection was limited to lineages with high EQs. Second, we explored the putative association between the evolutionary rate of *MCPH* genes and some morphological variables of cetacean brains. Third, the correlation between *MCPH* evolution and group size was examined to test support for the 'social brain hypothesis' at the molecular level.

Methods

MCPH genes and primary treatments

A total of 16 cetacean species (three mysticetes and 13 odontocetes) was used in our study (see Additional file 1: Table S1). Of them, samples of 13 species (two mysticetes and 11 odontocetes) were collected from dead individuals in the wild and no ethics statement was required in such occasions. We first downloaded the full-length coding sequence (CDS) of seven *MCPH* genes from the database of Orthologous Mammalian Markers (http://www.ortho-mam.univ-montp2.fr/orthomam/html/) and designed the primers to amplify each exon of seven *MCPH* genes. We then sequenced these exons in 13 samples and merged into the predicted full-length CDS. Species information, genomic DNA extraction, primer design, PCR amplification and sequencing were conducted as described in Xu et al. [20]. The *MCPH* orthologous gene sequences from the other three cetacean species (i.e. bowhead whale *Balaena mysticetus*, killer whale *Orcinus orca*, and sperm whale *Physeter macrocephalus*) and two terrestrial relatives (Hippopotamus *Hippopotamus amphibius* and cow *Bos taurus*) were available from their published genomes (see Additional file 1: Table S1). We used two alignment methods, i.e. CLUSTAL and MUSCLE, as implemented in MEGA 6.0 (Tamura et al. [22]) to align the nucleotide sequences of each *MCPH* gene and verified by visual inspection.

Molecular evolution analysis

The nonsynonymous (d_N) / synonymous substitution (d_S) rate ($\omega = d_N / d_S$) is a measure of selective pressure, with values of $\omega >1$, $= 1$, and <1 indicating positive selection, neutral selection, and purifying selection, respectively. The ω ratios were estimated using the codon-based maximum likelihood (ML) models implemented in the CODEML program in PAML 4.4 [23]. A well-accepted phylogeny of Cetacea [24] was used as input tree in our analysis for each gene. The phylogenetic trees were also reconstructed using the maximum-likelihood (ML) and Bayesian inference (BI, See the Additional file 2). The gene trees were similar to the well-accepted phylogeny with only some minor differences within Delphinidae (see the Additional file 2: Figure S1). According to Yang et al. [25] suggestions, the minor differences in the phylogeny do not make any significant difference in identification of positively selected sites. Hence, selection detection using the gene trees produced results similar to those obtained using the well-accepted phylogeny of Cetartiodactyla (see the Additional file 1: Table S2), only the latter result was reported here.

To examine the probabilities of sites under positive selection in the six intact genes, we first used two pair of site models: M7 (beta) versus M8 (beta & $\omega_2 > 1$) [26], and M8a (beta & $\omega_2 = 1$) versus M8 [27] implemented in the CODEML program of PAML 4.7. The nested models were compared using a likelihood ratio test (LRT) with a χ^2 distribution. Positively selected sites in the M8 were identified using a Bayes Empirical Bayes (BEB) analysis [28] with posterior probabilities ≥ 0.80. Positive selected sites were further detected by fixed effects likelihood (FEL) performed in HYPHY [29] (via the www.datamonkey.org web server), with the default settings with significance levels of 0.2. We then performed selective pressure detection using TreeSAAP v.3.2 [30], which detected selection based on 31 physicochemical amino acid properties. All magnitude category 6–8 changes with P values ≤ 0.05 were used as an index for the degree of radical amino acid substitution and positive selection.

To evaluate whether positive selection was restricted to specific cetacean lineages, we used branch models (including free-ratio model and two-ratio model [31, 32]) and branch-site model implemented in CODEML [23]. The free-ratio model (M1) that assumes an independent ω ratio for each branch was compared with the null one-ratio model (M0) with the same ω for all branches [31]. Two-ratio model and branch-site model require the foreground branches (lineages tested to be under positive selection) and background branches (rest of the lineages) to be defined a priori. Each cetacean lineage across the Cetartiodactyla phylogeny was used as the foreground branch, respectively, whereas the remaining branches were treated as background branches for each gene. We compared the two-ratio model where ω was allowed to differ in the background and a foreground branch with null M0 model [31, 32]. By contrast, the branch-site model appeared to be conservative but far more powerful than the branch-based model. The modified branch-site model A with ω varying among sites and among lineages [28, 33] was tested against the recommended null hypothesis of no selection in any of the foreground or background branches. According to Zhang et al. [33], sites identified by this method can still be evidence of positive selection

even if the BEB cannot be reliably inferred because the tested positive selection at any single site may not be strong enough for the BEB probability to reach high levels if positive selection has affected only one lineage or a very few lineages on the tree. A false discovery rate (FDR) correction for multiple tests was applied to the LRT P values for branch-site model analysis [34].

Association analysis between gene evolution and phenotypes

To explore potential relationships between the evolutionary rate (ω) of *MCPH* genes and brain size phenotypes we used the method of Montgomery et al. [16] whereby the root-to-tip ω is regarded as more suitable for regression against phenotypic data from extant species because the root-to-tip ω is more inclusive of the evolutionary history of a locus [35]. The following phenotypic traits including absolute brain mass and absolute body mass from 11 cetacean species were derived from published data [5, 32–39] (see Additional file 1: Table S3). EQ values for each species were derived from the eq. EQ = brain weight/0.12 (body weight) $^{0.67}$ from Jerison [40]. We then used phylogenetic generalized least squares (PGLS) regression, performed in R 3.1.2 using the packages Caper [41], to analyze the relationship between log-transformed (root-to-tip ω) and each log-transformed morphological variables. The detailed analytical procedures were provided in the supplementary material online (see Additional file 2).

Social complexity may be as a major force for brain evolution in cetaceans [12]. When group size is used as a measure of social complexity, one should expect to see strong relationships between group size and the evolutionary rates of *MCPH* genes in cetacean species. Thus, PGLS regression analyses were also used to test whether there was association between mean group size and gene evolution of such *MCPH* genes subject to positive selection.

Three-dimensional (3D) structure prediction

To provide further insights into the functional significance of these positively selected sites, they were mapped onto the three-dimensional (3D) structures of *MCPH* genes using PYMOL (http://pymol.sourceforge.net/). We first predicted the 3D structures of *MCPH* genes following homology modeling using the SWISS-MODEL (http://swissmodel.expasy.org). However, no significant amino acid sequence similarity with known proteins or no consistent results were detected using BLAST tools. Thus, an ab-initio 3D model of each *MCPH* gene was constructed by I-TASSER [42, 43], a state-of-the-art hierarchical protein structure modeling approach based on secondary-structure enhanced profile-profile threading alignment [42].

Functional assays of the *ASPM* gene

Bioinformatics analyses provided a series of support for the positive selection on *MCPH* genes, especially *CDK5RAP2* and *ASPM*, in the brain size enlargement or decrease of cetaceans. However, considering that these genes were first identified in humans, it would be better to provide some additional functional evidences which could not only suggest these genes do have function in cetaceans, but further support their important roles in brain size evolution that were revealed through bioinformatics analyses. In the present study, we chose *ASPM* gene as a representative to conduct functional experiments and expected to present partial and preliminary evidences.

The *ASPM* gene is a major determinant of cerebral cortical size [44]. Four distinguished regions were identified in the predicated *ASPM* gene in human, comprising a putative microtubule-binding domain, a calponin-homology domain, an IQ repeat domain containing 81 IQ motifs (CaM-binding motifs), and a C-terminal region [44]. Of these, CaM-binding IQ motifs were suggested to play an essential role in determining brain size [45]. Considering that the CDS of the cetacean *ASPM* gene was more than 10,300 bp in CDS with 28 exons of the bottlenose dolphin *T. truncatu*, the expression vector cannot carry such a large DNA fragment. Thus, only the one IQ motif (the 23rd IQ when the bottlenose dolphin *Tursiops truncates* as reference) including one positively selected site, i.e. 1684) and its adjacent 16 amino acid motif were cloned into the expression vector. Six species of odontocetes with increased EQ and three species of mysticetes with reduced EQ were chosen as representative samples in this study.

We then used GST pull-down assay to evaluate if CaM interact with IQ motif in cetacean lineages with brain enlargement or decrease. Binding affinities between CaM and IQ motif were further quantitatively determined by biolayer interferometry (BLI) using the FortoBio Octet Red system. Binding affinities were calculated using FortoBio Data Acquisition 6.3 software (FortoBio). Equilibrium dissociation constants (Kd) were calculated as the ratio of dissociation and association rate constants (Koff /Kon). All detailed experiment procedures were listed in the Additional file 2.

Results

Almost all exons of the seven *MCPH* genes were successfully amplified in 13 representative species of cetaceans. Newly obtained sequences for each *MCPH* gene (GenBank accession nos. KY011963- KY012055) covered at least 74.16% of the full CDS: 86.31% for *MCPH1*, 85.68% for *WDR62*, 74.78% for *CDK5RAP2*, 88.91% for *CEP152*, 90.34% for *ASPM*, 74.16% for

CENPJ and 87.86% for *STIL*. Evidence of intact gene was identified in cetaceans at six genes (*MCPH2–7*) because non-frameshift insertions/deletions and premature stop codons were observed in sequences of these genes. However, *MCPH1* has been pseudogenized in some cetacean lineages because premature stop codons were identified in three species (i.e. beluga *Delphinapterus leucas*, Risso's dolphin *Grampus griseus* and killer whale *O. orca*) and frameshift insertions/deletions were examined in five cetacean species (including Blainville's beaked whale *Mesoplodon densirostris*, dwarf sperm whale *Kogia sima*, Chinese white dolphin *Sousa chinensis*, *G. griseus*, and *P. macrocephalus*). We exclude the possibility that frameshift insertions/deletions and stop codons are the result of sequencing error because we have reamplified these pseudogenized fragment of *MCPH1* gene in different samples or using different primers and obtained the same result. Thus, the six intact genes of *MCPH* were used for our further analyses.

Selection on MCPH genes

We found M8 that incorporated selection fit the data better than the neutral model, M8a, at the four *MCPH* genes (*WDR62*, *CDK5RAP2*, *CEP152*, and *ASPM*; $P < 0.001$), suggesting that these genes were subjected to positive selection in cetaceans. However, for *CENPJ* and *STIL*, associated the LRT showed no significant difference between the models M8 and M8a (*CENPJ*: $P = 0.485$; *STIL*: $P = 0.195$), implying no positive selection. Using M8, the most stringent model carried out in PAML, a small proportion of codons (1.82–7.57%) were estimated to be under selection with average ω values of 4.438–10.203 at the four positively selected genes in cetaceans (Table 1). Seven, 39, 10, and 20 positively selected sites were identified by the BEB approach as having posterior probabilities ≥0.80 at *WDR62*, *CDK5RAP2*, *CEP152*, and *ASPM*, respectively (Table 1). When we used a significance threshold of 0.95 for posterior probabilities, the number of positive selected amino acids decreased to three, nine, six, and four at the four genes, respectively. FEL, performed in HYPHY [24], was also used to test for selection in the six intact *MCPH* genes. HYPHY can improve the estimation of the ω value by incorporating variation in d_S whereas d_S is fixed across sequences for all the PAML-based analysis [23, 24]. FEL analysis showed that significant signs of positive selection were detected at the six *MCPH* genes with many more positively selected sites than that identified by M8 (Table 1). Combining the two different maximum likelihood (ML) methods, a total of 36 positively selected sites (six at *WDR62*, 16 at *CDK5RAP2*, six at *CEP152*, and eight at *ASPM*) were picked out (Table 1). Sites identified to be under positive selection by two ML methods were regarded as robust candidates for sites under selection. Therefore, 36 robust sites under positive selection were

Table 1 Positively selected sites detected using two maximum likelihood (ML) methods across cetacean phylogeny

Gene	Test of Selection						Sites under Selection Identified by ML Methods		no. of Sites[c]	% of Sites
	-Ln (M8a)	-Ln (M7)	-Ln (M8)	-2ΔL (M7 vs.m8)	-2ΔL (M8a vs.m8)	ω value	PAML M8[a]	FEL[b]		
WDR62 (1298aa)	7703.049	7703.257	7690.821	24.872*	24.456*	10.203	**262**, **458**, **479**, **972**, **1008**, **1133**, 1244	207, **262**, **458**, **479**, 648, 705, 763, **972**, **1008**, 1076, 1092, 1107, **1133**, 1158	6	0.46%
CDK5RAP2 (1414aa)	10,316.782	10,317.746	10,285.732	64.082*	62.101*	4.438	**23**, **108**, **182**, **201**, **283**, 382, 394, **426**, **445**, 518, 523, **531**, 558, 615, 678, 686, 700, 717, **720**, 724, 726, 727, 740, **760**, 763, 786, **787**, **832**, **849**, **913**, 982, 1012, 1020, 1030, **1166**, 1167, **1284**, 1370, 1396	6, 10, **23**, 30, **108**, 174, **182**, **201**, 204, **283**, **426**, 427, **445**, **531**, **720**, **760**, **787**, **832**, **849**, 887, **913**, **1166**, 1280, **1284**, 1310, 1313	16	1.13%
CEP152 (1523aa)	9948.922	9949.080	9939.249	19.662*	19.347*	7.312	**79**, **305**, 477, **510**, 543, 613, **1147**, **1163**, 1360, **1398**	73, **79**, 239, **305**, 328, 401, **510**, 717, 941, 966, 1088, 1097, **1147**, **1163**, 1281, **1398**	6	0.39%
ASPM (3128aa)	20,714.804	20,714.924	20,700.708	28.432*	28.191*	5.236	**69**, **347**, **387**, 595, **692**, 805, 806, **1311**, **1602**, 1684, 1787, **1864**, 2096, 2098, 2350, **2619**, 2693, 2798, 2819, 3111	22, 23, **69**, **347**, **387**, 542, 677, **692**, 1178, **1311**, **1602**, 1657, 1660, 1682, 1737, 1741, 1758, **1864**, 2106, 2187, 2205, 2445, 2465, 2507, 2510, 2546, **2619**, 2817, 3098	8	0.26%
CENPJ (998aa)	7086.280	7086.473	7086.040	1.700 (NS)	0.488 (NS)	1.593	—	112, 366, 370, 451, 473, 488, 499, 513, 544, 567, 645	—	—
STIL (1125aa)	6173.275	6173.287	6172.437	0.866 (NS)	1.677 (NS)	4.921	—	471, 575, 991	—	—

The positively selected sites picked out by two methods are shown in bold

* indicates P < 0.05 whereas NS indicates 'No significant'

[a]Codons were detected by M8 model in PAML using a Bayes Empirical Bayes (BEB) analysis with posterior probabilities ≥0.80

[b]Condons was determined by FEL implemented in HYPHY with significance levels of 0.2

[c]No. of sites indicate positively selected sites identified by both ML methods

used in our next analyses. We further employed a complementary protein-level approach implemented in TreeSAAP [30] to evaluate destabilizing radical changes at each robust site. The result showed that 32 of 36 sites (88.89%) have radical changes in at least one property whereas 16 sites (44.44%) had at least three changes in properties at the four *MCPH* genes (see Additional file 1: Table S4). When an empirical threshold of $P \leq 0.05$ was applied we found 13 sites (36.11%) under strong positive selection at the protein-level (see Additional file 1: Table S4).

To evaluate whether positive selection is only limited to particular lineages at the six intact *MCPH* genes, we first used branch models (including free-ratio and two-ratio models) that allow the ω ratio to vary among branches across the phylogeny. The LRT tests showed that evidence of positive selection was detected at the *WDR62*, *CDK5RAP2* and *ASPM* genes whereas no selection was detected for the other three genes. Using a free-ratio model, we found that ω was greater than 1 in branches of odontocetes with increased EQ at *ASPM*: last common ancestor (LCA) of delphinids, LCA of the Baiji *Lipotes vexillifer*, and LCA of the Indo-Pacific finless porpoise *Neophocaena phocaenoides* and beluga *D. leucas* (Fig. 1). In contrast, evidence of positive selection

was examined at *CDK5RAP2* in lineages both with increased or decreased EQ. That is, a ω greater than 1 was found along lineages leading from the LCA of cetaceans to odontocetes with increased EQ (terminal branch of *P. macrocephalus*, LCA of delphinids, LCA of *M. densirostris*, branch leading to *L. vexillifer*, terminal branch of *D. leucas*, LCA of delphinids, and four terminal branches of delphinids) and the ancestral branch of mysticetes and branch leading to Omura's whale *B. omurai* with reduced EQ. In the two-ratio model, significant signs of positive selection were restricted to lineages with expanded relative brain size, such as the LCA of delphinids at both *CDK5RAP2* and *ASPM*, LCA of *L. vexillifer* at *ASPM*, the branch leading to *G. griseus* at *CDK5RAP2*, and the branch leading to *D. leucas* at *WDR62* (Fig. 1). Similar results were obtained with the stringent branch-site model, which revealed that two lineages with increased EQ (such as the terminal branch of *T. truncatus* and *P. macrocephalus*) at *CDK5RAP2* and one with reduced EQ (i.e. branch leading to *B. acutorostrata*) at *ASPM* were subject to selection after FDR correction, respectively (Fig. 1). In addition, seven and one positively selected sites were identified at *CDK5RAP2* in branches leading to *T. truncatus* and *P. macrocephalus*, respectively. However, none were found at *ASPM* even

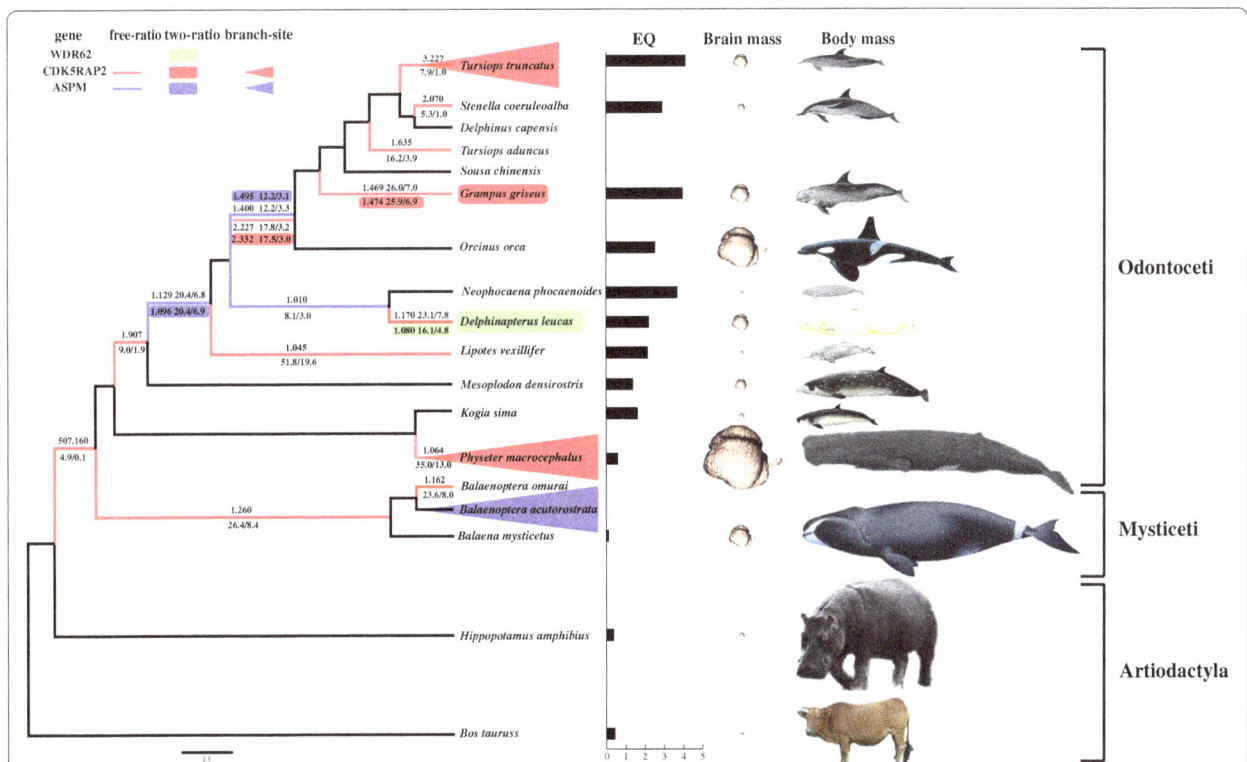

Fig. 1 Evidence of positive selection across the phylogeny of cetartiodactyla identified by branch and branch-site models. The three genes identified to be under positive selection are marked with different colors: *WDR62* (green), *CDK5RAP2* (pink), and *ASPM* (blue). Significant positive selection identified by free-ratio, two-ratio and branch-site models are highlighted and indicated by line, rectangle and triangle, respectively. The ω value greater than 1 for individual lineage according to free-ratio and two-ratio are shown

using the posterior probabilities ≥0.50 as the cutoff for the positively selected sites.

Association between *MCPH* evolution and morphological variables

To explore the associations between the evolution of each *MCPH* gene found to be under positive selection (represented by root-to-tip ω) and absolute brain mass, body mass and EQ (see Additional file 1: Table S3), we performed PGLS regressions, as implemented in the R 3.1.2 using the packages Caper [41]. Regression analyses revealed a positive association between log (root-to-tip ω) and log (EQ) at *CDK5RAP2* (R^2 = 0.521, P = 0.007) and *ASPM* (R^2 = 0.304, P = 0.046, Fig. 2; see Additional file 1: Table S5), whereas no such association was found for the other two *MCPH* genes under positive selection, i.e. *WDR62* (R^2 = 0.235, P = 0.074) and *CEP152* (R^2 = 0.005, P = 0.841). In addition, log (root-to-tip ω) was not related to brain mass and body mass for all of the four *MCPH* genes under positive selection (see Additional file 1: Table S5).

Relationship between ω and mean group size

To test whether social complexity drove cetacean brain size expansion we used mean group size as a measure of social complexity. Mean group size from 13 cetacean species used in our study was derived from May-Collado

et al. [46] (see Additional file 1: Table S3). Regression analyses showed a significant association between log (root-to-tip ω) and log (mean group size) for *ASPM* (R^2 = 0.267, P = 0.041) and *CDK5RAP2* (R^2 = 0.308, P = 0.029), whereas no significant association was found for *WDR62* (R^2 = 0.030, P = 0.574) and *CEP152* (R^2 = 0.086, P = 0.173) (Fig. 3; see Additional file 1: Table S6).

Spatial distribution of positively selected sites in 3D structures of MCPH genes

A total of 32 radical amino acid changes subjected to positive selection identified by two ML methods were mapped onto the 3D structures of four *MCPH* genes (six at *WDR62*, 14 at *CDK5RAP2*, four at *CEP152*, and eight at *ASPM*). We found that 16.67%–87.5% of positively selected sites were localized in the functional region in the predicted 3D structure of each *MCPH* gene (Fig. 4). For example, for *ASPM*, up to 87.50% of positively selected sites (7 / 8) were localized in two putative key functional domains: the CaM-binding IQ motifs (1311, 1602, 1864, and 2619) and microtubules domain (69, 347 and 387). For *CDK5RAP2*, eight sites (50%) were scattered over the three predicated functional regions: one (23) in the γTurc binding domain, three (182, 201, and 283) in the structural maintenance of chromosomes (SMC) domain, and another four (720, 760, 787, and 832) in SMC_N

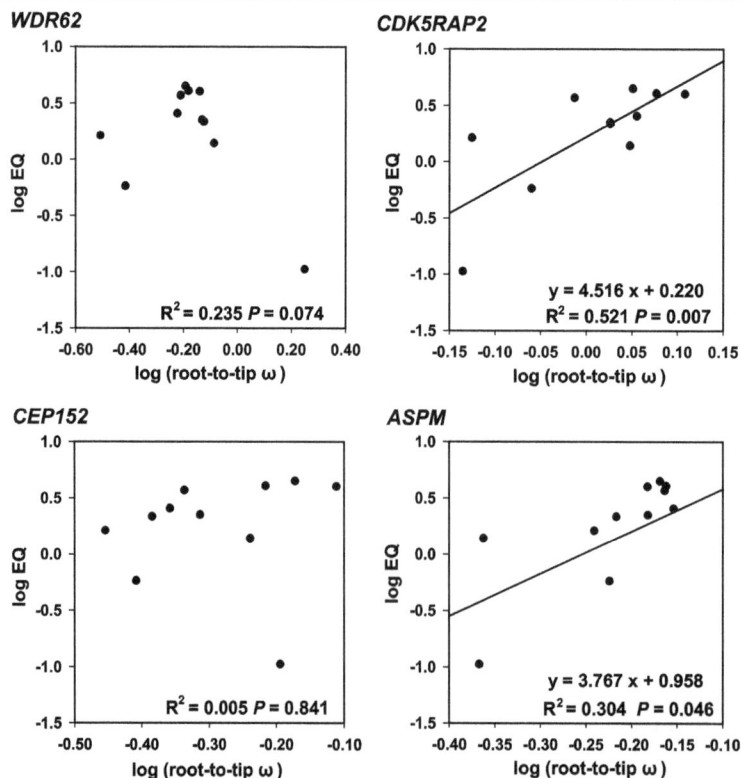

Fig. 2 Regression analyses between root-to-tip ω and EQ across cetacean phylogeny

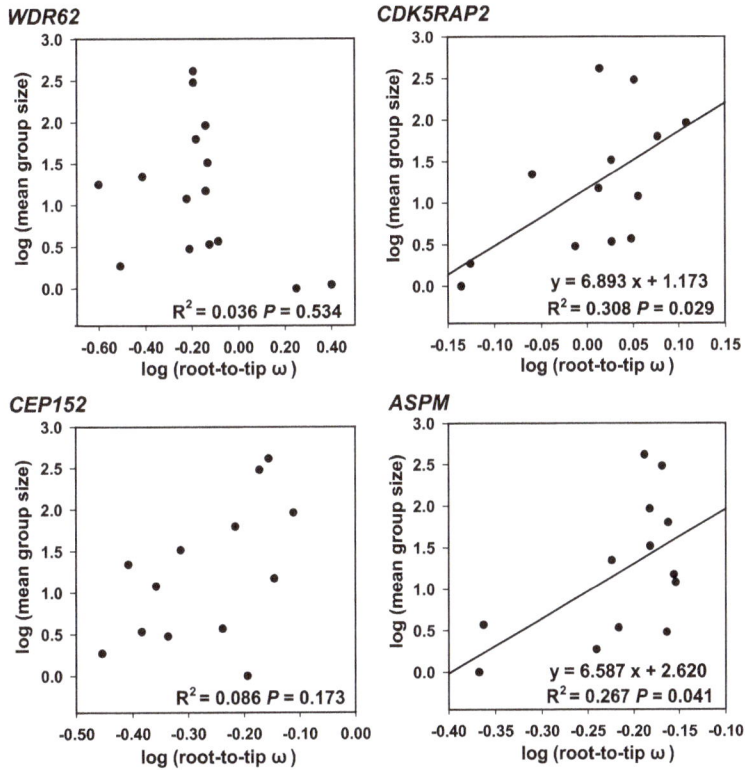

Fig. 3 Regression analysis between root-to-tip ω and mean group size in cetaceans

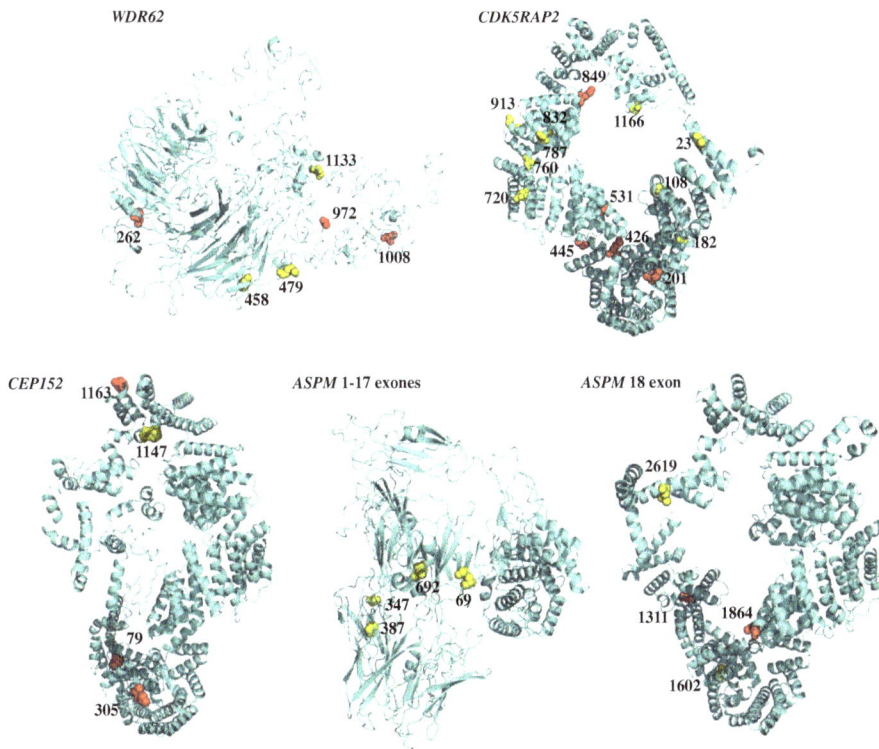

Fig. 4 Radical amino acid changes in selected sites mapped on the three-dimensional structure of four *MCPH* genes in cetaceans. Radical changes of positively selected sites identified by two ML methods are colored with yellow ball and red ball whereas those positively selected sites with significant radical changes are marked with red ball

domain, that were known to play a key role in the cohesion and condensation of chromosomes during mitosis. Only two (262, 458) or one (305) sites were localized in the functional region of *WDR62* (WD40 repeat domain) and *CEP152* (coiled_coil domain), respectively.

Functional assay of *ASPM*

Previous studies showed that the IQ motifs, which act as calmodulin-binding domains, were thought to be involved in increased cerebral cortical size in mammalian evolution (reviewed in ref. [15]). Considering that the coding sequence of the cetacean *ASPM* gene was more than 10,300 bp, the expression vector cannot carry such a large DNA fragment. Thus, only the 23rd IQ motif (66 bp) including one positively selected site, i.e. 1684) and its adjacent 16 amino acid motif were cloned into the expression vector. We performed a GST pull-down assay probed by immunoblotting with GST-IQ as bait and His-CaM as prey. The result showed that this IQ motif in the six odontocete species bound to CaM (Fig. 5a). In contrast, the IQ motif of mysticetes did not bind to CaM except for *B. acutorostrata*. Further, four odontocete and two mysticete species were chosen to measure the binding affinities of the GST-IQ protein and

CaM using BLI analysis that can quantitatively analyze protein interactions in real-time. BLI analysis revealed that the CaM could effectively bind to GST-IQ fusion protein of both odontocetes and mysticetes, but not the GST control (Fig. 5b). The affinities of the GST-IQ for the CaM were calculated with a 1:1 binding model. For odontocetes with increased EQ, the Kd was 27.6 nM. Specially, with the EQ increasing, the Kd values have a tendency to reduce, suggesting the interaction became weak with the EQ growth. For example, Kd of the *T. truncates* with highest EQ (4.14) was 17.7 nM, whereas Kd value increased 2.2 fold for the *P. macrocephalus* with lowest EQ (0.58), resulting from increased dissociation rate constants (K_{off}). By contrast, the GST-IQ binding CaM could still be measured in the EQ decreased mysticetes, but Kd value (mean 63.38 nM) increased dramatically due to increased K_{off} values. The result suggested that the affinity of this interaction in the odontocetes was stronger than for the mysticetes (Fig. 5c).

Discussion
Extensive adaptive evolution of cetacean *MCPH* genes
We systematically investigated the evolution of seven *MCPH* genes in cetaceans to explore the genetic basis of

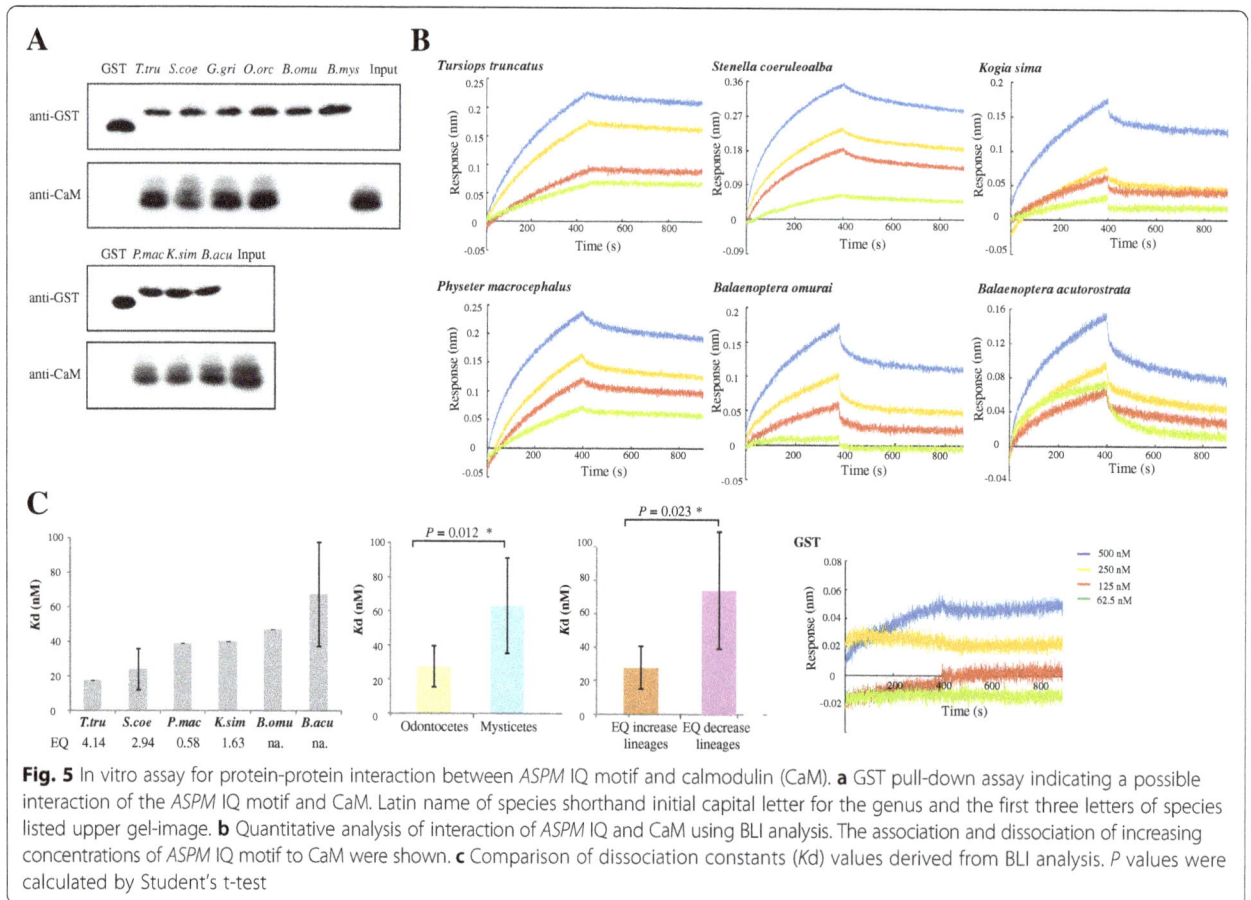

Fig. 5 In vitro assay for protein-protein interaction between *ASPM* IQ motif and calmodulin (CaM). **a** GST pull-down assay indicating a possible interaction of the *ASPM* IQ motif and CaM. Latin name of species shorthand initial capital letter for the genus and the first three letters of species listed upper gel-image. **b** Quantitative analysis of interaction of *ASPM* IQ and CaM using BLI analysis. The association and dissociation of increasing concentrations of *ASPM* IQ motif to CaM were shown. **c** Comparison of dissociation constants (*Kd*) values derived from BLI analysis. *P* values were calculated by Student's t-test

brain evolution in cetaceans. *MCPH1* genes appeared to be a pseudogene because premature stop codons and frameshift mutations were identified in three cetacean lineages. Evidence of pseudogene in *MCPH1* gene was also identified in other seven cetacean species by [19]. However, the sequences of the other six genes were intact. Strong evidence of positive selection was identified by site-specific modeling for four of six intact *MCPH* genes: *WDR62*, *CDK5RAP2*, *CEP152*, and *ASPM*. A total of 36 robust candidate sites under selection were identified by two ML methods (PAML and REL): six at *WDR62*, 16 at *CDK5RAP2*, six at *CEP152*, and eight at *ASPM*. Of these sites, 88.89% (32 / 36) were categorized as radical changes under positive selection at the protein-level. Notably, almost all of the radical amino acid changes subjected to positive selection were localized within or near the functional region of the predicted 3D structures of the four *MCPH* genes. For example, up to 87.50% positively selected sites (7 / 8) at *ASPM* were localized in the two putative key functional domains (i.e. calmodulin binding IQ motifs and microtubules domain), which play an essential role in orientation of mitotic spindles during embryonic neurogenesis. It has been reported that mutations in these putative functional domains would cause *MCPH* in humans [47, 48]. For *CDK5RAP2*, 50% (8 / 16) of sites under positive selection scattered over three predicated functional regions are known to play a role in the cohesion and condensation of chromosomes during mitosis [14].

According to ancestral state reconstruction of cetacean EQs [6, 49], almost all odontocetes have increased EQs whereas mysticetes have reduced EQ compared to their ancestor. Lineage-specific selection analyses found extensive positive selection at *CDK5RAP2* along cetacean lineages from the ancestor of cetaceans to descendant lineages, especially in lineages with both increased and decreased EQs. For *ASPM*, significant signs of positive selection were mainly determined in odontocete lineages with expanded EQ whereas the minke whale with contracted EQ was found to be under positive selection although none of the positively selected sites were identified. In contrast, only odontocetes with expanded EQ were subject to positive selection at *WDR62*. Such extensive adaptive evolution of *MCPH* genes in cetaceans may be well matched with complex evolution of their brain size, including EQ expansion in toothed whales and EQ reduction in baleen whales.

Significant association between brain phenotype and *MCPH* gene evolution

Statistical association between selection on a functional gene and changes in phenotype are an important indication for exploring the genetic basis of adaptive phenotypes [21, 48]. Regressions revealed a significant positive association between relative brain size (EQ) and evolutionary rates for *ASPM* and *CDK5RAP2* but not for *WDR62* and *CEP152* (Fig. 3). However, both absolute brain mass and absolute body mass were not related to selection rates of the four positively selected *MCPH* genes. By contrast, *CDK5RAP2* and *ASPM* are related to neonatal brain mass in anthropoid primates whereas no association was found between these genes and EQ or adult brain size [16]. The discrepancy between anthropoid primates and cetaceans may be the result of different evolutionary patterns in brain size enlargement, although relaxed constraints on brain-body allometry were examined in both groups. Primates have a directional trend in brain mass expansion but not body mass, leading to a wide pattern of EQ expansion through primate evolution. However, cetacean EQ changes are associated with body mass [49]. For example, the reduced EQ in mysticetes was mainly driven by a high rate of body mass enlargement whereas the EQ increase in odontocetes was due to body mass decrease at the origin of odontocetes according to ancestral state reconstruction. Body mass changes are a predominant influence in cetacean EQ evolution [6, 49]. Despite large differences in EQ, mysticetes and odontocetes evolved with similar patterns of brain mass that were generally increased over time.

Although odontocetes have a generally increased EQ over time, there are some exceptions. It is well known that *P. macrocephalus* has the largest absolute brain mass (up to 10 kg) among animals but the smallest EQ (0.58) among extant odontocetes due to its large body mass (more than 35,632 kg) [36]. *P. macrocephalus* had the second EQ reduction across cetacean lineages due to its 107-fold increase in body mass and only 6.5-fold increase in brain mass compared to their common ancestor [49]. Thus, we performed the association analyses after removing the sperm whale. The result revealed that *ASPM* and *CDK5RAP2* remain significantly related to EQ (*ASPM*: $R^2 = 0.409$, $P = 0.028$; *CDK5RAP2*: $R^2 = 0.477$, $P = 0.016$). The same pattern was also found after the removal of any one single species of odontocetes. This suggested that *P. macrocephalus* was not an outlier in the relationship between the genes' evolutionary rates and EQ. However, when *P. macrocephalus* was excluded, a marginally negative relationship was identified between *ASPM* and absolute body mass ($R^2 = 0.316$, $P = 0.053$). This pattern was not found after any other single odontocete species was excluded. Thus, the sperm whale appeared to be an outlier in the association between the selection and body mass at the *ASPM*, which might attribute to its largest body mass. Additionally, bowhead whale *B. mysticetus* is another species with reduced EQ used in our association analysis. When the two species with reduced EQ. (*P. macrocephalus* and *B. mysticetus*) were excluded, *ASPM* is still significantly related to EQ ($R^2 = 0.557$, $P = 0.013$), suggesting the both species are not outliers. These results further corroborated our previous work that selection on *ASPM* may contribute to relative brain size enlargement during cetacean evolution [20]. Only two species with reduced EQ were used in our study and we should test whether cetaceans have a similar pattern to primates regarding *ASPM* and increasing and decreasing EQ when more data becomes available.

Notably, no association was found in the *ASPM* gene when only the two exons (approx. 60% of the transcribed *ASPM* protein) were examined in our previous study [20], which was questioned by Montgomery et al. [21]. However, when the 22 exons, accounting for 90.34% of the transcribed *ASPM* protein, were used in this study, it was striking and interesting to find a significant association between root-to-tip ω of *ASPM* and cetacean EQ ($R^2 = 0.304$, $P = 0.046$, Fig. 2). Therefore, it is best to use the complete CDS to explore gene-phenotype associations in the future.

Is cetacean *ASPM* mainly linked to evolutionary increases in EQ?

ASPM plays a key role in mitotic spindle function including orientation of the cleavage plane [15]. To execute this function, *ASPM* must conjunct with CaM because *ASPM* is not detected at meiotic and mitotic spindles after RNAi of CaM [50]. Thus, it was suggested that CaM is needed for the localization of *ASPM*. More importantly, previous functional assays have proved that a minimal region of *ASPM*, such as the first IQ motif, can be sufficient for CaM binding [50]. Here, we examined binding affinities between CaM and the 23rd IQ (when the *T. truncates* as reference) of *ASPM* including one positively selected site in cetacean lineages to test whether there is a functional divergence between *ASPM* genes of EQ enlarged and contracted species.

The GST pull-down assay displayed that toothed species with the highest EQs had a strong effect on CaM binding to this IQ, whereas no such effect was found in baleen whales with decreased EQ (except for *B. acutorostrata*). The same result was found when the pull-down assay was repeated two times. *B. acutorostrata* is a special case for the pull-down assay of the baleen whales, likely because a significant sign of positive selection identified in this species but not in other baleen species at *ASPM*. In addition, *B. acutorostrata* is the smallest among the baleen whales with an average body length of 6.7–7.3 m and body mass of 11, 000 kg, similar to species of toothed whales. Specially, nearly significant negative association was identified between *ASPM* evolution and body mass ($P = 0.052$) in our association analysis when *P. macrocephalus* was excluded. Accordingly, the anomalous pattern in *B. acutorostrata* may be attributable it having the smallest body mass in the mysticetes.

Next, we quantitatively analyzed the binding affinities using BLI that revealed this IQ did bind to CaM in odontocetes with EQ expansion and mysticetes with EQ contraction. However, it was noted that the dissociation constants of mysticetes (mean 63.38 nM) were significantly higher than that of odontocetes (mean 27.6 nM), suggesting that the binding affinity of odontocetes with EQ expansion was stronger than for mysticetes. Notably, when only odontocetes were considered, there is a tendency that the Kd values decrease with EQ increasing, suggesting the binding affinity of CaM and GST-IQ greatly increase with EQ increasing (Fig. 5). Such discrepancy of binding affinities of mysticetes between the GST pull down experiment and BLI analysis may be due to its weak binding affinity making it easy to wash out in the pull down experiment. Of course, only one IQ motif was detected in our study, the complete CDS should be tested when fresh tissue is available. Collectively, our functional assay further supported that cetacean *ASPM* was mainly linked to brain size expansion, which was contrasted with the finding in primates that *ASPM* evolution was related to both increase and decrease EQ [18].

Molecular evidence to support social brain hypothesis in cetaceans

Field research shows that cetaceans, and particularly delphinids, live in large complex groups with highly differentiated relationships [51]. In such groups, cetaceans must identify their long-term bonds and higher-order alliances, and communicate, collaborate and compete among group members [12, 52]. It is widely accepted that brain size expansion in cetaceans is driven by complex social forces and cognitive demands for living in complex social groups [12, 52]. The 'social brain hypothesis' proposes that species living in a complex social group must manage a wide variety of information relevant to social living [8, 10, 53]. This hypothesis is supported by the correlation between high-level encephalization and sociality (particularly for stable groups) in mammalian species [54]. A positive relationship was found between the relative brain (or neocortex) size and group size in delphinids [13, 55], suggesting that relative brain size in delphinids enlarged in order to respond to cognitive demands, social complexity and group size. Similar results have been found in haplorhine primates [10, 56]. Group size is a proxy for social complexity, although it is not the driver of brain evolution [9, 53, 56].

In order to test whether the 'social brain hypothesis' is supported at the molecular level in cetaceans, we examined the association between positive selection on four *MCPH* genes and mean group size as summarized by May-Collado et al. [46]. Significant positive associations between evolutionary rate and mean group size were found for *ASPM* and *CDK5RAP2*. When only cetacean species with increased EQ were considered, a significant positive relationship remained for *CEP152* ($R^2 = 0.519$, $P = 0.007$). These findings confirm that cetaceans evolved large brains to manage their unusually complex social systems. Although mean group size is a crude measure of social complexity, group size data is relatively easy to obtain for wild mammals. Relationships between *MCPH* gene evolution and other ecological factors of sociality such as pair bonding, activity patterns and diet should be examined in cetaceans to further consolidate the social brain hypothesis at the molecular level.

Conclusions

Cetaceans evolved a dramatic brain size expansion but their body evolved in complex pattern, leading to odontocetes with increased EQ and mysticetes with decreased EQ. We comprehensively investigated seven *MCPH* genes associated with brain size development in representative cetacean lineages. Significant positive selection was examined at the four *MCPH* genes, special for *ASPM* and *CDK5RAP2* genes, selection identified along lineages with both increased and decreased EQ. The result is well matched with cetacean complex brain size evolution. Association analyses showed that *CDK5RAP2* and *ASPM* evolutionary rate (ω) were significantly related to EQ, suggesting that these two genes may have contributed to EQ expansion in cetaceans. This suggestion was further indicated by our preliminary function test that *ASPM* might be mainly linked to evolutionary increases in EQ. In addition, a positive association was determined between evolution rate of *ASPM* and *CDK5RAP2* and mean group size, which is consistent with 'social brain hypothesis' that that cetaceans evolved large brains to manage complex social systems.

Additional files

Additional file 1: Table S1. Sequence information of seven MCPH genes across the phylogeny of Cetartiodactyla used in this study. **Table S2.** Results for site model and free-ratio model analysis at the six MCPH genes using the gene tree and species tree. **Table S3.** Morphological variables of cetacean brain used in regression analyses. **Table S4.** Amino acid sites under positive selection identified by maximum likelihood (ML) methods and TreeSAAP. **Table S5.** Regression analyses between the root-to-tip ω and body mass, body mass, and EQ across cetacean lineages. **Table S6.** Phylogenetically controlled regression analyses between the root-to-tip ω and mean group size. (DOC 239 kb)

Additional file 2: Supplementary methods and results. **Figure S1.** Phylogeny of cetaceans based on ML and BI best topology; number above branches show bootstrap support and posterior probability value above 0.50. (ZIP 759 kb)

Additional file 3: Alignment sequences of seven MCPH genes used for this study. (DOCX 60 kb)

Abbreviations

3D: three-dimensional; BEB: Bayes Empirical Bayes; BI: Bayesian inference; BLI: biolayer interferometry; CaM: calmodulin; CDS: coding sequence; d_N: nonsynonymous substitution; d_S: synonymous substitution; EQ: encephalization quotients; FDR: false discovery rate; FEL: fixed effects likelihood; Kd: dissociation constants; Koff: dissociation rate constants; Kon: association rate constants; LRT: likelihood ratio test; M0: one-ratio model; M1: free-ratio model; Ma: million years ago; *MCPH*: microcephaly; ML: maximum likelihood; PGLS: phylogenetic generalized least squares

Acknowledgements

We thank Mr. Xinrong Xu for help with collecting samples for many years; Professors Yanfu Qu, Mei liu and Shan Lu for their technical supports. A special thank-you is also due from Professor Harold H Zakon for comments that helped to improve the manuscript.

Funding

This work was supported by the National Natural Science Foundation of China (NSFC, grant number 31570379 to SX), the National Key Programme of Research and Development, Ministry of Science and Technology (Grant number 2016YFC0503200 to GY and SX), the National Science Fund for Distinguished Young Scholars to GY (grant number 31325025), the State Key Program of National Natural Science of China to GY (grant number 31630071); the NSFC (grant number 31370401 to WR); the Priority Academic Program Development of Jiangsu Higher Education Institutions to GY and SX, the Natural Science Foundation of Jiangsu Province of China (grant number BK20141449) to SX.

Authors' contributions

SX and GY conceived the project and designed the experiments. XN performed functional assay. XN, XS, ZZ and RT performed the molecular evolution analysis. SX wrote the manuscript, and WR, KZ and GY improve the manuscript. All authors read and approved the final manuscript.

Competing interests

The authors declare that they have no competing interests.

References

1. Thewissen JG, Cooper LN, George JC, Bajpai S. From land to water, the origin of whales, dolphins, and porpoises. Evol Educ Outreach. 2009;2:272–88.
2. Uhen M. The origin(s) of whales. Annu Rev Earth Planet Sci. 2010;38:189–219.
3. Uhen M. Evolution of marine mammals: back to the sea after 300 million years. Anat Rec. 2007;290:514–22.
4. Marino L. A comparison of encephalization between odontocete cetaceans and anthropoid primates. Brain Behav Evol. 1998;51:230–8.
5. Marino L, McShea D, Uhen MD. The origin and evolution of large brains in toothed whales. Anat Rec. 2004;281:1247–55.
6. Boddy AM, McGowen MR, Sherwood CC, Grossman LI, Goodman M, Wildman DE. Comparative analysis of encephalization in mammals reveals relaxed constraints on anthropoid primate and cetacean brain scaling. J Evol Biol. 2012;25:981–94.
7. Goldbogen JA, Pyenson ND, Shadwick RE. Big gulps require high drag for fin whale lunge feeding. Mar Ecol Prog Ser. 2007;349:289–301.
8. Dunbar RI. The social brain hypothesis. Evol Anthropol. 1998;6:178–90.
9. Dunbar RI, Shultz S. Understanding primate brain evolution. Phil Trans R Soc B. 2007a;362:649–58.
10. Dunbar RI, Shultz S. Evolution in the social brain. Science. 2007;317:1344–7.
11. Marino L. Convergence in complex cognitive abilities in cetaceans and primates. Brain Behav Evol. 2002;59:21–32.
12. Marino L, Connor RC, Fordyce RE, Herman LM, Hof PR, Lefebvre L, et al. Cetaceans have complex brains for complex cognition. PLoS Biol. 2007;5:139.
13. Marino L. What can dolphins tell us about primate evolution. Evol Anthropol. 1996;5:81–6.

14. Cox J, Jackson AP, Bond J, Woods CG. What primary microcephaly can tell us about brain growth? Trends Mol Med. 2006;12:358–66.

15. Mahmood S, Ahmad W, Hassan MJ. Autosomal recessive primary microcephaly (MCPH): clinical manifestations, genetic heterogeneity and mutation continuum. Orphanet J Rare Dis. 2011;6:1–15.

16. Montgomery SH, Capellini I, Venditti C, Barton RA, Mundy NI. Adaptive evolution of four microcephaly genes and the evolution of brain size in anthropoid primates. Mol Biol Evol. 2011;28:625–38.

17. Ponting C, Jackson AP. Evolution of primary microcephaly genes and the enlargement of primate brains. Hemoglobin. 2005;15:13–40.

18. Montgomery SH, Mundy NI. Evolution of ASPM is associated with both increases and decreases in brain size in primates. Evolution. 2012;66:927–32.

19. McGowen MR, Montgomery SH, Clark C, Gatesy J. Phylogeny and adaptive evolution of the brain-development gene microcephalin (MCPH1) in cetaceans. BMC Evol Biol. 2011;11:98.

20. Xu SX, Chen Y, Cheng YF, Yang D, Zhou XM, Xu JX, et al. Positive selection at the ASPM gene coincides with brain size enlargements in cetaceans. Proc Roy Soc B. 2012;279:4433–40.

21. Montgomery SH, Mundy NI, Barton RA. ASPM and mammalian brain evolution: a case study in the difficulty in making macroevolutionary inferences about gene–phenotype associations. Proc R Soc B. 2014;281:380–93.

22. Tamura K, Stecher G, Peterson D, Filipski A, Kumar S. MEGA6: molecular evolutionary genetics analysis version 6.0. Mol Biol Evol. 2013;30:2725–9.

23. Yang Z. PAML 4, Phylogenetic analysis by maximum likelihood. Mol Biol Evol. 2007;24:1586–91.

24. Zhou X, Xu S, Yang Y, Zhou K, Yang G. Phylogenomic analyses and improved resolution of Cetartiodactyla. Mol Phylogenet Evol. 2011;61:255–64.

25. Yang Z*, Nielsen R, Goldman N, AMK P. Codon-substitution models for heterogeneous selection pressure at amino acid sites. Genetics. 2000;155:431–49.

26. Yang Z, Bielawski JP. Statistical methods for detecting molecular adaptation. Trends Ecol Evol. 2000;15:496–503.

27. Wong WSW, Yang Z, Goldman N, Nielsen R. Accuracy and power of statistical methods for detecting adaptive evolution in protein coding sequences and for identifying positively selected sites. Genetics. 2004;168:1041–51.

28. Yang Z, Wong WSW, Nielsen R. Bayes empirical Bayes inference of amino acid sites under positive selection. Mol Biol Evol. 2005;22:1107–18.

29. Pond SL, Frost SD. Datamonkey: rapid detection of selective pressure on individual sites of codon alignments. Bioinformatics. 2005;21:2531–3.

30. Woolley S, Johnson J, Smith MJ, Crandall KA, McClellan DA. TreeSAAP: selection on amino acid properties using phylogenetic trees. Bioinformatics. 2003;19:671–2.

31. Yang Z, Nielsen R. Synonymous and nonsynonymous rate variation in nuclear genes of mammals. J Mol Evol. 1998;46:409–18.

32. Yang Z. Likelihood ratio tests for detecting positive selection and application to primate lysozyme evolution. Mol Biol Evol. 1998;15:568–73.

33. Zhang J, Nielsen R, Yang Z. Evaluation of an improved branch–site likelihood method for detecting positive selection at the molecular level. Mol Biol Evol. 2005;22:2472–9.

34. Anisimova M, Yang Z. Multiple hypothesis testing to detect lineages under positive selection that affects only a few sites. Mol Biol Evol. 2007;24:1219–28.

35. Wolf JB, Künstner A, Nam K, Jakobsson M, Ellegren H. Nonlinear dynamics of nonsynonymous (dN) and synonymous (dS) substitution rates affects inference of selection. Genome Biol Evol. 2009;1:308–19.

36. Pilleri G, Gihr M. The central nervous system of the mysticete and odontocete whales. Invest Cetacea. 1970;2:87–135.

37. Stephan H, Frahm H, Baron G. New and revised data on volumes of brain structures in insectivores and primates. Folia Primatol (Basel). 1981;35:1–29.

38. Ridgway SH, Brownson RH. Relative brain sizes and cortical surface areas in odontocetes. Acta Zool Fenn. 1984;172:149–52.

39. Schwerdtfeger WK, Oelschläger HA, Stephan HC. Brain Struct Funct. 1984;170:11–9.

40. Jerison HJ. Evolution of the brain and intelligence. Trends Cogn Sci. 1973;45:250–7.

41. Orme CDL, Freckleton RP, Thomas GH, Petzoldt T, Fritz SA, Isaac N, et al. Caper: comparative analyses of phylogenetics and evolution in R. Methods Ecol Evol. 2012;3:145–51.

42. Zhang Y. I-TASSER server for protein 3D structure prediction. BMC Bioinformatics. 2008;9:297–315.

43. Roy A, Kucukural A, Zhang Y. I-TASSER: a unified platform for automated protein structure and function prediction. Nat Protoc. 2010;5:725–38.

44. Bond J, Scott S, Hampshire DJ, Springell K, Corry P, Abramowicz MJ, et al. Protein truncating mutations in ASPM cause variable reduction in brain size. Am J Hum Genet. 2003;73:1170–7.

45. Zhang J. Evolution of the human ASPM gene, a major determinant of brain size. Genetics. 2003;165:2063–70.

46. LJ M-C, Agnarsson I, Wartzok D. Phylogenetic review of tonal sound production in whales in relation to sociality. BMC Evol Biol. 2007;7:136.

47. Kaindl AM, Passemard S, Kumar P, Kraemer N, Issa L, Zwirner A, et al. Many roads lead to primary autosomal recessive microcephaly. Prog Neurobiol. 2010;90:363–83.

48. Mekel-Bobrov N, Lahn BT. Response to comments by Timpson et al. and Yu et al. Science. 2007;317:1036.

49. Montgomery SH, Geisler JH, Mcgowen MR, Fox C, Marino L, Gatesy J. The evolutionary history of cetacean brain and body size. Evolution. 2013;67:3339–53.

50. Vvan der Voet M, Berends CW, Perreault A, Nguyen-Ngoc T, Gönczy P, Vidal M, et al. NuMA-related LIN-5. ASPM-1, calmodulin and dynein promote meiotic spindle rotation independently of cortical LIN–5/GPR/Galpha. Nat Cell Biol. 2009;11:269–77.

51. Connor RC, Wells R, Mann J, Read A. The bottlenose dolphin: social relationships in a fission-fusion society. In: Mann J, Connor R, Tyack P, Whitehead H, editors. Cetacean societies: field studies of dolphins and whales. Chicago: The University of Chicago Press; 2000. p. 91–126.

52. Connor RC. Dolphin social intelligence: complex alliance relationships in bottlenose dolphins and a consideration of selective environments for extreme brain size evolution in mammals. Philos Trans R Soc Lond Ser B Biol Sci. 2007;362:587–602.

53. MacLean EL, Barrickman NL, Johnson EM, Wall CE. Sociality, ecology, and relative brain size in lemurs. J Human Evol. 2009;56:471–8.

54. Shultz S, Dunbar RI. Species differences in executive function correlate with hippocampus volume and neocortex ratio across nonhuman primates. J Comp Psychol. 2010;124:252–60.

55. Tschudin A J–P C. Relative neocortex size and its correlates in dolphins: comparisons with humans and implications for mental evolution. Pietermaritzburg; University of Natal. 1999.

56. Dunbar R. Neocortex size as a constraint on group size in primates. J Hum Evol. 1992;22:469–93.

10

A new species of *Xenoturbella* from the western Pacific Ocean and the evolution of *Xenoturbella*

Hiroaki Nakano[1*] (iD), Hideyuki Miyazawa[1], Akiteru Maeno[2], Toshihiko Shiroishi[2], Keiichi Kakui[3], Ryo Koyanagi[4], Miyuki Kanda[4], Noriyuki Satoh[5], Akihito Omori[6,7] and Hisanori Kohtsuka[6]

Abstract

Background: *Xenoturbella* is a group of marine benthic animals lacking an anus and a centralized nervous system. Molecular phylogenetic analyses group the animal together with the Acoelomorpha, forming the Xenacoelomorpha. This group has been suggested to be either a sister group to the Nephrozoa or a deuterostome, and therefore it may provide important insights into origins of bilaterian traits such as an anus, the nephron, feeding larvae and centralized nervous systems. However, only five *Xenoturbella* species have been reported and the evolutionary history of xenoturbellids and Xenacoelomorpha remains obscure.

Results: Here we describe a new *Xenoturbella* species from the western Pacific Ocean, and report a new xenoturbellid structure - the frontal pore. Non-destructive microCT was used to investigate the internal morphology of this soft-bodied animal. This revealed the presence of a frontal pore that is continuous with the ventral glandular network and which exhibits similarities with the frontal organ in acoelomorphs.

Conclusions: Our results suggest that large size, oval mouth, frontal pore and ventral glandular network may be ancestral features for *Xenoturbella*. Further studies will clarify the evolutionary relationship of the frontal pore and ventral glandular network of xenoturbellids and the acoelomorph frontal organ. One of the habitats of the newly identified species is easily accessible from a marine station and so this species promises to be valuable for research on bilaterian and deuterostome evolution.

Keywords: *Xenoturbella*, Acoels, Nemertodermatids, Acoelomorpha, Xenacoelomorpha, Frontal organ, Deuterostomes, Bilaterians, Metazoans, Evolution

Background

Xenoturbella is a group of marine benthic worms, first described in 1949 as a 'strange' platyhelminth [1]. It has a mouth but lacks an anus, hence the digestive organ is a sack rather than a tube. The nervous system of *Xenoturbella* is not centralized and is in the form of an intraepidermal nerve net [1–3]. Structures such as a coelom and reproductive organs are absent [1]. Various phylogenetic positions have been suggested for this animal based on different morphological characters - an early metazoan group based on its overall body plan [4], the sister group to

Bilateria based on its nervous system structure [2] and musculature [5], a member of the deuterostomes based on epidermal structure [6, 7] and a bivalve based on oocyte characteristics [8] (discussed in [9–11]). Recent molecular phylogenetic analyses support a close affinity with the Acoelomorpha [12–14], a group of marine worms also originally suggested to belong to the Platyhelminthes, but later suggested to be the sister group to the Nephrozoa (all remaining Bilateria) [15–21]. *Xenoturbella* and Acoelomorpha are suggested to form a new clade, the Xenacoelomorpha, and accordingly similarities in overall body plan [1], morphology of the free-swimming stage during development [22–24], ciliary ultrastructure [25–27] and degenerating epidermal cells [28] have been reported. However, diversity within the Xenacoelomorpha has been reported

* Correspondence: h.nakano@shimoda.tsukuba.ac.jp
[1]Shimoda Marine Research Center, University of Tsukuba, 5-10-1, Shimoda, Shizuoka 415-0025, Japan
Full list of author information is available at the end of the article

for some other features, such as the morphology of the digestive organ [1], statocyst [29, 30], sperm [31–33] and cleavage pattern [11, 34–36]. When a monophyletic clade consisting of *Xenoturbella* and Acoelomorpha was first proposed based on large scale molecular phylogenetic analyses, it was suggested to be a sister group to the Nephrozoa [12]. A later study, in which the name Xenacoelomorpha was first introduced, proposed that the group is a member of the deuterostomes [13]. However, a more recent study has again suggested a sister group relationship to the Nephrozoa [14]. Either way, it is clear that studies on *Xenoturbella* could provide important insights into the origins of bilaterian traits such as an anus, the nephron, feeding larvae and centralized nervous systems [24, 37–43].

The type species for *Xenoturbella*, *X. bocki*, is about 1–3 cm in body length and inhabits the seafloor of western Sweden coast at 50–200 m depth [1, 9, 11]. There is a considerable body of research on this species, and almost all knowledge of *Xenoturbella* comes from this animal. A second species, *X. westbladi*, was reported from the same habitat as *X. bocki* in 1999 [44], but a recent haplotype network analysis using cytochrome c oxidase subunit I (*cox1*) sequences showed that *X. bocki* and *X. westbladi* are a single species and that *X. westbladi* should be regarded as a junior synonym for the species [45]. Xenoturbellids were then reported from the Pacific Ocean when four species (*X. monstrosa*, *X. churro*, *X. profunda* and *X. hollandorum*) were reported from the west coast of USA and Mexico in 2016 [45]. This discovery revealed unknown diversity within the group: the body lengths of three of the species were over 10 cm, with *X. monstrosa* reaching about 20 cm; a structure called the ventral glandular network was described; and two sub-clades termed 'shallow' and 'deep' were identified. These four species all live on the sea floor at over 600 m deep, with some as deep as 3700 m, and require remotely operated vehicles (ROVs) equipped with a slurp gun for collection, making them difficult to work on as research organisms. Here we report the discovery of a new species of *Xenoturbella* off the Japanese coast and discuss the ancestral traits of *Xenoturbella* and Xenacoelomorpha. The new species can be collected using a marine biological dredge within an hour from a marine station, and therefore it promises to be a valuable model for further research on xenacoelomorphs.

Results
Xenoturbellida Bourlat et al., 2006 [46]
Genus *Xenoturbella* Westblad, 1949 [1]
Xenoturbella japonica sp. nov.
Etymology. Named for the locality where the specimens were collected.

Material examined. Holotype: NSMT-Xe 2, female (Figs. 1, 3, Additional file 1: Video S1 and Additional file 2:

Video S2). Paratype: NSMT-Xe 1, juvenile, sex unknown (Figs. 2, 3, Additional file 1: Video S1 and Additional file 3: Video S3).

Locality. Holotype: off Jogashima, Miura, Kanagawa, Japan, 35°06.93″ N 139°33.72″ E to 35°06.95″ N 139°33.33″ E; 380–554 m depth (Additional file 4: Figure S1). Specimen found from sediment obtained using a marine biology dredge (Rigo Co., Ltd., Tokyo, Japan) on December 9th, 2015 during a research survey (December 9th to 10th, 2015; JAMBIO application number 27–7) headed by Dr. Hiroshi Namikawa, National Museum of Nature and Science, using the RV *Rinkai-Maru* of the Misaki Biological Marine Station, The University of Tokyo.

Paratype: Sanriku coast, Iwate, Japan, 39°37.86″ N 142°18.22″ E to 39°37.00″ N 142°17.60″ E; 517–560 m depth (Additional file 4: Figure S1). Specimen found inside the inner small plankton net with a mesh size of 0.5 mm attached inside a larger beam trawl [47]. Collected on July 18th, 2013 during a research cruise (July 18th to 29th, 2013) headed by Dr. Ken Fujimoto, National Research Institute of Fisheries Science, Fisheries Research Agency, aboard FRV *Soyo-Maru* of the National Research Institute of Fisheries Science, Fisheries Research Agency, Japan.

Description of female. Based on holotype. Body 5.3 cm in length; pale orange with coloration getting darker toward the anterior (Fig. 1a). In live specimens, muscles hold the dorsal body wall in a W-shape (three ridges and two troughs). Body shape actively changes by contracting and elongating when alive. Ring furrow and side furrow are present (Fig. 1a,b). Ventral mouth present, oval-shaped, just anterior to ring furrow (Fig. 1b,c). Glandular network present over ventral surface, starting near anterior tip of body and ending just in front of ring furrow (Fig. 1b–d). Internally, body wall with epidermis, circular and longitudinal muscles, parenchyma and gastrodermis present (Fig. 1e). Oocytes present within intestine (Fig. 1f). Statocyst situated near anterior tip of body, just inside side furrow (Fig. 3a,b).

Description of juvenile. Based on paratype. Similar to female, but differs as follows: body 1.1 cm in length; pale orange in color (Fig. 2a); dorsal body surface in live specimen smooth, lacking longitudinal ridges and troughs, similar to that of *X. bocki*; gametes not observed. Ventral glandular network not detected externally, but observed with microCT imaging (Figs. 2 and 3).

Genetic information. Whole mitochondrial genome sequences (15,244 bp in holotype; 15,249 bp in paratype) and partial Histone H3 gene sequences (346 bp in holotype; 413 bp in paratype) were determined and deposited as INSD accession numbers LC228486, LC228485, LC228579 and LC228578, respectively. Exogenous mitochondrial and rSSU sequences of the following bivalves were detected: *Acila castrensis* from the holotype; *Nucula*

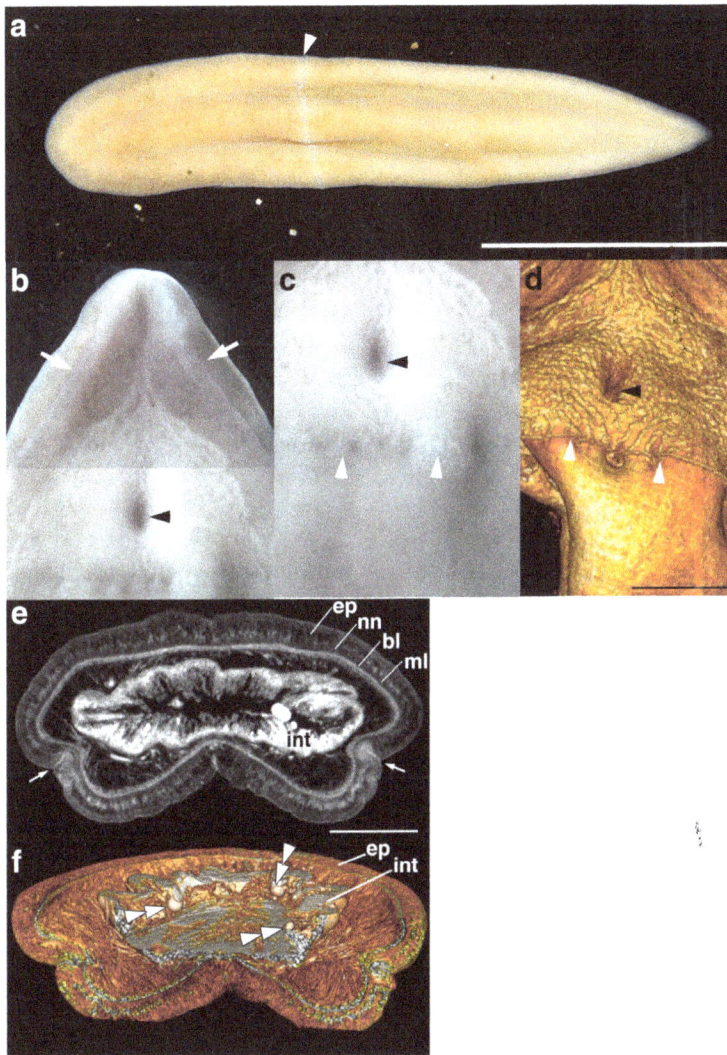

Fig. 1 Morphology of the holotype specimen (female) of *Xenoturbella japonica* sp. nov. **a** Live specimen with anterior to the left. **b** Antero-ventral part of a relaxed specimen, composed from two separate photographs. **c** Mid-ventral part of a relaxed specimen. The ventral glandular network ends at the ring furrow. **d** Volume rendering image from microCT scans showing the difference in epidermal composition between the anterior and posterior parts of the animal. **e** MicroCT scan showing a transverse section just anterior to the mouth. The epidermis (ep), intraepidermal nerve net (nn), basal lamina (bl) and muscle layer (ml) surround the intestine (int). **f** Volume rendering image from microCT scans showing oocytes inside the intestine (int). White arrowheads, ring furrow; white arrows, side furrow; black arrowheads; mouth, white double arrowheads, oocytes. Scale bars: a: 2 cm, d: 3 mm, e: 1 mm

nucleus, Ennucula cardara, A. castrensis and *Limaria fragilis* from the paratype.

Frontal pore and ventral glandular network of *X. japonica* and *X. bocki*

Volume rendering imaging of the anterior end of the holotype *X. japonica* specimen revealed that a frontal pore was present at the anterior tip of the body, ventral to side furrow (Fig. 3a). MicroCT imaging confirmed the presence of the pore, and also showed that the ventral glandular network continued posteriorly from the pore while branching (Fig. 3a,b, Additional file 2: Video S2). Analyses of the paratype juvenile specimen also showed that a frontal pore was present, and that the ventral glandular network, a linear structure, continued posteriorly from the pore along the basal lamina (Fig. 3c,d, Additional file 3: Video S3). Based on the discovery of these structures in the new species, we decided to reinvestigate the type species of *Xenoturbella, X. bocki*. MicroCT imaging showed that the frontal pore and the ventral glandular network are also both present in *X. bocki* (Fig. 3e,f, Additional file 1: Video S1 and Additional file 3: Video S3). Histological sections and light microscopic observations confirmed the presence of the structure (Fig. 3g,h).

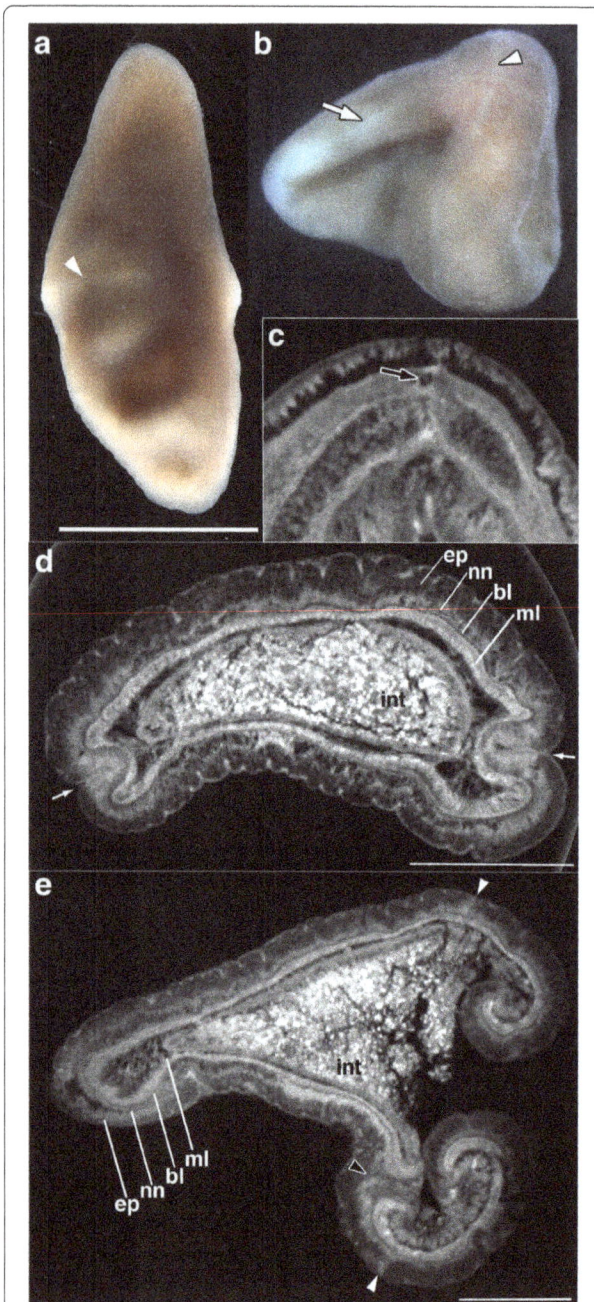

Fig. 2 Morphology of the paratype specimen (juvenile) of *Xenoturbella japonica* sp. nov. **a** Live specimen with anterior to the top. **b** Left-ventral view of a contracted specimen after fixation with anterior to the left. **c–e** MicroCT scans showing internal morphology. **c** Latitudinal section with anterior to the top. **d** Transverse section just anterior to the mouth. The epidermis (ep), intraepidermal nerve net (nn), basal lamina (bl) and muscle layer (ml) surround the intestine (int). **e** Longitudinal section with anterior to the left. The posterior epidermis, to the right, is curled due to contraction following fixation. White arrowheads, ring furrow; white arrows, side furrow; black arrow, statocyst; black arrowhead, mouth. Scale bars: a: 5 mm, d,e: 1 mm

Molecular phylogenetic analyses

The mitochondrial genome sizes of the *X. japonica* holotype and paratype were 15,244 bp and 15,249 bp, respectively. The gene content and gene order were identical with those of other *Xenoturbella* species (Additional file 5: Figure S2) [45, 48, 49].

Phylogenetic analyses using nucleotide alignments of the six reported mitochondrial genomes and those of the two new specimens showed that, in accordance with a previous study [45], *Xenoturbella* species are clustered into two major groups (Fig. 4a). *Xenoturbella japonica* is a sister group to a clade consisting of *X. bocki* and *X. hollandorum*, a group which was termed 'shallow' in Rouse et al. (2016) [45]. The other group, comprising *X. profunda*, *X. churro* and *X. monstrosa*, formed a clade that has previously been termed 'deep' [45]. All nodes in the tree were strongly (BP = 100) supported.

The pairwise genetic distances between the two collected specimens were compared with those of other xenoturbellids. Analyses using *cox1*, in which sequence data is available for the largest number of *Xenoturbella* specimens, showed that the genetic distance between the two new specimens are larger than almost all intra-species variations, but smaller than all inter-species variations (Fig. 4b, Additional file 6: Table S1). Analyses on either whole mitochondrial genomes, mitochondrial protein coding genes, or Histone H3 showed that the genetic distances between the two collected specimens were the smallest in all available pairwise genetic distance matrices (Additional file 7: Table S2).

Phylogenetic analyses of metazoans using mitochondrial genome sequences for the two new specimens resulted in different topology between data sets (Additional file 8: Figure S3, Additional file 9: Figure S4, Additional file 10: Figure S5 and Additional file 11: Figure S6). In agreement with previous studies [14, 45], this implies that mitochondrial genomes alone may not be suitable for resolving the phylogenetic positions of *Xenoturbella* and Xenacoelomorpha.

Discussion

The two new specimens collected off the Japanese coast were confirmed as xenoturbellids based on their morphology (Figs. 1, 2 and 3) and molecular phylogenetic analyses (Fig. 4, Additional file 8: Figure S3, Additional file 9: Figure S4, Additional file 10: Figure S5 and Additional file 11: Figure S6). The latter also showed that they do not belong to the five known species of *Xenoturbella* (Fig. 4, Additional file 6: Table S1 and Additional file 7: Table S2). But are the two specimens different species? Analyses of pairwise genetic distances showed that the distance between the two new specimens is larger than all xenoturbellid intra-species variations, except for three outliers seen among *X. bocki cox1*

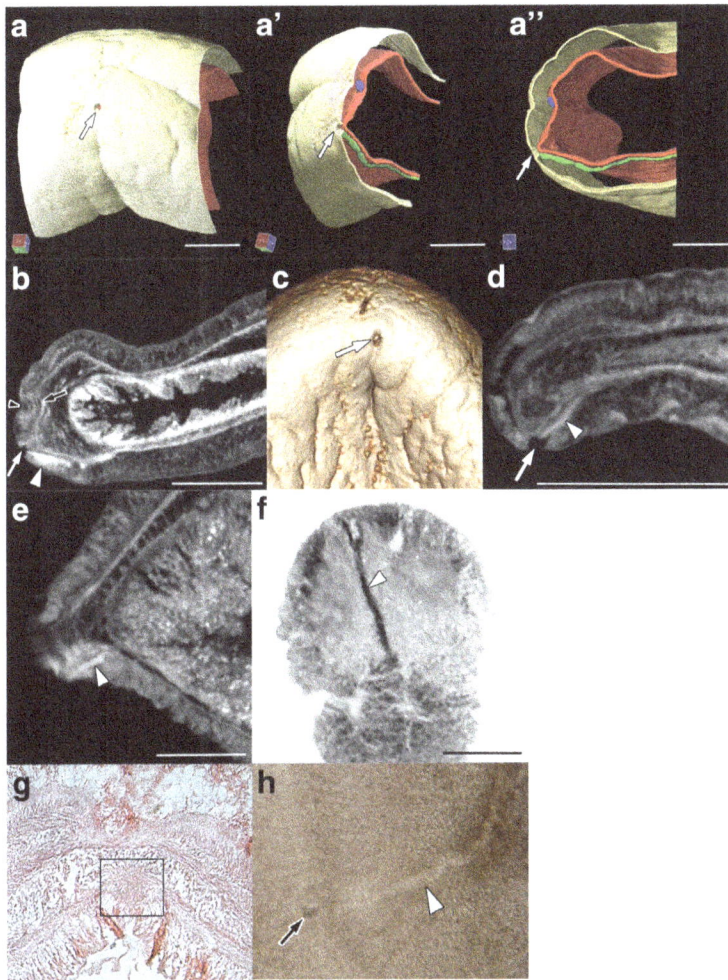

Fig. 3 Frontal pore and the ventral glandular network of *Xenoturbella*. a–b: *X. japonica* sp. nov., holotype female specimen. a–a″: Reconstructed images showing the relative positions of the frontal pore, statocyst (blue), ventral glandular network (green) and the basal lamina (red). b: Longitudinal section with anterior to the left. c–d: *X. japonica* sp. nov., paratype juvenile specimen. c: Volume rendering image of the anteroventral tip. d: Longitudinal section with anterior to the left. e–h: *X. bocki*. e: Longitudinal section with anterior to the left. f: Latitudinal section with anterior to the top. g: Transverse section of the anterior part of an animal, with dorsal to the top. The frontal organ is present in the mid-ventral region (black square). h: Anterior part of a specimen pressed under a cover glass. White arrowheads, ventral glandular network; white arrows, frontal pore; black arrowhead, side furrow; black arrows, statocyst. Scale bars: a–a″: 600 μm, b,d: 1 mm, e,f: 500 μm

sequences, but smaller than those of any inter-species pairs (Fig. 4b, Additional file 6: Table S1 and Additional file 7: Table S2). The two specimens were collected roughly 600 km apart, but *X. monstrosa* has been reported from two locations more than 2500 km apart, and *X. monstrosa* and *X. churro* have been shown to inhabit the same locality [45]. Therefore, the distance between the collection locations is not informative for species differentiation. There are morphological differences between the two specimens (Figs. 1, 2 and 3; Additional files 1: Video S1, Additional files 2: Video S2 and Additional files 3: Video S3), but these may be due to age differences. Therefore, until further specimens are collected, the two new xenoturbellids are considered as the same species, *X. japonica*, with the larger female

specimen as the holotype and the smaller specimen, probably a juvenile, as a paratype.

DNA extraction experiments have yielded contaminating bivalve DNA from four of the five known *Xenoturbella* species [45, 50, 51]. In this study, DNA sequences with similarities to those of the following bivalves were detected from *X. japonica*; *Nucula nucleus*, *Ennucula cardara*, *Acila castrensis* and *Limaria fragilis*, further supporting the theory that *Xenoturbella* feeds on bivalves. Since none of these species were collected together with *X. japonica*, further studies are needed to verify if it feeds on these exact species, or on closely related species in the area.

All Pacific xenoturbellid species, including *X. japonica* in this study, were collected from depths greater than 500 m. The type locality for *X. bocki* is around 100 m,

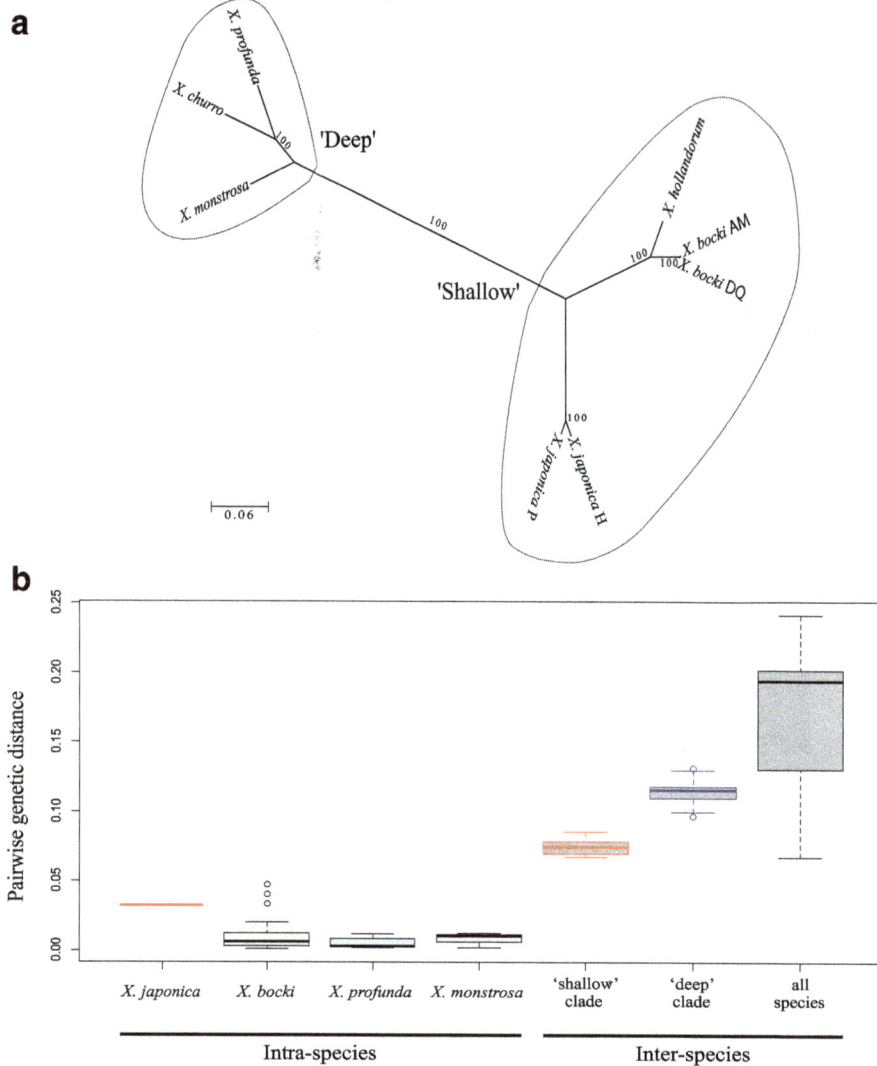

Fig. 4 Internal phylogeny of *Xenoturbella*. **a** Unrooted maximum likelihood tree of *Xenoturbella* species based on mitochondrial 13 protein-coding genes. Bootstrap values are shown at the nodes. H: holotype, P: paratype. AM, DQ: sequences deposited as AM296016 and DQ832701, respectively. **b** Pairwise genetic distances within and between *Xenoturbella* species calculated from *cox1* alignments. "'shallow' clade" and "'deep' clade" show inter-species differences between species belonging to each clade only. In "all species", inter-species differences between all xenoturbellid species are shown. Each box represents the inner 50% quantile (median marked by the solid line). Whiskers extend to 1.5 times the interquartile range, with circles indicating outliers. Original data are available in Additional file 6: Table S1

but the area is within the Gullmarsfjord, Sweden, known for its biodiversity of deep-sea animals living at a relatively shallow depth [11]. Therefore, *Xenoturbella* is probably a deep-sea fauna ancestrally. Considering Xenacoelomorpha, a clade inhabiting the deep-sea floor has been reported as a sister group to the rest of the acoels [52]. Comparison of acoels and *Xenoturbella* for morphological and physiological traits necessary for adapting to the deep-sea environment may provide insights into the ancestral habitat of Xenacoelomorpha.

We discovered a new structure in *Xenoturbella* - the frontal pore. It opens at the anterior tip of the animal, just below the side furrow, and connects with the ventral

glandular network (Fig. 3, Additional file 1: Video S1, Additional file 2: Video S2 and Additional file 3: Video S3). Although previously not described from the well-studied *X. bocki*, reinvestigation revealed that the frontal pore is also present in this species (Fig. 3e–h, Additional file 1: Video S1 and Additional file 3: Video S3). MicroCT scanning, a method previously not applied to *Xenoturbella*, proved to be instrumental in this discovery. Since collections of *Xenoturbella* can be rare (less than four specimens each for 4 of the 6 species [45]), this non-destructive method promises to be a powerful tool for investigating the internal morphology of this soft-bodied animal. A frontal organ, which consists of a frontal pore at the anterior end and a group of mucous glands (frontal glands) connected to the pore, is present in a number of acoel and nemertodermatid species [53–56], and has been suggested to be an ancestral trait for Acoelomorpha [57, 58]. Similarities in their overall morphologies and position tempts us to speculate that the frontal pore and ventral glandular network of xenoturbellids and the frontal organ of acoelomorphs are homologous and that the structure is a synapomorphy for Xenacoelomorpha. However, there are differences in their structures, with the acoelomorph frontal organ having a more elaborate morphology [53–56]. Comparisons of the frontal pore and ventral glandular network of xenoturbellids and the acoelo-morph frontal organ with respect to ultrastructure and the gene regulatory networks employed for morphogenesis may help to clarify the evolutionary relationship of these structures.

A previous study revealed that the five reported xenoturbellid species could be divided into two clades, 'shallow' and 'deep' [45]. Species of the first clade, *X. bocki* and *X. hollandorum*, live on the sea floor no deeper than 650 m and their body lengths are shorter than 4 cm. The ventral mouth is diamond shaped when animals are relaxed chemically. The ventral glandular network is not so conspicuous in this clade. The second clade consists of *X. monstrosa*, *X. churro* and *X. profunda*. These species have been reported from the seafloor between 1700 and 3700 m deep. They grow to be larger than the 'shallow' clade, reaching 10–20 cm in body length, with an obvious ventral glandular network and an oval mouth. Molecular phylogenetic analyses showed that *X. japonica* belong in the 'shallow' clade (Fig. 4a). The collected depth, body length, and the lack of an obvious ventral glandular network of the paratype specimen fit the characters of the 'shallow' clade. However, the holotype specimen shows characteristics of both clades. It was collected at 380–560 m depth, corresponding to the depth of the 'shallow' group. The body length of 5.3 cm falls between the two clades. Its oval mouth and conspicuous ventral glandular network suggest affinities to the 'deep' group. It is worth noting that the ventral epidermal network ends just anterior to the ring furrow in *X. japonica*, whereas the network passes the furrow posteriorly in *X. monstrosa*, *X. churro* and *X. profunda* [45]. Together with the lack of mature gametes, it is possible that the *X. japonica* holotype specimen is also still not fully grown, with the ventral epidermal network expanding posteriorly and the body lengthening with growth. Considering the phylogenetic position of *X. japonica*, it is parsimonious to regard characters common between this species and species in the 'deep' clade were present in the last common ancestor of *Xenoturbella*: body length over 5 cm, a well-developed ventral glandular network (and probably the frontal pore), and an oval mouth.

Conclusions

We have reported here the collection of two new specimens of *Xenoturbella*. Our discovery of *Xenoturbella* from the western Pacific Ocean greatly broadens the biogeographical range of the animal (Additional file 4: Figure S1). Molecular phylogenetic analyses of the two specimens revealed that they do not belong to the five previously described species. We propose that the new specimens belong to the same species, *X. japonica*, with the holotype specimen being female and the paratype specimen as juvenile, until further specimens are collected. MicroCT scanning established the presence of the frontal pore and ventral glandular network in *Xenoturbella*, and we suggest these structures as a new synapomorphy for *Xenoturbella*. *X. japonica* shows traits of both the 'shallow' and 'deep' groups, and therefore will be an important species for research on *Xenoturbella* and Xenacoelomorpha. This species can be collected using a marine biological dredge within a 1 h boat trip, and so promises to be a valuable source of information for investigating the evolution and diversity of deutero-stomes and bilaterians.

Methods
Collection and sample handling
The holotype specimen was collected as stated in the results. Pictures were taken using a Nikon Df fitted with AI AF Micro Nikkor 105 mm F2.8D or with a Leica DFC290 HD digital camera mounted on a Leica M205C stereomicroscope. 7% $MgCl_2$ in freshwater was used to relax the animal. Pieces of the animal were dissected and fixed in RNAlater for DNA extraction and then the partially dissected specimen was fixed in 4% paraformaldehyde in filtered seawater overnight, washed, and kept in 70% ethanol.

Collection of the paratype specimen was performed as described in the results. Pictures of the live animal were taken with an OLYMPUS OM-D E-M5, fitted with OLYMPUS M. ZUIKO DIGITAL ED 60 mm F2.8 Macro lens using a pair of Morris Hikaru Komachi Di flashes. The animal was anaesthetized in $MgCl_2$ solution isotonic to seawater and fixed in 10% neutral buffered formalin. The sample was stored in 10% neutral buffered formalin at room temperature for about 8 months. Pictures of the fixed specimen were taken with a Leica DFC290 HD digital camera mounted on a Leica M205C stereomicroscope. It was then transferred and preserved in 90% ethanol at 4 °C. Pieces of the animal were dissected and used for DNA extraction and then microCT scanning was performed on the partially dissected specimen.

Both specimens are deposited in the National Museum of Nature and Science, Tsukuba (NSMT), Japan.

Xenoturbella bocki was collected as previously described in [3, 11] at the Sven Lovén Centre for Marine Infrastructure, Gothenburg University, Sweden. Animals were fixed in 4% paraformaldehyde in filtered sea water at 4 °C overnight, dehydrated through a graded ethanol series and stored in 70% or 95% ethanol. Histological sections were made and photographed according to our previous studies [3, 33].

MicroCT scanning and image analysis
Samples stored in 70% or 95% ethanol were either stained with 1% PTA solution in 70% ethanol [59] or were rehydrated through a graded ethanol series and stained with 25% Lugol solution (1:3 mixture of Lugol's solution and deionized distilled water) [60]. The only exception was the paratype specimen, in which both staining methods were performed on a single sample. The staining time ranged from 3 to 48 h, depending on the size of the specimen. The stained samples were scanned using an X-ray microCT system (ScanXmate-E090S105; Comscan Techno) at a tube voltage peak of 60 kV and a tube current of 130 μA. During scanning, samples were rotated 360 degrees in steps of 0.1 to 0.18 degrees, generating 2000–3600 projection images of 992×992 pixels. Images were reconstructed using software provided with the X-ray microCT system (coneCTexpress; Comscan Techno) and volume data consisting of several hundred 8 bit TIFF files were obtained. For the *X. japonica* holotype and *X. bocki* No.2 samples, parts of the samples were scanned separately in order to obtain higher resolution images, and the reconstructed TIFF files were combined to generate data for the whole specimen. These data were used for analyses. 2D and 3D tomographic images were obtained using the OsiriX [61] and Tri/3D–BON (Ratoc System Engineering) software programs. Details of staining and scanning procedures are summarized in Additional file 12: Table S3.

DNA extraction and sequencing
Genomic DNA was extracted from a piece of the *X. japonica* holotype specimen stored in RNAlater solution (Ambion) using DNeasy Blood & Tissue Kit (Qiagen). Cytochrome-*c*-oxidase subunit I (*cox1*) and cytochrome B (*cob*) fragments were amplified by PCR, and purified with QIA quick Gel extraction Kit (Qiagen). Sequencing was carried out by a DNA sequencing service (Fasmac). Specific primers were designed from the sequences of the *cox1* and *cob*, and the full-length mitochondrial genome was amplified as two overlapping fragments by PCR. Barcode sequences were added to 5′ end of the

specific primers, and the PCR products were sequenced by Macrogen Japan using PacBio RS II Multiplexing Targeted Sequencing. PCR conditions (taq-polymerases, primers and amplification parameters) are shown in Additional file 13: Table S4. PacBio RS II reads were classified according to their barcode sequences using standalone BLAST (blastn, version 2.2.29), and aligned with MAFFT v7.221 [62] with a gap opening penalty of 0.1. Barcode sequences and ambiguous sites in the alignments were excluded using in-house Perl scripts. The mitochondrial genome of the *X. japonica* holotype specimen was obtained by concatenating two fragments, assembled from 573 reads (Average Coverage: 428) and 743 reads (Average Coverage: 558), respectively, from PacBio library.

Genomic DNA of the *X. japonica* paratype specimen was extracted using Allprep DNA/RNA FFPE Kit (QIAGEN). The extracted genomic DNA was fragmented to a target length of 600 bp by S220 Focused-Ultrasonicator (Covaris). Before preparing the library, the fragmented DNA was repaired using PreCR Repair Mix (New England Biolabs). Using the repaired DNA, a paired-end sequencing library was prepared with KAPA Hyper Prep Kit (KAPA Biosystems) according to the manufacturer's instructions. *Xenoturbella japonica* paratype specimen genome sequence data were obtained using a 300 bp paired-end protocol on an Illumina Miseq instrument (Illumina). BLAST searches against mitochondrial sequences registered in Refseq and Silva SSU 119 were performed to extract contigs with similarities to mitochondrial genomes and ribosomal small subunit RNAs, respectively. The mitochondrial genome of the paratype was reconstructed by assembling 40,879 reads from a MiSeq library (Average Coverage: 465) using MITObim v1.8 [63], the *X. bocki* mitochondrial genome as an initial reference sequence.

The mitochondrial genomes of both *X. japonica* specimens were annotated using the MITOS web server, employing translation table 5 [64].

The nucleotide sequence of Histone H3 from the *X. japonica* holotype specimen was amplified by PCR, purified with QIA quick Gel extraction Kit (Qiagen), and ligated into pMD20-T vector (Takara Bio). The nucleotide sequences of the vectors were directly amplified from clones by PCR, purified using Exonuclease I and Calf intestine alkaline phosphatase (both Takara Bio), and sequenced using a DNA sequencing service (Fasmac). The PCR conditions (taq-polymerases, primers and amplification parameters) are shown in Additional file 13: Table S4. The nucleotide sequence of the paratype specimen Histone H3 was reconstructed from reads of MiSeq library: 182 reads showing high similarity (>85% identity) to the nucleotide sequence of *X. bocki* Histone H3 were identified in MiSeq library using a standalone BLASTn search and were aligned using MAFFT v7.221 with gap opening penalty of 0.5.

Sequences obtained were deposited in the International Nucleotide Sequence Database (INSD) through the DNA Data Bank of Japan (DDBJ).

Molecular phylogenetic analyses

For phylogenetic analyses of xenoturbellids, nucleotide sequences of the eight available mitochondrial genomes were aligned using MAFFT L-INS-i, and refined by Gblocks using the following parameters: b2 = 75%, b3 = 5, b4 = 5, b5 = half [65]. Phylogenetic relationships were inferred using RAxML (v8.1.11) under the GTR + gamma model with 100 bootstrap replicates.

For calculating pairwise genetic distances, amino acid sequences of 13 concatenated mitochondrial protein-coding genes and nucleotide sequences of whole mitochondrial genomes, nuclear histone H3 and *cox1* genes were aligned using MAFFT L-INS-i. Pairwise distances were calculated using the R package 'phangorn' [66], employing the MtMam model for amino acid sequences and the JC69 model for nucleotide sequences. A nuclear histone H3 sequence was not available for *X. churro*. Among *cox1* alignments, genetic distances between two non-overlapping sequences were not calculated.

Four different phylogenetic analyses of bilaterian mitochondrial genomes were performed. In three analyses (*i–iii*), each dataset was aligned using MAFFT L-INS-i and ambiguous sites were removed with Gblocks using stringent parameters (b2 = 75%, b3 = 5, b4 = 5, b5 = half) [65]. RAxML was run under the LG4X + gamma model with the data partitioned by genes. In a fourth analysis (*iv*), genomes were aligned with MAFFT FFT-NS-i (faster but less accurate than MAFFT L-INS-i) and refined by Gblocks 0.91b [67] using less stringent condition parameters (b2 = 65%, b3 = 10, b4 = 5, b5 = all) [68]. The maximum likelihood (ML) analysis was carried out with RAxML v8.1.1 [69] under the GTR + gamma model with the data partitioned (−q option) by genes. Datasets used in the four analyses were amino acid sequences of 13 mitochondrial protein-coding genes (*atp6* and *8, cob, cox1–3, nad1–6* and *nad4L*) from 31 metazoans and six *Xenoturbella* sequences (same dataset as ref. [45]) with the addition of two new sequences acquired in this study, with the following differences between analyses. (*i*) all four acoelomorph species were excluded; (*ii*) all 31 metazoans and eight *Xenoturbella* data were used; (*iii*) non-bilaterian metazoans (3 sponges and 3 cnidarians) were excluded; (*iv*) all eight *Xenoturbella* and 31 metazoans mitochondrial genomes were used. The alignment lengths of the four phylogenetic analyses were: (*i*) 2179 aa; (*ii*) 2208 aa; (*iii*) 2202 aa; (*iv*) 2791 aa.

Additional files

Additional file 1: Video S1. Volume rendering image reconstructed from micoCT scans showing external morphology of *X. japonica* holotype (H), paratype (P) and *X. bocki*. (MP4 19263 kb)

Additional file 2: Video S2. MicroCT sections showing internal structures of *X. japonica* holotype female. vgn, ventral glandular network. (MP4 19089 kb)

Additional file 3: Video S3. MicroCT sections showing internal structures of *X. japonica* paratype juvenile and *X. bocki*. vgn, ventral glandular network. (MP4 19407 kb)

Additional file 4: Figure S1. Distribution of *Xenoturbella*. a: Collection sites of the two specimens of *X. japonica* from the western Pacific. H: holotype, P: paratype. b: Worldwide distribution of *Xenoturbella*. Only sites where the species of the collected specimens were confirmed by molecular phylogenetic analyses are shown. The map and plots were generated with GMT5 software [71]. Xb: *X. bocki*, Xc: *X. churro*, Xh: *X. hollandorum*, Xj: *X. japonica* sp. nov., Xm: *X. monstrosa*, Xp: *X. profunda*. (PDF 1477 kb)

Additional file 5: Figure S2. Linearized mitochondrial genome maps of *X. japonica* sp. nov. holotype, paratype and *X. bocki*. Red; protein coding genes, blue; tRNA, green; rRNA. Gene orders of both *X. japonica* specimens were identical with that of *X. bocki*. (PDF 904 kb)

Additional file 6: Table S1. Pairwise genetic distances of nucleotide *cox1* alignments. Intra-species genetic distances of *X. japonica, X. bocki, X. profunda* and *X. monstrosa* are colored red, light green, light blue and light purple, respectively. Inter-species genetic distances are shown in gray. Inter-species genetic distances between species in the 'shallow' clade are surrounded by an orange square, and those in the 'deep' clade are surrounded by cobalt squares. (PDF 74 kb)

Additional file 7: Table S2. Pairwise genetic distances of *Xenoturbella* species. (PDF 52 kb)

Additional file 8: Figure S3. Maximum likelihood tree of metazoans (excluding Acoelomorpha) based on 13 mitochondrial protein-coding genes. Bootstrap values are shown at the nodes. Bilaterian taxon names are indicated to the right of the tree. H: holotype, P: paratype. AM, DQ: sequences deposited as AM296016 and DQ832701, respectively. (PDF 1200 kb)

Additional file 9: Figure S4. Maximum likelihood tree of metazoans based on 13 mitochondrial protein-coding genes. Bootstrap values are shown at the nodes. Bilaterian taxon names are indicated to the right of the tree. H: holotype, P: paratype. AM, DQ: sequences deposited as AM296016 and DQ832701, respectively. (PDF 1249 kb)

Additional file 10: Figure S5. Unrooted maximum likelihood tree of bilaterians based on 13 mitochondrial protein-coding genes. Bootstrap values are shown at the nodes. Bilaterian taxon names are marked with dashed lines. H: holotype, P: paratype. AM, DQ: sequences deposited as AM296016 and DQ832701, respectively. (PDF 920 kb)

Additional file 11: Figure S6. Maximum likelihood tree of metazoans based on 13 mitochondrial protein-coding genes with less stringent conditions. Bootstrap values are shown at the nodes. Bilaterian taxon names are indicated to the right of the tree. H: holotype, P: paratype. AM, DQ: sequences deposited as AM296016 and DQ832701, respectively. (PDF 1458 kb)

Additional file 12: Table S3. Details for microCT scans. (PDF 52 kb)

Additional file 13: Table S4. Primers list and PCR conditions. (PDF 57 kb)

Abbreviations

Cob: Cytochrome B; cox1: Cytochrome C oxidase subunit I; DDBJ: DNA Data Bank of Japan; FRV: Fisheries research vessel; INSD: International Nucleotide Sequence Database; JAMBIO: Japanese Association for Marine Biology; microCT: micro computed tomography; NSMT: National Museum of Nature and Science, Tsukuba, Japan; ROVs: Remotely operated vehicles; rSSU: Ribosomal small subunit; RV: Research vessel

Acknowledgements

We thank Ken Fujimoto, National Research Institute of Fisheries Science, Fisheries Research Agency, Hiroshi Namikawa, National Museum of Nature and Science, and all the people that participated in the research survey onboard FRV *Soyo-Maru* and RV *Rinkai-Maru* for the sample collections. We are grateful for Mamoru Sekifuji, Junko Inoue, Nao Oumi, and Mariko Kondo for collections and discussions. We also thank Shunsuke Yaguchi for reagents, Munetsugu Bam for discussions, and Atsuko Suzuki for taking photographs of the holotype specimen. We thank the staff at Sven Lovén Centre for Marine Infrastructure, Gothenburg University, Sweden for their help in the *X. bocki* collections. We thank Maurice R. Elphick for critical reading of the manuscript. We are grateful to Yasunori Saito for equipment and discussions, who sadly passed away in December 2015.

Funding

This work was supported by the NIG Collaborative Research Program (2014-A36, 2015-A128, and 2016-A127), the KVA fund for 2015 and 2016, JSPS Grant-in-Aid for Young Scientists (A) (JP26711022), and JAMBIO, Japanese Association for Marine Biology, as a program of Joint Usage/Research Center by the Ministry of Education, Culture, Sports, Science and Technology.

Authors' contributions

HN conceived the project. For the *X. japonica* holotype specimen, HN, AO and HK made the initial external observations. HN and KK made the initial external observations for the paratype specimen. HN, HK, AM and TS performed the microCT analyses, and HN, HM, RK, MK and NS analyzed the sequence data. HN drafted the paper with HM and AM, and all authors read and approved the manuscript.

Competing interests

The authors declare that they have no competing interests.

Author details

[1]Shimoda Marine Research Center, University of Tsukuba, 5-10-1, Shimoda, Shizuoka 415-0025, Japan. [2]Mammalian Genetics Laboratory, National Institute of Genetics, 1111 Yata, Mishima, Shizuoka, 411-8540, Japan. [3]Faculty of Science, Hokkaido University, N10 W8, Kita-ku, Sapporo, Hokkaido 060-0810, Japan. [4]DNA Sequencing Section, Okinawa Institute of Science and Technology Graduate University, Onna, Okinawa 904-0495, Japan. [5]Marine Genomics Unit, Okinawa Institute of Science and Technology Graduate University, Onna, Okinawa 904-0495, Japan. [6]Misaki Marine Biological Station, The University of Tokyo, 1024 Koajiro, Misaki, Miura, Kanagawa 238-0225, Japan. [7]Present address: Sado Marine Biological Station, Faculty of Science, Niigata University, Sado, Niigata 952-2135, Japan.

References

1. Westblad E. *Xenoturbella bocki* n.g, n.sp, a peculiar, primitive turbellarian type. Ark Zool. 1949;1:3–29.
2. Raikova O, Reuter M, Jondelius U, Gustafsson M. An immunocytochemical and ultrastructural study of the nervous and muscular systems of *Xenoturbella westbladi* (Bilateria inc. sed.). Zoomorphology. 2000;120:107–18. doi:10.1007/s004350000028
3. Stach T, Dupont S, Israelsson O, Fauville G, Nakano H, Kånneby T, et al. Nerve cells of *Xenoturbella bocki* (phylum uncertain) and *Harrimania kupfferi* (Enteropneusta) are positively immunoreactive to antibodies raised against echinoderm neuropeptides. J Mar Biol Assoc UK. 2005;85:1519–24. doi:10.1017/S0025315405012725.
4. Jagersten G. Further remarks on the early phylogeny of Metazoa. Zool Bidr Upps. 1959;33:79–108.
5. Ehlers U, Sopott-Ehlers B. Ultrastructure of the subepidermal musculature of *Xenoturbella bocki*, the adelphotaxon of the Bilateria. Zoomorphology. 1997; 117:71–9. doi:10.1007/s004350050032.

6. Reisinger E. Was ist *Xenoturbella*? Z Wiss Zool. 1960;164:188–98.
7. Pedersen KJ, Pedersen LR. Fine structural observations on the extracellular matrix (ECM) of *Xenoturbella bocki* Westblad, 1949. Acta Zool. 1986;67:103–13. doi:10.1111/j.1463-6395.1986.tb00854.x.
8. Israelsson O. And molluscan embryogenesis. Nature. 1997;390:32. 10.1038/36246.
9. Telford MJ. Xenoturbellida: the fourth deuterostome phylum and the diet of worms. Genesis. 2008;46:580–6. doi:10.1002/dvg.20414.
10. Edgecombe GD, Giribet G, Dunn CW, Hejnol A, Kristensen RM, Neves RC, et al. Higher-level metazoan relationships: recent progress and remaining questions. Org Divers Evol. 2011;11:151–72. doi:10.1007/s13127-011-0044-4.
11. Nakano H. What is *Xenoturbella*? Zoolog Lett. 2015;1:22. doi:10.1186/s40851-015-0018-z.
12. Hejnol A, Obst M, Stamatakis A, Ott M, Rouse GW, Edgecombe GD, et al. Assessing the root of bilaterian animals with scalable phylogenomic methods. Proc Biol Sci. 2009;276:4261–70. doi:10.1098/rspb.2009.0896.
13. Philippe H, Brinkmann H, Copley RR, Moroz LL, Nakano H, Poustka AJ, et al. Acoelomorph flatworms are deuterostomes related to *Xenoturbella*. Nature. 2011;470:255–8. 10.1038/nature09676.
14. Cannon JT, Vellutini BC, Smith J III, Ronquist F, Jondelius U, Hejnol A. Xenacoelomorpha is the sister group to Nephrozoa. Nature. 2016;530:89–93. doi:10.1038/nature16520.
15. Ehlers U. On the fine structure of *Paratomella rubra* Rieger & Ott (Acoela) and the position of the taxon *Paratomella* Dorjes in a phylogenetic system of the Acoelomorpha (Plathelminthes). Microfauna Mar. 1992;7:265–93.
16. Ehlers U. Dermonephridia- modified epidermal cells with a probable excretory function in *Paratomella rubru* (Acoela, Plathelminthes). Microfauna Mar. 1992;7:253–64.
17. Katayama T, Yamamoto M, Wada H, Satoh N. Phylogenetic position of acoel turbellarians inferred from partial 18S rDNA sequences. Zool Sci. 1993;10:529–36.
18. Haszprunar G. The Mollusca: Coelomate turbellarians or mesenchymate annelids? In: Taylor JD, editor. Origin and evolutionary radiation of the Mollusca. Oxford: Oxford University Press; 1995. p. 1–28.
19. Haszprunar G. Plathelminthes and Plathelminthomorpha- paraphyletic taxa. J Zool Syst Evol Res. 1996;34:41–8.
20. Ruiz-Trillo I, Riutort M, Littlewood DTJ, Herniou EA, Baguna J. Acoel flatowrms: earliest extant bilaterian metazoans, not members of Platyhelminthes. Science. 1999;283:1919–23. 10.1126/science.283.5409.1919.
21. Philippe H, Brinkmann H, Martinez P, Riutort M, Baguna J. Acoel flatworms are not Platyhelminthes: evidence from phylogenomics. PLoS One. 2007;2: e717. doi:10.1371/journal.pone.0000717.
22. Gardiner EG. Early development of *Polychoerus caudatus*, Mark. J Morphol. 1895;11:155–76. doi: 10.1002/jmor.1050110104 .
23. Ramachandra NB, Gates RD, Ladurner P, Jacobs DK, Hartenstein V. Embryonic development in the primitive bilaterian *Neochildia fusca*: normal morphogenesis and isolation of POU genes *Brn-1* and *Brn-3*. Dev Genes Evol. 2002;212:55–69. doi:10.1007/s00427-001-0207-y.
24. Nakano H, Lundin K, Bourlat SJ, Telford MJ, Funch P, Nyengaard JR, et al. *Xenoturbella bocki* exhibits direct development with similarities to Acoelomorpha. Nat Commun. 2013;4:1537. doi:10.1038/ncomms2556.
25. Franzen A, Afzelius BA. The ciliated epidermis of *Xenoturbella bocki* (Platyhelminthes Xenoturbellida) with some phylogenetic considerations. Zool Scripta. 1987;16:9–17. doi:10.1111/j.1463-6409.1987.tb00046.x.
26. Rohde K, Watson N, Cannon LRG. Ultrastructure of epidermal cilia of *Pseudactinoposthia* sp. (Platyhelminthes, Acoela): implications for the phylogenetic status of the Xenoturbellida and Acoelomorpha. J Submicrosc Cytol Pathol. 1988;20:759–67.
27. Lundin K. The epidermal ciliary rootlets of *Xenoturbella bocki* (Xenoturbellida) revisited: new support for a possible kinship with the Acoelomorpha (Platyhelminthes). Zool Scripta. 1998;27:263–70. doi:10.1111/j.1463-6409.1998.tb00440.x.
28. Lundin K. Degenerating epidermis cells in *Xenoturbella bocki* (phylum uncertain), Nemertodermatida and Acoela (Platyhelminthes). Belgian. J Zool. 2001;131:153–7.
29. Ehlers U. Comparative morphology of statocysts in the Platyhelminthes and the Xenoturbellida. Hydrobiologia. 1991;227:263–71. doi:10.1007/BF00027611.
30. Israelsson O. Ultrastructural aspects of the 'statocyst' of *Xenoturbella* (Deuterostomia) cast doubt on its function as a georeceptor. Tissue Cell. 2007;39:171–7. doi:10.1016/j.tice.2007.03.002.
31. Lundin K, Hendelberg J. Is the sperm type of the Nemertodermatida close to that of the ancestral Platyhelminthes? Hydrobiologia. 1998;383:197–205. doi:10.1023/A:1003439512957.

32. Petrov A, Hooge M, Tyler S. Ultrastructure of sperms in Acoela (Acoelomorpha) and its concordance with molecular systematics. Invert Biol. 2004;123:183–97. doi:10.1111/j.1744-7410.2004.tb00154.x.

33. Obst M, Nakano H, Bourlat SJ, Thorndyke MC, Telford MJ, Nyengaard JR, et al. Spermatozoon ultrastructure of Xenoturbella bocki (Westblad 1949). Acta Zool. 2011;92:109–15. doi:10.1111/j.1463-6395.2010.00496.x.

34. Henry JQ, Martindale MQ, Boyer BC. The unique developmental program of the acoel flatworm, Neochildia fusca. Dev Biol. 2000;220:285–95. doi:10.1006/dbio.2000.9628.

35. Jondelius U, Larsson K, Raikova O. Cleavage in Nemertoderma westbladi (Nemertodermatida) and its phylogenetic significance. Zoomorphology. 2004;123:221–5. doi:10.1007/s00435-004-0105-8.

36. Børve A, Hejnol A. Development and juvenile anatomy of the nemertodermatid Meara stichopi (Bock) Westblad 1949 (Acoelomorpha). Front Zool. 2014;11:50. doi:10.1186/1742-9994-11-50.

37. Lowe CJ, Pani AM. Animal evolution: a soap opera of unremarkable worms. Curr Biol. 2011;21:R151–3. doi:10.1016/j.cub.2010.12.017.

38. Hejnol A, Martin-Duran JM. Getting to the bottom of anal evolution. Zool Anz. 2015;256:61–74. doi:10.1016/j.jcz.2015.02.006.

39. Gavilan B, Perea-Atienza E, Martinez P. Xenoacoelomorpha: a case of independent nervous system centralization? Phil Trans R Soc B. 2016;371:20150039. doi:10.1098/rstb.2015.0039.

40. Haszprunar G. Review of data for a morphological look on Xenacoelomorpha (Bilateria incertae sedis). Org Divers Evol. 2016;16:363–89. doi:10.1007/s13127-015-0249-z.

41. Hejnol A, Pang K. Xenacoelomorpha's significance for understanding bilaterian evolution. Curr Opin Genet Dev. 2016;39:48–54. doi:10.1016/j.gde.2016.05.019.

42. Telford MJ, Copley RR. Zoology: war of the worms. Curr Biol. 2016;26:R319–37. doi:10.1016/j.cub.2016.03.015.

43. Robertson HE, Lapraz F, Egger B, Telford MJ, Schiffer PH. The mitochondrial genomes of the acoelomorph worms Paratomella rubra, Isodiametra pulchra and Archaphanostoma ylvae. Sci Rep. 2017;7:1847. doi:10.1038/s41598-017-01608-4.

44. Israelsson O. New light on the enigmatic Xenoturbella (phylum uncertain): ontogeny and phylogeny. Phil Trans R Soc B. 1999;266:835–41. doi:10.1098/rspb.1999.0713.

45. Rouse GW, Wilson NG, Carvajal JI, Vrijenhoek RC. New deep-sea species of Xenoturbella and the position of Xenacoelomorpha. Nature. 2016;530:94–7. doi:10.1038/nature16545.

46. Bourlat SJ, Juliusdottir T, Lowe CJ, Freeman R, Aronowicz J, Kirschner M, et al. Deuterostome phylogeny reveals monophyletic chordates and the new phylum Xenoturbellida. Nature. 2006;444:85–8. doi:10.1038/nature05241.

47. Akiyama T, Shimomura M, Nakamura K. Collection of deep-sea small arthropods: gears for collection and processing of samples on deck. TAXA. 2008;24:27–32.

48. Pereske M, Hankeln T, Weich B, Fritzsch G, Stadler PF, Israelsson O, et al. The mitochondrial DNA of Xenoturbella bocki: genomic architecture and phylogenetic analysis. Theory Biosci. 2007;126:35. doi:10.1007/s12064-007-0007-7.

49. Bourlat SJ, Rota-Stabelli O, Lanfear R, Telford MJ. The mitochondrial genome structure of Xenoturbella bocki (phylum Xenoturbellida) is ancestral within the deuterostomes. BMC Evol Biol. 2009;9:107. doi:10.1186/1471-2148-9-107.

50. Bourlat SJ, Nielsen C, Lockyer AE, Littlewood DTJ, Telford MJ. Xenoturbella is a deuterostome that eats molluscs. Nature. 2003;424:925–8. doi:10.1038/nature01851.

51. Bourlat SJ, Nakano H, Akerman M, Telford MJ, Thorndyke MC, Obst M. Feeding ecology of Xenoturbella bocki (phylum Xenoturbellida) revealed by genetic barcoding. Mol Ecol Resour. 2008;8:18–22. doi:10.1111/j.1471-8286.2007.01959.x.

52. Arroyo AS, López-Escardó D, de Vargas C, Ruiz-Trillo I. Hidden diversity of Acoelomorpha revealed through metabarcoding. Biol Lett. 2016;12:20160674. doi:10.1098/rsbl.2016.0674.

53. Klauser MD, Smith IIIJPS, Tyler S. Ultrastructure of the frontal organ in Convoluta and Macrostomum spp.: significance for models of the turbellarian archetype. Hydrobiologia. 1985;132:47–52. doi:10.1007/BF00046227.

54. Smith JPS III, Tyler S. Fine-structure and evolutionary implications of the frontal organ in Turbellaria Acoela: 1. Diopisthoporus gymnopharyngeus sp.n. Zool Scripta. 1985;14:91–102. doi:10.1111/j.1463-6409.1985.tb00180.x.

55. Smith JPS III, Tyler S. Frontal organs in the Acoelomorpha (Turbellaria): ultrastructure and phylogenetic significance. Hydrobiologia. 1986;132:71–8. doi:10.1007/BF00046231.

56. Ehlers U. Frontal glandular and sensory structures in Nemertoderma (Nermertodermatida) and Paratomella (Acoela): ultrastructure and phylogenetic implications for the monophyly of the Euplathelminthes (Plathelminthes). Zoomorphology. 1992;112:227–36. doi:10.1007/BF01632820.

57. Jondelius U, Wallberg A, Hooge M, Raikova OI. How the worm got its pharynx: phylogeny, classification and Bayesian assessment of character evolution in Acoela. Syst Biol. 2011;60:845–71. doi:10.1093/sysbio/syr073.

58. Achatz JG, Chiodin M, Salvenmoser W, Tyler S, Martinez P. The Acoela: on their kind and kinships, especially with nemertodermatids and xenoturbellids (Bilateria incertae sedis). Org Divers Evol. 2013;13:267–86. doi:10.1007/s13127-012-0112-4.

59. Metscher BD. MicroCT for comparative morphology: simple staining methods allow high-contrast 3D imaging of diverse non-mineralized animal tissues. BMC Physiol. 2009;9:11. doi:10.1186/1472-6793-9-11.

60. Degenhardt K, Wright AC, Horng D, Padmanabhan A, Epstein JA. Rapid 3D phenotyping of cardiovascular development in mouse embryos by micro-CT with iodine staining. Circ Cardiovasc Imaging. 2010;3:314–22. doi:10.1161/CIRCIMAGING.109.918482.

61. Rosset A, Spadola L, Ratib O. OsiriX: an open-source software for navigating in multidimensional DICOM images. J Digit Imaging. 2004;17:205–16. doi:10.1007/s10278-004-1014-6.

62. Katoh K, Standley DM. MAFFT multiple sequence alignment software version 7: improvements in performance and usability. Mol Biol Evol. 2013;30:772–80. doi:10.1093/molbev/mst010.

63. Hahn C, Bachmann L, Chevreux B. Reconstructing mitochondrial genomes directly from genomic next-generation sequencing reads—a baiting and iterative mapping approach. Nucl Acids Res. 2013;41:e129. doi:10.1093/nar/gkt371.

64. Bernt M, Donath A, Juhling F, Externbrink F, Florentz C, Frizsch G, et al. MITOS: Improved de novo metazoan mitochondrial genome annotation. Mol Phylogenet Evol. 2013;69:313–9. doi:10.1016/j.ympev.2012.08.023.

65. Philippe H, Brinkmann H, Lavrov DV, Littlewood DTJ, Manuel M, Worheide G, et al. Resolving difficult phylogenetic questions: why more sequences are not enough. PLoS Biol. 2011;9:e1000602. doi:10.1371/journal.pbio.1000602.

66. Schliep KP. Phangorn: phylogenetic analysis in R. Bioinformatics. 2011;27:592–3. doi:10.1093/bioinformatics/btq706.

67. Castresana J. Selection of conserved blocks from multiple alignments for their use in phylogenetic analysis. Mol Biol Evol. 2000;17:540–52. doi:10.1093/oxfordjournals.molbev.a026334.

68. Dunn CW, Hejnol A, Matus DQ, Pang K, Browne WE, Smith SA, et al. Broad phylogenomic sampling improves resolution of the animal tree of life. Nature. 2008;452:745–9. doi:10.1038/nature06614.

69. Stamatakis A. RAxML version 8: a tool for phylogenetic analysis and post-analysis of large phylogenies. Bioinformatics. 2014;30:1312–3. doi:10.1093/bioinformatics/btu033.

70. Nakano H, Miyazawa H, Maeno A, Shiroishi T, Kakui K, Koyanagi R, et al. Micro CT files from 'A new species of Xenoturbella from the western Pacific Ocean and the evolution of Xenoturbella'. figshare. 2017. doi:10.6084/m9.figshare.5330908.

71. Wessel P, Smith WHF, Scharroo R, Luis J, Wobbe F. Generic mapping tools: improved version released. EOS Trans Am Geophys Union. 2013;94:409–10. doi:10.1002/2013EO450001.

11

Evolution of cytokinesis-related protein localization during the emergence of multicellularity in volvocine green algae

Yoko Arakaki[1], Takayuki Fujiwara[2], Hiroko Kawai-Toyooka[1], Kaoru Kawafune[1,3], Jonathan Featherston[4,5], Pierre M. Durand[4,6], Shin-ya Miyagishima[2] and Hisayoshi Nozaki[1*] ⓘ

Abstract

Background: The volvocine lineage, containing unicellular *Chlamydomonas reinhardtii* and differentiated multicellular *Volvox carteri*, is a powerful model for comparative studies aiming at understanding emergence of multicellularity. *Tetrabaena socialis* is the simplest multicellular volvocine alga and belongs to the family Tetrabaenaceae that is sister to more complex multicellular volvocine families, Goniaceae and Volvocaceae. Thus, *T. socialis* is a key species to elucidate the initial steps in the evolution of multicellularity. In the asexual life cycle of *C. reinhardtii* and multicellular volvocine species, reproductive cells form daughter cells/colonies by multiple fission. In embryogenesis of the multicellular species, daughter protoplasts are connected to one another by cytoplasmic bridges formed by incomplete cytokinesis during multiple fission. These bridges are important for arranging the daughter protoplasts in appropriate positions such that species-specific integrated multicellular individuals are shaped. Detailed comparative studies of cytokinesis between unicellular and simple multicellular volvocine species will help to elucidate the emergence of multicellularity from the unicellular ancestor. However, the cytokinesis-related genes between closely related unicellular and multicellular species have not been subjected to a comparative analysis.

Results: Here we focused on dynamin-related protein 1 (DRP1), which is known for its role in cytokinesis in land plants. Immunofluorescence microscopy using an antibody against *T. socialis* DRP1 revealed that volvocine DRP1 was localized to division planes during cytokinesis in unicellular *C. reinhardtii* and two simple multicellular volvocine species *T. socialis* and *Gonium pectorale*. DRP1 signals were mainly observed in the newly formed division planes of unicellular *C. reinhardtii* during multiple fission, whereas in multicellular *T. socialis* and *G. pectorale*, DRP1 signals were observed in all division planes during embryogenesis.

Conclusions: These results indicate that the molecular mechanisms of cytokinesis may be different in unicellular and multicellular volvocine algae. The localization of DRP1 during multiple fission might have been modified in the common ancestor of multicellular volvocine algae. This modification may have been essential for the re-orientation of cells and shaping colonies during the emergence of multicellularity in this lineage.

Keywords: Multicellularity, Volvocine algae, *Tetrabaena socialis*, DRP1

* Correspondence: nozaki@bs.s.u-tokyo.ac.jp
[1]Department of Biological Sciences, Graduate School of Science, University of Tokyo, 7-3-1 Hongo, Bunkyo-ku, Tokyo 113-0033, Japan
Full list of author information is available at the end of the article

Background

The transition to multicellularity is one of the most compelling events during the evolution of life and has occurred more than 25 times in distinct eukaryotic lineage [1]. To fully understand these transitions comparative biological studies are essential. Major complex multicellular groups such as metazoans and land plants emerged from unicellular ancestors that existed approximately 600 to 1000 million years ago [2]. There are few extant species that represent the initial features of multicellular ancestors or transitional forms from unicellular to multicellular within the lineages closely related to metazoans and land plants. In contrast, the multicellular volvocine green algae (Fig. 1) diverged from a unicellular ancestor only 200 million years ago [3]. In addition, there are many extant species representing various steps in the transition to multicellularity, which makes the volvocines unique as a model lineage.

The volvocine lineage (Fig. 1) includes unicellular *Chlamydomonas reinhardtii*, undifferentiated multicellular species like *Tetrabaena socialis* and *Gonium pectorale*, and differentiated multicellular species such as *Volvox carteri* [3–5]. The simplest multicellular species is four-celled *T. socialis* belonging to the Tetrabaenaceae, which is sister to the large clade composed of the remaining, more complex colonilal/multicellular volvocine algae (Goniaceae and Volvocaceae) (Fig. 1) [5–7]. *T. socialis* shares at least four common features with more complex multicellular volvocine members: incomplete cytokinesis, rotation of basal bodies, transformation of the cell wall to extracellular matrix, and modulation of cell number [3, 5]. Recently, whole nuclear genome analyses of *G. pectorale* [8] suggested that modifications of cell cycle regulation genes (duplication of *cyclin D1* gene and alterations in the retinoblastoma protein) occurred in the common ancestor of

G. pectorale and *V. carteri* and that these modifications were the basis for genetic modulation of cell number. However, there is less information about more downstream molecules that actually participate in the formation of integrated multicellular individuals during embryogenesis.

During the asexual life cycles of unicellular and multicellular volvocine algae, reproductive cells perform successive divisions (rapid S/M phase alternating without G2 phase) known as multiple fission [9, 10]. The unicellular species *C. reinhardtii* forms 2^n (n: number of rounds of cell divisions) daughter cells depending on the size of the mother cell [11], whereas reproductive cells of multicellular volvocine members form 2^n daughter protoplasts that are regulated by mother cell size and genetic control [4]. These daughter protoplasts are connected to one another by cytoplasmic bridges, which are important for the arrangement of cells within the daughter colony of multicellular volvocine algae [4] like *T. socialis* [5], *G. pectorale* [12] and *V. carteri* [13]. Considering that both of these multicellular member-specific traits (modulation of daughter cell number and incomplete cytokinesis) are recognized in the tetrabaenacean species *T. socialis*, comparative molecular analyses of multiple fission between unicellular and multicellular forms is essential to understand the initial steps to multicellularity in this lineage.

Various cytokinesis-related genes are characterized in metazoans [14] and land plants [15]. Modes of cytokinesis are variable: mother cells divide into two daughter cells by fission in metazoans, whereas two daughter cells are produced by a cell plate newly formed in land plant mother cells [16]. However, there are some common molecules associated with cytokinesis in metazoans and land plants [14, 15] such as dynamin-related protein

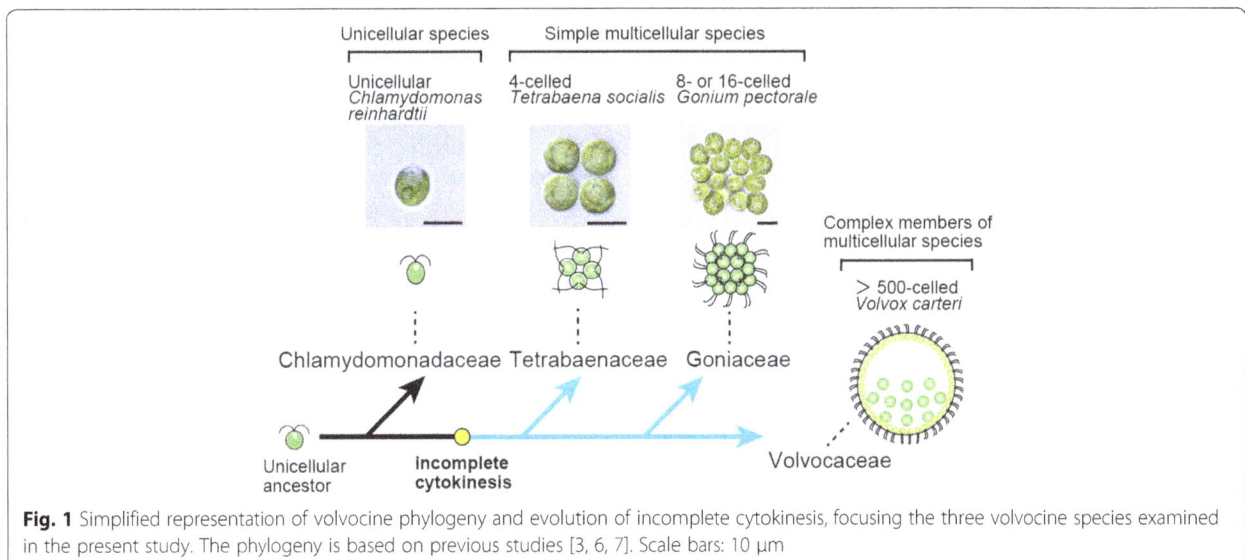

Fig. 1 Simplified representation of volvocine phylogeny and evolution of incomplete cytokinesis, focusing the three volvocine species examined in the present study. The phylogeny is based on previous studies [3, 6, 7]. Scale bars: 10 μm

(DRP). Dynamin was originally described as a microtubule-binding protein that was isolated from bovine brain extracts [17]. Dynamin homologs are often categorized as classical or conventional dynamins, which have five distinct domains: GTPase domain, middle domain, pleckstrin-homology domain, GTPase effecter domain (GED), and proline-rich domain [18, 19]. The dynamin superfamily contains additional members that lack a pleckstrin-homology domain and/or proline-rich domain, or have additional domains that are not present in classical dynamins. These members are defined as dynamin-related proteins (DRPs) [19]. In metazoans and land plants, several dynamins and DRPs play important roles in cytokinesis [18, 19]. For example, in *Arabidopsis thaliana*, DRP1A and DRP1E are localized in the cell plate [20], DRP2B is co-localized with DRP1A in the cell plate during cytokinesis and functions in vesicle formation [21], and DRP5A is localized in the cell plate at the end of cell division [22]. The double mutant line *drp1a/drp1e* of *A. thaliana* is unable to accomplish the embryogenesis because of defects in cell wall formation [20].

In this study, we focused on the DRP1 homologs in the volvocine lineage to examine the contribution of cytokinesis-related genes to the initial stages of multicellularity. We determined the complete coding region of the *DRP1* homolog of the simplest multicellular species *T. socialis* (*TsDRP1*), and performed immunofluorescence microscopy with a newly raised anti-TsDRP1 antibody in unicellular *C. reinhardtii* and two colonial multicellular species *T. socialis* and *G. pectorale* (Fig. 1).

Methods
Strains and culture conditions
Three algal strains were used in this study: *C. reinhardtii* strain cw92 (CC-503, cell-wall deficient, distributed by the *Chlamydomonas* Resource Center) [23], *T. socialis* strain NIES-571 [5], and *G. pectorale* strain 2014–0520-F1–4 (a sibling strain of *plus* and *minus* strains used previously [24]). *C. reinhardtii* strain cw92 was cultured synchronously in 300 mL tris-acetate-phosphate medium [25] in a silicon-capped 500 mL flask with aeration at 25 °C, on a light: dark cycle 12 h: 12 h under cool-white fluorescent lamps at an intensity of 110–150 μmol·m^{-2}·s^{-1}. *T. socialis* strain NIES-571 was cultured synchronously in 300 mL standard *Volvox* medium [26] in a silicon-capped 500 mL flask with aeration at 20 °C, on a light: dark cycle 12 h: 12 h under cool-white fluorescent lamps at an intensity of 110–150 μmol·m^{-2}·s^{-1} [5]. *G. pectorale* strain 2014–0520-F-4 was cultured synchronously in 300 mL standard *Volvox* medium [26] in a silicon-capped 500 mL flask with aeration at 20 °C, on a light: dark cycle 12 h: 12 h under cool-white fluorescent lamps at an intensity of 130–180 μmol·m^{-2}·s^{-1}. To evaluate the synchrony, percentages of dividing cells were monitored 15–16 times during 24 h.

Identification of *TsDRP1* gene
To determine the complete coding region of *TsDRP1*, partial coding sequences were obtained from our ongoing *T. socialis* strain NIES-571 genome assembly by a TBLASTN search using *C. reinhardtii* DRP1 (CrDRP1) as a query, and *TsDRP1* specific primers (Additional file 1: Table S1) were designed based on the sequences. Polyadenylated mRNA of *T. socialis* was isolated with Dynabeads oligo (dT)$_{25}$ (Thermo Fisher Scientific, Waltham, MA, USA) and reverse transcribed with Superscript III reverse transcriptase (Thermo Fisher Scientific) [24]. PCR was carried out using the synthesized cDNA as follows: 94 °C for 1 min, 35 cycles of 94 °C for 30 s, 60 °C for 30 s, and 72 °C for 2 min, followed by 72 °C for 5 min with *TaKaRa LA-Taq* with GC Buffer (Takara Bio Inc., Otsu, Japan) by using a thermal cycler GeneAmp PCR System 9700 (Thermo Fisher Scientific). Amplified DNA was purified with illustra GFX PCR DNA and Gel Band Purification Kits (GE Healthcare, Buckinghamshire, UK) and sequenced directly by an ABI PRISM 3100 Genetic Analyzer (Thermo Fisher Scientific) using a Big-Dye Terminator cycle sequencing ready reaction kit, v.3.1 (Thermo Fisher Scientific). The 5′ and 3′ ends of *TsDRP1* were determined by RACE using the GeneRacer™ kit (Thermo Fisher Scientific) according to the manufacturer's protocol. Each synthesized RACE product was amplified with KOD FX Neo DNA polymerase (TOYOBO, Osaka, Japan). PCR was carried out as follows: 94 °C for 2 min, 5 cycles of 98 °C for 10 s and 74 °C for 1.5 min, 5 cycles of 98 °C for 10 s and 72 °C for 1.5 min, 5 cycles of 98 °C for 10 s and 70 °C for 1.5 min, 15 cycles of 98 °C for 10 s and 68 °C for 1.5 min, and followed by 68 °C for 7 min. Amplified DNA was purified and sequenced as described above. Domains of DRP1proteins were searched by using Pfam program (http://pfam.xfam.org) [27].

Phylogenetic analyses
The amino acid sequences of DRP1 and DRP2 from six streptophytes (land plants and *Klebsormidium flaccidum*) and six chlorophytes (Additional file 1: Table S2) were collected from the genome databases in Phytozome (https://phytozome.jgi.doe.gov/pz/portal.html) and National Center for Biotechnology Information (https://www.ncbi.nlm.nih.gov/) by BLASTP and TBLASTN searches [28] using CrDRP1 and *A. thaliana* DRP2A as queries and aligned with the newly determined TsDRP1 by MAFFTv7 [29] server (http://mafft.cbrc.jp/alignment/server/). Phylogenetic analyses were performed using the maximum-likelihood and neighbor joining methods with PhyML [30] and MEGA 5.2.2 programs [31], respectively. The LG + I + G + F model was selected using ProtTest3 [32] optimized using Akaike information criteria. The bootstrap analyses were performed with 1000 replicates. Because DRP1 clade

is sister to DRP2 clade [22], DRP2 sequences were treated as the outgroup in the present study.

Preparation of antibodies

The antibody against *T. socialis* DRP1 (anti-TsDRP1 antibody) was raised in rabbits using the recombinant polypeptide. The cDNA sequence encoding the protein (corresponding to 11–597 positions of TsDRP1 amino acids; Additional file 2: Figure S1) was amplified by PCR using the primers listed in Additional file 1: Table S1. The PCR product was cloned into a pET100 expression vector (Thermo Fisher Scientific) and 6xHis fusion polypeptide was expressed in Rosetta (DE3) *Escherichia coli* cells, purified using a HisTrap HP column (GE healthcare). The purified polypeptide was subjected to SDS-PAGE, and gel slices containing the recombinant polypeptide were homogenized and injected into rabbits for antibody production (Kiwa Laboratory Animals. Co., Ltd., Wakayama, Japan). Antibodies were affinity-purified from the antisera by using the recombinant polypeptide coupled to a HiTrap NHS-activated HP column (GE Healthcare). Evaluation of the specificity of the anti-TsDRP1 antibody in *T. socialis*, *G. pectorale* and *C. reinhardtii* was performed by western blot analyses as described below (Additional file 2: Figure S2, Information S1).

Western blot analyses

Expression of DRP1 proteins in the three volvocine species was analyzed by SDS-PAGE and western blot modified from a previous study [33]. Cells were harvested, suspended in SDS-sample buffer (100 mM dithiothreitol, 2% SDS, 10% glycerol, 0.005% Bromophenol blue in 62.5 mM Tris-HCl) and boiled for three minutes. The prepared samples were separated on an Any kD Mini-PROTEAN TGX precast gel (Bio-Rad, Hercules, CA, USA) and transferred onto a Hybond-P membrane (GE Healthcare, Uppsala, Sweden). The blotted membrane was blocked with 3% skim milk in TPBS [0.1% Tween 20 (Sigma Aldrich) in phosphate-buffered saline (PBS)] at 4 °C overnight. The blot was incubated with an anti-TsDRP1 antibody diluted 1: 2000 with 3% skim milk in TPBS for 1 h at room temperature and washed in 3% skim milk in TPBS. The membrane was incubated with a goat anti-rabbit IgG antibody conjugated to horseradish peroxidase (Jackson ImmunoResearch, WestGrove, PA, USA) diluted 1:2000 with 3% skim milk in TPBS for 1 h at room temperature and washed with TPBS. The protein signals were detected with Amersham ECL prime Western blotting detection reagent (GE Healthcare). Images were obtained by ChemiDoc XRS system (Bio-Rad) with Quantity One software (Bio-Rad).

To examine the expression of DRP1 at the protein level, five time-course samples were obtained from each synchronous culture of *C. reinhardtii*, *T. socialis*, and *G. pectorale*: a sample with the greatest number of dividing cells during 24 h (0 point), three hours before the 0 point (–3 point), six hours before the 0 point (–6 point), three hours after the 0 point (+3 point), and six hours after the 0 point (+6 point). The time-course samples were analyzed by SDS-PAGE and western blot as described above. For Coomassie brilliant blue staining as loading control, each duplicated SDS-PAGE gel was stained by EzStain AQua (ATTO, Tokyo, Japan).

Indirect immunofluorescence microscopy

Fixation of *C. reinhardtii* was performed using a modified method in a previous study [34]. *C. reinhardtii* cells were attached to polyethyleneimine coated coverslips and fixed in –20 °C methanol for 5 min, transferred to fresh –20 °C methanol for 5 min and air-dried. The dried cells were incubated in PBS for 10 min. Subsequent blocking and antibody reactions were performed based on a previous study [5]. Immunostaining of *T. socialis* and *G. pectorale* were performed as described previously [5]. The anti-TsDRP1 antibody and a monoclonal anti-tubulin alpha antibody (clone YL1/2, Bio-Rad, Hercules, CA, USA), used as primary antibodies, were diluted 1: 500 with blocking buffer (0.11% Gelatin [Sigma Aldrich], 0.05% NaN_3, 0.25% bovine serum albumin [Sigma Aldrich] in TPBS). Alexa Fluor 488 goat anti-rabbit IgG (H + L) (# A11008, Invitrogen, Carlsbad, CA, USA) and Alexa Fluor 568 goat anti-rat IgG (H + L) (# A11077, Invitrogen) were also diluted 1: 500 with the blocking buffer. Confocal and differential interference contrast (DIC) images were obtained with an FV-1200 (Olympus, Tokyo, Japan) and three serial images were merged by using Adobe Photoshop CS6 software (Adobe Systems Inc., San Jose, CA, US).

Results
Identification and characterization of *TsDRP1*

The full-length coding region of *TsDRP1* (1890 bp) was determined and the deduced amino acid sequence (629 amino acids) aligned with DRP1 homologs of *A. thaliana* (AtDRP1A), *C. reinhardtii* (CrDRP1), *G. pectorale* (GpDRP1), and *V. carteri* (VcDRP1) (Additional file 2: Figure S1). The DRP1 sequences were highly conserved within the volvocine algae: the identity of TsDRP1 with CrDRP1, GpDRP1, and VcDRP1 was 92%, 92%, and 91%, respectively. The GC contents of the volvocine *DRP1* coding region were higher (*CrDRP1*: 64.18%, *TsDRP1*: 65.12%, *GpDRP1*: 62.22%, and *VcDRP1*: 57.55%) than that of *AtDRP1A* (46.78%), which is consistent with their GC-rich genome compositions [8, 23, 35]. Three domains that characterize DRP1, GTPase domain, middle domain, and GED [19], were found in all DRP1 sequences of volvocine algae as well as AtDRP1A. The motifs for interactions to

GTP (G1–4) [18] were conserved in AtDRP1A and volvocine DRP1 sequences (Additional file 2: Figure S1).

To confirm that DRP1 sequences of volvocine algae are orthologous to the DRP1 proteins of land plants, phylogenetic analyses were performed. All five land plants possessed several paralogs of *DRP1* genes, whereas each of the green algae (six chlorophytes and streptophyte *K. flaccidum*) possessed a single *DRP1* gene in the nuclear genome. The DRP1 clade was subdivided into two monophyletic groups corresponding to streptophytes and chlorophytes (Fig. 2). However, while the streptophyte clade was robustly resolved (with 92–100% bootstrap values), the monophyly of the chlorophytes was much less supported (with 60% bootstrap values in only maximum-likelihood method). The chlorophytes

were composed of two robust clades, Chlorophyceae/Trebouxiophyceae and Mamiellophyceae with 99–100% bootstrap values. Within the former clade, four chlorophycean or volvocine DRP1 sequences formed a robust monophyletic group with 100% bootstrap values. These results were consistent with the phylogeny using multiple chloroplast genes [36–38], and indicate that *DRP1* genes of volvocine algae are orthologs of *DRP1* of streptophytes.

Expression patterns of CrDRP1, TsDRP1 and GpDRP1 during the asexual life cycle

To examine the relationship between volvocine DRP1 and cytokinesis at a protein expression level, we performed western blot analyses using the anti-TsDRP1 antibody.

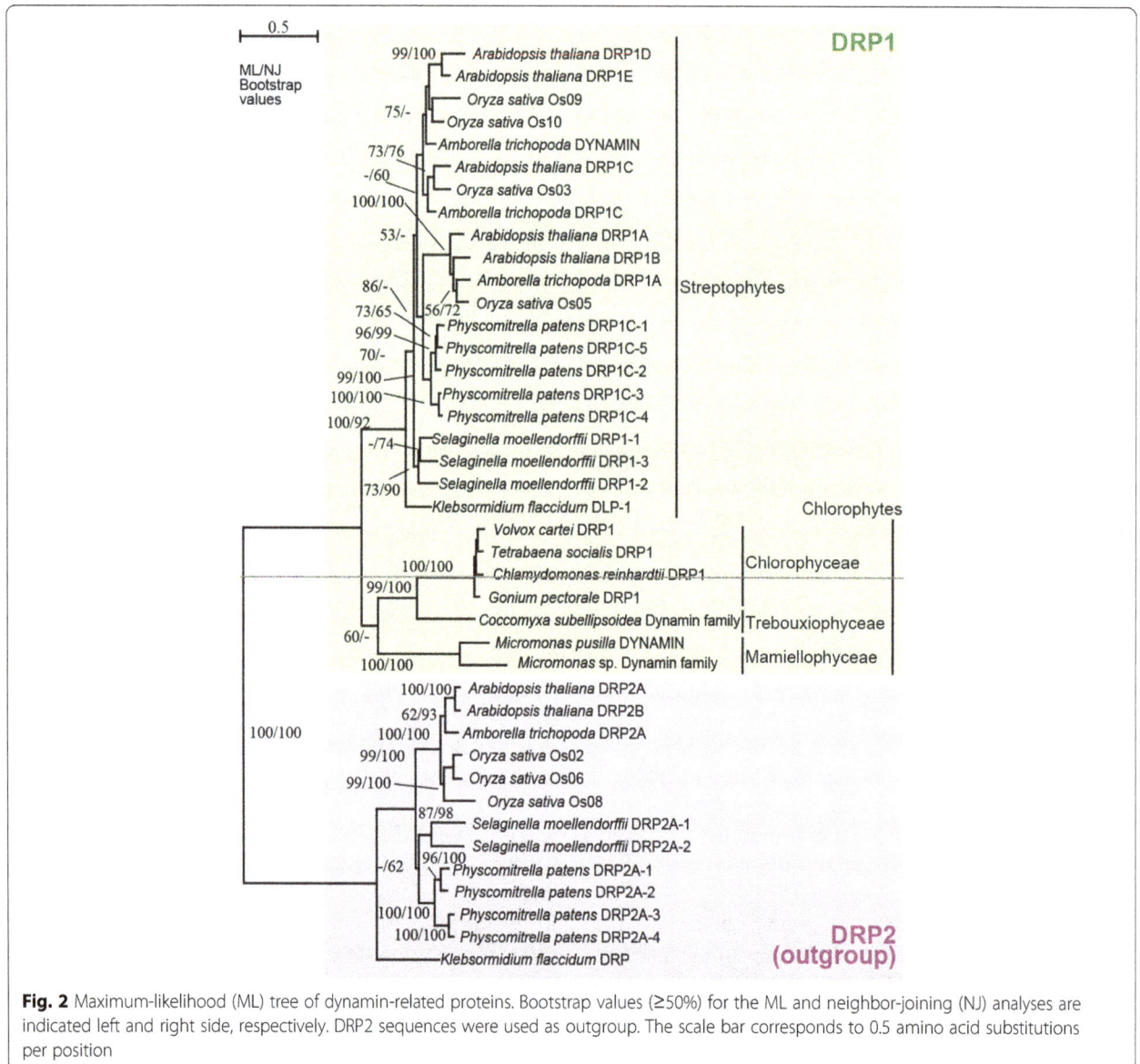

Fig. 2 Maximum-likelihood (ML) tree of dynamin-related proteins. Bootstrap values (≥50%) for the ML and neighbor-joining (NJ) analyses are indicated left and right side, respectively. DRP2 sequences were used as outgroup. The scale bar corresponds to 0.5 amino acid substitutions per position

The CrDRP1, TsDRP1, and GpDRP1 signals were detected as major bands at ~75 kDa with the anti-TsDRP1 antibody (Additional file 2: Figure S2). DRP1 signals were constitutively detected from all time-course samples of the three species (Additional file 2: Figure S3).

Subcellular localization of DRP1 during the asexual life cycle

To verify the subcellular localization of DRP1, immunofluorescence microscopy was carried out using the anti-TsDRP1 and anti-tubulin alpha antibodies. Immunofluorescences of tubulin were used as the division plane marker, because microtubule structures (phycoplast) are observed in division planes of volvocine algae [39–41]. In vegetative cells of *C. reinhardtii*, *T. socialis* and *G. pectorale*, DRP1 signals were observed as many speckles in their cytoplasm (Figs. 3a-d, 4a-d, 5a-d). In two-celled stage, microtubules were observed in the cleavage furrows and DRP1 signals were localized in the vicinity of the division plane of *C. reinhardtii* (white arrowheads in Fig. 3e-h), *T. socialis* (white arrowheads in Fig. 4e-h) and *G. pectorale* (white arrowheads in Fig. 5e-h). While tubulin and DRP1 signals were not perfectly co-localized in the two-celled stage, these appeared to be partially overlapped (Figs. 3h, 4h, 5h). The DRP1 fluorescence in cytoplasm at this stage was less than that of vegetative cell (Figs. 3f, 4f, 5f).

DRP1 localizations of the four-celled stage were different between the unicellular and multicellular species examined here. In the four-celled stage of *C. reinhardtii*, DRP1 was mainly localized in second division planes (Fig. 3i-l). In contrast, DRP1 in four-celled embryos of *T. socialis* and *G. pectorale* were observed clearly in both first and second division planes (Figs. 4i-l, 5i-l). Eight-celled embryo of *G. pectorale* also exhibited clear DRP1 localization in the first, second and third division planes (Fig. 5m-p).

Discussion

In *C. reinhardtii*, *T. socialis* and *G. pectorale*, DRP1 signals were mainly observed in their cytoplasm during the vegetative phase (Figs. 3b, 4b, 5b). However, the signals were mainly accumulated in the division plane during multiple fission of the three volvocine species (Figs. 3f, j, 4f, j, 5f, j, n). DRP1 expression was always detected and did not drastically change during the time courses examined in the three species (Additional file 2: Figure S3). Thus, volvocine DRP1 might change localization during the life cycle. The volvocine DRP1 localizations in the division plane were similar to those of some DRP1 proteins of land plants: DRP1A (ADL1) in *A. thaliana* [20] and tobacco BY-2 cells [42], and DRP1C in tobacco BY-2 cells [42]. Therefore, DRP1 proteins of volvocine algae may act as cytokinesis-related molecules.

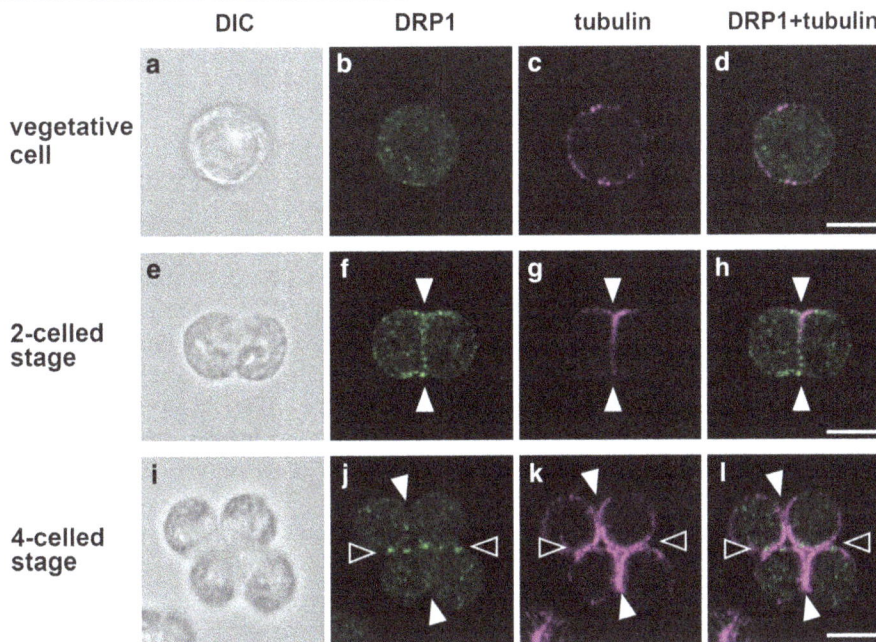

Fig. 3 Immunofluorescence images of DRP1 localization in *Chlamydomonas reinhardtii*. Differential interference contrast (DIC) images (**a**, **e**, **i**), immunofluorescence images labeled with an anti-TsDRP1 (**b**, **f**, **j**) and an anti-tubulin alpha (**c**, **g**, **k**), and merged immunofluorescence images of DRP1 and tubulin (**d**, **h**, **l**) are shown. In vegetative cells, DRP1 was localized to cytoplasm (**b-d**). DRP1 was localized to a first division plane of two-celled embryo (white arrowheads in **f-h**) and second division planes of four-celled embryo (black arrowheads in **j-l**) mainly. Note that DRP1 signals localized to the first division plane of the two-celled embryo (**f**) become weak in the four-celled stage (**j**). The first and second division planes in the four-celled stage were estimated by relative positions of pyrenoids. Scale bars: 5 μm

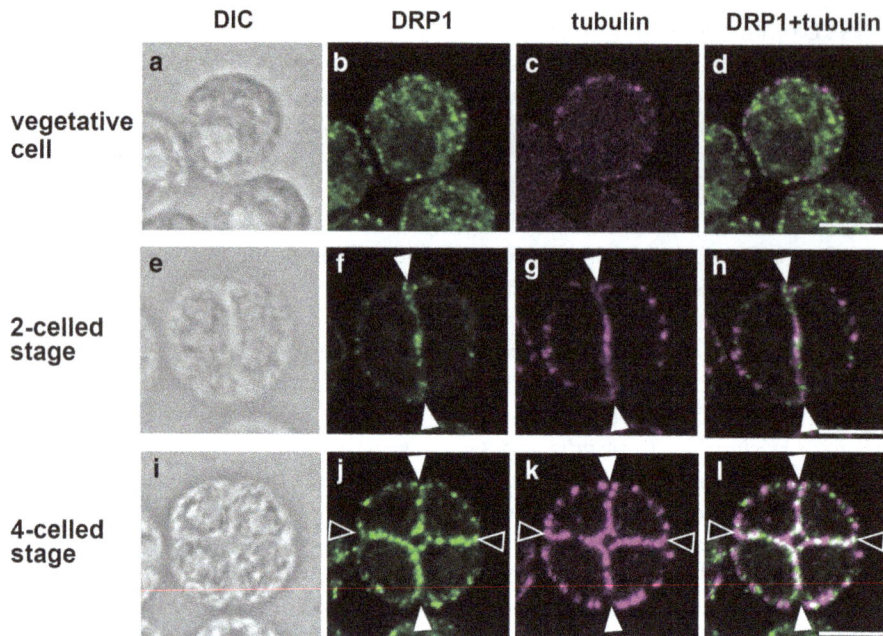

Fig. 4 Immunofluorescence images of DRP1 localization in *Tetrabaena socialis*. Differential interference contrast (DIC) images (**a**, **e**, **i**), immunofluorescence images labeled with an anti-TsDRP1 (**b**, **f**, **j**) and an anti-tubulin alpha (**c**, **g**, **k**), and merged immunofluorescence images of DRP1 and tubulin (**d**, **h**, **l**) are shown. In vegetative cells, DRP1 was localized to cytoplasm (**b-d**). DRP1 was localized to a first division plane (white arrowheads in **f-h**) of two-celled embryo and both first (white arrowheads in **j-l**) and second (blank arrowheads in **j-l**) division planes of four-celled embryo. The first and second division planes in the four-celled stage were estimated by directions of the planes within the parental colony. Scale bars: 10 μm

In the second division of multiple fission (4-celled stage), DRP1 of *C. reinhardtii* was mainly localized in the second division plane (Fig. 3j), while DRP1 signals of *T. socialis* (Fig. 4j) and *G. pectorale* (Fig. 5j) were clearly observed in the both division planes, which were formed by the first and second divisions. Moreover, in the third division of *G. pectorale* (8-celled stage), DRP1 signals were observed in all division planes (Fig. 5n). Thus, DRP1 may have a role mainly in last division plane of unicellular *C. reinhardtii* (Fig. 3f, j), whereas DRP1 may equally work in all division planes of multicellular *T. socialis* and *G. pectorale* (Figs. 4f, j,5f, j, n). Each cell division during multiple fission of multicellular *T. socialis* [5], *G. pectorale* [12], and other multicellular volvocine algae [13, 43–45] proceeds through incomplete cytokinesis and the resulting daughter protoplasts are connected to one another by cytoplasmic bridges, whereas daughter protoplasts of *C. reinhardtii* are completely separated from one another by means of complete cytokinesis [9, 10]. The division planes of multicellular volvocine species are therefore different from those of unicellular *C. reinhardtii* at subcellular and molecular levels.

Dynamins and DRPs are involved in various membrane remodeling events such as budding and trafficking of vesicles, fission or fusion of organelles, and cytokinesis [22, 46]. Particularly in cytokinesis, dynamins are associated with newly formed membranes of metazoans such as *Caenorhabditis elegans* [47] and zebrafish [48]. In *Dictyostelium discoideum*, the relationships between cytokinesis and dynamin A [49], dynamin-like protein (Dlp) A, DlpB and DlpC [22] were reported. In *A. thaliana*, DRP1A [20], DRP1E [20], DRP2B [21], and DRP5A [22] are localized in the division plane, and function in cytokinesis. Those dynamins and DRPs play important roles for cytokinesis such as membrane fission and vesicle formation for cytokinesis. Therefore, volvocine DRP1 proteins might be related to membrane remodeling during the multiple fission. Hence, the few DRP1 signals in the first division plane of *C. reinhardtii* 4-celled stage (Fig. 3j) may indicate that membrane fission/formation mediated by DRP1 has been finished in this plane after the first division (white arrowheads in Fig. 3j), but has been occurring at the second division plane (black arrowheads in Fig. 3j). The presence of DRP1 signals in all division planes of *T. socialis* (Fig. 4j) and *G. pectorale* (Fig. 5j, n) may be due to the continuous membrane remodeling during the incomplete cytokinesis in these two multicellular species.

Conclusion

This study demonstrated that localization patterns of DRP1 are different between unicellular *C. reinhardtii* and two multicellular species *T. socialis* and *G. pectorale*.

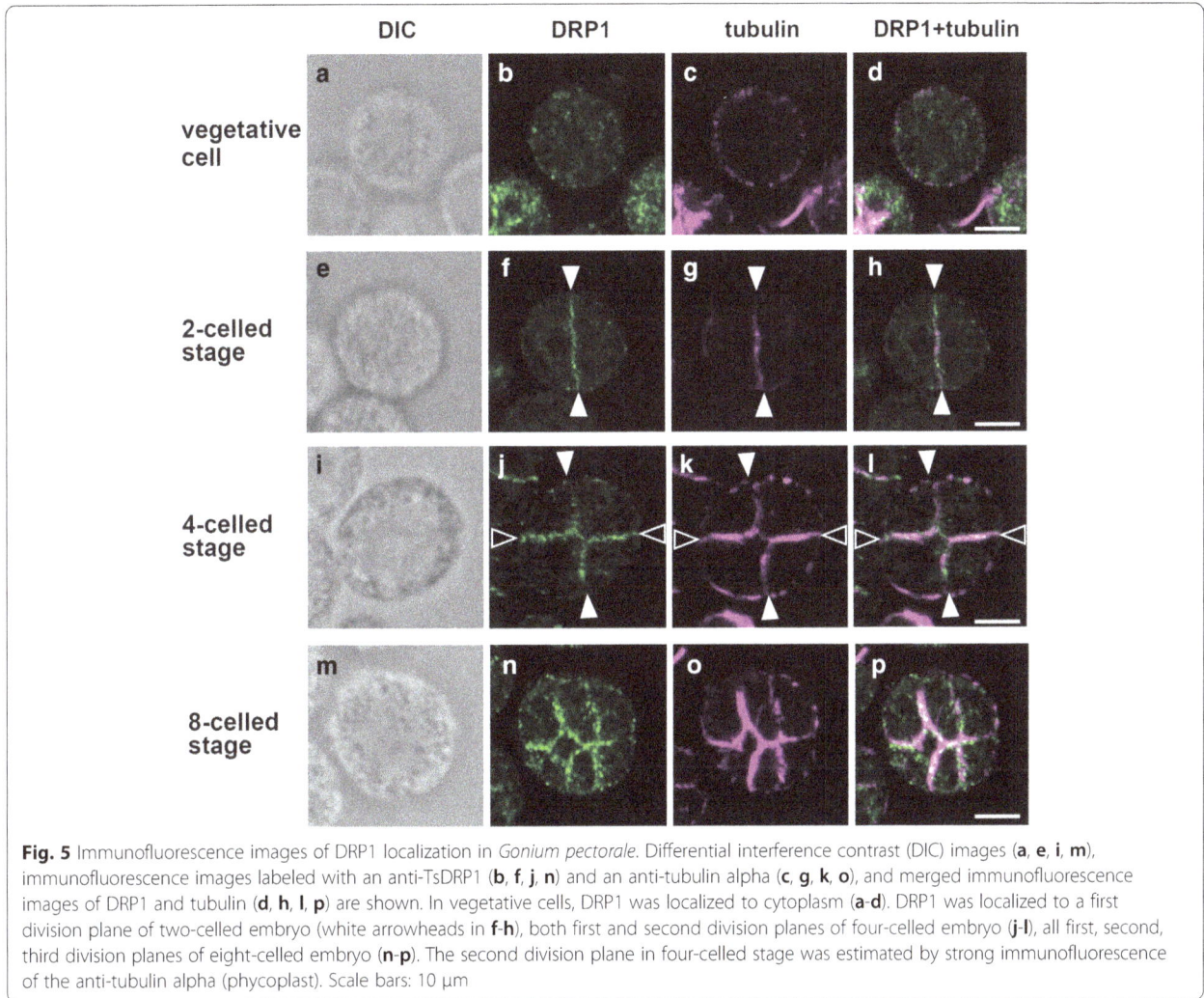

Fig. 5 Immunofluorescence images of DRP1 localization in *Gonium pectorale*. Differential interference contrast (DIC) images (**a**, **e**, **i**, **m**), immunofluorescence images labeled with an anti-TsDRP1 (**b**, **f**, **j**, **n**) and an anti-tubulin alpha (**c**, **g**, **k**, **o**), and merged immunofluorescence images of DRP1 and tubulin (**d**, **h**, **l**, **p**) are shown. In vegetative cells, DRP1 was localized to cytoplasm (**a-d**). DRP1 was localized to a first division plane of two-celled embryo (white arrowheads in **f-h**), both first and second division planes of four-celled embryo (**j-l**), all first, second, third division planes of eight-celled embryo (**n-p**). The second division plane in four-celled stage was estimated by strong immunofluorescence of the anti-tubulin alpha (phycoplast). Scale bars: 10 μm

Given that DRP1 may function in volvocine cytokinesis, the different DRP1 localization patterns between *C. reinhardtii* and two multicellular species *T. socialis* and *G. pectorale* (Fig. 6) indicate differences in the molecular mechanisms of cytokinesis during multiple fission between unicellular forms and multicellular forms. *T. socialis* and *G. pectorale* are considered to represent ancestral (plesiomorphic) multicellular morphology based on the cladistic analysis of morphological data and molecular phylogenetic analyses (Fig. 1) [3–7]. These data indicate therefore, that the localization patterns of DRP1 during multiple fission might have been modified from unicellular *Chlamydomonas*-type to multicellular *Tetrabaena/Gonium*-type (Fig. 6) in the

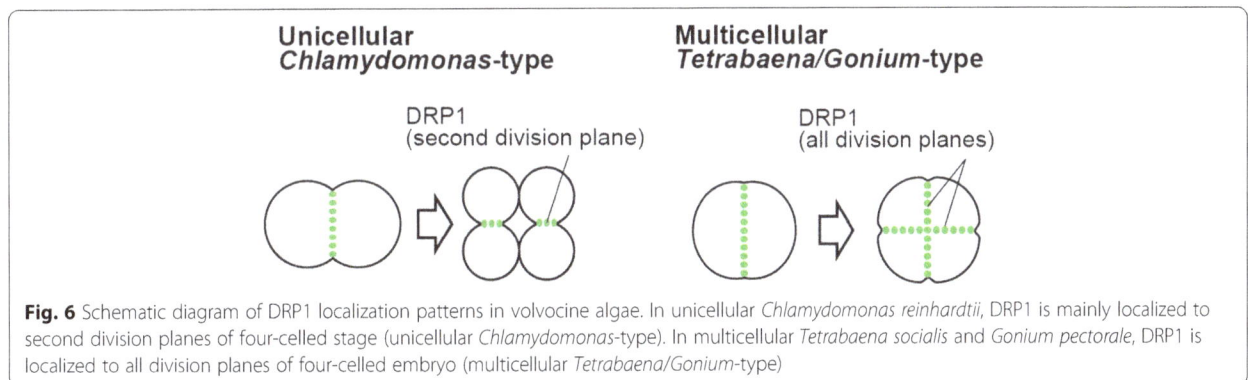

Fig. 6 Schematic diagram of DRP1 localization patterns in volvocine algae. In unicellular *Chlamydomonas reinhardtii*, DRP1 is mainly localized to second division planes of four-celled stage (unicellular *Chlamydomonas*-type). In multicellular *Tetrabaena socialis* and *Gonium pectorale*, DRP1 is localized to all division planes of four-celled embryo (multicellular *Tetrabaena/Gonium*-type)

common ancestor of multicellular volvocine algae (Fig. 1). This modification was essential for the initial stages of colonial living in this lineage.

Additional files

Additional file 1 **Table S1.** List of primers used for amplification and sequencing of *TsDRP1*. **Table S2.** List of DRP1 and DRP2 proteins used in this study. (PDF 84 kb)

Additional file 2 **Figure S1.** Alignment of DRP1A from *Arabidopsis thaliana* (At) and DRP1 from *Chlamydomonas reinhardtii* (Cr), *Tetrabaena socialis* (Ts), *Gonium pectorale* (Gp) and *Volvox carteri* (Vc). Black and gray background indicates identical or similar amino acid, respectively. GTPase domain, dynamin middle domain, and GTPase effector domain are indicated by pink, green, and yellow background color, respectively. The region corresponding to the antigen for an anti-TsDRP1 antibody is showed under the alignment (gray bar). **Figure S2.** Specificity of the affinity-purified anti-TsDRP1 antibody. The specificity of the anti-TsDRP1 antibody was validated in three volvocine algae by western blotting. A single band was detected in each lane (~75 kDa) with the antibody that was incubated with acetone powder of *E. coli* with the empty vector (left) while no signal was detected with the antibody that was incubated with acetone powder of *E. coli* expressing TsDRP1 (middle). For details of the methods, see Information S1 (Additional file 2). **Figure S3.** Western blot analyses of DRP1 proteins of *Chlamydomonas reinhardtii* (CrDRP1), *Tetrabaena socialis* (TsDRP1), and *Gonium pectorale* (GpDRP1) using anti-TsDRP1 antibody. Time-course of synchronous culture and western blot (WB) of *C. reinhardtii*, *T. socialis*, and *G. pectorale* are shown in **a**, **b**, and **c**, respectively. Time-course samples were obtained from five points (arrows in each line graph): the greatest number of dividing cells (0), three (−3) and six (−6) hours before 0 point, and three (+3) and six (+6) hours after 0 point. Coomassie brilliant blue (CBB) staining of a duplicate gel shows the equal protein loading in each lane. **Information S1.** Methods for specificity of the affinity-purified anti-TsDRP1 antibody (Additional file 2: Figure S2). (PDF 1785 kb)

Abbreviations
DIC: Differential interference contrast; Dlp: Dynamin-like protein; DRP: Dynamin-related protein; GED: GTPase effector domain; GTP: Guanosine triphosphate; GTPase: Guanosine triphosphatease; PBS: Phosphate-buffered saline; TPBS: 0.1% Tween 20 in PBS

Acknowledgements
We would like to thank Dr. Kenji Kimura (National Institute of Genetics) for fluorescence observation using the confocal laser microscopy.

Funding
This work was supported by NIG-JOINT (2016-B to HN), Grants-in-Aid for JSPS Fellows (No. 25–9234 to YA) and Scientific Research (A) (grant number 16H02518 to HN) from MEXT/JSPS KAKENHI, and the National Research Foundation, South Africa (grant number RA151217156515 to PD).

Authors' contributions
YA, HKT, KK, SM and HN conceived and designed the experiments. YA, JF, PD, and HN constructed the draft genome assembly of *Tetrabaena*. YA, TF and SM performed the experiments and analyzed the data. YA and HN wrote the manuscript. All authors read and approved the manuscript.

Competing interests
The authors declare that they have no competing interests.

Author details
[1]Department of Biological Sciences, Graduate School of Science, University of Tokyo, 7-3-1 Hongo, Bunkyo-ku, Tokyo 113-0033, Japan. [2]Department of Cell Genetics, National Institute of Genetics, 1111 Yata, Mishima, Shizuoka 411-8540, Japan. [3]Department of Life Science and Technology, School of Life Science and Technology, Tokyo Institute of Technology, 2-12-1 Ookayama, Meguro-ku, Tokyo 152-8550, Japan. [4]Evolutionary Studies Institute, University of the Witwatersrand, Johannesburg 2000, South Africa. [5]Agricultural Research Council, Biotechnology Platform, Pretoria 0040, South Africa. [6]Department of Ecology and Evolutionary Biology, University of Arizona, Tucson, AZ 85721, USA.

References
1. Grosberg RK, Strathmann RR. The evolution of multicellularity: a minor major transition? Annu Rev Ecol Evol Syst. 2007;38:621–54.
2. Sharpe SC, Eme L, Brown MW, Roger AJ. Timing the origins of multicellular eukaryotes through Phylogenomics and relaxed molecular clock analyses. In: Trillo IR, Nedelcu AM, editors. Evolutionary transitions to multicellular life. Dordrecht: Springer Netherlands; 2015. p. 3–29.
3. Herron MD, Hackett JD, Aylward FO, Michod RE. Triassic origin and early radiation of multicellular volvocine algae. Proc Natl Acad Sci U S A. 2009; 106:3254–8.
4. Kirk DLA. Twelve-step program for evolving multicellularity and a division of labor. BioEssays. 2005;27:299–310.
5. Arakaki Y, Kawai-Toyooka H, Hamamura Y, Higashiyama T, Noga A, Hirono M, et al. The simplest integrated multicellular organism unveiled. PLoS One. 2013;8:e81641.
6. Nozaki H, Ito M. Phylogenetic relationships within the colonial Volvocales (Chlorophyta) inferred from cladistic analysis based on morphological data. J Phycol. 1994;30:353–65.
7. Nozaki H, Misawa K, Kajita T, Kato M, Nohara S, Watanabe MM. Origin and evolution of the colonial Volvocales (Chlorophyceae) as inferred from multiple, chloroplast gene sequences. Mol Phylogenet Evol. 2000;17:256–68.
8. Hanschen ER, Marriage TN, Ferris PJ, Hamaji T, Toyoda A, Fujiyama A, et al. The *Gonium pectorale* genome demonstrates co-option of cell cycle regulation during the evolution of multicellularity. Nat Commun. 2016;7: 11370.
9. Harris EH. The *Chlamydomonas* sourcebook: introduction to *Chlamydomonas* and its laboratory use. 2nd ed. Oxford: Academic Press; 2009.
10. Kirk DL. *Volvox*: molecular-genetic origins of multicellularity and cellular differentiation. Cambridge: Cambridge University Press; 1998.
11. Umen JG, Goodenough UW. Control of cell division by a retinoblastoma protein homolog in *Chlamydomonas*. Genes Dev. 2001;15:1652–61.
12. Iida H, Ota S, Inouye I. Cleavage, incomplete inversion, and cytoplasmic bridges in *Gonium pectorale* (Volvocales, Chlorophyta). J Plant Res. 2013; 126:699–707.
13. Green KJ, Viamontes GI, Kirk DL. Mechanism of formation, ultrastructure, and function of the cytoplasmic bridge system during morphogenesis in *Volvox*. J Cell Biol. 1981;91:756–69.
14. Eggert US, Mitchison TJ, Field CM. Animal cytokinesis: from parts list to mechanisms. Annu Rev Biochem. 2006;75:543–66.
15. Jürgens G. Plant cytokinesis: fission by fusion. Trends Cell Biol. 2005;15:277–83.
16. Field C, Li R, Oegeme K. Cytokinesis in eukaryotes: a mechanistic comparison. Curr Opin Cell Biol. 1999;11:68–80.
17. Shpetner HS, Vallee RB. Identification of dynamin, a novel mechanochemical enzyme that mediates interactions between microtubules. Cell. 1989;59: 421–32.
18. Praefcke GJK, McMahon HT. The dynamin superfamily: universal membrane tubulation and fission molecules? Nat Rev Mol Cell Biol. 2004;5:133–47.
19. Konopka CA, Schleede JB, Skop AR, Bednarek SY. Dynamin and cytokinesis. Traffic. 2006;7:239–47.
20. Kang B-H, Busse JS, Bednarek SY. Members of the *Arabidopsis* dynamin-like gene family, ADL1, are essential for plant cytokinesis and polarized cell growth. Plant Cell. 2003;15:899–913.
21. Fujimoto M, Arimura S, Ueda T, Takahashi H, Hayashi Y, Nakano A, et al. *Arabidopsis* dynamin-related proteins DRP2B and DRP1A participate together in clathrin-coated vesicle formation during endocytosis. Proc Natl Acad Sci U S A. 2010;107:6094–9.

22. Miyagishima S, Kuwayama H, Urushihara H, Nakanishi H. Evolutionary linkage between eukaryotic cytokinesis and chloroplast division by dynamin proteins. Proc Natl Acad Sci U S A. 2008;105:15202–7.

23. Merchant SS, Prochnik SE, Vallon O, Harris EH, Karpowicz SJ, Witman GB, et al. The *Chlamydomonas* genome reveals the evolution of key animal and plant functions. Science. 2007;318:245–51.

24. Kawai-Toyooka H, Mori T, Hamaji T, Suzuki M, Olson BJSC, Uemura T, et al. Sex-specific posttranslational regulation of the gamete fusogen GCS1 in the isogamous volvocine alga *Gonium pectorale*. Eukaryot Cell. 2014;13:648–56.

25. Gorman DS, Levine RP. Cytochrome *f* and plastocyanin: their sequence in the photosynthetic electron transport chain of *Chlamydomonas reinhardtii*. Proc Natl Acad Sci U S A. 1965;54:1665–9.

26. Kirk DL, Kirk MM. Protein synthetic patterns during the asexual life cycle of *Volvox carteri*. Dev Biol. 1983;96:493–506.

27. Finn RD, Coggill P, Eberhardt RY, Eddy SR, Mistry J, Mitchell AL, et al. The Pfam protein families database: towards a more sustainable future. Nucleic Acids Res. 2016;44:D279–85.

28. Altschul SF, Madden TL, Schäffer AA, Zhang J, Zhang Z, Miller W, Lipman DJ, Gapped BLAST. PSI-BLAST: a new generation of protein database search programs. Nucleic Acids Res. 1997;25:3389–402.

29. Katoh K, Standley DMMAFFT. Multiple sequence alignment software version 7: improvements in performance and usability. Mol Biol Evol. 2013;30:772–80.

30. Guindon S, Dufayard JF, Lefort V, Anisimova M, Hordijk W, Gascuel O. New algorithms and methods to estimate maximum-likelihood phylogenies: assessing the performance of PhyML 3.0. Syst Biol. 2010;59:307–21.

31. Tamura K, Peterson D, Peterson N, Stecher G, Nei M, Kumar S. MEGA5: Molecular evolutionary genetics analysis using maximum likelihood, evolutionary distance, and maximum parsimony methods. Mol Biol Evol. 2011;28:2731–9.

32. Darriba D, Taboada GL, Doallo R, Posada D. ProtTest 3: fast selection of best-fit models of protein evolution. Bioinformatics. 2011;27:1164–5.

33. Nakazawa Y, Hiraki M, Kamiya R, Hirono M. SAS-6 is a cartwheel protein that establishes the 9-fold symmetry of the centriole. Curr Biol. 2007;17:2169–74.

34. Lechtreck K-F, Luro S, Awata J, Witman GB. HA-tagging of putative flagellar proteins in *Chlamydomonas reinhardtii* identifies a novel protein of intraflagellar transport complex B. Cell Motil Cytoskeleton. 2009;66:469–82.

35. Prochnik SE, Umen J, Nedelcu AM, Hallmann A, Miller SM, Nishii I, et al. Genome analysis of organismal complexity in the multicellular green alga *Volvox carteri*. Science. 2010;329:223–6.

36. Smith SA, Beaulieu JM, Donoghue MJ. Mega-phylogeny approach for comparative biology: an alternative to supertree and supermatrix approaches. BMC Evol Biol. 2009;9:37.

37. Leliaert F, Smith DR, Moreau H, Herron MD, Verbruggen H, Delwiche CF, Clerck OD. Phylogeny and molecular evolution of green algae. CRC Crit Rev Plant Sci. 2012;31:1–46.

38. Ruhfel BR, Gitzendanner MA, Soltis PS, Soltis D, Burleigh JG. From algae to angiosperms-inferring the phylogeny of green plants (*Viridiplantae*) from 360 plastid genomes. BMC Evol Biol. 2014;14:23.

39. Doonan JH, Grief C. Microtubule cycle in *Chlamydomonas reinhardtii*: an immunofluorescence study. Cell Motil Cytoskeleton. 1987;7:381–92.

40. Kirk DL, Kaufman MR, Keeling RM, Stamer KA. Genetic and cytological control of the asymmetric divisions that pattern the *Volvox* embryo. Dev Suppl. 1991;1:67–82.

41. Dymek EE, Goduti D, Kramer T, Smith EFA. Kinesin-like calmodulin-binding protein in *Chlamydomonas*: evidence for a role in cell division and flagellar functions. J Cell Sci. 2006;119:3107–16.

42. Hong Z, Geisler-Lee CJ, Zhang Z, Verma DPS. Phragmoplastin dynamics: multiple forms, microtubule association and their roles in cell plate formation in plants. Plant Mol Biol. 2003;53:297–312.

43. Fulton AB. Colonial development in *Pandorina morum*. II. Colony morphogenesis and formation of the extracellular matrix. Dev Biol. 1978;64:236–51.

44. Marchant HJ. Colony formation and inversion in green alga *Eudorina elegans*. Protoplasma. 1977;93:325–39.

45. Iida H, Nishii I, Inouye I. Embryogenesis and cell positioning in *Platydorina caudata* (Volvocaceae, Chlorophyta). Phycologia. 2011;50:530–40.

46. Antonny B, Burd C, De Camilli P, Chen E, Daumke O, Faelber K, et al. Membrane fission by dynamin: what we know and what we need to know. EMBO J. 2016;35:2270–84.

47. Thompson HM, Skop AR, Euteneuer U, Meyer BJ, McNiven MA. The large GTPase dynamin a associates with the spindle midzoe and is required for cytokinesis. Curr Biol. 2002;12:2111–7.

48. Feng B, Schwarz H, Jesuthasan S. Furrow-specific endocytosis during cytokinesis of zebrafish blastomeres. Exp Cell Res. 2002;279:14–20.

49. Wienke DC, Knetsch MLW, Neuhaus EM, Reedy MC, Manstein DJ. Disruption of a dynamin homologue affects endocytosis, organelle morphology, and cytokinesis in *Dictyostelium discoideum*. Mol Biol Cell. 1999;10:225–43.

12

Musculoskeletal networks reveal topological disparity in mammalian neck evolution

Patrick Arnold[1,2]* (iD), Borja Esteve-Altava[3] and Martin S. Fischer[1]

Abstract

Background: The increase in locomotor and metabolic performance during mammalian evolution was accompanied by the limitation of the number of cervical vertebrae to only seven. In turn, nuchal muscles underwent a reorganization while forelimb muscles expanded into the neck region. As variation in the cervical spine is low, the variation in the arrangement of the neck muscles and their attachment sites (i.e., the variability of the neck's musculoskeletal organization) is thus proposed to be an important source of neck disparity across mammals. Anatomical network analysis provides a novel framework to study the organization of the anatomical arrangement, or connectivity pattern, of the bones and muscles that constitute the mammalian neck in an evolutionary context.

Results: Neck organization in mammals is characterized by a combination of conserved and highly variable network properties. We uncovered a conserved regionalization of the musculoskeletal organization of the neck into upper, mid and lower cervical modules. In contrast, there is a varying degree of complexity or specialization and of the integration of the pectoral elements. The musculoskeletal organization of the monotreme neck is distinctively different from that of therian mammals.

Conclusions: Our findings reveal that the limited number of vertebrae in the mammalian neck does not result in a low musculoskeletal disparity when examined in an evolutionary context. However, this disparity evolved late in mammalian history in parallel with the radiation of certain lineages (e.g., cetartiodactyls, xenarthrans). Disparity is further facilitated by the enhanced incorporation of forelimb muscles into the neck and their variability in attachment sites.

Keywords: Anatomical network analysis, Network theory, Forelimb evolution, Mammalian cervical spine, Sloths, Meristic constraints, Modularity

Background

The increase in locomotor and metabolic performance was one of the most important innovations in the evolution of mammals [1–6]. This innovation, however, was accompanied by an exceptionally low variability in the number of presacral vertebrae compared to other tetrapods (e.g., [7–14]). In fact, the number of cervical vertebrae in mammals is limited to seven, except in extant manatees and sloths [12]. As mammals evolved a new

locomotor mode based on an increase in sagittal axial motions, their back and nuchal muscles underwent an anatomical reorganization [15, 16]. The epaxonic muscles (particularly the iliocostalis system) were reduced along with the decrease of lateral axial motion [15, 16]. With the predominance of girdle-limb system as the main propeller in mammals, pectoral muscles also expanded into the dorsal region [15] and were integrated into the head/neck functional unit. Studies on the evolution of the mammalian neck usually have focused on the role of those muscles emigrating from the cervical region during early development [17–20]. In contrast, the muscles that expanded into the neck have been solely investigated for their impact on shoulder and forelimb mechanics (e.g., [21–24]).

* Correspondence: patrick_arnold@eva.mpg.de
[1]Department of Human Evolution, Max Planck Institute for Evolutionary Anthropology, Leipzig, Germany
[2]Institut für Spezielle Zoologie und Evolutionsbiologie mit Phyletischem Museum, Friedrich-Schiller-Universität Jena, Jena, Germany
Full list of author information is available at the end of the article

Differences in ecology and size resulted in interspecific differences in the posture and mobility of the head in mammals during standing, locomotion, foraging, oral grooming, and other daily activities (e.g., [25–30]). The morphological basis of these differences, however, is poorly understood. Variation of the cervical column length as a whole has recently been shown to be an important factor in generating morphological disparity of the neck in mammals [31]. As a consequence of the limited variability in the number of vertebrae [9, 12] and in vertebral shape [17, 32–35], the disparity of the cervical skeleton alone is still low. Hence, we suggest that interspecific variation in the arrangement of the neck muscles plus their attachment sites on the cervical vertebrae, the skull and other bones (i.e., the variability of the musculoskeletal organization) should be an important source of morphological disparity of the neck across mammals. Although there are numerous descriptions of the myology of the neck region for almost every mammalian family, only a few studies compared the neck muscle arrangement interspecifically in an evolutionary context (e.g., [36–39]). Moreover, these studies compared neck muscles only qualitatively, which prevents the quantification of the differentiation of the neck muscles arrangement. As a consequence, it is currently unknown whether the interspecific variation in muscle attachments actually affected the changes of the musculoskeletal organization of the neck across mammals.

Anatomical Network Analysis (AnNA) provides a novel framework to study the organization of the anatomical arrangement of bones and muscles of anatomical structures (i.e., the connectivity pattern) [40, 41]. Within this framework, bones and muscles are formalized as the nodes of the network, and the physical contacts among them are formalized as the links that connect the network's nodes. Anatomical network models thus offer a mathematical description of the organization of the body [41]. Through such mathematical formalism we can identify and quantify structural patterns, such as anatomical modules, without a priori assumptions about the developmental or functional factors causing them [40] (for a recent review on morphological modularity, see [42]). This allows direct phylogenetic comparisons. In this context, AnNA has been used, for example, to infer evolutionary trends in the skull of tetrapods [43] and the phylogenetic relation between morphological complexity and modularity in the skull of primates [44]. AnNA formalization also allows combining information on skeletal and muscular tissues as it was done to study congenital musculoskeletal malformations [45], secondary injuries [46] in humans, and hindlimb functional integration in frogs [47].

We can also use network parameters as proxies to infer the morphological organization of the body. Table 1 summarizes the network parameters used in this study and their most common morphological interpretation. Further

details on the interpretation of network concepts in a morphological context and their historical roots have been given elsewhere (see, e.g., [41, 43, 48–50]). In short, every interpretation derives from the biological role of connections in the network model. Broadly speaking, the connections we have modeled among the bones and muscles of the neck embody functional interactions (e.g., determining the range of neck motion), as well as developmental factors (e.g., those inducing muscles to attach to a specific vertebrae and not to other). Thus, for example, the number of such connections for a given element, or for the entire network (i.e., K), represents the amount of functional and developmental dependences of this element or of the whole network. Functional and developmental dependences are often associated to Rupert Riedl's concept of burden, or more generally, to the concept of constrain of body parts [40, 51–53]. Morphological interpretation of more elaborated network parameters, such as the density of connections (D), the mean clustering coefficient (C), and the mean shortest path length (L), follow a similar logic. Because connections represent biological interactions among anatomical parts, their relative amount (D) serves as a proxy of the complexity available to the system (e.g., to perform complex functions). In addition to quantifying the amount of connections, the way connections are set (e.g., creating intertwined patterns such as 3-node loops (C)) and their effects on topology (e.g., increasing the effective proximity of elements to interact together (L)) also have consequences in the overall integration of the anatomy. Thus, the greater the intertwining, the greater the integration; the closer the elements, the greater the integration. Moreover, differences in the amount of connections among the elements of the network (some have many, most have a few) introduces heterogeneity in the organization of the network. Such heterogeneity can be related before to structural disparity (or anisomerism sensu Gregory [43, 54]). Finally, the overall patterns of integration and heterogeneity among the parts of a network often results in the emergence of new properties, for example, modularity [48] (see [55–58] for general reviews on the origin and macroevolution of modularity at a morphological level). The more specific details of the modular organization of a network need a closer observation, but the overall degree of parcellation of the network into large, uniform modules (P) captures how much modular the neck is.

Here we applied AnNA to a phylogenetically broad dataset of mammalian necks, including all bones (cervical vertebrae, cranium, sternum, hyoid and pectoral girdle) and muscles involved in the motion of the head and neck (Fig. 1a-f). First, we modeled the neck of each species as a network in which nodes represented the aforementioned bones and muscles and links represented their physical arrangement or contacts. Then, we

Table 1 Summary of network parameters used in this study

Network parameter	Mathematical definition	Morphological interpretation
Number of nodes (N)	Direct count of the number of nodes in the network	Number of anatomical elements
Number of links (K)	Direct count of the number of links in the network	Number of anatomical relations (burden or constrain), connectivity
Density of connections (D)	Relative amount of links: $D = \frac{2K}{N(N-1)}$	Morphological complexity
Mean clustering coefficient (C)	Relative amount of 3-node loops: $C = \frac{1}{N}\sum \frac{\sum loops_i}{k_i(k_i-1)}$ where e_i is the existing number of links among the neighbors of node i and k_i is the total number of links of a node i	Co-dependency (integration)
Mean shortest path length (L)	Average distance between every pair of nodes: $L = \frac{1}{N-1}\sum d_{n_i,n_j}$ where d is the shortest distance in number of links between a given pair of nodes n_i and n_j	Effective proximity (integration)
Heterogeneity of connections (H)	Disparity in the number of links per node: $H = \sigma_K/\mu_K$ where σ_K and μ_K are the standard deviation and mean of K, respectively	Anisomerism
Parcellation (P)	Extent and uniformity of the modular division: $P = 1 - \sum (N_m/N)^2$ where N_m is the number of nodes in module m	Degree of modularity

quantified seven network parameters that serve as proxies for the morphological organization of the neck anatomy (Table 1). Finally, we explored the disparity of neck musculoskeletal organization through time to answer, specifically: 1) whether the musculoskeletal organization (as captured by network parameters) of the neck really differs among mammals; 2) whether closely related species share similar network organization; 3) whether there is a consistent pattern of modularity across mammalian necks; 4) how extreme elongation and deviating vertebral numbers alter neck organization; and 5) how disparity in neck anatomical organization changed during mammalian evolutionary history.

Results

Network parameters and phylogenetic signal
The values of the network parameters used as proxies for the musculoskeletal organization of the neck for individual taxa are listed in Table 2. The phylogenetic signal is statistically significant for the multivariate dataset of all network parameters (Kmult = 0.89, $p < 0.001$) as well as for five of the seven parameters (N, K, D, H, P; Table 3, Additional file 1: AF1). This suggests similar variation in neck organization in closely related species (Fig. 2). A Brownian motion model of evolution best explains trait evolution of the network parameters (see Additional file 1: AF1). Relative variability in connectivity K, complexity D, and integration by co-dependency C is high among the species examined (significant higher

coefficients of variation; Msrl = 139.58, $p < 0.001$) (Table 3, Additional file 1: AF1). In contrast, relative variability of integration by effective proximity L, anisomerism H, and the degree of modularity P are low across all mammals (Table 3, Additional file: AF1). There is no significant relationship between the network parameters and either body mass (F = 0.170, $p = 0.68$), absolute neck length (F = 0.005, $p = 0.94$), relative neck length (F = 0.661, $p = 0.42$), or predatory behavior (i.e., predatory vs. non-predatory; F = 0.191, p = 0.68).

The number of anatomical elements N is low in monotremes compared to the general pattern of therians (Fig. 2a). However, most xenarthrans, the chiropterans, and the Pygmy sperm whale (*Kogia breviceps*) also have a decreased number of elements in their neck network. The number of anatomical connections K is uniformly high in marsupials in contrast to most other mammals (Fig. 2b). K is decreased in monotremes, xenarthrans, chiropterans, the Pygmy sperm whale and the Bactrian camel (*Camelus bactrianus*). Morphological complexity D is high in monotremes, intermediate in marsupials and xenarthrans and tends to decrease in most of the other placental lineages (Fig. 2c). For H, the largest contrast can be found between monotremes (very low H) and therians in general, whereas the pattern within therians is not uniform (Fig. 2d). The degree of modularity P is relatively invariable and only slight decreases can be found in diprotodonts, Pen-tailed treeshrew (*Ptilocercus lowii*), chiropterans and eulipotyphlans (Fig. 2e). The

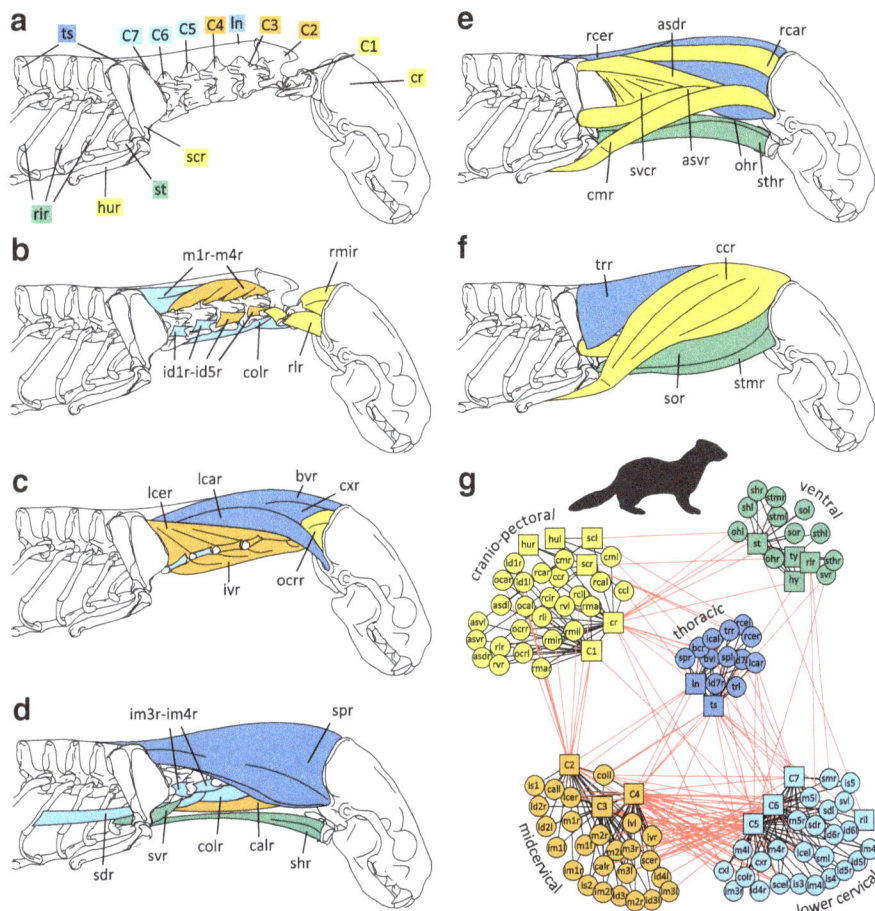

Fig. 1 Musculoskeletal anatomy of the mammalian neck and its translation into the anatomical network. **a** Skeletal and (**b-f**) muscular elements of the neck included in the analysis (from deep to superficial) exemplified for the lesser grison (*Galictis cuja*). **g** Anatomical network representing the same topological information. Colors code for the identified connectivity modules. Red links between-modules. Black links within modules. **a-f** adopted and modified after [138, 139]

phylomorphospace (Fig. 3; Additional file 1: AF1) highlights the differences in neck musculoskeletal organization between monotreme and therian mammals (F = 1.695, *p* < 0.001). Marsupials cluster closely together. They slightly overlap with the distribution of placental mammal, which occupy a huge part of the morphospace. The Pygmy sperm whale is far from the other placental mammals.

Community structure and phenotypic modules in neck networks

The number and constitution of the connectivity modules in the neck varies among the mammalian species examined (a summary is given in Table 4; for detailed information see Additional file 1: AF1). In many cases however, module number varies based on a left-right split of modules that are united in other species (e.g., the pectoral elements are separated in left and right modules). Overall, five principal connectivity modules were detected: 1) cranio-pectoral, 2) ventral, 3) mid-cervical, 4) lower cervical, and 5) thoracic.

This pattern is exemplified here for the lesser grison (*Galictis cuja*) (Fig. 1g). The cranio-pectoral module (present in 32 out of 48 taxa) groups the cranium, the C1, and the bones of the pectoral girdle (scapulae, claviculae, and, if included, humeri), as well as the suboccipital, cleidocephalic (or cephalohumeralis), atlantoscapularis, capital longus, and capital rhomboid muscles. The ventral module groups the sternum, hyoid, thyroid, mandible (when included), and the sternocephalic and infrahyoid muscles. In 32 out of 48 cases, ribs and the related scalenii muscles are also included in this module. The ventral module is combined with (parts of) the pectoral bones and muscles in 15 species, none of which are aclaviculate (Fig. 4b). The mid-cervical module groups C2 to C4 as well as the longus cervicis, spinalis, and their related interspinal, intertransversarii, and multifidii muscles. The lower cervical module groups C5 to C7 with the cervical longissimus, spinalis, and their related interspinal, intertransversarii, and multifidi muscles. These mid-cervical and lower cervical modules are present in all

Table 2 Network parameters of the musculoskeletal organization of the neck of 48 mammalian species

Order	Species	N	K	D	C	L	H	P
Afrosoricida	Chrysospalax trevelyani	112	328	0.053	0.456	2.81	1.448	0.820
	Micropotamogale ruwenzorii	108	329	0.057	0.482	2.771	1.449	0.784
Carnivora	Canis lupus	123	368	0.049	0.351	2.714	1.529	0.784
	Civettictis civetta	113	357	0.056	0.468	2.706	1.504	0.767
	Felis silvestris	103	320	0.061	0.452	2.65	1.401	0.784
	Galictis cuja	118	322	0.047	0.448	2.928	1.416	0.780
	Zalophus californianus	118	338	0.049	0.479	2.795	1.43	0.801
Cetartiodactyla	Babyrousa babyrussa	106	344	0.062	0.398	2.734	1.409	0.784
	Bos taurus	108	329	0.057	0.41	2.777	1.395	0.789
	Camelus bactrianus	96	232	0.051	0.295	2.903	1.268	0.771
	Giraffa camelopardalis	106	309	0.056	0.435	2.731	1.421	0.793
	Kogia breviceps	96	219	0.048	0.488	3.099	1.316	0.825
Chiroptera	Pteropus vampyrus	92	260	0.062	0.452	2.778	1.329	0.732
	Vespertilio murinus	96	264	0.058	0.476	2.821	1.358	0.729
Cingulata	Dasypus novemcinctus	95	279	0.062	0.524	2.769	1.331	0.844
Dasyuromorpha	Sarcophilus harrisii	118	390	0.056	0.449	2.714	1.502	0.779
Didelphimorphia	Didelphis virginiana	108	374	0.065	0.385	2.668	1.413	0.795
Diprotodontia	Macropus rufus	112	355	0.057	0.446	2.726	1.425	0.725
	Phascolarctos cinereus	119	397	0.057	0.429	2.699	1.538	0.710
	Trichosurus vulpecula	109	348	0.059	0.402	2.743	1.432	0.725
Eulipotyphla	Erinaceus europaeus	104	330	0.062	0.441	2.747	1.419	0.734
	Scalopus aquaticus	108	322	0.056	0.425	2.788	1.439	0.728
	Suncus murinus	104	304	0.057	0.411	2.835	1.418	0.753
Hyracoidea	Procavia capensis	115	340	0.052	0.451	2.84	1.507	0.815
Lagomorpha	Oryctolagus cuniculus	122	343	0.046	0.422	2.831	1.556	0.786
Monotremata	Ornithorhynchus anatinus	84	282	0.081	0.382	2.575	1.179	0.765
	Tachyglossus aculeatus	85	260	0.073	0.389	2.709	1.183	0.791
Notoryctemorphia	Notoryctes typhlops	112	363	0.058	0.455	2.736	1.485	0.806
Paucituberculata	Caenolestes fuliginosus	122	383	0.052	0.435	2.767	1.544	0.770
Peramelemorphia	Macrotis lagotis	120	396	0.055	0.434	2.72	1.541	0.790
Perissodactyla	Equus caballus	124	327	0.043	0.454	2.874	1.555	0.774
	Tapirus indicus	114	321	0.05	0.463	2.862	1.463	0.815
Pholidota	Manis pentadactyla	101	292	0.058	0.456	2.873	1.375	0.791
Pilosa	Bradypus tridactylus	110	318	0.053	0.548	2.909	1.293	0.782
	Choloepus didactylus	90	255	0.064	0.489	2.708	1.339	0.762
	Cyclopes didactylus	96	260	0.057	0.456	2.917	1.319	0.820
Primates	Homo sapiens	113	334	0.053	0.512	2.712	1.442	0.793
	Loris tardigradus	114	344	0.053	0.352	2.731	1.47	0.808
	Macaca mulatta	122	369	0.05	0.344	2.719	1.504	0.812
Proboscidea	Elephas maximus	110	273	0.046	0.439	2.829	1.424	0.797
Rodentia	Chinchilla lanigera	108	320	0.055	0.404	2.761	1.423	0.831
	Heteromys desmarestianus	96	294	0.064	0.356	2.719	1.256	0.803
	Neotoma fuscipes	120	333	0.047	0.409	2.825	1.521	0.814
	Pedetes capensis	112	323	0.052	0.434	2.775	1.474	0.802

Table 2 Network parameters of the musculoskeletal organization of the neck of 48 mammalian species *(Continued)*

Order	Species	N	K	D	C	L	H	P
	Sciurus vulgaris	130	364	0.043	0.421	2.807	1.575	0.780
Scandentia	*Ptilocercus lowii*	114	336	0.052	0.483	2.785	1.49	0.719
Sirenia	*Dugong dugon*	106	310	0.056	0.505	2.782	1.413	0.795
Tubulidentata	*Orycteropus afer*	101	272	0.054	0.504	2.86	1.308	0.783

N Number of elements, *K* Number of connections, *D* Density of connections, *C* Mean clustering coefficient, *L* Mean shortest path length, *H* Heterogeneity of connections, *P* Parcellation index

networks (Fig. 4a-d). The border between them is, however, shifted in some species (e.g., C4/C5 to C3/C4). If the attachments of scalenii muscles are limited to few specific cervical vertebrae, these muscle and the ribs are also included in the mid or lower cervical module. The thoracic module groups the thoracic spine, nuchal ligament, semispinalis (complexus + biventer cervicis), capital longissimus, cervical rhomboid and trapezius muscle.

In several species, the pectoral bones (plus the related muscles) are not grouped together with the cranium and C1 but separated; otherwise they are included in the ventral or thoracic module, respectively. For instance, in the long-necked camel the pectoral bones and muscles constitute a distinct module together with the nuchal ligament (Fig. 4c). In the giraffe (*Giraffa camelopardalis*), attachment sites of the pectoral muscles are shifted proximally to the midcervical module. In contrast, the pectoral bones and muscles are combined with the ventral elements into a ventro-pectoral unit in the particolored bat (*Vespertilio murinus*) (Fig. 4b). The sloths differ in the organization of their neck due to their aberrant number of cervical vertebrae. In the two-toed sloth (*Choloepus didactylus*; six cervical vertebrae), the pectoral elements form a separate module. The C5, C6, thoracic spine, and related muscles are grouped within one module. In the three-toed sloth (*Bradypus tridactylus*; nine cervical vertebrae; Fig. 4d), there is an upper (C2-C5) and lower (C6-C7) midcervical module. The evolutionary 'new' vertebrae C8 and C9, however, are

grouped with the thoracic spine, scapulae, and related muscles. The claviculae are included in the ventral module. The ribs are grouped with the cranium and atlas in the Pygmy sperm whale.

Disparity in neck organization through time

The mean subclade disparity values for the observed and simulated data were plotted against node age (Fig. 2f). Subclade disparity through time is low, which is particularly obvious in the first two-thirds of mammalian evolution (i.e., during the Mesozoic). However, it is not significantly different from the expectation under a Brownian motion model of neck organizational evolution (morphological disparity index = 0.016, $p = 0.65$). Nevertheless, major shifts in disparity rate occurred in the middle to late Paleocene and in the middle to late Eocene. These shifts resulted in disparity peaks exceeding the 95% confidence interval of the simulated data. The disparity in neck musculoskeletal organization decreased after the Eocene-Oligocene border. A larger sample size would be required to infer significant results for the post-Eocene ages (i.e., more divergence events are needed).

Discussion
Variation in neck organization across phylogeny
The more conserved network parameters (*L*, *H*, *P*) represent measurements of the neck's integration by effective proximity, anisomerism, and degree of modularity

Table 3 Variability and phylogenetic signal in neck network parameters

	N	K	D	C	L	H	P
Minimum	84	219	0.043	0.295	2.575	1.179	0.71
1st Quartil	102.5	293.5	0.052	0.41	2.72	1.371	0.769
Mean	108.6	321.5	0.056	0.437	2.782	1.421	0.782
3rd Quartil	115.8	345	0.058	0.464	2.83	1.493	0.802
Maximum	130	397	0.081	0.548	3.099	1.575	0.844
Coefficient of Variation	0.097	0.132	0.128	0.114	0.031	0.066	0.04
95% Confidence Intervalls	0.077/0.115	0.104/0.155	0.092/0.16	0.085/0.138	0.022/0.04	0.051/0.079	0.032/0.047
Abouheif's Cmean	0.372***	0.421***	0.366***	0.133	0.172	0.359***	0.354***
Blomberg's K	0.995***	0.879***	1.479***	0.559	0.762	1.108***	0.719**

N Number of elements, *K* Number of connections, *D* Density of connections, *C* Mean clustering coefficient, *L* Mean shortest path length, *H* Heterogeneity of connections, *P* Parcellation index. Significance levels of the tests for phylogenetic signal: $p < 0.01$**; $p < 0.001$***

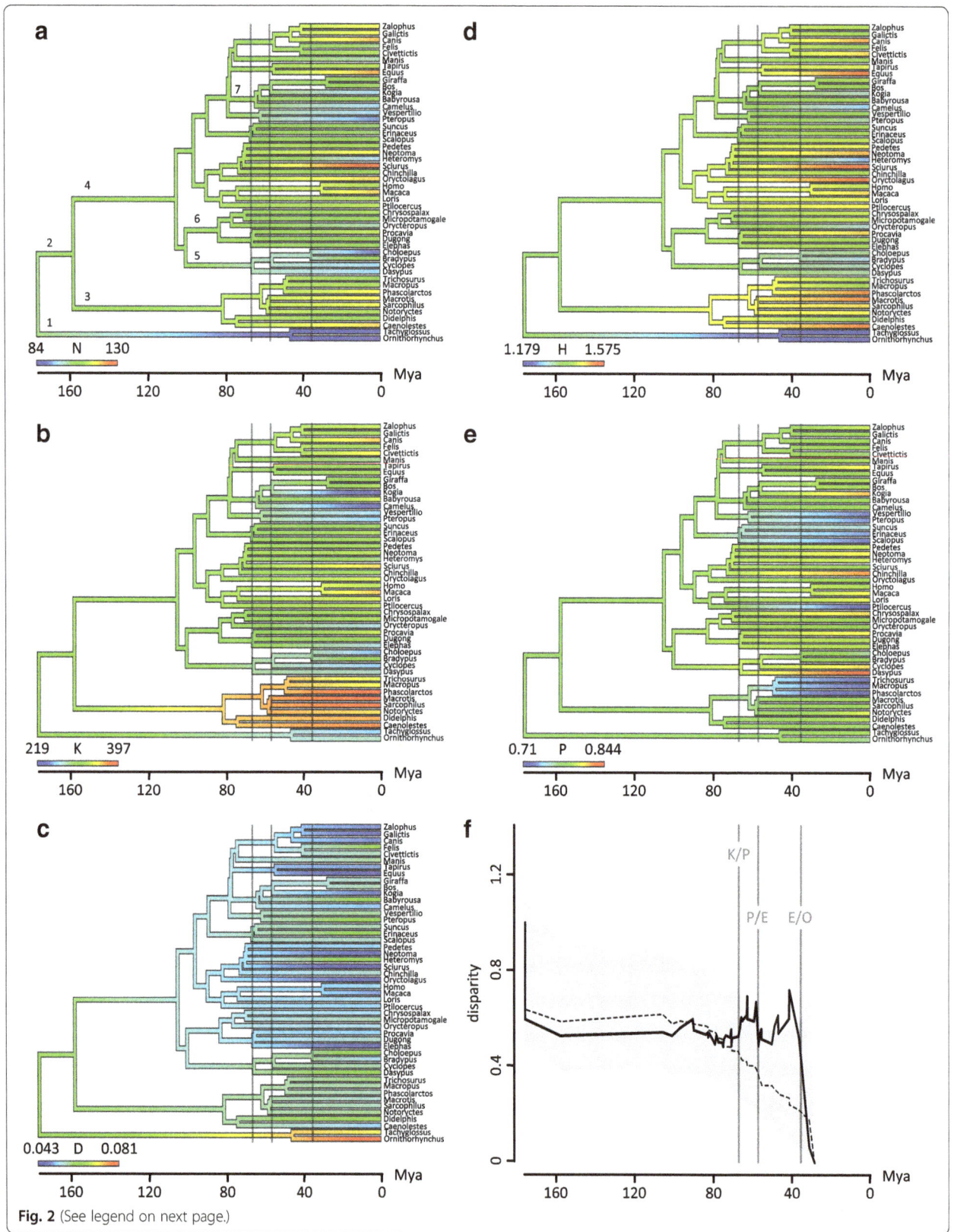

Fig. 2 (See legend on next page.)

(See figure on previous page.)
Fig. 2 Phylogenetic signal and disparity through time (DTT) for neck network parameters across mammals. **a** Number of elements *N*; (**b**) Number of connections *K*; (**c**) Density of connections *D*; (**d**) Heterogeneity of connections *H*; (**e**) Parcellation index *P*; (**f**) mean subclade disparity through time plot. Gray vertical lines indicate Cretaceous-Tertiary (K/T) Paleocen-Eocene (P/E), and Eocene-Oligocene (E/O) boundary. Numbers indicate selected taxa: 1 Monotremata: 2 Theria; 3 Marsupialia; 4 Placentalia; 5 Xenarthra; 6 Afrotheria; 7 Cetartiodactyla. In (**f**). the solid line indicates actual median subclade DTT of the sample. The dashed line indicates the median subclade DTT based on 10,000 simulations of character evolution under Brownian motion. The shaded area indicates the 95% DTT range for the simulated data

[41, 43]. These measurements capture how distant parts of the neck (e.g., head and trunk, lower and upper cervical column) are integrated, that anatomical connections are not evenly distributed across bone (e.g. vertebrae) and the way the neck is modularized. Thus, our results indicate a basic constructional set-up of the neck across mammals determined by morphological regionalization (see further below). These conserved features likely arise due to shared developmental [8, 9, 17, 19, 59–61] and/or biomechanical/constructional constraints [31, 62, 63]. However, the musculoskeletal organization of the neck is not uniform for other morphological features captured by network proxies. Specifically, mammalian necks considerably vary in terms of their morphological burden, complexity, and integration by co-dependency (quantified by *K*, *D*, and *C*, respectively). These features describe the grade of specialization in the neck due to reduction of elements or

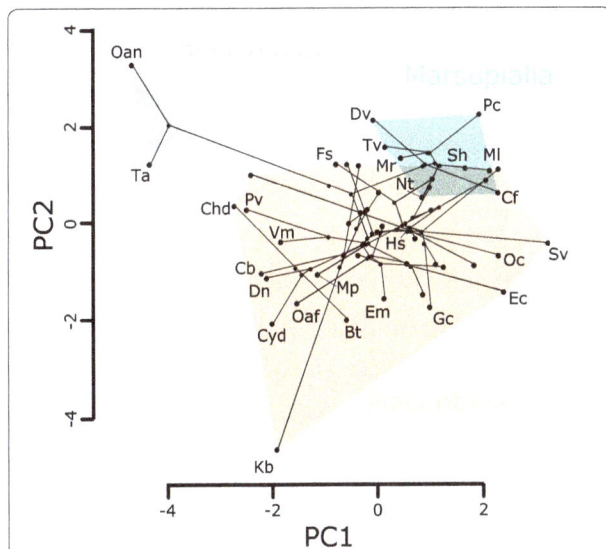

Fig. 3 Phylomorphospace of the network parameters. Monotreme, marsupial and selected placental species are labeled. PC1 and PC2 represent 44% and 27% of the total variation, respectively. Species abbreviations: Bt *Bradypus tridactylus;* Cb *Camelus bactrianus;* cf *caenolestes fuliginosus;* Chd *Choloepus didactylus;* Cyd *Cyclopes didactylus;* Dn *Dasypus novemcinctus;* Dv *Didelphis virginiana;* Ec *Equus caballus;* Em *Elephas maximus;* Fs *Felis silvestris;* Gc *Galictis cuja;* Hs *Homo sapiens;* Kb *Kogia breviceps;* Ml *Macrotis lagotis;* Mp *Manis pentadactyla;* Mr. *Macropus rufus;* Nt *Notoryctes typhlops;* Oaf *Orycteropus afer;* Oan *Ornithorhynchus anatinus;* Oc *Oryctolagus cuniculus;* PC *Phascolarctos cinereus;* Pv *Pteropus vampyrus;* Sh *Sarcophilus harrisii;* Sv *Sciurus vulgaris;* Ta *Tachyglossus aculeatus;* Tv *Trichosurus vulpecula;* Vm *Vespertilio murinus*

enhancement of passive structures (e.g., the nuchal ligament) and the way the neck is structurally constrained by the setup of its muscular connections (i.e., its evolvability) [43, 49]. The variation arises from the major trends of epaxonic muscle modification during mammalian evolution, leading to differences in nuchal muscle organization among monotremes, marsupials and therians. Our findings confirm the plesiomorphic pattern of epaxonic neck muscle arrangement in monotremes [64–67]. It results in a musculoskeletal organization that is distinctively different from that of therian mammals (Fig. 3). Their low differentiation within the three longitudinal systems (longissimus, iliocostalis, transversospinalis; in particular the deep intervertebral muscles; low number of muscles) and muscle attachments that are evenly distributed among the vertebrae (high complexity but low irregularity) suggest low specialization to specific neck motion patterns. In marsupials, epaxonic muscles are more differentiated in deep and superficial layers. Moreover, most of the superficial muscles are attached to every cervical vertebra [66, 68–72]. This high connectivity results in structural constraints in the neck (i.e., morphological burdens) and low musculoskeletal disparity among marsupials in comparison to placentals (Fig. 3). Attachments of epaxonic neck muscles are very variable among placental mammals [70] and thus network parameters are as well. However, two major trends in neck evolution have been shown: First, there is a reduction of attachment sites of the neck muscles to only a few vertebrae/the skull; and second, there is an increased bracing of the head-trunk distance by ligamentous structures to accommodate for increasing head weight and neck length [15, 73–78]. This results in placental mammals generally having necks that are less complex compared to monotreme and marsupial mammals. For instance, few but specialized muscles have the small attachment sites and are able to induce a similar motion or the superficial epaxonic muscles attach only secondarily to the head via the nuchal ligament. In addition, neck variation in placental mammals is also highly influenced by variation in the organization of pectoral bones and muscles (see below).

The phylogenetic signal of most of the network parameters reveals that phylogenetic relationship accounts for much of the variation in neck organization. At the same time, network parameters also discriminate between monotreme, marsupial, and placental mammals. Within placental mammals, however, variation of network parameters

Table 4 Summary of network modules

Order	Species	M1	M2	M3	M4	M5
Afrosoricida	*Chrysospalax trevelyani*	**cranio-atlantal pectoral**	midcervcial	lower cervical - thoracic	ventral	
	Micropotamogale ruwenzorii	cranio-atlantal	midcervcial[a]	lower cervical - thoracic	ventral	**pectoral**
Carnivora	*Canis lupus*	**cranio-atlantal pectoral**	midcervical	lower cervical	ventral	thoracic
	Civettictis civetta	**cranio-atlantal pectoral**	midcervcial	lower cervical - thoracic	ventral	
	Felis silvestris	**cranio-atlantal pectoral**	midcervcial	lower cervical	ventral	thoracic
	Galictis cuja	**cranio-atlantal pectoral**	midcervcial	lower cervical	ventral	thoracic
	Zalophus californianus	cranio-atlantal	midcervcial	lower cervical - thoracic	ventral	**pectoral**
Cetartiodactyla	*Babyrousa babyrussa*	**cranio-atlantal humeral**	midcervical	lower cervical - thoracic	ventral	**scapular**
	Bos taurus	**cranio-atlantal humeral**	midcervcial	lower cervical - thoracic	ventral	**scapular**
	Camelus bactrianus	cranio-atlantal	midcervical	lower cervical - thoracic	ventral	**pectoral**
	Giraffa camelopardalis	cranio-atlantal	**midcervcial-pectoral**	lower cervical	ventral	thoracic
	Kogia breviceps	cranio-atlantal costal	midcervcial	lower cervical - thoracic	ventral	**pectoral**
Chiroptera	*Pteropus vampyrus*	cranio-atlantal	midcervcial	lower cervical - thoracic	**ventro-pectoral**	
	Vespertilio murinus	cranio-atlantal	midcervcial	lower cervical - thoracic	**ventro-pectoral**	
Cingulata	*Dasypus novemcinctus*	cranio-atlantal axial	midcervcial	lower cervical	ventral	thoracic
Dasyuromorpha	*Sarcophilus harrisii*	cranio-atlantal	midcervcial	lower cervical - thoracic	ventral	pectoral
Didelphimorphia	*Didelphis virginiana*	**cranio-atlantal pectoral**	midcervcial	lower cervical - thoracic	ventral	
Diprotodontia	*Macropus rufus*	**cranio-atlantal pectoral**	midcervcial	lower cervical - thoracic	ventral[d]	
	Phascolarctos cinereus	**cranio-atlantal pectoral**	midcervcial	lower cervical - thoracic	ventral	
	Trichosurus vulpecula	**cranio-atlantal pectoral**	midcervcial	lower cervical - thoracic	ventral	
Eulipotyphla	*Erinaceus europaeus*	**cranio-atlantal pectoral**	midcervcial	lower cervical - thoracic	ventral	
	Scalopus aquaticus	**cranio-atlantal pectoral**	midcervcial	lower cervical - thoracic	ventral	
	Suncus murinus	atlantal	midcervcial	lower cervical - thoracic	**cranio-ventro-pectoral**	costal
Hyracoidea	*Procavia capensis*	**cranio-atlantal pectoral**	midcervcial	lower cervical	ventral	thoracic
Lagomorpha	*Oryctolagus cuniculus*	**cranio-atlantal pectoral**	midcervical[b]	lower cervical - thoracic	ventral	
Monotremata	*Ornithorhynchus anatinus*	cranio-atlantal	midcervcial	lower cervical - thoracic	**ventro-pectoral**	
	Tachyglossus aculeatus		midcervcial		ventral	**pectoral**

Table 4 Summary of network modules *(Continued)*

Order	Species	M1	M2	M3	M4	M5
		cranio-atlantal axial		lower cervical - thoracic		
Notoryctemorphia	*Notoryctes typhlops*	cranio-atlantal	midcervcial	lower cervical	**ventro-pectoral**	thoracic
Paucituberculata	*Caenolestes fuliginosus*	**cranio-atlantal pectoral**	midcervical[c]	lower cervical - thoracic	ventral	
Peramelemorphia	*Macrotis lagotis*	**cranio-atlantal pectoral**	midcervcial[a]	lower cervical - thoracic	ventral	
Perissodactyla	*Equus caballus*	**cranio-atlantal pectoral**	midcervcial	lower cervical	ventral	thoracic
	Tapirus indicus	**cranio-atlantal pectoral**	midcervcial	lower cervical	ventral	thoracic
Pholidota	*Manis pentadactyla*	**cranio-atlantal pectoral**	midcervical[b]	lower cervical - thoracic	ventral	
Pilosa	*Bradypus tridactylus*	cranio-atlantal	upper midcervical	lower midcervical	**ventro-clavicular**	**lower cervical - thoracic scapular**
	Choloepus didactylus	cranio-atlantal	midcervcial	lower cervical - thoracic	ventral	**pectoral**
	Cyclopes didactylus	cranio-atlantal	midcervcial - thoracic	lower cervical C5&rest	ventral	**pectoral**
Primates	*Homo sapiens*	cranio-atlantal	midcervical[b]	lower cervical - thoracic	ventral	**pectoral**
	Loris tardigradus	**cranio-atlantal pectoral**	midcervcial[a]	lower cervical - thoracic	ventral	thoracic
	Macaca mulatta	**cranio-atlantal pectoral**	midcervical[c]	lower cervical - thoracic	ventral	costal
Proboscidea	*Elephas maximus*	**cranio-atlantal pectoral**	midcervical[c]	lower cervical - thoracic	ventral	
Rodentia	*Chinchilla lanigera*	**cranio-atlantal scapular**	midcervical	lower cervical	**ventro-clavicular**	thoracic
	Heteromys desmarestianus	cranio-atlantal	midcervcial	lower cervical - thoracic	ventral	**pectoral**
	Neotoma fuscipes	**cranio-atlantal pectoral**	midcervical	lower cervical - thoracic	ventral	
	Pedetes capensis	**cranio-atlantal pectoral**	midcervcial[a]	lower cervical - thoracic	ventral	
	Sciurus vulgaris	**cranio-atlantal scapular**	midcervcial	lower cervical - thoracic	**ventro-clavicular**	
Scandentia	*Ptilocercus lowii*	**cranio-atlantal pectoral**	midcervcial	lower cervical - thoracic	ventral	
Sirenia	*Dugong dugon*	**cranio-atlantal pectoral**	midcervcial	lower cervical	ventral	thoracic
Tubulidentata	*Orycteropus afer*	**cranio-atlantal pectoral**	midcervical[b]	lower cervical - thoracic	ventral	

Contribution of the pectoral elements to different modules marked in bold. Left-right division of pectoral and costal elements is not considered in this summary table. Scapular, humeral, and clavicular elements are separately indicated when the pectoral bones and associated muscles are not group within the same module
[a] no clear assignment of C5 to the midcervical or lower cervical module
[b] potential subdivision of the midcervical module into C2/C3 and C4/C5
[c] no clear assignment of C2 to the cranio-atlantal or midcervical module
[d] no clear division between the cranio-atlantal and ventral module

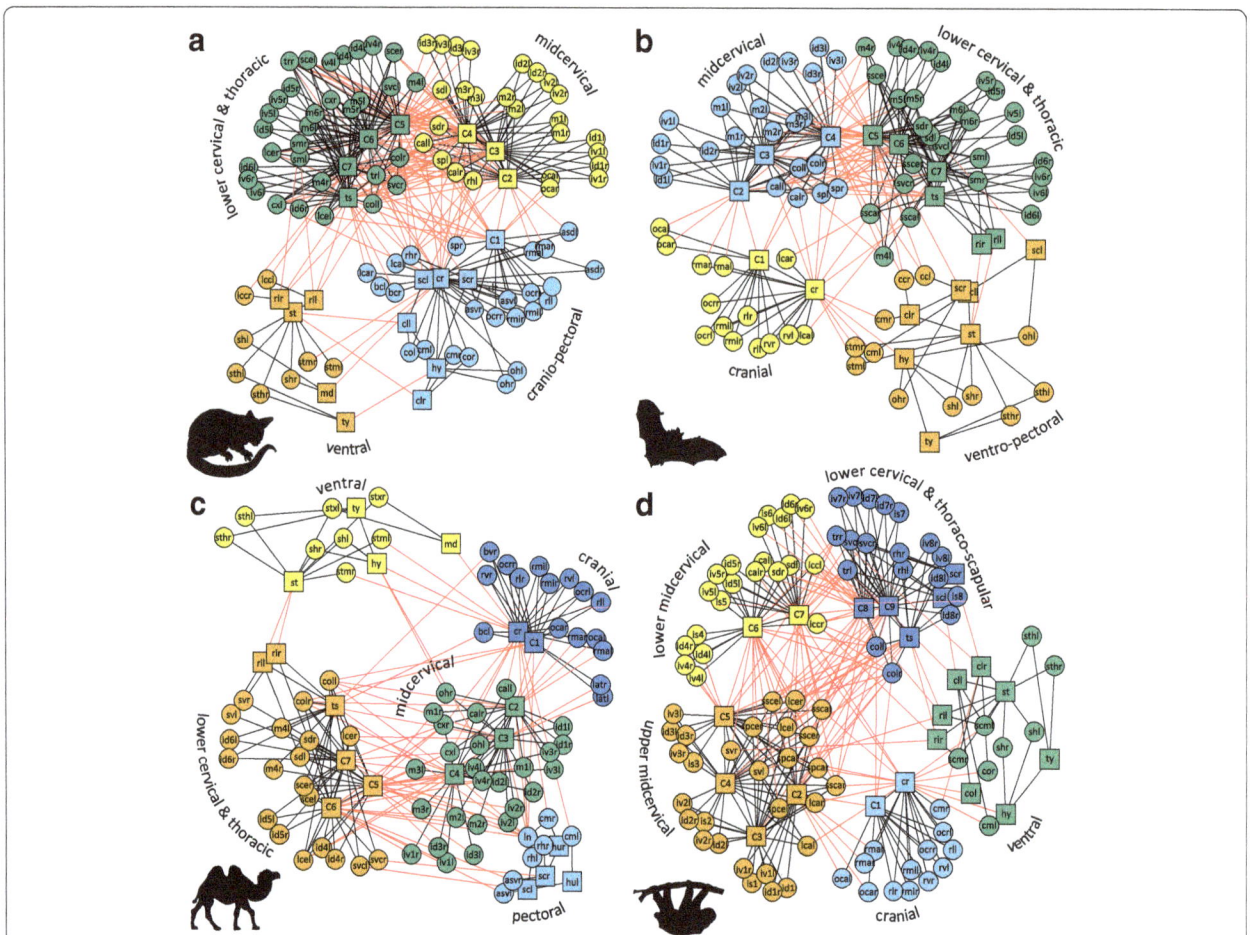

Fig. 4 Network representations and connectivity modules of the neck of different mammals. **a** Common brushtail possum (*Trichosurus vulpecula*); (**b**) Parti-coloured bat (*Vespertilio murinus*); (**c**) Bactrian camel (*Camelus bactrianus*); (**d**) Three-toed sloth (*Bradypus tridactylus*). Colors code for the identified connectivity modules. Red links between-modules. Black links within modules

is mostly limited to xenarthrans, chiropterans, and some cetartiodactyls. Although relatively species-poor, xenarthrans, show highly specialized neck morphologies related to their diverse fossorial or arboreal ecologies [79–81] and unique development [33, 82, 83]. In chiropterans, back muscles contribute only marginally to the stabilization of the head because of the lack of most cranial and cervical attachments [84–86]. Sagittal stability is instead achieved by the modified morphology of the cervical vertebrae in accordance with roosting behaviors [87]. Cetartiodactyls have recently been show to exhibit the highest disparity in neck morphology across mammals [31]. It ranges from the very short necks of cetaceans up to the extreme long ones of camelids and giraffids. As a consequence, neck musculoskeletal organization is similarly diverse. Several muscles with cervical attachment are reduced in the Pygmy sperm whale (and other cetaceans) or their attachment is shifted to the skull (e.g. scaleni muscles) and thus head stabilization is increased [88, 89]. On the other hand, cranial attachments of the dorsal neck muscle are mainly reduced in long

necked species, such as the camel and giraffe. Muscle force is instead transferred by the modified nuchal ligament [74, 75, 77]. Surprisingly, neck network parameters in the dugong (*Dugong dugon*), although also being fully aquatic, do not show a similar alteration as in the Pygmy sperm whale. Instead, it closely resembles the Asian elephant (*Elephas maximus*) and other afrotherians (Table 1, Fig. 2).

In accordance with our findings, [90, 91] also showed that the effect of size and prey capture behavior is low in the neck compared to the thoracolumbar region. However, functional interpretations of the results of the analysis of the topological arrangement of parts needs to be inferred on a one to one basis and taking into account the specific ecological context of each taxa.

Regionalization and modularity in the mammalian neck
Despite the relative low and invariant number of neck vertebrae in mammals, several studies have uncovered a tripartite regionalization of the cervical spine based on

developmental, morphological, allometric, and functional evidence [17, 25, 31, 32, 91]. Our results have now uncovered a corresponding regionalization of the musculoskeletal organization of the mammalian neck into an upper (cranium, C1), mid (C2-C4), and lower cervical module (C5-C7, in some species also the thoracic spine). This modularity pattern is conserved across mammals despite variations in size, feeding mechanisms, and locomotor modes (indicated by a uniform grade of modularity). This conserved pattern probably arose from the high number of connections between the vertebrae of the same module (or the cranium and C1) resulting in an increase of structural constraints and integration of these elements [40, 49, 52]. However, the boundaries between adjacent modules/regions are not consistent across different studies analyzing the morphology of the neck using different criteria. For example, vertebrae C1 and C2 are not part of the same connectivity module despite their close developmental, functional, and evolutionary relationship [25, 92, 93]. A similar dissociation of C1 and C2 into different regions has been shown for their scaling properties [31] and highlights the role of C2 as a functional mediator between the head joint and the postaxial column (see also [94]). In addition to the three 'inner' axial modules, there are two additional 'outer' modules bridging the distance between the trunk and the head (or the hyoid or upper vertebrae), with a muscular cuff on the dorso-lateral (pectoral) and ventro-lateral side. Many of these muscles were crucial for the evolutionary origin of the vertebrate neck [95, 96].

Neck organization in sloths

In general, a similar regionalization of the neck is observed in both genera of sloths, despite their variation in the number of cervical vertebrae. The evolutionary new C8 and C9 in *Bradypus* and their associated muscles are grouped together with the thoracic spine. This agrees with their thoracic origin and ossification sequence [82]. Conversely, the evolutionary new Th1 provides the basis for the close association of the thoracic vertebral region to C5 and C6 in *Choloepus* neck organization. Divergence of the sloths' necks becomes obvious when including their pectoral bones and muscles in the comparison. Their neck-shoulder arrangement represents two different solutions of locomotor possibilities under common functional constraints [80, 97]. Neck organization and modularity of *Bradypus* resembles those of other long necked species. The resemblance stems from its nearly complete lack of cervical and cranial attachments of the neck/shoulder muscles and its unusual clavicular and shoulder morphology [80, 98–100] (see the unusual lower cervical-thoracic-scapular module in Table 4). The muscles of *Choloepus*, in contrast, are so placed as to offer the greatest possible support dorsally and ventrally to the head as well as to the scapula [80, 98, 101,

102]. Thus, neck organization and modularity is closer to the general pattern as seen in the lesser grison (i.e., pronounced head support, functional connection of head and forelimb) [15].

Evolutionary integration of the neck and the forelimb

The enduring evolutionary and developmental relationship between the neck and the forelimb in mammals (and other amniotes) is well documented (e.g., [17, 19, 20, 95, 96, 103]). This relationship is most obvious in the brachial plexus innervating shoulder and forelimb muscles [104, 105]. However, there is also a strong functional integration between the neck and the forelimb, with several muscles connecting the pectoral girdle and the head/neck (often with repeated slips). Based on this functional connection, the posture and movements of the neck have a crucial influence on the mechanics of the forelimb in terms of gait efficiency, balance, stabilization, ground reaction forces, and kinematics [30, 106–112]. Our findings now highlight the consequences of this integration on the musculoskeletal organization of the neck. Although there is a conserved tripartition of the cervical spine and its associated muscles, the varying contribution of pectoral bones and muscles to different connectivity modules accounts for much of the observable neck disparity across mammals (e.g., see results on sloths). Major shifts in forelimb morphology and function (e.g., mobilization of the pectoral girdle, reduction of the claviculae) [113–115] are associated with increasing decoupling of the pectoral elements from the ventral module and their connection to the cranium and upper cervical region. This coincides with the increased role of head/neck movements on forelimb mechanics during fast and endurance running (i.e., cursoriality) [30, 108]. In mammals with extreme long necks (camel, giraffe), although also being capable of enduring walking, the pectoral bones and muscles are separated from the cranial module. However, it has recently been shown that neck pendulum mechanics and function in long-necked mammals is different compared to actual cursors [108].

Implications for the evolution and disparity of the mammalian neck

The differences in neck organization between monotremes and therian mammals is one of the striking findings of this study. They result in high disparity between them whereas their within-subclade disparity is low during the first two-thirds of mammalian evolution. Accordingly, the disparity of the neck of mammals was low during the Mesozoic (Fig. 2f) [116]. Figure 5 illustrates the major grades of musculoskeletal organization during mammalian neck evolution. Monotreme, marsupial, and placental mammals differ in their degree of epaxonic muscle differentiation and the varying integration of the

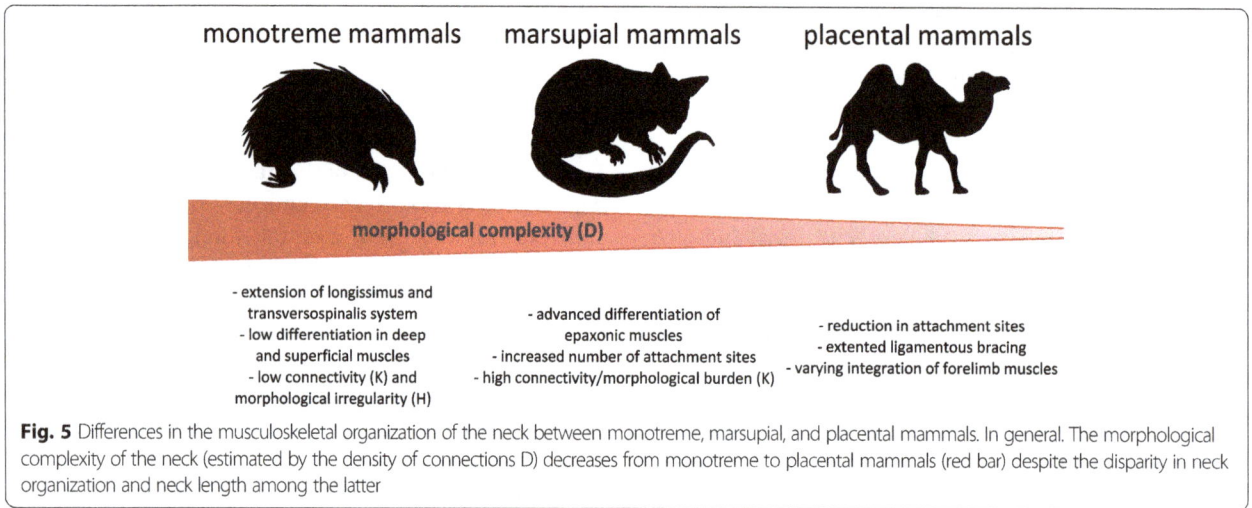

Fig. 5 Differences in the musculoskeletal organization of the neck between monotreme, marsupial, and placental mammals. In general. The morphological complexity of the neck (estimated by the density of connections D) decreases from monotreme to placental mammals (red bar) despite the disparity in neck organization and neck length among the latter

forelimb muscles. In general, the morphological complexity of the neck decreases from monotremes to placentals (Fig. 5) but disparity increases (Fig. 3).

The therian radiation that followed the K/T mass extinction and the appearance of most of the therian and marsupial (supra)orders during the Paleocene was accompanied by an abrupt increase in neck musculoskeletal disparity after a long period of low neck disparity. This was even associated with the appearance of locomotor and foraging specializations [117] and the diversification in body size (e.g., [118]). The second disparity peak during the Eocene coincides with the radiation and increased diversity of modern placental orders, like cetartiodactyls, perissodactyls, carnivorans, and xenarthrans (see [117] and references therein). Thus, disparity in neck organization emerged relatively late in the long mammalian history and is associated with the origin and radiation of specific lineages.

Conclusion

One of our crucial findings is that the musculoskeletal organization of the neck differs between monotreme, marsupial, and placental mammals (Fig. 5). Moreover, particularly the necks of placental mammals are characterized by a reduced complexity despite their increased disparity in musculoskeletal organization and length. Our network analyses revealed a mosaic complexity and disparity in the musculoskeletal organization of the mammalian neck despite the more obvious meristic (and other) constraints on the cervical spine. Musculoskeletal irregularity, effective proximity, degree of modularity, and the occurrence of three inner/axial regions are conserved features among mammalian necks. Thus, a shared biomechanical construction and common developmental interrelationships not only constrain variation in the cervical spine, but are similarly likely to limit musculoskeletal variability in the neck. The conservation

of these traits contrasts, however, with the high variability in morphological burden, integration by co-relation, morphological complexity, and the configuration of the ventral and (cranio-)pectoral module in the neck. The expansion of limb muscles in the cervical region not only facilitated enhanced forelimb mechanics but also increased structural disparity (and thus derived motor patterns and mechanics) in the neck. Thus, we highlight the close integration of the neck and the forelimb during mammalian evolution. The disparity in neck musculoskeletal organization evolved late in mammalian history and in parallel with the radiation of some lineages (e.g., cetartiodactyls, xenarthrans). Finally, our findings show that the limited number of vertebrae in the cervical spine does not necessarily result in low musculoskeletal disparity during mammalian evolutionary diversification.

Methods
Data collection and anatomical network modeling

We collected the topographic data of the musculoskeletal system of the neck in 48 mammalian species through an extensive literature review (Table 2, Additional file 2: AF2). The sample represents all major monotreme, marsupial, and placental clades, as well a diversity of locomotor and feeding strategies. Rodent diversity is represented by members of all suborders (sciuromorphs, myomorphs, hystricomorphs, castorimorphs, and anumaluromorphs, respectively). Representatives of both genera of extant sloth (*Bradypus* and *Choloepus*) were included to examine the influence of their deviating number of cervical vertebrae on neck organization. We documented the number and specific connections/attachments of all skeletal structures and muscles constituting the neck motion system in these taxa.

In contrast to the domestic mammals, for which veterinary textbooks as well as anatomical publications are available, information for exotic species is scarce and literature descriptions are in many cases old. To overcome this problem, in all but two cases we always consulted two or more references to compare the data. We made sure that at least one detailed anatomical monograph was included. As the network models used here focus on the presence/absence of attachments among bones and muscles rather than more specific information on the nature and area of such attachments, the literature reviewed was of enough quality for our modeling approach.

We included all muscles originating from the cervical vertebrae, skull (cranium or mandible), nuchal ligament, or hyoid/thyroid, and inserting on the (cervical or thoracic) vertebrae, sternum, pectoral girdle (scapulae, claviculae, humeri), or ribs (see details in Additional file 3: AF3). Accordingly, we excluded the masticatory, facial, laryngeal, pharyngeal, and suprahyoid muscles from the analysis. A calibrated phylogenetic tree was constructed using the data from the Timetree of Life database [119, 120] (Fig. 2).

We built anatomical network models of the necks' musculoskeletal systems, which comprised all anatomical units as well as the different types of physical interaction among them. Network nodes represented all bones (cranium, cervical vertebrae, thoracic spine, left and right ribs, hyoid, left and right claviculae, left and right scapulae, sternum, and, if involved, left and right humeri, and mandible), other passive elements (nuchal ligament, if present, thyroid), and all cervical muscles, as described above. Network connections represented all physical articulations between bones and other passive elements described, as well as the fleshy and tendinous attachments of the muscles onto the bones (see adjacency matrices in Additional file 4: AF4). Network models were analyzed using the *igraph* package [121] in R [122].

Network parameter analyses

The mathematical definitions and calculations of the network parameters examined here (N, K, D, C, L, H, P) are provided in Table 1. The degree of modularity (parcellation P) was measured from the connectivity modules identified using a spin-glass model and simulated annealing algorithm implemented in the R package *netcarto* [123, 124]. A connectivity module is defined as a group of nodes highly connected among them and poorly connected to nodes outside the group. P is 0 when all nodes are in a same module, and tends to 1 when nodes are evenly distributed within many modules. We tested the phylogenetic signal in the multivariate dataset of all network parameters by calculating K_{mult} [125] using R package *phylocurve* [126]. We additionally tested the phylogenetic signal of the individual parameters using the Abouheif's test with 1000 permutations [127] and Blomberg's test [128] in the R package *phytools* [129]. Mode of trait evolution was explored by comparing multivariate fits of Brownian motion, Ornstein-Uhlenbeck (single adaptive optimum), and Early burst models using Akaike information criterion (AIC) weights in R package *mvMORPH*

[130]. Distribution of the network parameters was visualized with a phylomorphospace of the first two principal components using R package *phytools* [129].

Relative variability of the network parameters was analyzed by statistical comparison of their coefficients of variation (CVs). 95% confidence intervals of the CVs were calculated by 10,000 bootstrap resampling. Significant differences among the parameters' CVs were tested using the modified signed-likelihood ratio test (MSLRT) for equality of CVs (all parameters) and the asymptotic test for the equality of CVs (pair-wise comparisons, Bonferroni corrected) in the R package *cvequality* [131]. In order to test for allometric effects on network parameters they were regressed against logtransformed body mass, absolute neck length, and relative neck length. Body mass and neck length data were taken from [31]. Relative neck lengths were calculated by dividing absolute neck length by body mass$^{1/3}$. Allometric analyses were done using phylogenetic generalized least square regressions in the R package *caper* [132]. The effect of predatory behavior on logtransformed network parameters was examined by testing for significant differences between predatory and non-predatory mammals using a multivariate distance-based phylogenetic generalized least square regression (D-PGLS) with 1000 permutations in the R package *geomorph* [133]. Species were classified as predatory when food intake involves head-neck movements to hold the food counteracting its resisting movements (carnivorous, insectivorous, piscivorous species). In contrast, species were classified as non-predatory when food intake does not involve such head-neck movements (food is just picked or harvested, e.g., browsers, grazers, but also myrmecophagous species). Differences in network parameters between monotreme, marsupial, and placental mammals were tested using a multivariate D-PGLS regression with 1000 permutations.

Modularity analysis

We calculated the quality of the partitions identified by the community detection algorithm using the optimization function Q [134]. According to Newman and Girvan [134], if the number of connections within modules is not different from that expected at random, then Q will be close to 0. The higher the Q the stronger the modular pattern of the network ($Q_{max} = 1$). In practice, strongly modular networks show Q values ranging from 0.3 to 0.7 [134]. Thus, we considered that an anatomical network has a strongly modular structure if $Q - Q_{error} > 0.3$. The expected error of Q was calculated using a jackknife procedure, where every link was taken as an independent observation [134] (more details are provided in Additional file 3: AF3). Finally, we estimated the statistical significance of each module using a two-sample Wilcoxon rank-sum test on the internal vs. external links of the module's nodes. The null hypothesis was that the number of connections is the same inside as outside the module (i.e., as expected if the module were created at random); the alternative hypothesis was that the number of connections is higher inside than outside the module (i.e., the definition of connectivity module). An extensive account of these methods has been given elsewhere [41, 48, 135].

Disparity through time analysis

We carried out a disparity through time (DTT) analysis using the R package *geiger* [136] to trace the variation in neck organization through the evolution of mammals. First, we performed a principal component analysis (PCA) of the network parameters used as proxies of the morphological organization of the neck (i.e., N, K, C, D, L, H, and P) to account for their co-variation structure. Mean subclade disparity through time for the PC scores were calculated [116, 137]. Observed disparity in neck organization across our phylogeny was compared with that expected under a Brownian motion process performing 10,000 iterations. High disparity values indicate high variance within subclades; low disparity values indicate conservation within subclades and high variance among subclades. Finally, we calculated the morphological disparity index to quantify the overall difference in relative disparity of a clade compared to that expected under the null Brownian motion model [116, 137].

Abbreviations of network elements

Left and right side are indicated by *l* and *r*, respectively, added to the abbreviations.

asd atlantoscapularis dorsalis; *asv* atlantoscapularis ventralis; *bc* biventer cervicis; *C1-C9* cervical vertebrae; *cal* longus capitis; *cc* cleidocervicalis; *cl* clavicle; *cm* cleidomastoideus; *co* cleidooccipitalis; *col* longus colli; *cx* complexus; *cr* cranium; *hu* humerus; *hy* hyoid; *icc* iliocostalis cervicis; *id1-id8* intertransversarii cervicis dorsales; *im1-im4* intertransversarii cervicis mediales; *is1-is8* interspinalis; *iv/iv1-iv8* intertransversarii cervicis ventrales (fused/separate); *lat* longus atlantis; *lca* longissimus capitis; *lce* longissimus cervicis; *ln* nuchal ligament; *m1-m6* multifidi; *md* mandible; *oca* obliquus capitis caudalis; *ocr* obliquus capitis cranialis; *oh* omohyoideus; *rca* rhomboideus capitis; *rce* rhomboideus cervicis; *rci* rectus capitis dorsalis intermedius; *rh* rhomboideus (undifferentiated); *ri* ribs; *rl* rectus capitis lateralis; *rma* rectus capitis dorsalis major; *rmi* rectus capitis dorsalis minor; *rv* rectus capitis ventralis; *sce* spinalis cervicis; *sc* scapula; *scm* sternocleidomastoideus (sternal and clavicle part not separate); *sd* scalenus dorsalis; *sh* sternohyoideus; *sm* scalenus medius; *so* sternooccipitalis; *spca* splenius capitis; *spce* splenius cervicis; *sp* splenius (undifferentiated); *ssca* semispinalis capitis; *ssce* semispinalis cervicis; *st* sternum; *sth* sternothyroideus; *stm* sternomastoideus; *stx* sternomaxillaris; *svc* serratus ventralis cervicis; *sv* scalenus ventralis; *tr* trapezius; *ts* thoracic spine; *ty* thyroid

Additional files

Additional file 1: AF1 Additional information on the results of phylogenetic, network, and modularity analyses. (PDF 1108 kb)

Additional file 2: AF2 Systematics and references of investigated species. (PDF 414 kb)

Additional file 3: AF3 Additional information on methods. (PDF 210 kb)

Additional file 4: AF4 Adjacency matrices coding for the topological information of neck's musculoskeletal organization of investigated species. (XLS 677 kb)

Acknowledgements
We thank Heiko Stark for helpful discussions, Jan Wölfer for assistance with the phylogentic comparative methods, and Mikaela Lui for language improvement. We are grateful to associated editor Leandro Monteiro and two anonymous reviewers who improved the manuscript with their helpful suggestions.

Funding
PA is funded by the Max Planck Society. BE-A is funded by the European Union's Horizon 2020 research and innovation program under the Marie Skłodowska-Curie grant agreement No 654155. The funding supported the experimental design and execution of the work described in this study. The funding institution had no influence on the analysis and interpretation of the data.

Authors' contributions
PA and MSF designed the study. PA collected all data and prepared the manuscript and figures. PA and BE-A analyzed the data. All authors contributed to the interpretation of the results. All authors have read approved the manuscript.

Competing interests
The authors declare that they have no competing interests.

Author details
[1]Department of Human Evolution, Max Planck Institute for Evolutionary Anthropology, Leipzig, Germany. [2]Institut für Spezielle Zoologie und Evolutionsbiologie mit Phyletischem Museum, Friedrich-Schiller-Universität Jena, Jena, Germany. [3]Structure & Motion Lab, Department of Comparative Biomedical Sciences, Royal Veterinary College, Hatfield, UK.

References
1. Alexander R, Dimery NJ, Ker R. Elastic structures in the back and their role in galloping in some mammals. J Zool. 1985;207(4):467–82.
2. Bramble DM, Carrier DR. Running and breathing in mammals. Science. 1983;219(4582):251–6.
3. Fischer MS. Crouched posture and high fulcrum, a principle in the locomotion of small mammals: the example of the rock hyrax (*Procavia capensis*)(Mammalia: Hyracoidea). J Hum Evol. 1994;26(5-6):501–24.
4. Gambaryan PP, Hardin H. How mammals run: anatomical adaptations. New York: Wiley; 1974.

5. Hildebrand M. Motions of the running cheetah and horse. J Mammal. 1959;40(4):481–95.

6. Magne de la Croix P. The evolution of locomotion in mammals. J Mammal. 1936;17(1):51–4.

7. Asher R, Lin K, Kardjilov N, Hautier L. Variability and constraint in the mammalian vertebral column. J Evol Biol. 2011;24(5):1080–90.

8. Buchholtz EA: Flexibility and constraint: patterning the axial skeleton in mammals. In: From clone to bone: the synergy of morphological and molecular tools in Palaeobiology. Edited by Asher R, Müller J. Cambridge: Cambridge University Press; 2012: 230-256.

9. Galis F. Why do almost all mammals have seven cervical vertebrae? Developmental constraints, Hox genes, and cancer. J Exp Zool. 1999;285(1):19–26.

10. Galis F, Carrier DR, Van Alphen J, Van Der Mije SD, Van Dooren TJ, Metz JA, ten Broek CM. Fast running restricts evolutionary change of the vertebral column in mammals. Proc Natl Acad Sci. 2014;111(31):11401–6.

11. Müller J, Scheyer TM, Head JJ, Barrett PM, Werneburg I, Ericson PG, Pol D, Sánchez-Villagra MR. Homeotic effects, somitogenesis and the evolution of vertebral numbers in recent and fossil amniotes. Proc Natl Acad Sci. 2010;107(5):2118–23.

12. Narita Y, Kuratani S. Evolution of the vertebral formulae in mammals: a perspective on developmental constraints. J Exp Zool B Mol Dev Evol. 2005;304(2):91–106.

13. Sánchez-Villagra MR, Narita Y, Kuratani S. Thoracolumbar vertebral number: the first skeletal synapomorphy for afrotherian mammals. Syst Biodivers. 2007;5(1):1–7.

14. Todd TW. Numerical significance in the thoracicolumbar vertebrae of the mammalia. Anat Rec. 1922;24(5):260–86.

15. Jouffroy F. Evolution of the dorsal muscles of the spine in light of their adaptation to gravity effects. In: Berthoz A, Graf W, Vidal PP, editors. The head-neck sensory motor system. Oxford: Oxford University Press; 1992. p. 22–35.

16. Slijper EJ. Comparative biological-anatomical investigations of the vertebral column and spinal musculature. K Ned Akad Wet, Verh (Tweede Sectie). 1946;42:1–128.

17. Buchholtz EA, Bailin HG, Laves SA, Yang JT, Chan MY, Drozd LE. Fixed cervical count and the origin of the mammalian diaphragm. Evol Dev. 2012;14(5):399–411.

18. Burke AC, Nowicki J. A new view of patterning domains in the vertebrate mesoderm. Dev Cell. 2003;4(2):159–65.

19. Hirasawa T, Fujimoto S, Kuratani S. Expansion of the neck reconstituted the shoulder–diaphragm in amniote evolution. Develop Growth Differ. 2016;58(1):143–53.

20. Hirasawa T, Kuratani S. A new scenario of the evolutionary derivation of the mammalian diaphragm from shoulder muscles. J Anat. 2013;222(5):504–17.

21. Carrier DR, Deban SM, Fischbein T. Locomotor function of forelimb protractor and retractor muscles of dogs: evidence of strut-like behavior at the shoulder. J Exp Biol. 2008;211(1):150–62.

22. Deban SM, Schilling N, Carrier DR. Activity of extrinsic limb muscles in dogs at walk, trot and gallop. J Exp Biol. 2012;215(2):287–300.

23. English AWM. An electromyographic analysis of forelimb muscles during overground stepping in the cat. J Exp Biol. 1978;76(1):105–22.

24. Payne R, Veenman P, Wilson A. The role of the extrinsic thoracic limb muscles in equine locomotion. J Anat. 2004;205(6):479–90.

25. Graf W, De Waele C, Vidal P. Functional anatomy of the head-neck movement system of quadrupedal and bipedal mammals. J Anat. 1995;186(Pt 1):55.

26. Graf W, de Waele C, Vidal P, Wang D, Evinger C. The orientation of the cervical vertebral column in unrestrained awake animals. II. Movement strategies. Brain Behav Evol. 1995;45(4):209–31.

27. Hart BL, Hart LA, Mooring MS, Olubayo R. Biological basis of grooming behaviour in antelope: the body-size, vigilance and habitat principles. Anim Behav. 1992;44(4):615–31.

28. Mooring MS, Blumstein DT, Stoner CJ. The evolution of parasite-defence grooming in ungulates. Biol J Linn Soc. 2004;81(1):17–37.

29. Vidal PP, Graf W, Berthoz A. The orientation of the cervical vertebral column in unrestrained awake animals. Exp Brain Res. 1986;61(3):549–59.

30. Zsoldos RR, Licka TF. The equine neck and its function during movement and locomotion. Zoology. 2015;118(5):364–76.

31. Arnold P, Amson E, Fischer MS. Differential scaling patterns of vertebrae and the evolution of neck length in mammals. Evolution. 2017;71(6):1587–1599.

32. Arnold P, Forterre F, Lang J, Fischer MS. Morphological disparity, conservatism, and integration in the canine lower cervical spine: insights

into mammalian neck function and regionalization. Mamm Biology-Zeitschrift für Säugetierkunde. 2016;81(2):153–62.

33. Buchholtz EA, Stepien CC. Anatomical transformation in mammals: developmental origin of aberrant cervical anatomy in tree sloths. Evol Dev. 2009;11(1):69–79.

34. Buchholtz EA, Wayrynen KL, Lin IW. Breaking constraint: axial patterning in Trichechus (Mammalia: Sirenia). Evol Dev. 2014;16(6):382–93.

35. Johnson D, McAndrew T, Oguz Ö. Shape differences in the cervical and upper thoracic vertebrae in rats (Rattus norvegicus) and bats (Pteropus poiocephalus): can we see shape patterns derived from position in column and species membership? J Anat. 1999;194(2):249–53.

36. Bekele A. The comparative functional morphology of some head muscles of the rodents Tachyoryctes splendens and Rattus rattus. II. Cervical muscles. Mammalia. 1983;47(4):549–72.

37. Diogo R. The head and neck muscles of the Philippine colugo (Dermoptera: Cynocephalus volans), with a comparison to tree-shrews, primates, and other mammals. J Morphol. 2009;270(1):14–51.

38. Diogo R, Wood B. Soft-tissue anatomy of the primates: phylogenetic analyses based on the muscles of the head, neck, pectoral region and upper limb, with notes on the evolution of these muscles. J Anat. 2011;219(3):273–359.

39. Filan SL. Myology of the head and neck of the bandicoot (Marsupialia, Peramelemorphia). Aust J Zool. 1990;38(6):617–34.

40. Esteve-Altava B, Marugán-Lobón J, Botella H, Bastir M, Rasskin-Gutman D. Grist for Riedl's mill: a network model perspective on the integration and modularity of the human skull. J Exp Zool B Mol Dev Evol. 2013;320(8):489–500.

41. Rasskin-Gutman D, Esteve-Altava B. Connecting the dots: anatomical network analysis in morphological EvoDevo. Biol Theory. 2014;9(2):178–93.

42. Esteve-Altava B. In search of morphological modules: a systematic review. Biol Rev. 2017;92(3):1332–47.

43. Esteve-Altava B, Marugán-Lobón J, Botella H, Rasskin-Gutman D. Structural constraints in the evolution of the tetrapod skull complexity: Williston's law revisited using network models. Evol Biol. 2013;40(2):209–19.

44. Esteve-Altava B, Boughner JC, Diogo R, Villmoare BA, Rasskin-Gutman D. Anatomical network analysis shows decoupling of modular lability and complexity in the evolution of the primate skull. PLoS One. 2015;10(5):e0127653.

45. Diogo R, Esteve-Altava B, Smith C, Boughner JC, Rasskin-Gutman D. Anatomical network comparison of human upper and lower, newborn and adult, and normal and abnormal limbs, with notes on development, pathology and limb serial homology vs. homoplasy. PLoS One. 2015;10:e0140030.

46. Murphy AC, Muldoon SF, Baker D, Lastowka A, Bennett B, Yang M, Bassett DS. Structure, function, and control of the musculoskeletal network. In: arXiv preprint arXiv:161206336; 2016.

47. Dos Santos DA, Fratani J, Ponssa ML, Abdala V. Network architecture associated with the highly specialized hindlimb of frogs. PLoS One. 2017;12(5):e0177819.

48. Esteve-Altava B. Challenges in identifying and interpreting organizational modules in morphology. J Morphol. 2017;278(7):960–74.

49. Rasskin-Gutman D. Boundary constraints for the emergence of form. In: Müller GB, Newman SA, editors. Origination of Organismal form: beyond the gene in developmental and evolutionary biology. Camebridge: MIT Press; 2003. p. 305.

50. Rasskin-Gutman D, Buscalioni AD. Theoretical morphology of the Archosaur (Reptilia: Diapsida) pelvic girdle. Paleobiology. 2001;27(1):59–78.

51. Riedl R, Jefferies RPS. Order in living organisms: a systems analysis of evolution. New York: Wiley; 1978.

52. Schoch RR. Riedl's burden and the body plan: selection, constraint, and deep time. J Exp Zool B Mol Dev Evol. 2010;314(1):1–10.

53. Wagner GP, Laubichler MD. Rupert Riedl and the re-synthesis of evolutionary and developmental biology: body plans and evolvability. J Exp Zool B Mol Dev Evol. 2004;302(1):92–102.

54. Gregory WK. Polyisomerism and anisomerism in cranial and dental evolution among vertebrates. Proc Natl Acad Sci. 1934;20(1):1–9.

55. Eble GJ. Morphological modularity and macroevolution: conceptual and empirical aspects. In: Callebaut W, Rasskin-Gutman D, editors. Modularity: understanding the development and evolution of natural complex systems. Camebridge: The MIT Press; 2005. p. 221–38.

56. Goswami A, Smaers J, Soligo C, Polly P. The macroevolutionary consequences of phenotypic integration: from development to deep time. Philos Trans R Soc B: Biol Sci. 2014;369(1649):20130254.

57. Klingenberg CP. Morphological integration and developmental modularity. Annu Rev Ecol Evol Syst. 2008;39:115–32.

58. Wagner GP, Pavlicev M, Cheverud JM. The road to modularity. Nat Rev Genet. 2007;8(12):921–31.

59. Buchholtz EA. Crossing the frontier: a hypothesis for the origins of meristic constraint in mammalian axial patterning. Zoology. 2014;117(1):64–9.

60. Galis F, Metz JA. Anti-cancer selection as a source of developmental and evolutionary constraints. BioEssays. 2003;25(11):1035–9.

61. Galis F, Van Dooren TJ, Feuth JD, Metz JA, Witkam A, Ruinard S, Steigenga MJ, Wunaendts LC. Extreme selection in humans against homeotic transformations of cervical vertebrae. Evolution. 2006;60(12):2643–54.

62. Kummer B. Bauprinzipien des Säugerskeletes Stuttgart. Germany: Georg Thieme Verlag; 1959.

63. Kummer B: Biomechanik des Säugetierskeletts. In: Handbuch der Zoologie, Band 8 (6). Edited by Helmcke J-G, Legerken Hv, Starck D, vol. 24. Berlin: Walter de Gruyter & Co. Verlag; 1959.

64. Coues E. On the myology of the Ornithorhynchus. Proc Essex Inst. 1871;6:127–73.

65. Jouffroy F, Lessertisseur J, Saban R. Particularités musculaires des Monotrémes – musculature post-craniénne. In: Grassé P-P, editor. Traité de Zoologie Mamifères, Tome XVI, Fac III. Paris: Masson et Cie editeurs; 1971. p. 679–836.

66. Nishi S. Zur vergleichenden Anatomie der eigentlichen (genuinen) Rückenmuskeln. Gegenbaurs Morphol Jahrb. 1916;50:167–318.

67. Virchow H. Die tiefen Rückenmuskeln des Ornithorhynchus. Gegenbaurs Morphol Jahrb. 1929;60:481–559.

68. Barbour RA. The musculature and limb plexuses of Trichosurus vulpecula. Aust J Zool. 1963;11(4):488–610.

69. Coues E, Wyman J. On the osteology and myology of Didelphys virginiana. Mem Boston Soc Nat Hist. 1872;2:41–154.

70. Jouffroy FK. Musculature épisomatique. In: Grassé PP, editor. Traité de Zoologie Mamifères, Tome XVI, Fac II. Paris: Masson et Cie editeurs; 1968. p. 479–548.

71. Jüschke S. Untersuchungen zur funktionellen Anpassung der Rückenmuskulatur und der Wirbelsäule quadrupeder Affen und Känguruhs. Anat Embryol. 1972;137(1):47–85.

72. Osgood WH, Herrick CJ, Obenchain JB. A monographic study of the American marsupial Caenolestes, with a description of the brain of Caenolestes by C. Judson Herrick. Field Mus Nat Hist (Zool Ser). 1921;14:1–162.

73. Bianchi M. The thickness, shape and arrangement of elastic fibres within the nuchal ligament from various animal species. Anat Anz. 1988;169(1):53–66.

74. Dimery NJ, Alexander R, Deyst KA. Mechanics of the ligamentum nuchae of some artiodactyls. J Zool. 1985;206(3):341–51.

75. Endo H, Yamagiwa D, Fujisawa M, Kimura J, Kurohmaru M, Hayashi Y. Modified neck muscular system of the giraffe (Giraffa camelopardalis). Ann Anat-Anat Anz. 1997;179(5):481–5.

76. Gellman K, Bertram J. The equine nuchal ligament 1: structural and material properties. Vet Comp Orthop Traumatol. 2002;15(1):1–6.

77. Mobarak A, Fouad S. A study on Lig. Nuchae of the one-humped camel (Camelus dromedarius). Anat Histol Embryol. 1977;6(2):188–90.

78. Preuschoft H, Klein N. Torsion and bending in the neck and tail of sauropod dinosaurs and the function of cervical ribs: insights from functional morphology and biomechanics. PLoS One. 2013;8:e78574.

79. Galliari FC, Carlini AA, Sánchez-Villagra MR. Evolution of the axial skeleton in armadillos (Mammalia, Dasypodidae). Mamm Biology-Zeitschrift für Säugetierkunde. 2010;75(4):326–33.

80. Miller RA. Functional adaptations in the forelimb of the sloths. J Mammal. 1935;16(1):38–51.

81. VanBuren CS, Evans DC. Evolution and function of anterior cervical vertebral fusion in tetrapods. Biol Rev. 2017;92:608–26.

82. Hautier L, Weisbecker V, Sánchez-Villagra MR, Goswami A, Asher RJ. Skeletal development in sloths and the evolution of mammalian vertebral patterning. Proc Natl Acad Sci. 2010;107(44):18903–8.

83. Varela-Lasheras I, Bakker AJ, van der Mije SD, Metz JA, van Alphen J, Galis F. Breaking evolutionary and pleiotropic constraints in mammals: on sloths, manatees and homeotic mutations. EvoDevo. 2011;2:11.

84. Macalister A. The myology of the Cheiroptera. Philos Trans R Soc Lond. 1872;162:125–71.

85. Maisonneuve P. Traité de l'ostéologie et de la myologie du Vespertilio murinus, précédé d'un exposé de la classification des chéiroptèi et de considérations sur les moeurs de ces animaux. Paris: O. Doin; 1878.

86. Mori M. Muskulatur des Pteropus edulis. Okajimas Folia Anat Jpn. 1960;36(3-4):253–307.

87. Fenton MB, Crerar LM. Cervical vertebrae in relation to roosting posture in bats. J Mammal. 1984;65(3):395–403.

88. Howell A. Aquatic mammals. Springfield: Charles C Thomas; 1930.

89. Schulte HW, Smith MDF. The external characters, skeletal muscles, and peripheral nerves of Kogia breviceps (Blainville). Bull Am Mus Nat Hist. 1918;38:7–72.

90. Randau M, Cuff AR, Hutchinson JR, Pierce SE, Goswami A. Regional differentiation of felid vertebral column evolution: a study of 3D shape trajectories. Organ Divers Evol. 2017;17(1):305–19.

91. Randau M, Goswami A, Hutchinson JR, Cuff AR, Pierce SE. Cryptic complexity in felid vertebral evolution: shape differentiation and allometry of the axial skeleton. Zool J Linnean Soc. 2016;178:183–202.

92. Evans FG. The morphology and functional evolution of the atlas-axis complex from fish to mammals. Ann N Y Acad Sci. 1939;39(1):29–104.

93. Jenkins FA. The evolution and development of the dens of the mammalian axis. Anat Rec. 1969;164(2):173–84.

94. Bogduk N, Mercer S. Biomechanics of the cervical spine. I: normal kinematics. Clin Biomech. 2000;15(9):633–48.

95. Matsuoka T, Ahlberg PE, Kessaris N, Iannarelli P, Dennehy U, Richardson WD, McMahon AP, Koentges G. Neural crest origins of the neck and shoulder. Nature. 2005;436(7049):347–55.

96. Ericsson R, Knight R, Johanson Z. Evolution and development of the vertebrate neck. J Anat. 2013;222(1):67–78.

97. Nyakatura JA. The convergent evolution of suspensory posture and locomotion in tree sloths. J Mamm Evol. 2012;19(3):225–34.

98. Humphry G. The myology of the limbs of the Unau, the Ai, the two-toed anteater, and the pangolin. J Anat Physiol. 1869;3:2–78.

99. Macalister A. On the myology of Bradypus Tridactylus; with remarks on the general mucular anatomy of the Edentata. Ann Mag Nat Hist. 1869;4(19):51–67.

100. Mackintosh H. On the myology of the genus Bradypus. Proc R Ir Acad. 1870;1:517–29.

101. Lucae JCG. Statik und Mechanik der Quadrupeden an dem Skelett und den Muskeln des Lemur und eines Choloepus. Abhandlungen der Senckenbergischen Naturforschenden Gesellschaft Frankfurt. 1884;13:1–92.

102. Mackintosh H. On the muscular anatomy of Choloepus didactylus. Proc R Ir Acad. 1875;1:66–67^69.

103. Gross MK, Moran-Rivard L, Velasquez T, Nakatsu MN, Jagla K, Goulding M. Lbx1 is required for muscle precursor migration along a lateral pathway into the limb. Development. 2000;127(2):413–24.

104. Howell AB. Morphogenesis of the shoulder architecture. Part VI. Therian Mammalia. Q Rev Biol. 1937;12(4):440–63.

105. Miller RA. Comparative studies upon the morphology and distribution of the brachial plexus. Am J Anat. 1934;54(1):143–75.

106. Dunbar DC, Macpherson JM, Simmons RW, Zarcades A. Stabilization and mobility of the head, neck and trunk in horses during overground locomotion: comparisons with humans and other primates. J Exp Biol. 2008;211(24):3889–907.

107. Hirasaki E, Kumakura H. Head movements during locomotion in a gibbon and Japanese macaques. Neuroreport. 2004;15(4):643–7.

108. Loscher DM, Meyer F, Kracht K, Nyakatura JA. Timing of head movements is consistent with energy minimization in walking ungulates. Proc R Soc B. 2016;283:20161908.

109. Runciman RJ, Richmond FJ. Shoulder and forelimb orientations and loading in sitting cats: implications for head and shoulder movement. J Biomech. 1997;30(9):911–9.

110. Weishaupt M, Wiestner T, Peinen K, Waldern N, Roepstorff L, Weeren R, Meyer H, Johnston C. Effect of head and neck position on vertical ground reaction forces and interlimb coordination in the dressage horse ridden at walk and trot on a treadmill. Equine Vet J. 2006;38(S36):387–92.

111. Zsoldos R, Kotschwar A, Kotschwar A, Groesel M, Licka T, Peham C. Electromyography activity of the equine splenius muscle and neck kinematics during walk and trot on the treadmill. Equine Vet J. 2010;42(s38):455–61.

112. Zubair HN, Beloozerova IN, Sun H, Marlinski V. Head movement during walking in the cat. Neuroscience. 2016;332:101–20.

113. Eaton TH. Modifications of the shoulder girdle related to reach and stride in mammals. J Morphol. 1944;75(1):167–71.

114. Jenkins FA. The movement of the shoulder in claviculate and aclaviculate mammals. J Morphol. 1974;144(1):71–83.

115. Schmidt M, Voges D, Fischer MS. Shoulder movements during quadrupedal locomotion in arboreal primates. Z Morphol Anthropol. 2002;83:235–42.

116. Harmon LJ, Schulte JA, Larson A, Losos JB. Tempo and mode of evolutionary radiation in iguanian lizards. Science. 2003;301(5635):961–4.

117. Janis C. Tertiary mammal evolution in the context of changing climates, vegetation, and tectonic events. Annu Rev Ecol Syst. 1993;24(1):467–500.

118. Slater GJ. Phylogenetic evidence for a shift in the mode of mammalian body size evolution at the cretaceous-Palaeogene boundary. Methods Ecol Evol. 2013;4(8):734–44.

119. Timetree of Life Database [http://timetreebeta.igem.temple.edu/]. Accessed 22.11.2016.
120. Hedges SB, Marin J, Suleski M, Paymer M, Kumar S. Tree of life reveals clock-like speciation and diversification. Mol Biol Evol. 2015;32:835–45.
121. Csardi G, Nepusz T. The igraph software package for complex network research. Int J Complex Syst. 2006;1695(5):1–9.
122. R Core Team. R: a language and environment for statistical computing. Vienna, Austria: R Foundation for Statistical Computing; 2016.
123. Network cartography - Netcarto [http://seeslab.info/downloads/network-cartography-netcarto/]. Accessed 02.10.2016.
124. Guimera R, Amaral LAN. Functional cartography of complex metabolic networks. Nature. 2005;433(7028):895–900.
125. Adams DC. A generalized K statistic for estimating phylogenetic signal from shape and other high-dimensional multivariate data. Syst Biol. 2014;63(5):685–97.
126. Goolsby EW. Phylogenetic comparative methods for evaluating the evolutionary history of function-valued traits. Syst Biol. 2015;64(4):568–78.
127. Abouheif E. A method for testing the assumption of phylogenetic independence in comparative data. Evol Ecol Res. 1999;1(8):895–909.
128. Blomberg SP, Garland T Jr, Ives AR, Crespi B. Testing for phylogenetic signal in comparative data: behavioral traits are more labile. Evolution. 2003;57(4):717–45.
129. Revell LJ. Phytools: an R package for phylogenetic comparative biology (and other things). Methods Ecol Evol. 2012;3(2):217–23.
130. Clavel J, Escarguel G. Merceron G: mvMORPH: an R package for fitting evolutionary models to morphometric data. Methods Ecol Evol. 2015;6(11):1311–9.
131. Marwick B, Krishnamoorthy K: cvequality: Tests for the Equality of Coefficients of Variation from Multiple Groups. 2016. R package version 0.1.1. https://CRAN.R-project.org/package=cvequality.
132. Orme D, Freckleton R, Thomas G, Petzoldt T, Fritz S, Isaac N, Pearse W: caper: Comparative Analyses of Phylogenetics and Evolution in R. 2013. R package version 0.5.2. https://CRAN.R-project.org/package=caper.
133. Adams DC, Collyer ML, Kaliontzopoulou A, Sherratt E: Geomorph: Software for geometric morphometric analyses. 2013. R package version 3.0.5. https://CRAN.R-project.org/package=geomorph.
134. Newman ME, Girvan M. Finding and evaluating community structure in networks. Phys Rev E. 2004;69(2):026113.
135. Esteve-Altava B, Marugán-Lobón J, Botella H, Rasskin-Gutman D. Network models in anatomical systems. J Anthropol Sci. 2011;89:1–10.
136. Harmon LJ, Weir JT, Brock CD, Glor RE, Challenger W. GEIGER: investigating evolutionary radiations. Bioinformatics. 2008;24(1):129–31.
137. Slater GJ, Price SA, Santini F, Alfaro ME. Diversity versus disparity and the radiation of modern cetaceans. Proc R Soc Lond B Biol Sci. 2010;277(1697):3097–104.
138. Ercoli MD, Álvarez A, Busker F, Morales MM, Julik E, Smith HF, Adrian B, Barton M, Bhagavatula K, Poole M. Myology of the head, neck, and thoracic region of the lesser Grison (Galictis cuja) in comparison with the red panda (Ailurus fulgens) and other carnivorans: phylogenetic and functional implications. J Mamm Evol. 2017;24(3):289–322.
139. Ercoli MD, Álvarez A, Stefanini MI, Busker F, Morales MM. Muscular anatomy of the forelimbs of the lesser grison (Galictis cuja), and a functional and phylogenetic overview of Mustelidae and other Caniformia. J Mamm Evol. 2015;22(1):57–91.

13

Improved mitochondrial amino acid substitution models for metazoan evolutionary studies

Vinh Sy Le[1*†], Cuong Cao Dang[1] and Quang Si Le[2*†]

Abstract

Background: Amino acid substitution models play an essential role in inferring phylogenies from mitochondrial protein data. However, only few empirical models have been estimated from restricted mitochondrial protein data of a hundred species. The existing models are unlikely to represent appropriately the amino acid substitutions from hundred thousands metazoan mitochondrial protein sequences.

Results: We selected 125,935 mitochondrial protein sequences from 34,448 species in the metazoan kingdom to estimate new amino acid substitution models targeting metazoa, vertebrates and invertebrate groups. The new models help to find significantly better likelihood phylogenies in comparison with the existing models. We noted remarkable distances from phylogenies with the existing models to the maximum likelihood phylogenies that indicate a considerable number of incorrect bipartitions in phylogenies with the existing models. Finally, we used the new models and mitochondrial protein data to certify that Testudines, Aves, and Crocodylia form one separated clade within amniotes.

Conclusions: We introduced new mitochondrial amino acid substitution models for metazoan mitochondrial proteins. The new models outperform the existing models in inferring phylogenies from metazoan mitochondrial protein data. We strongly recommend researchers to use the new models in analysing metazoan mitochondrial protein data.

Keywords: Mitochondrial amino acid substitution models, Metazoa, Vertebrates, Invertebrates

Background

An amino acid substitution model (model for short) includes a 20 × 20 matrix and an amino acid frequency vector. The matrix represents the instantaneous substitution rates among amino acids while the amino acid frequency vector serves as the equilibrium frequencies of the 20 amino acids. The substitution rates characterise the biological, chemical, and physical correlations among amino acids [1]. Amino acid substitution models are the key to infer phylogenies from protein data. Distance-based methods use amino acid substitution models to estimate pairwise distances among sequences, while maximum likelihood or Bayesian methods require amino acid substitution models to calculate the likelihood of data [2].

Estimating amino acid substitution models is much more challenging than estimating nucleotide substitution models due to a large number of parameters to be optimised. For example, the general time reversible model for nucleotides contains 8 parameters in comparing to 208 parameters for models of amino acid substitutions. Thus, amino acid substitution models are typically estimated from large datasets.

It is well established that models of different species or protein types would be diverse [3–5]. For example, Dang et al. showed that the model for influenza proteins is highly different from general models [3]. Note that protein structures also contribute to amino acid evolution patterns [6, 7].

Mitochondria (mt) are energy factories and play an essential role in supplying cellular energy [8]. The mitochondrial genome encodes 13 proteins that are widely used to infer phylogenies [7, 9–12]. Few groups have estimated empirical models from mt protein data (mt models). Adachi and Hasegawa were the first to estimate an mt model, named

---
* Correspondence: vinhls@vnu.edu.vn; lsquang@gmail.com
†Equal contributors
[1]University of Engineering and Technology, Vietnam National University Hanoi, Hanoi, Vietnam
[2]School of Pharmacy and Biomedical Sciences, University of Portsmouth, Winston Churchill Avenue Portsmouth, Portsmouth, PO1 2UP, UK

mtREV, from 20 complete vertebrate sequences [13]. They argued that the difference between the universal code and the mitochondrial code might be partially responsible to the difference between amino acid substitution patterns from nuclear and mitochondrial-encoded proteins. Abascal et al. built another mt model, mtArt, from 36 arthropod species to analyse the data of invertebrate species [14]. Note that although invertebrates are paraphyletic, the term *invertebrates* is widely used as a convenient shorthand in communication [5, 13–15]. Neither mtREV nor mtArt is appropriate for datasets consisting of diverse metazoan lineages, as they were specifically estimated from either vertebrate or invertebrate protein data. Rota-Stabelli et al. solved the problem by introducing an mt model (mtZoa) estimated from 117 general metazoan species [5]. They recommended to use mtZoa for analysing datasets from diverse or basal metazoan groups. The existing mt models (mtREV, mtArt, and mtZoa) outperform general models (e.g., LG [16] and WAG [17]) in inferring phylogenies from mt protein data, even though they were estimated from small datasets.

The main issue of the existing mt models comes from their small training datasets of at most 117 species. This was due to the limited mt protein data available and the capability of estimation methods at the time these studies were carried out. Consequently, the models might over-fit to training data due to a large number of free parameters of the amino acid substitution model (precisely 208 free parameters). In other words, the existing models may fit too well to training sequences but poorly represent others. Above all, the existing mt models cannot appropriately represent nearly a million available mt protein sequences of more than 34 thousands metazoan species, as they were estimated from only a limited number of species.

In this paper, we introduce new mt models for metazoan and vertebrates. Although invertebrates are not monophyletic, their mitochondria have the same genetic codes. The genetic codes of invertebrate mitochondria are different from that of vertebrate mitochondria. The difference might result in different amino acid substitution patterns from invertebrate and vertebrate mitochondrial-encoded proteins [5, 13, 14]. Therefore, we also introduce a new mt model for invertebrates. To this end, we created three datasets from 125,935 mt sequences of 13 proteins from 34,448 metazoan species. Then, we implemented the fast and accurate method, FastMG [18], to estimate three new mt models from these three datasets.

We validated the new models by assessing the likelihood of phylogenies with the new models for both training and testing data. We summarised the experimental results to show the advantage of the new models in inferring the maximum likelihood phylogenies (called the best phylogenies) in comparison to existing mt models. Experimental results revealed remarkable distances from the phylogenies with the existing models to the best phylogenies. We proved that

the remarkable distances imply a considerable number of incorrect bipartitions in the phylogenies with the existing models. Although we could not evaluate the topological quality of phylogenies with the new models, as they were often the best phylogenies, we would expect significant topological improvement due to their large likelihood advantage over the phylogenies with the existing models.

Finally, we applied the new models to tackle a debated question about the location of Testudines within amniotes. We used IQ-TREE with the new models to build the maximum likelihood phylogeny of 993 amniotes from their mt protein data. We learned from the phylogeny that Testudines, Aves, and Crocodylia form one separated clade within amniotes.

Results and discussion
Data preparation
We downloaded all mt protein sequences of 34,448 species in the metazoan kingdom from NCBI (National Center for Biotechnology Information, 2016) and then mapped them onto 13 mt proteins. We selected one sequence per species to eliminate bias on intensively studied species (e.g., 30,000 human sequences). As the result, we obtained 125,935 sequences to form three datasets for metazoan, vertebrate, and invertebrate categories. We kept all sites, as removing sites with missing data would lead to worse phylogenies [19]. We divided each dataset into a training dataset and a testing dataset containing 90% and 10% of sequences, respectively.

We implemented the fast and accurate method, FastMG [18], to estimate three new mt models, *mtMet, mtVer*, and *mtInv* from metazoan, vertebrate, and invertebrate training datasets, respectively. As FastMG is infeasible for alignments of several thousands sequences, we split alignments based on the taxonomy tree to obtain sub-alignments of at most one thousand sequences. Then we divided these sub-alignments into smaller sub-alignments of at most 128 sequences using the tree-based splitting algorithm in FastMG. In addition, we removed branches with lengths equal to zero or larger than two in order to eliminate data noise. The data are summarised in Tables 1 and 2. Note that the FastMG algorithm starts from an initial model and iteratively optimises the model until the likelihood improvement is insignificant.

The fit of new models to training datasets
We measured the fit of new models to the training datasets. Table 3 shows significant likelihood improvements of the new models over the initial model, mtZoa, for metazoan, vertebrate, and invertebrate training datasets. The first iteration contributed about 99% of the total likelihood improvement. The optimisation process was terminated after the third iteration, as the gain from the third iteration was insignificant.

Table 1 The number of sequences of 13 mt proteins for metazoan, vertebrate, and invertebrate datasets

Protein	Metazoan		Vertebrate		Invertebrate	
	Training	Testing	Training	Testing	Training	Testing
ATP6	8493	938	5752	636	2741	302
ATP8	8412	928	5726	632	2686	296
COX1	7090	784	4633	512	2457	272
COX2	9363	1033	5023	555	4340	478
COX3	6867	759	4208	466	2659	293
CYTB	12,894	1422	10,326	1139	2569	282
ND1	8280	912	5355	590	2926	321
ND2	14,541	1597	11,885	1306	2655	292
ND3	9074	997	6262	687	2812	310
ND4	7191	793	4567	503	2625	289
ND4L	7274	803	4498	496	2776	307
ND5	6975	769	4409	487	2566	282
ND6	6977	769	4360	480	2617	289
Total	125,935		85,493		40,442	

Each dataset is divided into a training dataset and a testing dataset with a 9 to 1 ratio

Table 3 Total log-likelihood of the target function (Eq. 1) on training datasets

	Metazoan	Vertebrate	Invertebrate
mtZoa (initial model)	−1.23427e + 07	−5.50036e + 06	−6.85299e + 06
First iteration	−1.21987e + 07	−5.32959e + 06	−6.77590e + 06
Second iteration	−1.21987e + 07	−5.32671e + 06	−6.77536e + 06
Third iteration (final model)	−1.21987e + 07	−5.32671e + 06	−6.77536e + 06
AIC/site	0.795	1.456	1.232
BIC/site	0.790	1.430	1.220

AIC/site (BIC/site) is the AIC (BIC) improvement per site of the final model in comparison to the initial model mtZoa
There is no likelihood improvement after two iterations

The better Akaike and Bayesian information criterion scores [20, 21] of the new models in comparison to the initial model, mtZoa, confirm the better fit of the new models to the training data. The scores guarantee that the likelihood gain of the new models comes from their genuine fit and overwhelm the penalty of free parameters.

Model analysis

Figures 1 and 2 show significant differences in exchangeability patterns between amino acids among the four models: mtZoa, mtMet, mtVer, and mtInv (see). For example, the exchangeability rate between *methionine* and *glutamine* in mtMet is about 10 times greater than that in mtZoa (0.155 vs 0.0016). The exchangeability rate between these two amino acids in mtVer is a third of that in mtInv (0.075 vs 0.228). Figure 3 shows a clear variety of amino acid frequencies among the four models, especially between mtVer and mtInv). For instance, the frequency of *Threonine* in mtVer is about three times as much as that in mtInv (0.146 vs 0.0428).

The low pairwise correlations of exchangeability rate matrices (or frequency vectors) of the mt models confirm high varieties among the models (Table 4). The mtInv and

mtVer models are the most diverse pair with the smallest correlation of exchangeability rates (0.775). Note that the correlation between the two popular general models LG and WAG is 0.912. As expected, mtMet is the closest model to mtZoa in terms of exchangeability rates, with a 0.929 correlation score, as both were trained from the metazoan data. Interestingly, mtMet is closer to mtInv than mtVer, although the metazoan training dataset consists of less invertebrate data than vertebrate data. The results indicate diverse evolutionary processes among lineages in the metazoan kingdom.

We observed remarkably low correlations between mt models and general models (e.g., the 0.46 correlation score between mtInv and LG). The low correlations imply considerably diverse evolutionary patterns between mt proteins and general proteins. Thus, general models are not an appropriate choice in inferring phylogenies from mt protein data.

Likelihood improvement on testing alignments

We assessed the performance of the new mt models (mtMet, mtVer, and mtInv) and the existing mt models (mtZoa, mtREV, and mtArt) on building maximum likelihood phylogenies. To this end, we used IQ-TREE [22] to build phylogenies with different models on the metazoan, vertebrate, and invertebrate testing datasets. For each testing alignment D and a model M, we optimised parameters of the rate heterogeneity model (i.e., proportion of invariable sites and shape of Gamma distribution with 4 categories), but fixed the exchangeability rates and base frequencies of the model M.

Table 2 The number of sequences, alignments, and sites in metazoan, vertebrate, and invertebrate training and testing datasets

	Training			Testing		
	#Sequences	#Alignments	#Sites	#Sequences	#Alignments	#Sites
Metazoan	103,637	1155	362,062	12,701	139	47,477
Vertebrate	68,536	772	238,429	8878	95	29,999
Invertebrate	35,089	390	125,849	3908	48	17,792

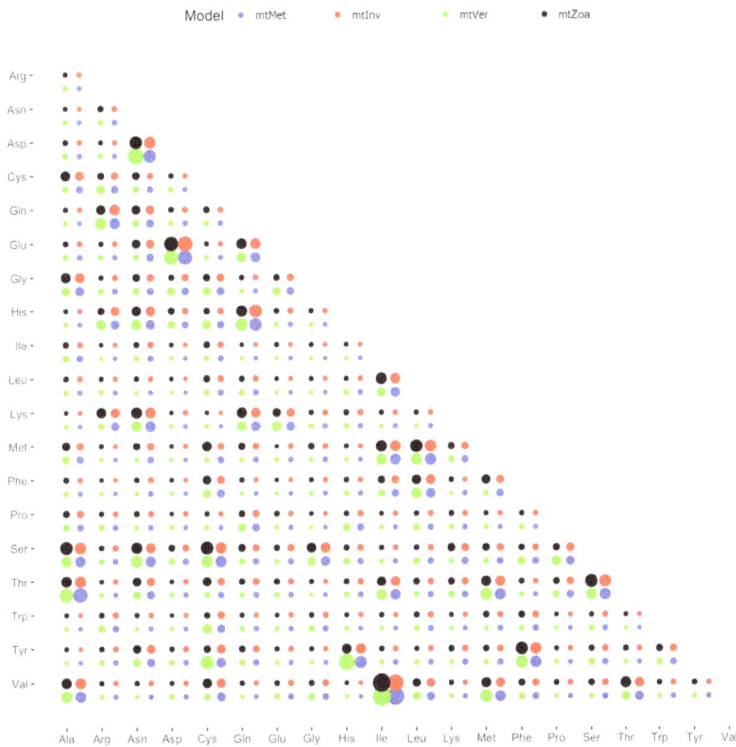

Fig. 1 Amino acid exchangeability rates of the mtMet, mtInv, mtVer, and mtZoa models. There are some considerable difference between mtZoa and the new models

It is clear from Fig. 4 that the new models outperform the existing models for all three testing datasets. They are the best-fit models for their corresponding testing data (e.g., mtMet is the best-fit model for the metazoan testing data). Note that the second-best fit model for a certain testing dataset is the existing model estimated from the training data of the same category as the testing dataset (e.g., mtZoa is the second-best fit model for the metazoan testing data). The log-likelihoods of the phylogenies with the new models are significantly higher than those of the existing models. For example, the likelihood advantage of mtMet to the second-best model, mtZoa, on the metazoan testing data is about 0.41 log points per site (or 1640 log points for a concatenated alignment of 4000 sites). This improvement is about four times as much as the improvements of LG from WAG [16]. In short, the three new models outperform the three existing models in their corresponding categories.

We analysed the performance of the mt models at the individual alignment level. We used the approximately unbiased SH test [23] to compute confidence levels for phylogenies with the models. Given a testing alignment D, we estimated the maximum likelihood tree T_i according to model M_i where M_i is one of the six mt models. We computed the site-wise log likelihoods for every $(T_i, M_i|D)$, and subsequently used the CONSEL program [24] for assessing

their confidence levels. The approximately unbiased SH test helps us to confirm whether the likelihood improvement comes from models and trees or from artefacts of numerical analyses in IQ-TREE. Figure 5 confirms the advantage of the new models in inferring phylogenies for all three testing datasets. The new models demonstrate a better fit for almost all testing alignments in comparison with the existing models (e.g., 85 out of 95 vertebrate alignments). The approximately unbiased SH test also confirms the superiority of the new models with high confidence levels (e.g., 67 out of 95 vertebrate alignments at the 0.9 confidence level). The existing models are still the best-fit models for some alignments, but only significantly better than the new models in a few cases. For example, the existing models are the best-fit models for 10 out of 95 vertebrate alignments, but only significantly better for one alignment at the 0.9 confidence level.

More specifically, we examined the performance of the six mt models individually (see Fig. 6). We highlight some following findings:

- The best-fit model for a certain testing alignment is typically the one estimated from the training data of the same category as the testing alignment. For example, 85 out of 95 vertebrate testing alignments

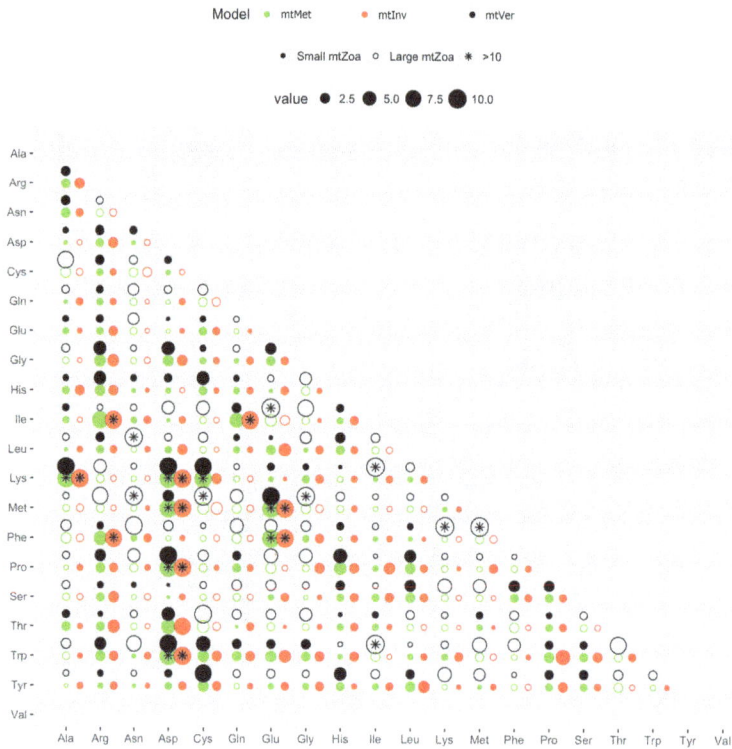

Fig. 2 The ratio of exchangeability rates between mtZoa and mtMet/mtVer/mtInv models. The size of one circle represents the exchangeability rate between mtZoa and other models. The solid (unfilled) circles represent exchangeability rates where mtZoa is smaller (bigger) than the three models. For visualization, the large ratios are trimmed at 10 and marked with '*'

fit best with mtVer, which was estimated from the vertebrate training data.

- The mtVer model outperforms the mtInv model for all vertebrate testing alignments and vice versa. This is explainable, as the two models are highly diverse. The

mtMet model is usually the best-fit model for metazoan testing alignments. However, some metazoan testing alignments are biased on vertebrate or invertebrate species, therefore, mtVer or mtInv might fit better than mtMet for those diverse metazoan alignments.

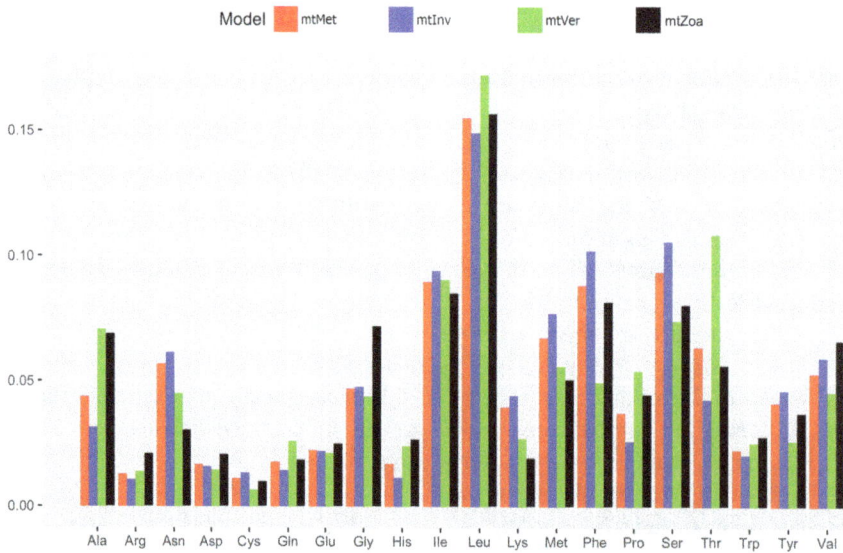

Fig. 3 Amino acid frequencies of the mtMet, mtInv, mtVer, and mtZoa models. There are some considerable difference between mtZoa and the new models

Table 4 Correlations between four models: mtMet, mtInv, mtVer, and mtZoa

	mtMet	mtInv	mtVer	mtZoa	LG	WAG
mtMet		0.976	0.89	0.929	0.527	0.439
mtInv	0.959		0.775	0.875	0.457	0.363
mtVer	0.94	0.866		0.893	0.591	0.529
mtZoa	0.92	0.956	0.829		0.619	0.587
LG	0.837	0.887	0.787	0.894		0.912
WAG	0.825	0.878	0.778	0.85	0.961	

The values in the top triangle represent the correlations between exchangeability matrices, while values in the low triangle are the correlations between frequency vectors

Finally, we compared the performance of new mt models to LG4X, C60 (site-heterogeneous models) [25] and PHAT (a transmembrane-specific amino acid substitution model) models [26]. Table 5 shows that the new mt models outperformed LG4X, C60 and PHAT models in terms of AIC and BIC.

Phylogeny topology differentiation on testing alignments
We investigated the topological quality of phylogenies with the six mt models by measuring their topological distances from the best phylogenies. Specifically, we used the RobinsonFoulds (RF) metric to measure the distance

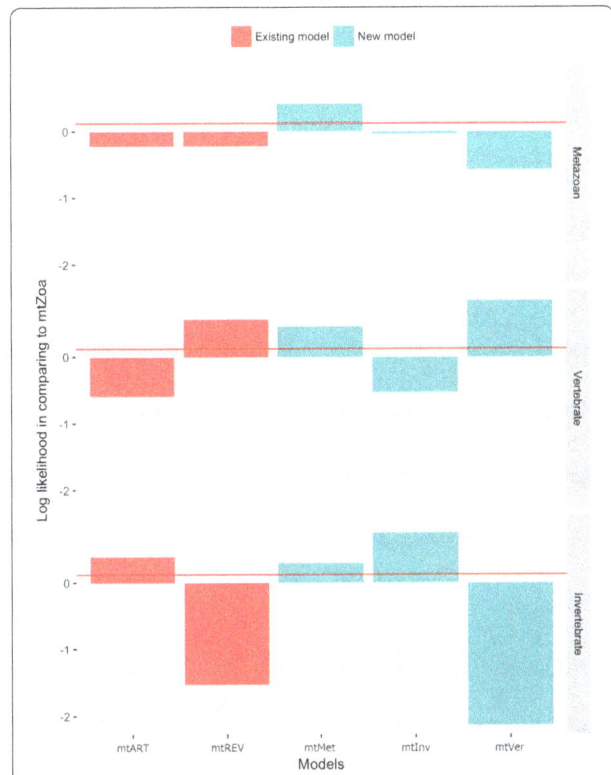

Fig. 4 Difference per site between log-likelihood of phylogenies with mtZoa and that with the existing models (mtREV and mtArt), and the new models (mtMet, mtVer, and mtVer). The red line represents the improvement of LG from WAG

between two phylogenies, as it represents the number of unique bipartitions in two phylogenies [27]. We learn from Lemma 1 that the lower-bound number of incorrect bipartitions in a phylogeny can be approximated as a quarter of its RF distance from the best phylogeny.

Lemma 1. Given two binary unrooted trees T and T' inferred from the same alignment of n taxa. The number of incorrect bipartitions in the worse likelihood phylogeny is at least a quarter of the RF distance between T and T'.

Proof: Let T_0 be the true binary unrooted tree. It is true that T, T', and T_0 have the same number of bipartitions, $2n - 3$ [28].

Let p be the number of shared bipartions in both T and T'. Let x and y be the number of unique bipartitions in T and T', respectively. As $x = (2n - 3) - p$ and $y = (2n - 3) - p$, x must be equal to y.

The RF distance between T and T' is $x + y$ or $2x$.

Let S be the set of all bipartitions in T and T', and S consists of $(2n - 3) + x$ bipartitions. Since the true tree T_0 has $(2n - 3)$ bipartitions, S must consist of at least x (half of the RF distance) incorrect bipartitions.

Let T be the worse likelihood phylogeny. Then, T should include at least half of the incorrect bipartitions ($x/2$) as T is considered the worse phylogeny. In other words, T includes at least a quarter of RF distance between T and T'. Figure 7 illustrates an example with five taxa.

Table 6 discloses remarkable topological distances from the phylogenies with the three existing models to the best phylogenies. The distances imply a considerable number of incorrect bipartitions in the phylogenies. For example, the phylogenies with mtZoa for metazoan testing alignments contain at least 6.37% incorrect bipartitions (i.e., a quarter of their normalised RF distance from the best phylogenies, 0.255). The results reconfirm the essential role of model selections in inferring phylogenies as a poor model selection (i.e., model and testing data coming from different categories) would lead to low quality phylogenies. The lower-bound numbers of incorrect bipartitions of phylogenies with the new models are indeterminable as they are often the best phylogenies. However, the significant likelihood improvement would expectedly lead to better phylogenies with fewer incorrect bipartitions.

We also applied the approximately unbiased SH test to examine the tree topologies under the best-fit models. Given a testing alignment D and its best-fit model M_b, we fixed tree topologies, but reoptimised other parameters (i.e., branch lengths, parameters of rate heterogeneity model) under the best-fit model M_b. Then we used the CONSEL program for assessing their confidence

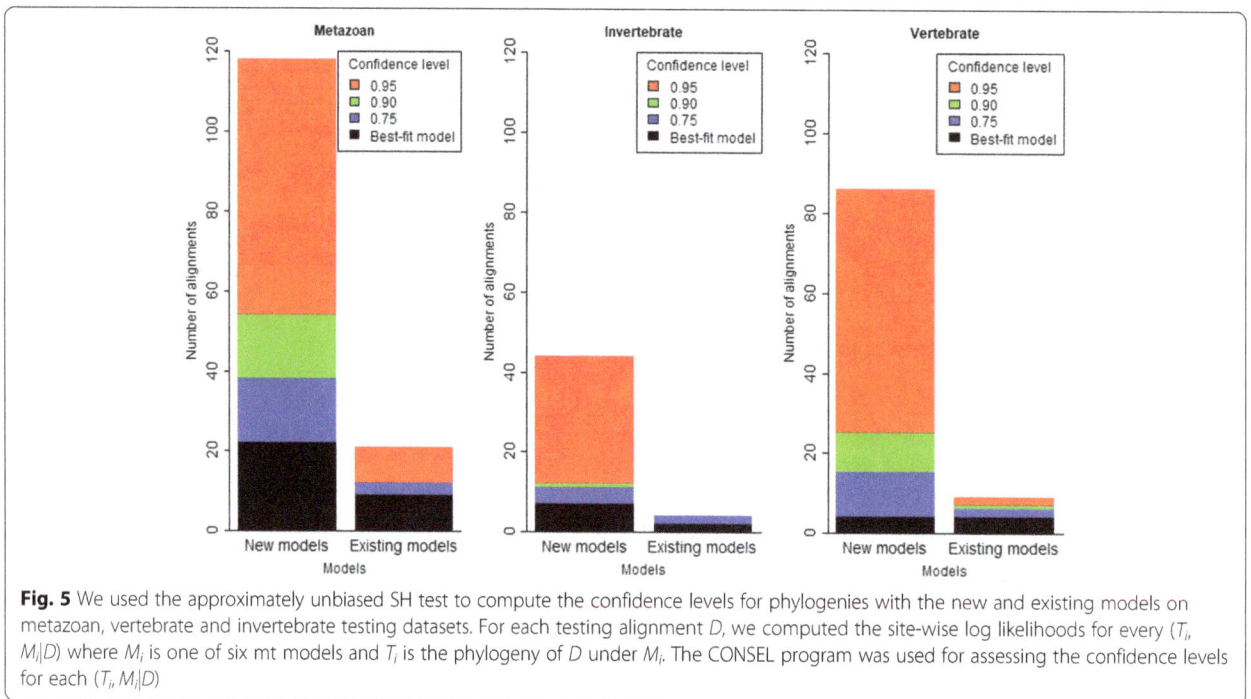

Fig. 5 We used the approximately unbiased SH test to compute the confidence levels for phylogenies with the new and existing models on metazoan, vertebrate and invertebrate testing datasets. For each testing alignment D, we computed the site-wise log likelihoods for every (T_i, $M_i|D$) where M_i is one of six mt models and T_i is the phylogeny of D under M_i. The CONSEL program was used for assessing the confidence levels for each (T_i, $M_i|D$)

levels. The test shows that the tree topologies built with the new models are better than that with the existing models in term of likelihood but with lower confidence (Fig. 8). The significant drop of confidence levels reveals that a large proportion of likelihood gain is due to the new models other than tree topologies.

Location of Testudines within amniotes

We applied the new models to tackle a question about the phylogenetic position of Testudines within amniotes. The question has a long history of debate with at least four hypothesises [29]. To this end, we built a concatenated alignment of 13 proteins for 993 amniotes and used IQ-TREE with all mt, LG4X, and C10 models to infer the best phylogeny, named T_a(Fig. 9). As expected, mtVer resulted in a huge likelihood advantage over other models (i.e., 18,351 log-likelihood advantage over the second-best model, mtMet). We also used a bootstrap method [30] to estimate the reliability of clades in T_a.

In general, T_a strongly supports the main clades of the NCBI taxonomy at the family, subfamily, and genus levels. However, the low bootstrap values of some clades at more high levels show the limitation of mt protein data in resolving ambiguous relationships among high level clades.

Specifically, T_a shows strong support (100% bootstrap values) for the clades of the Testudines order, Crocodylia order, and Aves class. In other words, mt proteins contain sufficient phylogenetic signals to correctly place a Testudines, Crocodylia, or Aves species into its

corresponding order or class. Moreover, T_a also displays a strong support (100% bootstrap value) for the clade of all Testudines, Crocodylia, and Aves. This means that Testudines, Crocodylia, and Aves form one separated group within amniotes. We validated the finding by moving Testudines out of the clade of Crocodylia and Aves to other positions around. We found that T_a was much better than other phylogenies examined (i.e., better than the second-best phylogeny with 76 log-likelihood points). In other words, Testudines is unlikely to be located elsewhere, rather than within the clade of Crocodilian and Aves. The finding agrees with the conclusion by Crawford et al. [31].

Although T_a shows a strong support for the position of Testudines within the clade of Crocodylia and Aves, unfortunately it cannot determine the exact relationships among them. The low bootstrap value of the clade including Testudines and Aves suggests the uncertainty of the ((Testudines,Aves),Crocodylia) topology. We examined this hypothesis by comparing the topology to two other possible topologies ((Crocodylia,Testudines),Aves) and ((Crocodylia, Aves),Testudines). The tiny likelihood difference among the three topologies implies that none of these topologies really outweighs the others (Table 7). For example, the 0.467 log-likelihood advantage of ((Testudines,Aves),Crocodylia) to ((Crocodylia,Aves),Testudines) is likely caused by the limits of numerical optimisation in IQ-TREE rather than by topological differentiation. The approximately unbiased SH test shows no evidence in favour of any topology (Table 7).

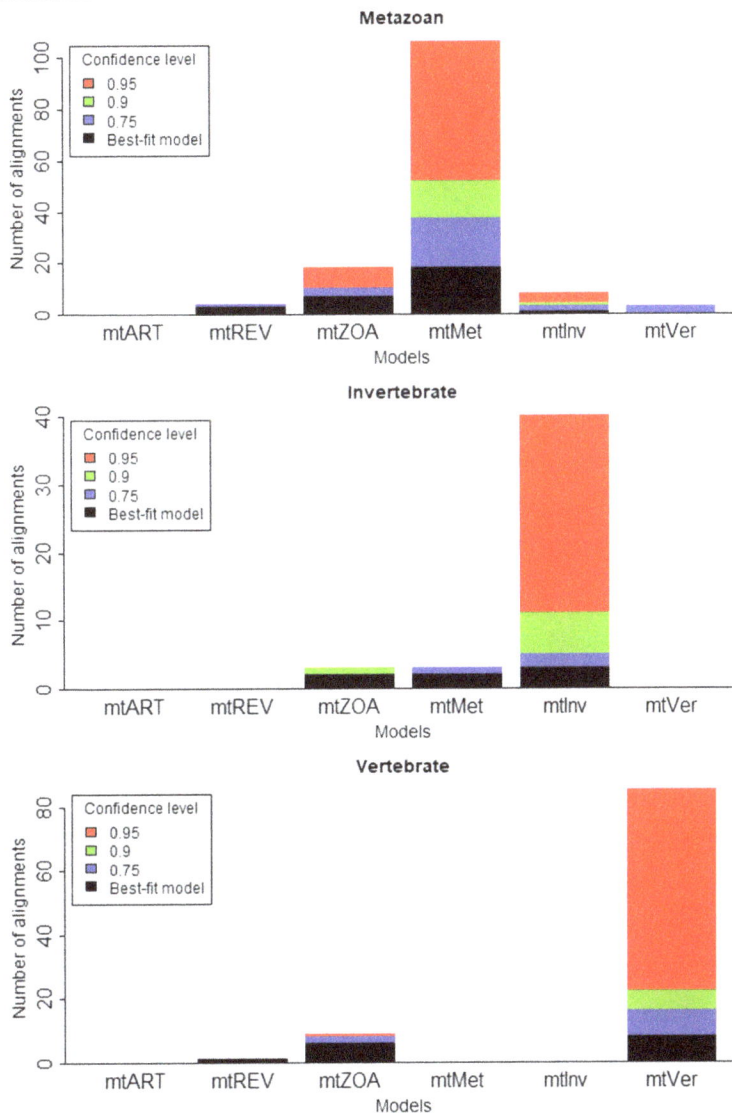

Fig. 6 We used the approximately unbiased SH test (explanations are given in Fig. 5) to compute the confidence levels for phylogenies with six mt models (mtMet, mtVer, mtInv, mtArt, mtREV, and mtZoa) on metazoan, vertebrate and invertebtate testing datasets

Conclusions

We introduced three new mt models estimated from large mt protein datasets of metazoan, vertebrate, and invertebrate species. Experimental results showed the advantage of the mt new models in inferring phylogenies

for both training and testing data in comparison to the existing mt models. The significant likelihood improvement for almost all testing alignments suggests that the new mt models would help find better phylogenies. The phylogenies with the existing mt models may consist of

Table 5 The AIC (BIC) per site of nine models on three testing datasets (the smaller AIC (BIC) the better model)

	mtZOA	mtREV	mtArt	LG4X	C60	PHAT	mtMet	mtInv	mtVer
Metazoan	120.049 (122.011)	120.478 (122.440)	120.476 (122.438)	124.613 (126.629)	124.748 (126.710)	132.966 (134.928)	119.216 (121.178)	120.125 (122.087)	120.769 (122.731)
Invertebrate	133.182 (134.831)	136.229 (137.878)	132.394 (134.044)	138.975 (140.675)	137.979 (139.628)	146.924 (148.573)	132.587 (134.236)	131.674 (133.324)	137.432 (139.082)
Vertebrate	97.129 (99.249)	95.979 (98.099)	98.301 (100.421)	99.851 (102.028)	99.040 (101.159)	107.722 (109.842)	96.195 (98.315)	98.180 (100.299)	95.435 (97.555)

Nine models include six mt models, two site-heterogeneous models (i.e., LG4X, C60), and PHAT model (a transmembrane-specific substitution model)

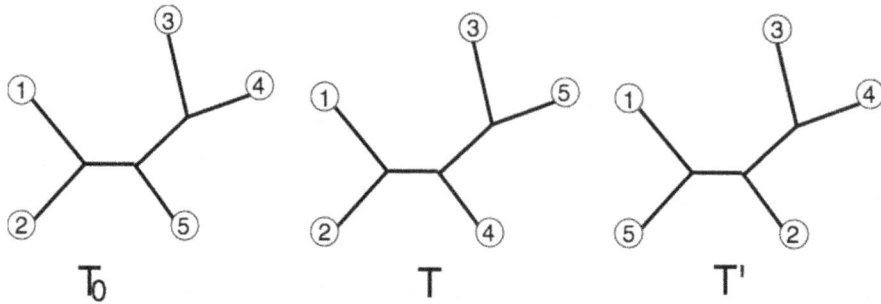

Fig. 7 Unrooted binary trees T, T', and true tree T_0 each has 7 bipartitions. The bipartitions that in T but not in T' is $\{(12|345),(124|35)\}$. The bipartitions that in T' but not in T is $\{(15|234),(152|34)\}$. The Robinson and Foulds distance between T and T' is four. The set S of all bipartitions in T and T' is $\left\{ \begin{array}{l} (12|345),(124|35),(15|234),\ (152|34), \\ (1|2345),(2|1345),(3|1245),(4|1235),(5|1234) \end{array} \right\}$. As the set S consists of 2 incorrect bipartitions (i.e., $(124|35)$ and $(15|234)$), the worse tree must contain at least one incorrect bipartition (a quarter of the Robinson and Foulds distance between T and T')

a considerable number of incorrect bipartitions due to their large distances from the best phylogenies.

The low pairwise correlations among mt models for both amino acid frequency vectors and exchangeability rate matrices suggest remarkable varieties of evolutionary processes of different metazoan lineages. This is particularly true for vertebrates and invertebrates, where their models are the most diverse pair. The new mt models are highly specified to the category of the training data and significantly different from the general models. Note that we also applied the approach to

Table 6 Normalised RobinsonFoulds (RF) distances between phylogenies with six mt models

		mtArt	mtREV	mtZoa	mtMet	mtInv	mtVer
Metazoan	mtREV	0.323					
	mtZoa	0.243	0.286				
	mtMet	0.307	0.281	0.28			
	mtInv	0.299	0.318	0.293	0.239		
	mtVer	0.353	0.277	0.313	0.276	0.332	
	Best	0.304	0.269	0.255	0.058	0.242	0.277
Vertebrate	mtREV	0.115					
	mtZoa	0.087	0.103				
	mtMet	0.109	0.099	0.100			
	mtInv	0.098	0.104	0.095	0.093		
	mtVer	0.124	0.098	0.114	0.1	0.115	
	Best	0.122	0.096	0.104	0.099	0.112	0.012
Invertebrate	mtREV	0.087					
	mtZoa	0.067	0.082				
	mtMet	0.082	0.075	0.076			
	mtInv	0.08	0.08	0.079	0.064		
	mtVer	0.094	0.076	0.089	0.076	0.087	
	Best	0.081	0.081	0.075	0.064	0.006	0.088

The distances are normalised by dividing by $(2n-3)$, where n is the number of taxa

estimate mtPro and mtDeu models for Protostomia and Deuterostomia clades, respectively.

Experimental results confirmed the essential role of model selections in inferring phylogenies from mt protein data. As a general rule, the best-fit model for a certain alignment is the new model estimated from the training data of the same category as the alignment. However, we recommend testing all three new mt models for the study of datasets containing diverse metazoan groups, as mtVer and mtInv might fit better than mtMet for the diverse metazoan alignments.

An alternative approach for model selection is to use model averaging method that allows the estimation of phylogenies and model parameters using all available mt models [32]. In addition, the new empirical mt models can be used as prior probability distribution of amino acid substitution rates in Bayesian analyses [33]. As the new empirical models do not explicitly encode site-specific biological constrains, it is worth testing site-heterogeneous models (e.g., LG4X or C60). Finally, mitochondrially encoded proteins are transmembrane proteins with non stationary evolutions, researchers should consider to test transmembrane-specific amino acid substitution models (e.g. PHAT [26]) and non stationary models (e.g. Coala [34]).

The phylogeny of 993 amniote species inferred from mt proteins with the new models shows strong support for the hypothesis that Testudines, Crocodylia, and Aves form one separated clade within amniotes. However, we could not determine precise relationships among Testudines, Crocodylia, and Aves.

Methods
Model

We assume the amino acid substitution process to be a general time-reversible process and that the substitution processes of amino acid sites are independent [16]. The amino acid substitution model is characterised by a

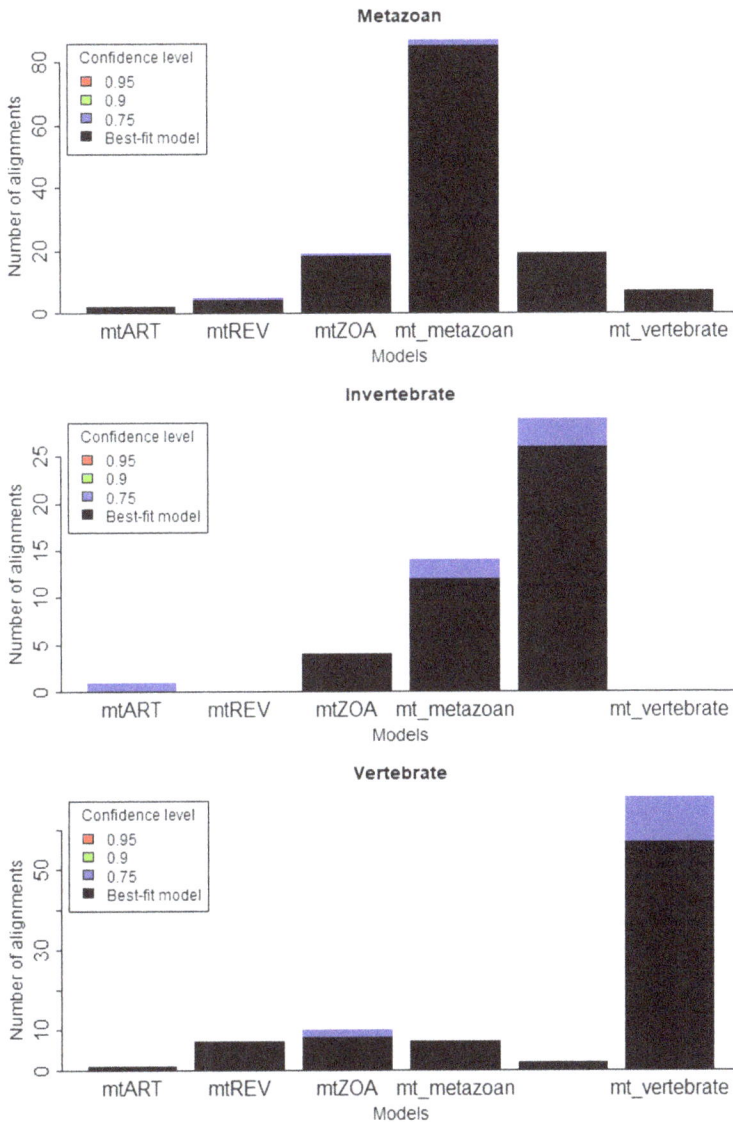

Fig. 8 We used the approximately unbiased SH test to examine tree topologies on metazoan, vertebrate and invertebrate testing datasets. For each testing alignment D, we determined its best-fit model M_b. We fixed tree topologies, but reoptimised other parameters (i.e., branch lengths, parameters of rate heterogeneity model) under the best-fit model M_b. Then we used the CONSEL program to assess the confidence levels for every tree topologies

Markovian substitution matrix, $Q = \{q_{x,y}\}$, that is unchanged during the evolution across all sites. The distribution of amino acid frequencies, $\pi = \{\pi_x\}$, is also assumed to be stationary (or in equilibrium) and fixed across sites and evolution histories. Moreover, Q and π are dependent, where $Q\pi = 0$. Since the process is time-reversible, $Q = \{q_{x,y}\}$ can be rewritten as:

$$q_{x,y} = \pi_y r_{x,y} \text{ and } q_{x,x} = -\Sigma_{x \neq y} q_{x,y},$$

where $r_{x,y} = r_{y,x}$ is the exchangeability coefficient between amino acids x and y.

Since time and branch lengths are normally measured by the number of mutations, matrix Q is normalised such that a time unit is equivalent to one amino acid mutation as follows:

$$\dot{Q} = \frac{Q}{\mu} \text{ where } \mu = -\Sigma_x q_{x,x}.$$

The normalisation of Q would not affect likelihood values or tree topologies but branch lengths only.

Given normalised matrix Q, the probability of amino acid substitutions over the course of time t is calculated as:

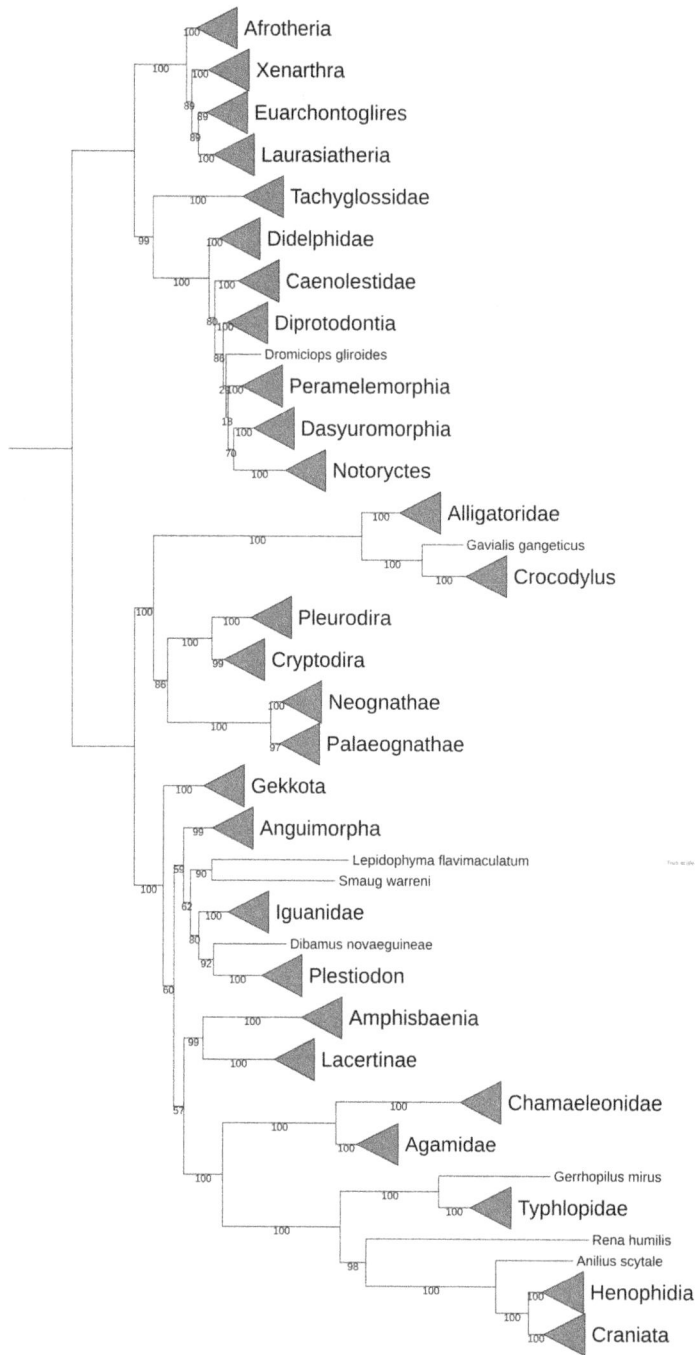

Fig. 9 Location of Turtles in Amiphiona. The Testudines clade including two clades (Pleurodira and Cryptodira) is located within the clade of Crocodylia and Aves

$$P(t) = e^{Qt},$$

where the right term, e^{Qt}, denotes the matrix exponential.

The likelihood of phylogeny T and matrix Q of a given alignment D is calculated as:

$$LK(T, Q; D) = \prod_i LK(T, Q; D_i),$$

where D_i is the data at site i of alignment D. In addition, $LK(T, Q; D_i)$ can be calculated using the pruning algorithm [35].

Table 7 Log-likelihoods and confidence levels of three different tree topologies for Aves, Testudines, and Crocodylia

	Log likelihood	Au	Np
((Aves,Testudines), Crocodylia)	−1,266,499.097	0.569	0.511
((Aves, Crocodylia),Testudines)	−1,266,499.559	0.511	0.471
((Crocodylia,Testudines), Aves)	−1,266,511.883	0.043	0.017

The abbreviations Au and Np stand for the approximately unbiased SH test and the bootstrap probability of the selection

It is well known that evolution rates among sites are variant and are best described by a gamma distribution with parameter α [36]. The proportion of invariant sites also contributes to the likelihood of a phylogeny. The likelihood of phylogeny T, matrix Q, rate variants α,and the proportion of invariant sites, v, with given alignment D can be calculated as follows:

$$LK(T,Q,\alpha,v;D) = v\prod_i LK(\text{Invariant};D_i)$$
$$+ (1-v)\prod_i \frac{1}{C}\Sigma_c LK\left(\rho_c T,Q;D_i\right)$$

where ρ_c is the rate of category c of the gamma distribution with parameter α, and $\rho_c T$ is tree T with branch lengths multiplied by the factor ρ_c.

Many software applications have been developed to estimate T, Q, α, and v for a given alignment D [22, 37, 38].

Given a set of alignments, $\mathbf{D} = \{D^i\}$, matrix Q can be estimated from $\mathbf{D}$ by maximising the likelihood function as follows:

$$LK(Q;\mathbf{D}) = \prod_i LK(T^i,Q,\alpha^i,v^i;D^i). \tag{1}$$

Le and Gascuel [16] proposed a method to estimate matrix Q. First, T^i, α^i, and v^i are estimated using an initial matrix Q, and subsequently matrix Q is estimated based on the newly estimated parameters T^i, α^i, and v^i. The optimising process is repeated until the likelihood improvement is insignificant.

Abbreviations
Mt.: Mitochondrial; NCBI: National Center for Biotechnology Information; RF: Robinson-Foulds

Acknowledgements
This work is financially supported by Vietnam National Foundation for Science and Technology Development (102.01-2013.04).

Funding
This work is financially supported by Vietnam National Foundation for Science and Technology Development (102.01–2013.04). The funding was used for the design of the study and collection, analysis, and interpretation of data and writing the manuscript.

Authors' contributions
VSL and QSL discussed ideas, conducted experiments, wrote the manuscript. All authors revised and approved the final manuscript. CCD participated in additional experiments for the revised version.

Competing interests
The authors declare that they have no competing interests.

References
1. Gray IC, Barnes MR. Amino acid properties and consequences of substitutions. Bioinforma. Genet. Chichester, UK: John Wiley & Sons. Ltd. 2003;4:289–304.
2. Benner S a, Cohen MA, Gonnet GH. Amino acid substitution during functionally constrained divergent evolution of protein sequences. Protein Eng. 1994, p. 1323–32.
3. Dang CC, Le QS, Gascuel O, Le VS. FLU, an amino acid substitution model for influenza proteins. BMC Evol Biol. 2010;10:99.
4. Nickle DC, Heath L, Jensen MA, Gilbert PB, Mullins JI, Kosakovsky Pond SL. HIV-Specific Probabilistic Models of Protein Evolution. Pybus O, editor. PLoS One 2007, 2:e503.
5. Rota-Stabelli O, Yang Z, Telford MJ. MtZoa: a general mitochondrial amino acid substitutions model for animal evolutionary studies. Mol Phylogenet Evol. 2009;52:268–72.
6. Le SQ, Gascuel O. Accounting for solvent accessibility and secondary structure in protein phylogenetics is clearly beneficial. Syst Biol. 2010;59: 277–87.
7. Dunn KA, Jiang W, Field C, Bielawski JP. Improving Evolutionary Models for Mitochondrial Protein Data with Site-Class Specific Amino Acid Exchangeability Matrices. Salamin N, editor. PLoS One 2013, 8:e55816.
8. Taanman J-W. The mitochondrial genome: structure, transcription, translation and replication. Biochim. Biophys. Acta - Bioenerg 1999, 1410: 103–123.
9. Carapelli A, Liò P, Nardi F, van der Wath E, Frati F. Phylogenetic analysis of mitochondrial protein coding genes confirms the reciprocal paraphyly of Hexapoda and Crustacea. BMC Evol. Biol. 2007, 7 Suppl 2:S8.
10. Eo SH, DeWoody JA. Evolutionary rates of mitochondrial genomes correspond to diversification rates and to contemporary species richness in birds and reptiles. Proc Biol Sci. 2010;277:3587–92.
11. Cook CE, Yue Q, Akam M. Mitochondrial genomes suggest that hexapods and crustaceans are mutually paraphyletic. Proc Biol Sci. 2005;272:1295–304.
12. Spinks PQ, Shaffer HB, Iverson JB, McCord WP. Phylogenetic hypotheses for the turtle family Geoemydidae. Mol Phylogenet Evol. 2004;32:164–82.
13. Adachi J, Hasegawa M. Model of amino acid substitution in proteins encoded by mitochondrial DNA. J Mol Evol. 1996;42:459–68.
14. Abascal F, Posada D, Zardoya R. MtArt: a new model of amino acid replacement for Arthropoda. Mol Biol Evol. 2007;24:1–5.
15. Donoghue PCJ, Purnell MA. Genome duplication, extinction and vertebrate evolution. Trends Ecol. Evol. 2005, p. 312–9.
16. Le SQ, Gascuel O. An improved general amino acid replacement matrix. Mol Biol Evol. 2008;25:1307–20.
17. Whelan S, Goldman N. A general empirical model of protein evolution derived from multiple protein families using a maximum-likelihood approach. Mol Biol Evol. 2001;18:691–9.
18. Dang CC, Le VS, Gascuel O, Hazes B, Le QS. FastMG: a simple, fast, and accurate maximum likelihood procedure to estimate amino acid replacement rate matrices from large data sets. BMC Bioinformatics. 2014;15:341.
19. Tan G, Muffato M, Ledergerber C, Herrero J, Goldman N, Gil M, et al. Current methods for automated filtering of multiple sequence alignments frequently worsen single-gene phylogenetic inference. Syst Biol. 2015;64:778–91.
20. Akaike H. A new look at the statistical model identification. IEEE Trans Autom Control. 1974;19:716–23.
21. Schwarz G. Estimating the dimension of a model. Ann Stat. 1978;6:461–4.
22. Nguyen LT, Schmidt HA, Von Haeseler A, Minh BQ. IQ-TREE: a fast and effective stochastic algorithm for estimating maximum-likelihood phylogenies. Mol Biol Evol. 2015;32:268–74.
23. Shimodaira H. An approximately unbiased test of phylogenetic tree selection. Syst Biol. 2002;51:492–508.
24. Shimodaira H, Hasegawa M. CONSEL: for assessing the confidence of phylogenetic tree selection. Bioinformatics. 2001;17:1246–7.
25. Le SQ, Dang CC, Gascuel O. Modeling protein evolution with several amino acid replacement matrices depending on site rates. Mol Biol Evol. 2012;29: 2921–36.
26. Ng PC, Henikoff JG, Henikoff S. PHAT: a transmembrane-specific substitution matrix. Predicted hydrophobic and transmembrane. Bioinformatics. 2000;16:760–6.

27. Robinson DF, Foulds LR. Comparison of phylogenetic trees. Math Biosci. 1981;53:131–47.
28. Felsenstein J. The number of evolutionary trees. Syst Zool. 1978;27:27–33.
29. Fong JJ, Brown JM, Fujita MK, Boussau B. A Phylogenomic approach to vertebrate phylogeny supports a turtle-archosaur affinity and a possible paraphyletic Lissamphibia. PLoS One. 2012;7
30. Minh BQ, Nguyen MAT, Von Haeseler A. Ultrafast approximation for phylogenetic bootstrap. Mol Biol Evol. 2013;30:1188–95.
31. Crawford NG, Faircloth BC, McCormack JE, Brumfield RT, Winker K, Glenn TC. More than 1000 ultraconserved elements provide evidence that turtles are the sister group of archosaurs. Biol Lett. 2012;8:783–6.
32. Posada D, Buckley TR. Model selection and model averaging in phylogenetics: advantages of akaike information criterion and bayesian approaches over likelihood ratio tests. Syst Biol. 2004;53:793–808.
33. Huelsenbeck JP, Joyce P, Lakner C, Ronquist F. Bayesian analysis of amino acid substitution models. Philos. Trans. R. Soc. Lond. B. Biol. Sci 2008, 363: 3941–3953.
34. Groussin M, Boussau B, Gouy M. A branch-heterogeneous model of protein evolution for efficient inference of ancestral sequences. Syst Biol. 2013;62: 523–38.
35. Felsenstein J. Evolutionary trees from DNA sequences: a maximum likelihood approach. J Mol Evol. 1981;17:368–76.
36. Yang Z. Maximum-likelihood estimation of phylogeny from DNA sequences when substitution rates differ over sites. Mol Biol Evol. 1993;10:1396–401.
37. Guindon S, Dufayard JF, Lefort V, Anisimova M, Hordijk W, Gascuel O. New algorithms and methods to estimate maximum-likelihood phylogenies: assessing the performance of PhyML 3.0. Syst Biol 2010, 59:307–321.
38. Yang Z. PAML 4: phylogenetic analysis by maximum likelihood. Mol Biol Evol. 2007;24:1586–91.

14

The effect of body size evolution and ecology on encephalization in cave bears and extant relatives

Kristof Veitschegger ⓘD

Abstract

Background: The evolution of larger brain volumes relative to body size in Mammalia is the subject of an extensive amount of research. Early on palaeontologists were interested in the brain of cave bears, *Ursus spelaeus*, and described its morphology and size. However, until now, it was not possible to compare the absolute or relative brain size in a phylogenetic context due to the lack of an established phylogeny, comparative material, and phylogenetic comparative methods. In recent years, many tools for comparing traits within phylogenies were developed and the phylogenetic position of cave bears was resolved based on nuclear as well as mtDNA.

Results: Cave bears exhibit significantly lower encephalization compared to their contemporary relatives and intraspecific brain mass variation remained rather small. Encephalization was correlated with the combined dormancy-diet score. Body size evolution was a main driver in the degree of encephalization in cave bears as it increased in a much higher pace than brain size. In *Ursus spelaeus*, brain and body size increase over time albeit differently paced. This rate pattern is different in the highest encephalized bear species within the dataset, *Ursus malayanus*. The brain size in this species increased while body size heavily decreased compared to its ancestral stage.

Conclusions: Early on in the evolution of cave bears encephalization decreased making it one of the least encephalized bear species compared to extant and extinct members of Ursidae. The results give reason to suspect that as herbivorous animals, cave bears might have exhibited a physiological buffer strategy to survive the strong seasonality of their environment. Thus, brain size was probably affected by the negative trade-off with adipose tissue as well as diet. The decrease of relative brain size in the herbivorous *Ursus spelaeus* is the result of a considerable increase in body size possibly in combination with environmental conditions forcing them to rest during winters.

Keywords: Physiological buffer, Dormancy, Diet, *Ailuropoda, Helarctos, Melursus, Tremarctos, Ursus*

Background

Cave bears, *Ursus spelaeus*, were a common faunal element during the Pleistocene of Europe and Asia [1]. The habitat of *U. spelaeus* was Eurasia with an east-west extension ranging from Spain to the Altai Region of Russia [1–3]. The ancestral species of *U. spelaeus*, *U. deningeri*, was even more widespread, with a habitat ranging from Spain to Siberia and even reaching the British Isles [1, 3–5]. At the end of the Pleistocene, cave bears shared the same fate as most other elements of the Pleistocene megafauna and became extinct [6–8]. Their time of extinction was

Correspondence: kristof.veitschegger@pim.uzh.ch
Palaeontological Institute and Museum, University of Zurich, Karl Schmid-Strasse 4, 8006 Zürich, Switzerland

proposed to be around 27.800–25.000 years BP [9, 10]. Based on molecular data, the sister group to cave bears are brown bears, *U. arctos*, and polar bears, *U. maritimus*, together (Fig. 2). The evolutionary lineage of *U. spelaeus* split from these two bear species sometime between 2.75 to 1.2 Ma years ago [11–13]. Traditionally, cave bears were considered to be predominantly or exclusively herbivorous based on the morphology of their teeth and jaws [1, 14–18]. Several studies presented isotopic as well as morphometric evidence confirming this hypothesis [2, 19–25]. However, the predominantly herbivorous diet of cave bears was questioned based on isotopic [26, 27], morphometric [28, 29], microwear [30, 31], and taphonomic evidence [32]. In recent years, many of these studies were dismissed based on methodological errors or repeated with the result that cave bears were indeed herbivorous [2, 19, 20, 33].

Cave bear brains are among the earliest ones of an extinct species to be investigated and several studies discuss different aspects of its evolution [34–42]. Many of these studies focus on the external morphology of artificial, fossil, or virtual endocasts [34, 35, 39–41]. Conflicting statements were presented concerning the overall size of the cave bear brain. Some authors suggested a small brain size compared to body size and speculated that the increase of skull size in the evolution of *U. spelaeus* outpaced brain size [35, 36]. Others suggested high brain volumes for cave bears and an opposite scenario with brain size outpacing body size [37, 38, 42]. Many factors affect the size of brains. Brain tissue itself is known to be expensive to produce and maintain [43–45]. Absolute as well as relative brain size can be influenced by social structure [46–48], environment [48–52], sensory systems [53], evolutionary history [54–57], body size evolution [42], and different physiological as well as life history trade-offs [43, 52, 57–66].

Diet can have a profound effect on brain size as was exemplified in bats and primates [67]. Recently, it was even suggested that diet had a bigger effect on brain size than sociality in primates [68]. The diet of bears is diverse with varying amounts of plant and animal matter within and among species [2]. It ranges from hypercarnivorous in polar bears, *U. maritimus*, to folivorous in giant pandas, *Ailuropoda melanoleuca* [2, 69]. Thus, diet of bears might exhibit a link to brain size.

Some bear species survive the cold seasons with extended resting periods, whereas especially tropical species are active

year-round [69]. Resting periods in bears are different from deep hibernation as movement still can occur [70]. Thus, these periods are better described as dormancy in bears. Previous to dormancy, bears increase the amount of stored body fat [70]. The storage of high amounts of adipose tissue was linked to a decreased brain size [60]. Bears represent a good study object to investigate the effect of dormancy on brain size because some species are active year round whereas others increase the amount of adipose tissue annually [69].

In this study, I investigate the absolute and relative brain size of *U. spelaeus* and all extant bear species in a phylogenetic context and add remarks on *U. deningeri*. For this, I created a comprehensive brain size dataset for all extant bear species and cave bears. Additionally, I examine potential variables which could introduce energetic constraints affecting brain size evolution such as dormancy, diet, and body size. These variables were chosen because they can be reconstructed for cave bears with some measure of certainty.

Methods
Data collection
Altogether, I measured 412 skulls of 10 extant and extinct bear species (Table 1). *U. spelaeus* samples cover a time period of about 20.000 years based on radiocarbon dating [9]. Brain volume was measured using the glass bead method [71]. I used 6 mm diameter soda lime glass beads. The individual body mass (g) was inferred using the basicranial length (SKL) as described by van Valkenburgh: body mass (kg) = 2.02*Log10(SKL)-2.80 (least squares regression) [72]. Brain volume was converted into brain mass (g) using the specific weight of brain substance 1.036

Table 1 Results of body mass (g) and brain mass (g) estimates as well as residuals and investigated ecological scores

Species	n	average body mass (g)	StD body mass (g)	average body mass literature (g)	average brain mass (g)	StD brainmass (g)	average residuals	StD average residuals	diet score	dormancy score	d*d
Ailuropoda melanoleuca	5	118'637 (105'324–135'094)	10,748.36	97'500 (70'000–125'000)	281.79 (238.28–331.52)	33.89	−0.0029	0.0548	1.000	3.000	3.000
Tremarctos ornatus	8	80'918 (64'223–110'621)	15,049.56	117'500 (60'000–175,000)	227.92 (176.12–279.72)	31.33	0.0373	0.0320	1.814	3.000	5.443
Ursus americanus	28	117'116 (83'885–155'600)	20,168.42	170'000 (40'000–300'000)	256.78 (186.48–352.24)	38.39	−0.0373	0.0422	1.884	1.000	1.884
Ursus arctos	93	177'628 (92'655–320'042)	40,696.57	390'000 (55'000–725'000)	378.08 (207.20–538.72)	61.38	−0.0080	0.0464	1.637	1.000	1.637
Ursus deningeri	1	254,996	-	-	341.88	-	−0.1770	-	-	-	-
Ursus malayanus	50	82'379 (56'333–108'841)	13,709.85	52'500 (25'000–80'000)	340.43 (227.92–435.12)	47.31	0.2047	0.0403	2.684	3.000	8.051
Ursus maritimus	82	211'265 (144'141–277'270)	33,275.87	402'500 (150'000–655'000)	498.80 (393.68–611.24)	53.75	0.0525	0.0320	2.970	2.000	5.940
Ursus spelaeus	99	322'764 (209'553–425'411)	57,207.28	362'500 (225'000–500'000)	430.10 (321.16–569.80)	52.36	−0.1550	0.0443	1.000	1.000	1.000
Ursus thibetanus	29	113'424 (78'533–166'402)	21,401.65	120'000 (40'000–200'000)	282.58 (186.48–414.40)	45.66	0.0155	0.0577	1.920	1.000	1.920
Ursus ursinus	17	147'081 (124'439–183'291)	18,122.18	100'000 (50'000–150'000)	292.52 (248.64–352.24)	26.04	−0.0573	0.0360	2.606	3.000	7.818

(g/cm3) [73]. The collected data is presented in Additional file 1: Table S1. To assess the validity of previously published cranial volumes of cave bears, I additionally created a data subset predicting endocranial volume based on external skull measurements for *U. spelaeus*, *U. arctos*, and *U. malayanus* [74]. Raw data for this analysis can be found in Additional file 2: Table S2.

The materials examined in this study are from the following collections: Biologiezentrum Linz (BZL), Geology School of Aristotle University Thessaloniki (AUTH), Institut für Paläontologie Wien (PIUW), Naturalis Biodiversity Center Leiden (NBC), Naturhistorisches Museum der Burgergemeinde Bern (NMBE). Naturhistorisches Museum Wien (NHM), Naturmuseum St. Gallen (NMSG), Naturmuseum Südtirol Bozen (PZO), Muséum National d'Histoire Naturelle Paris (MNHN), Museum für Naturkunde Berlin (MfN), Paleontological Institute and Museum University of Zurich (PIMUZ), and Zoological Museum University of Zurich (ZMUZH).

Data analyses

Data were log10-transformed and examined using ordinary least squares (OLS) and phylogenetic generalized least squares (PGLS) (Fig. 1, Additional file 3: Supplementary Information). I used OLS to investigate the differences in intercepts and slopes between species. Residuals from a PGLS based on brain/body mass (g) were used to investigate the differences in relative brain size. With this, the data were corrected for the effect of size. An initial investigation revealed that the data were heavily skewed by *U. malayanus* and *U. spelaeus* because of the uneven sampling (Additional file 3: Supplementary Information). All other bear species were more similar in body mass (g)/brain mass (g). This was supported by the multiple and adjusted R^2 (Additional file 3: Supplementary Information). Thus, the

basis for brain/body mass (g) residuals was the slope (0.78069) and intercept (−1.50995) as retrieved by a PGLS excluding *U. malayanus* and *U. spelaeus*. For PGLS, the species-averaged brain mass and body mass were used. Analyses were performed in R, version 3.2.3 [75]. PGLS was executed as implemented in the packages ape and caper [76, 77]. Results from OLS regressions on all data points as well as a PGLS regression with all species are presented in the Additional file 3: Supplementary Information.

The phylogenetic relationships among Ursidae is not completely understood as there are clear discrepancies between trees based on nuclear and mitochondrial DNA (mtDNA), mirroring a complex evolutionary history with introgression and incomplete lineage sorting [78]. Complete phylogenies of Ursidae including cave bears are based on mtDNA [11, 12], therefore, I use mtDNA topology as basis for the phylogenetic analyses. The relationship between cave bears and brown bear as well as polar bear was also confirmed by nuclear DNA [79]. Recently, several new, former unrecognized species and subspecies of *U. spelaeus* were described based on morphological and genetic data [3, 80–83]. However, some of these taxa are polyphyletic [84, 85]. Here, I include all these proposed cave bear species and subspecies in *U. spelaeus*, but exclude the well-established ancestral species *U. deningeri* [17]. *U. deningeri* is considered an anagenetic ancestor to *U. spelaeus* [1, 80] and thus was excluded from all analyses as it would either represent a duplication or cannot be properly placed in phylogeny. Branch lengths for phylogenetically informed analyses were retrieved from Nyakatura and Bininda-Emonds [86] and Bon [12].

Due to uneven sampling and small sample sizes in species-averaged datasets, I use non-parametric analyses. A Kruskal-Wallis test followed by a Dunn's test with Bonferroni adjustment was used on the resulting

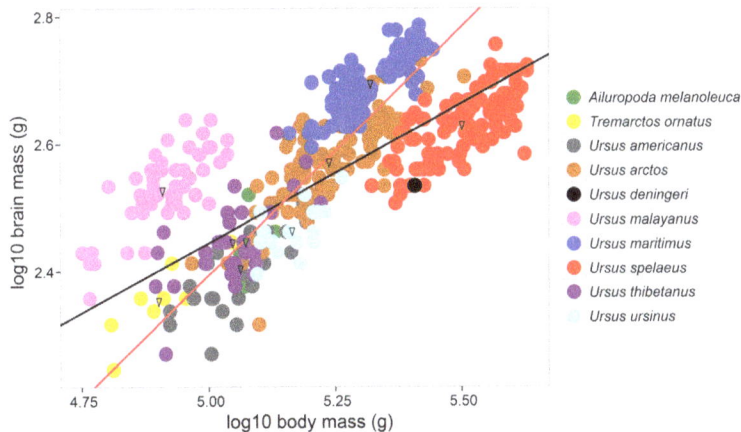

Fig. 1 Scatterplot of log10 brain mass (g) against log10 body mass (g) with a PGLS regression lines (phylogenetic generalized least squares). In black is the PGLS regression line for all data points (*p* value: 0.0148, slope: 0.43978, intercept: 0.24623, adjusted R^2: 0.5378), in red the PGLS regression line without *Ursus malayanus* and *Ursus spelaeus* (*p* value: 0.0016, slope: 0.78069, intercept: −1.50995, adjusted R^2: 0.8606). The triangles represent the mean for each species on which PGLS was calculated

residuals to test for significant differences. This was performed in R, version 3.2.3 [75], using the packages pgirmess and PMCMR [87, 88]. The subset of different brain volume estimations was analysed using a Wilcoxon signed-rank test for paired samples in R, version 3.2.3. [75]. Boxplots were created in the package ggplot2 [89].

I used squared-change parsimony [90] to reconstruct ancestral stages for log 10 average body mass (g), log 10 average brain mass (g), and averaged residuals respectively. This analysis follows a Brownian motion model of evolution [91]. The resulting ancestral character states were then used to investigate the relative mass change (in percent) from one node to the following within the tree. These analyses were performed for each variable separately in Mesquite software (version 3.01) [92].

To test for a possible effect of dormancy and diet on relative and absolute brain size, I scored each of these variables between 1 and 3: 1 represents states where a smaller brain size is expected and 3 the opposite. Dormancy was scored as 1 (dormancy), 2 (fasting periods), and 3 (no dormancy) [69]. Dietary preferences were scored using the compilation from van Heteren et al. [2]. The diet was scored between 1 (completely folivorous/low caloric diet) to 3 (completely faunivorous (high caloric diet) using the formula:

$$\text{Diet score} = (\text{percent folivory/overall percent})^*1$$
$$+(\text{percent frugivory/overall percent})^*2$$
$$+(\text{percent faunivory/overall percent})^*3$$

The scoring enables to multiply both scores to one under the assumption that unidirectional or opposing trends show a combined effect on brain size. This is possible because the array of possible variables is constrained among three states. I performed the Kendall's tau correlation analysis in R, version 3.2.3, using the package Kendall [93].

Results

The resulting averaged reconstructed body mass (g) and brain mass (g) with standard deviation as well as the ecological scores are given in Table 1.

The slopes of the OLS regression lines of the different bear species were not significantly different from each other. Intercepts, however, were in many cases significantly different among species (Table 2, Additional file 4: Table S3). The intercept of cave bears was not significantly different from that of U. americanus and U. ursinus.

U. spelaeus and U. deningeri have the lowest average residuals within the dataset, followed by U. ursinus and U. americanus (Fig. 2, Table 1). The highest average residuals were found in U. malayanus and U. maritimus. The Kruskal-Wallis test followed by a Dunn's test with Bonferroni adjustment revealed that the residuals of U. spelaeus are significantly smaller than of most other bear species, except for U. ursinus and A. melanoleuca (Table 3).

The biggest documented cave bear brain volume is 1.8 times bigger than the smallest. In comparison, in U. arctos it is 2.6 times bigger and in U. thibetanus 2.2 times. Polar bears, however, exhibit low variation with the biggest brain volume being 1.6 times bigger than the smallest (Table 1).

The comparison between different methods to estimate brain volumes revealed that external measurements produced results significantly different from brain volume measured directly with glass beads (Fig. 3). In U. spelaeus, brain volumes inferred by external measurements were significantly higher than those measured with soda lime glass beads ($n = 15$, median glass beads = 410 ml, median external measurements = 480 ml, V = 120, p-value = <0.0001). The opposite is true for U. arctos and U. malayanus. Here, brain volumes were significantly higher when measured with glass beads (U. arctos: $n = 34$, median glass beads = 370 ml, median external measurements = 312 ml, V = 66, p-value = <0.0001; U. malayanus: $n = 9$, median glass beads = 310 ml, median external measurements = 191 ml, V = 0, p-value = 0.0039).

The ancestral stage reconstruction based on squared-change parsimony revealed that the small relative brain size of U. spelaeus and U. ursinus represent a secondarily derived condition, as their respective ancestral stages exhibit a higher relative brain size (Fig. 2, Additional file 5: Table S4). The comparison between the relative change of body mass (g) and brain mass (g) shows that the evolution of a bigger body size in U. spelaeus outpaced brain size evolution. Both increased size compared to their ancestral stages, respectively; however, body size increased at a much higher pace. The reverse was found in U. maritimus, in which brain size evolution outpaced body size increase. Nonetheless, in U. maritimus and U. spelaeus brain as well as body size evolution are unidirectional towards increasing. In U. americanus the trend is unidirectional towards decrease. These cases contrast with the decoupling trend recorded for U. malayanus. In this species, the body size decreases where the brain size increases leading to the high relative brain size found in this species. At the basis of the tree, the analysis retrieved an ancestral body mass of 112,052 g and a brain mass of 277 g.

Using Kendall's tau to find correlations between ecological scores and brain mass (g) revealed no significant results. Residuals were not significantly correlated with dormancy or diet scores. However, residuals were correlated with the combined score (Table 4).

Discussion
Encephalization in Ursidae
U. spelaeus had a significantly smaller relative brain size than most extant bear species. The brain size variation in cave bears over time, between males and females [1] as well as high altitude and lowland populations [81] did not exceed the intraspecific variation in extant U. americanus,

Table 2 Results of the pairwise comparisons of slopes and intercepts among different bear species

	Ailuropoda melanoleuca	Tremarctos ornatus	Ursus americanus	Ursus arctos	Ursus malayanus	Ursus maritimus	Ursus spelaeus	Ursus thibetanus	Ursus ursinus
Ailuropoda melanoleuca		+/− 0.0034	+/− 0.0366	+/− 0.0306	+/− 0.1725*****	+/− 0.1086*****	+/− 0.0596***	+/− 0.0129	+/− 0.0347
Tremarctos ornatus	+/− 0.4053		+/− 0.0400**	+/− 0.0273	+/− 0.1691*****	+/− 0.1052*****	+/− 0.0630***	+/− 0.0096	+/− 0.0380*
Ursus americanus	+/− 0.3957	+/− 0.0096		+/− 0.0672****	+/− 0.2091*****	+/− 0.1452*****	+/− 0.0230	+/− 0.0495*****	+/− 0.0019
Ursus arctos	+/− 0.3362	+/− 0.0691	+/− 0.0595		+/− 0.1419*****	+/− 0.0779*****	+/− 0.0903*****	+/− 0.0177	+/− 0.0653*****
Ursus malayanus	+/− 0.3895	+/− 0.0158	+/− 0.0062	+/− 0.0533		+/− 0.0639*****	+/− 0.2321*****	+/− 0.1596*****	+/− 0.2072*****
Ursus maritimus	+/− 0.2864	+/− 0.1189	+/− 0.1093	+/− 0.0498	+/− 0.1031		+/− 0.1682*****	+/− 0.0956*****	+/− 0.1432*****
Ursus spelaeus	+/− 0.2161	+/− 0.1892	+/− 0.1796	+/− 0.1200	+/− 0.1734	+/− 0.0702		+/− 0.0726*****	+/− 0.0250
Ursus thibetanus	+/− 0.2497	+/− 0.1556	+/− 0.1460	+/− 0.0865	+/− 0.1398	+/− 0.0367	+/− 0.0335		+/− 0.0476*****
Ursus ursinus	+/− 0.1606	+/− 0.2447	+/− 0.2351	+/− 0.1756	+/− 0.2289	+/− 0.1258	+/− 0.0555	+/− 0.0891	

Significant results are marked with stars (p-value: *< 0.5, **< 0.1, ***< 0.01, ****< 0.001, *****< 0.0001)
Upper triangle shows intercept comparisons and lower triangle shows slope comparisons

U. arctos, *U. malayanus*, and *U. thibetanus*. Especially, the relative brain size of *U. arctos* and *U. thibetanus* exhibits a considerable amount of variation. The study of brain size evolution often focuses on the evolution of increased encephalization and intelligence [38, 94–99]. Animals with bigger relative brain size often show more flexibility in behaviour and are potentially more adaptable [100–104]. Nonetheless, brain tissue is expensive and producing it comes at the cost of a slower life history [43–45, 57, 64, 105]. Therefore, in some species a secondary reduction of relative or absolute brain size was described [106]. Especially, islands represent a challenging habitat for many mammals and several species exhibit a secondary decrease in encephalization [107, 108]. Dormancy and diet, separately, were not correlated with brain size; however, the combination of both variables showed a significant effect. A possible explanation for this correlation could be that cave bears underwent a change in diet in a habitat in which they were still forced to rest during winters [1, 9] limiting the possibility of so called cognitive buffering [66, 109]. Under the Cognitive Buffer hypothesis, it is expected that relative brain size of mammals in highly seasonal environment increases due to the necessity of behavioural flexibility. This, however, also implies an active reaction towards the environmental change. In contrast, dormancy does not require this high level of behavioural flexibility but relies on body fat storage, which additionally has a negative trade-off with brain size [60, 66]. This suggests that brain size in cave bears might exhibit a physiological buffering effect [66] partly constraining relative brain size. Other bear species such as *U. arctos* and *U. americanus*

would also exhibit this physiological buffering effect but their food quality or life history might lessen the constraint on relative brain size.

In Ursidae, three life history variables have been demonstrated to correlate with encephalization: gestation time (negative), newborn mass (positive), and litter size (negative) [57]. In *A. melanoleuca*, a combination of these variable with a year-round active strategy [69] is potentially the reason why the second herbivorous species in the dataset exhibits an encephalization higher than found in cave bears. Nonetheless, the life history correlates with encephalization are not unidirectional in the giant panda. In contrast, the highest encephalized species, *U. malayanus*, shows unidirectional trends towards increased encephalization in most variables with heavy newborns, small litter size, non-resting strategy, and 68% faunivory [2, 69, 110]. Gestation time and litter size are not known for *U. spelaeus*. However, cave bears were about the same size as *U. arctos* at birth [14, 111], contributing to its small relative brain size. A small relative brain size can already be traced in *U. deningeri*. This ancestor of *U. spelaeus* also exhibits low encephalization and is usually considered a herbivorous species with winter resting behaviour as well [25, 112].

The effect of diet alone on brain size in Ursidae remains elusive. In other groups such as primates and bats the link is more apparent. Fruit, blood, and meat eating bats tend to be more encephalized than their insect-eating relatives and in primates leaf-eaters are the least encephalized [67, 68]. Although a comparable link was proposed for Carnivora, it is hypothesized to be more

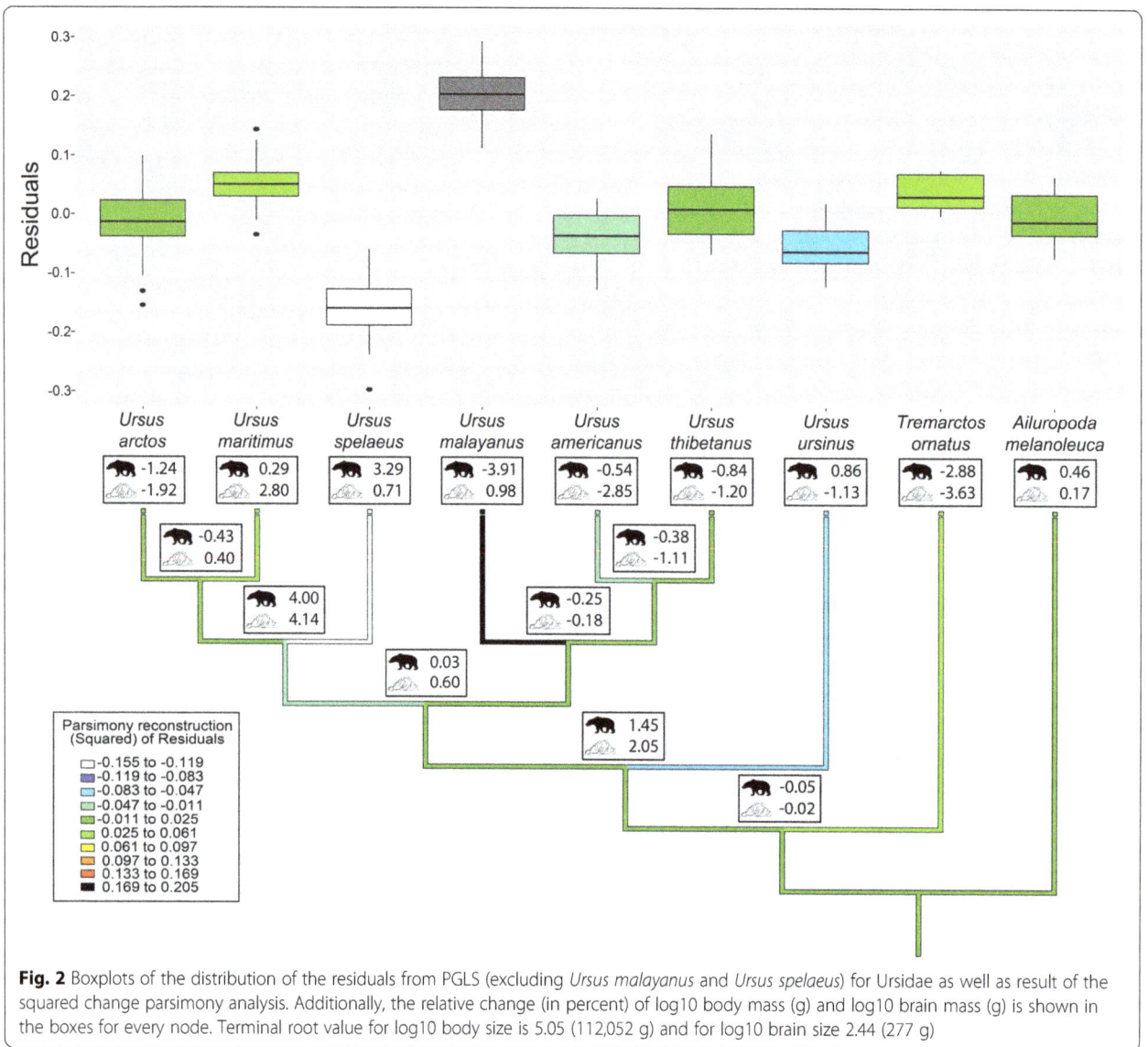

Fig. 2 Boxplots of the distribution of the residuals from PGLS (excluding *Ursus malayanus* and *Ursus spelaeus*) for Ursidae as well as result of the squared change parsimony analysis. Additionally, the relative change (in percent) of log10 body mass (g) and log10 brain mass (g) is shown in the boxes for every node. Terminal root value for log10 body size is 5.05 (112,052 g) and for log10 brain size 2.44 (277 g)

Table 3 Results of the Kruskal-Wallis rank sum test on the residuals of investigated bear species

Kruskal-Wallis rank sum test								
K-W chi-squared: 338.89 df: 8, *p*-value: <0.0001	*Ailuropoda melanoleuca*	*Tremarctos ornatus*	*Ursus americanus*	*Ursus arctos*	*Ursus malayanus*	*Ursus maritimus*	*Ursus thibetanus*	*Ursus ursinus*
Tremarctos ornatus	1.0000	-	-	-	-	-	-	-
Ursus americanus	1.0000	0.7560	-	-	-	-	-	-
Ursus arctos	1.0000	1.0000	1.0000	-	-	-	-	-
Ursus malayanus	0.0699	0.4557	**<0.0001**	**<0.0001**	-	-	-	-
Ursus maritimus	1.0000	1.0000	**<0.0001**	**<0.0001**	0.0006	-	-	-
Ursus thibetanus	1.0000	1.0000	0.8645	1.0000	**<0.0001**	0.7134	-	-
Ursus ursinus	1.0000	0.2586	1.0000	1.0000	**<0.0001**	**<0.0001**	0.2484	-
Ursus spelaeus	0.1133	**<0.0001**	**0.0005**	**<0.0001**	**<0.0001**	**<0.0001**	**<0.0001**	0.2602

In bold are significant results

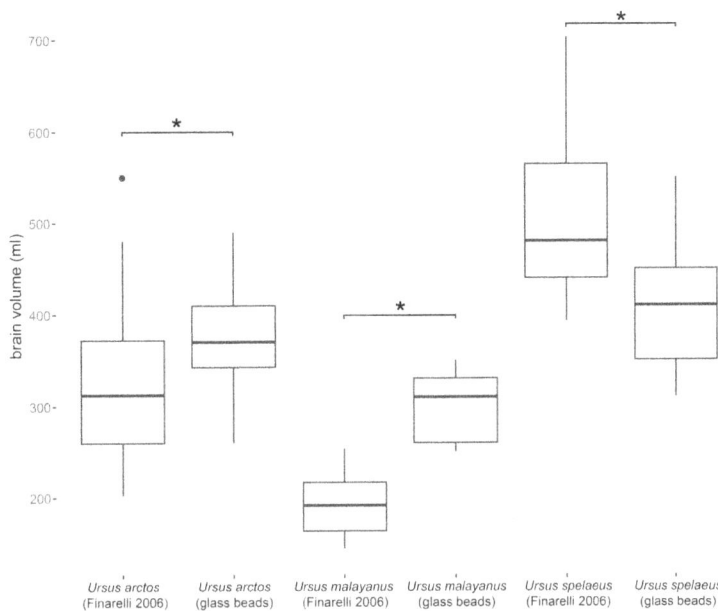

Fig. 3 Comparison between two methods for estimating brain volumes of Ursidae (asterisks mark significant differences based on a Wilcoxon signed-rank test)

associated with the process of acquiring food rather than the energetics of the diet itself [54, 67]. The change in diet in cave bears and associated smaller relative brain size is reminiscent of the often mentioned evolutionary arms-race between Carnivora and Ungulates in which Carnivora had to be more encephalized to outsmart their (herbivorous) prey [98]. This scenario, however, was later found to be unsubstantiated [113].

Smaers et al. [42] suggested that absolute brain size in the evolution of *U. spelaeus* was outpacing body size. This pattern was based on brain size estimates obtained by external measurements [37, 38, 74]. Although external measurements can predict brain volume with a certain confidence [74, 114], they can also have considerable prediction error [114]. The results of this study show that external measurements overestimate the endocranial volume of *U. spelaeus* (Fig. 3). The reason for this might be the frontal bossing found in cave bears likely caused by an extension of the frontal sinuses [16, 17, 35, 36, 41]. My results show that in cave bears body size evolution out-paced brain size evolution. Thus attesting to a remark by Marinelli [36]. Smaers et al. [42] also published brain and body size variables for three other extinct bear species *Arctodus simus* (3 Ma – 0.01 Ma), *Cephalogale ursinus*

(23.8 Ma – 22.8 Ma), and *Indarctos oregonensis* (10.3 Ma – 5.3 Ma). With these values *C. ursinus* would be placed high above the regression line (residual: 0.33), *A. simus* close to the line (residual: 0.04), and *I. oregonensis* below the line (residual: –0.16). Fossil evidence has shown to change the results of suggested bidirectional evolution in brain size [99, 106]. However, in ursids, the cave bear lineage represents one of the least encephalized compared to extant and most extinct relatives.

On the methodology of body mass reconstruction
I calculated the mass of every specimen individually based on skull length [72]. My body mass estimates, generally, were well within the range of known body mass distribution for each species (Table 1) [69]. However, the estimations for polar bears, *U. maritimus*, are generally small. Thus, this animal might be closer to the range of other bear species such as *U. arctos* in the scatterplot (Fig. 1). It is, nonetheless, possible that the measured skulls are from individuals from the lower range of mass distribution of this species. The opposite is true for the two smaller bear species *U. malayanus* and *U. ursinus*. *U. malayanus* potentially could have even bigger brains compared to body size than in the presented dataset. *U. ursinus* would

Table 4 Results of Kendall's tau on different scores as well as the combination of both

	Diet score (d)		Dormancy score (d)		d*d	
	tau	p-value	tau	p-value	tau	p-value
Average brain mass (g)	0.1970	0.5294	−0.1360	0.7285	0.0000	1.0000
Residuals	0.3660	0.2084	0.3400	0.2976	**0.5560**	**0.0476**

In bold are significant correlations

be within the range of other extant bear species in the scatterplot (Fig. 1) such as *U. arctos*. *U. spelaeus* is considered to be one of the biggest carnivorans [115], with some estimates suggesting it to have surpassed the size of the polar bear or the Kodiak brown bear, *U. a. middendorffi*, by reaching a body mass of about 1'500 kg [1]. Based on this, the cave bear could have had an exceptionally small relative brain size. Considering the possible bias body mass estimations based on skull length had on the dataset, encephalization in Ursidae could be more even with two strong outliers, *U. malayanus* towards increased encephalization and *U. spelaeus* towards decreased one.

Conclusion

The aim of this study was to examine the encephalization in cave bears and comparing it with living and extinct members of Ursidae. *U. spelaeus*, and subsequently all potential species associated with this taxon, exhibit one of the lowest encephalization in Ursidae because body size increase outpaced brain size increase in its evolution. This is a trend observable early on in the cave bear lineage as is evidenced by the low encephalization of *U. deningeri*. My results stand in contrast to previous interpretations of cave bear brain evolution [42]. I showed that this study has used overestimated brain volumes due to the shape of cave bear skulls. Bear species, which do not exhibit dormancy and have a high caloric diet, showed a weak but significant correlation with bigger relative brain size. This would be in accordance with the trait-off between brain size and adipose tissue as well as studies on diet and brain size [60, 66–68]. The ecological shift towards a plant based diet alone did not affect encephalization in cave bears. However, a more general link associated with food acquirement strategy might still exist [67]. The herbivorous *U. spelaeus* has a small relative brain size possibly due to the combined effect of unequal body/brain size evolution and a seasonal environment in which dormancy was necessary for survival.

Additional files

Additional file 1:Table S1. Basicranial length, body mass estimates, brain volumes and brain mass for all examined bear skulls. (XLSX 33 kb)

Additional file 2: Table S2. Data subset of brain volume estimates based on external measurements by Finarelli [74] and glass bead method. (XLSX 12 kb)

Additional file 3: Supplementary Information. Results for different linear models and corresponding graphical output as well as boxplot on residuals based on PGLS with all species. (PDF 658 kb)

Additional file 4: Table S3. Additional results for slope and intercept pairwise comparisons. (XLSX 31 kb)

Additional file 5: Table S4. Node values for ancestral stage reconstructions. (XLSX 13 kb)

Acknowledgments

This paper is part of a doctoral thesis at the University of Zurich supervised by Marcelo R. Sánchez-Villagra, Evolutionary Morphology and Palaeobiology of Vertebrates (PIMUZ). I want to acknowledge the many institutions and people giving us access to their collections. Thus, I want to thank Björn Berning (BZL), Toni Bürgin (NMSG), Stephanie Fassl (PIUW), Christine Frischauf (PIUW), Christiane Funk (MfN), Ursula B. Göhlich (NHM), Marianne Haffner (ZMUZH), Oliver Hampe (MfN), Lars van den Hoek Ostende (NBC), Evelyn Kustatscher (PZO), Marc Nussbaumer (NMBE), Barbara Oberholzer (ZMUZH), Natasja den Ouden (NBC), Gernot Rabeder (PIUW), Thomas Schossleitner (MfN), Martin Studeny (BZL), Evangelia Tsoukala (AUTH), Géraldine Veron (MNHN) for their support. Additionally, I am grateful to Karin Isler (AIMZH) for fruitful discussions on methodology and design of this study and Marcelo R. Sánchez-Villagra (PIMUZ) for discussions and editions to the text. I also want to thank Vera Weisbecker (UQ) and two anonymous reviewers for their helpful comments, which considerably improved this study. Funding for this research was provided by the Swiss National Science Foundation (SNF) grant no. 31003A-149605 to Marcelo R. Sánchez-Villagra.

Competing interests

I declare no competing interests of any kind.

References

1. Rabeder G, Nagel D, Pacher M. Der Höhlenbär. Stuttgart: Thorbecke Verlag; 2000.
2. van Heteren AH, MacLarnon A, Soligo C, Rae TC. Functional morphology of the cave bear (*Ursus spelaeus*) mandible: a 3D geometric morphometric analysis. Org Divers Evol. 2016;16(1):299–314.
3. Knapp M, Rohland N, Weinstock J, Baryshnikov G, Sher A, Nagel D, et al. First DNA sequences from Asian cave bear fossils reveal deep divergences and complex phylogeographic patterns. Mol Ecol. 2009;18(6):1225–38.
4. Stuart AJ. Insularity and quaternary vertebrate faunas in Britain and Ireland. Geol Soc Spec Publ. 1995;96(1):111–25.
5. Garcia N, Arsuaga JL, Torres T. The carnivore remains from the Sima de los Huesos Middle Pleistocene site (sierra de Atapuerca, Spain). J Hum Evol. 1997;33(2):155–74.
6. Barnosky AD, Koch PL, Feranec RS, Wing SL, Shabel AB. Assessing the causes of Late Pleistocene extinctions on the continents. Science. 2004;306(5693):70–5.
7. Stuart AJ. Late quaternary megafaunal extinctions on the continents: a short review. Geol J. 2015;50(3):338–63.
8. Stiller M, Baryshnikov G, Bocherens H, Grandal d'Anglade A, Hilpert B, Münzel SC, et al. Withering away – 25,000 years of genetic decline preceded cave bear extinction. Mol Biol Evol 2010;27(5):975-978.
9. Pacher M, Stuart AJ. Extinction chronology and palaeobiology of the cave bear (*Ursus spelaeus*). Boreas. 2009;38(2):189–206.
10. Baca M, Popović D, Stefaniak K, Marciszak A, Urbanowski M, Nadachowski A, et al. Retreat and extinction of the Late Pleistocene cave bear (*Ursus spelaeus* Sensu Lato). Sci Nat. 2016;103(11):92.
11. Krause J, Unger T, Noçon A, Malaspinas A-S, Kolokotronis S-O, Stiller M, et al. Mitochondrial genomes reveal an explosive radiation of extinct and extant bears near the Miocene-Pliocene boundary. BMC Evol Biol. 2008;8(1):1–12.
12. Bon C, Caudy N, de Dieuleveult M, Fosse P, Philippe M, Maksud F, et al. Deciphering the complete mitochondrial genome and phylogeny of the extinct cave bear in the Paleolithic painted cave of Chauvet. PNAS. 2008; 105(45):17447–52.
13. Loreille O, Orlando L, Patou-Mathis M, Philippe M, Taberlet P, Hanni C. Ancient DNA analysis reveals divergence of the cave bear, *Ursus spelaeus*, and brown bear, *Ursus arctos*, lineages. Curr Biol. 2001;11(3):200–3.
14. Ehrenberg K. Die Variabilität der Backenzähne beim Höhlenbären. In: Abel O, Kyrle G, editors. Die Drachenhöhle bei Mixnitz. Speläologische Monographien 7/8. Wien: Österreichische Staatsdruckerei; 1931. p. 537-573.
15. Abel O. Das Lebensbild der eiszeitlichen Tierwelt der Drachenhöhle bei Mixnitz. In: Abel O, Kyrle G, editors. Die Drachenhöhle bei Mixnitz. Speläologische Monographien 7/8. Wien: Österreichische Staatsdruckerei; 1931. p. 885-920.
16. Kurtén B. The cave bear story. New York: Columbia University Press; 1976.
17. Kurtén B. Pleistocene mammals of Europe. Chicago: Aldine; 1968.
18. Thenius E. Zähne und Gebiß der Säugetiere. Berlin: W. de Gruyter; 1989.
19. van Heteren AH, MacLarnon A, Soligo C, Rae TC. Functional morphology of the cave bear (*Ursus spelaeus*) cranium: a three-dimensional geometric morphometric analysis. Quat Int. 2014;339:209–16.

20. Naito YI, Germonpré M, Chikaraishi Y, Ohkouchi N, Drucker DG, Hobson KA, et al. Evidence for herbivorous cave bears (Ursus spelaeus) in Goyet cave, Belgium: implications for palaeodietary reconstruction of fossil bears using amino acid δ15N approaches. J Quat Sci. 2016;31(6):598–606.

21. Münzel SC, Rivals F, Pacher M, Döppes D, Rabeder G, Conard NJ, et al. Behavioural ecology of Late Pleistocene bears (Ursus spelaeus, Ursus ingressus): Insight from stable isotopes (C, N, O) and tooth microwear. Quat Int. 2014;339–340:148–63.

22. Bocherens H, Stiller M, Hobson KA, Pacher M, Rabeder G, Burns JA, et al. Niche partitioning between two sympatric genetically distinct cave bears (Ursus spelaeus and Ursus ingressus) and brown bear (Ursus arctos) from Austria: isotopic evidence from fossil bones. Quat Int. 2011;245(2):238–48.

23. Bocherens H, Drucker DG, Billiou D, Geneste J-M, van der Plicht J. Bears and humans in Chauvet cave (Vallon-Pont-d'Arc, Ardèche, France): insights from stable isotopes and radiocarbon dating of bone collagen. J Hum Evol. 2006;50(3):370–6.

24. Bocherens H, Billiou D, Patou-Mathis M, Bonjean D, Otte M, Mariotti A. Paleobiological implications of the isotopic signatures (13C,15N) of fossil mammal collagen in Scladina cave (Sclayn, Belgium). Quat Res. 1997;48(3):370–80.

25. Bocherens H, Fizet M, Mariotti A. Diet, physiology and ecology of fossil mammals as inferred from stable carbon and nitrogen isotope biogeochemistry: implications for Pleistocene bears. Palaeogeogr Palaeoclimatol Palaeoecol. 1994;107(3):213–25.

26. Richards MP, Pacher M, Stiller M, Quilès J, Hofreiter M, Constantin S, et al. Isotopic evidence for omnivory among European cave bears: Late Pleistocene Ursus spelaeus from the Peştera cu Oase, Romania. PNAS. 2008;105(2):600–4.

27. Robu M, Fortin JK, Richards MP, Schwartz CC, Wynn JG, Robbins CT, et al. Isotopic evidence for dietary flexibility among European Late Pleistocene cave bears (Ursus spelaeus). Can J Zool. 2013;91(4):227–34.

28. Meloro C. Feeding habits of Plio-Pleistocene large carnivores as revealed by the mandibular geometry. J Vert Paleontol. 2011;31(2):428–46.

29. Figueirido B, Palmqvist P, Pérez-Claros JA. Ecomorphological correlates of craniodental variation in bears and paleobiological implications for extinct taxa: an approach based on geometric morphometrics. J Zool. 2009;277(1):70–80.

30. Jones BD, DeSantis LRG. Dietary ecology of the extinct cave bear: evidence of omnivory as inferred from dental microwear textures. Acta Palaeontol Pol. 2016;61(4):735–41.

31. Peigné S, Goillot C, Germonpré M, Blondel C, Bignon O, Merceron G. Predormancy omnivory in European cave bears evidenced by a dental microwear analysis of Ursus spelaeus from Goyet, Belgium. PNAS. 2009; 106(36):15390–3.

32. Rabal-Garcés R, Cuenca-Bescós G, Ignacio Canudo J, De Torres T. Was the European cave bear an occasional scavenger? Lethaia. 2012;45(1):96–108.

33. Bocherens H. Isotopic tracking of large carnivore palaeoecology in the mammoth steppe. Quat Sci Rev. 2015;117:42–71.

34. Edinger T. Über einige fossile Gehirne. Paläont Z. 1928;9(4):379–402.

35. Über DH. Hirnschädelausgüsse von Ursus spelaeus. In: Abel O, Kyrle G, editors. Die Drachenhöhle bei Mixnitz. Speläologische Monographien 7/8. Wien: Österreichischen Staatsdruckerei; 1931. p. 498–536.

36. Marinelli W. Bericht über die Untersuchung der Höhlenbärenschädel. In: Abel O, Kyrle G, editors. Die Drachenhöhle bei Mixnitz. Speläologische Monographien 7/8. Wien: Österreichischen Staatsdruckerei; 1931. p. 332–497.

37. Finarelli JA, Flynn JJ. Brain-size evolution and sociality in Carnivora. PNAS. 2009;106(23):9345–9.

38. Finarelli JA, Flynn JJ. The evolution of encephalization in caniform carnivorans. Evolution. 2007;61(7):1758–72.

39. Groiss JT. Untersuchungen der Gehirnmorphologie von Ursus deningeri v. REICHENAU und von Ursus spelaeus ROSENMÜLLER (Mammalia, Ursidae) an Schädelausgüssen quartärer Funde aus österreichischen Höhlen. Abh Geol B-A. 1994;50:115–23.

40. Santos E, Garcia N, Carretero JM, Arsuaga JL, Tsoukala E. Endocranial traits of the Sima de los Huesos (Atapuerca, Spain) and Petralona (Chalkidiki, Greece) Middle Pleistocene ursids. Phylogenetic and biochronological implications. Ann Paleontol. 2014;100(4):297–309.

41. García N, Santos E, Arsuaga JL, Carretero JM. Endocranial morphology of the Ursus deningeri von Reichenau 1904 from the Sima de los Huesos (sierra de Atapuerca) Middle Pleistocene site. J Vert Paleontol. 2007;27(4):1007–17.

42. Smaers JB, Dechmann DKN, Goswami A, Soligo C, Safi K. Comparative analyses of evolutionary rates reveal different pathways to encephalization in bats, carnivorans, and primates. PNAS. 2012;109(44):18006–11.

43. Isler K, van Schaik CP. The expensive brain: a framework for explaining evolutionary changes in brain size. J Hum Evol. 2009;57(4):392–400.

44. Mink JW, Blumenschine RJ, Adams DB. Ratio of central nervous system to body metabolism in vertebrates: its constancy and functional basis. Am J Physiol Regul Integr Comp Physiol. 1981;241(3):R203–R12.

45. Aiello LC, Bates N, Joffe T. In defense of the expensive tissue hypothesis. In: Falk D, Gibson KR, editors. Evolutionary Anatomy of the primate Cerebral Cortex. Cambridge: Cambridge University Press; 2001. p. 57–78.

46. Dunbar RIM. The social brain hypothesis. Evol Anthropol. 1998;6(5):178–90.

47. Pérez-Barbería FJ, Shultz S, Dunbar RIM, Janis C. Evidence for coevolution of sociality and relative brain size in three orders of mammals. Evolution. 2007; 61(12):2811–21.

48. Shultz S, Dunbar RIM. Both social and ecological factors predict ungulate brain size. Proc Biol Sci. 2006;273(1583):207–15.

49. Taylor AB, van Schaik CP. Variation in brain size and ecology in Pongo. J Hum Evol. 2007;52(1):59–71.

50. van Woerden JT, van Schaik CP, Isler K. Effects of seasonality on brain size evolution: evidence from strepsirrhine primates. Am Nat. 2010; 176(6):758–67.

51. van Woerden JT, Willems EP, van Schaik CP, Isler K. Large brains buffer energetic effects of seasonal habitats in Catarrhine primates. Evolution. 2012;66(1):191–9.

52. Weisbecker V, Blomberg S, Goldizen AW, Brown M, Fisher D. The evolution of relative brain size in marsupials is energetically constrained but not driven by behavioral complexity. Brain Behav Evol. 2015;85(2):125–35.

53. Garamszegi LZ, Møller AP, Erritzøe J. Coevolving avian eye size and brain size in relation to prey capture and nocturnality. Proc Biol Sci. 2002; 269(1494):961–7.

54. Gittleman JL. Carnivore brain size, behavioral ecology, and phylogeny. J Mammal. 1986;67(1):23–36.

55. Mace GM, Harvey PH, Clutton-Brock TH. Brain size and ecology in small mammals. J Zool. 1981;193(3):333–54.

56. Eisenberg JF, Wilson DE. Relative brain size and feeding strategies in the Chiroptera. Evolution. 1978;32(4):740–51.

57. Finarelli JA. Does encephalization correlate with life history or metabolic rate in Carnivora? Biol Lett. 2010;6(3):350–3.

58. Raichlen DA, Gordon AD. Relationship between exercise capacity and brain size in mammals. PLoS One. 2011;6(6):e20601.

59. Eisenberg JF, Wilson DE. Relative brain size and demographic strategies in didelphid marsupials. Am Nat. 1981;118(1):1–15.

60. Navarrete A, van Schaik CP, Isler K. Energetics and the evolution of human brain size. Nature. 2011;480(7375):91–3.

61. Barton RA, Capellini I. Maternal investment, life histories, and the costs of brain growth in mammals. PNAS. 2011;108(15):6169–74.

62. Isler K, van Schaik CP. Metabolic costs of brain size evolution. Biol Lett. 2006; 2(4):557–60.

63. Weisbecker V, Goswami A. Brain size, life history, and metabolism at the marsupial/placental dichotomy. PNAS. 2010;107(37):16216–21.

64. Barrickman NL, Bastian ML, Isler K, van Schaik CP. Life history costs and benefits of encephalization: a comparative test using data from long-term studies of primates in the wild. J Hum Evol. 2008;54(5):568–90.

65. Western D. Size, life history and ecology in mammals. Afr J Ecol. 1979;17(4): 185–204.

66. Heldstab SA, van Schaik CP, Isler K. Being fat and smart: a comparative analysis of the fat-brain trade-off in mammals. J Hum Evol. 2016;100:25–34.

67. Striedter GF. Principles of brain evolution. Sunderland: Sinauer Associates; 2005.

68. DeCasien AR, Williams SA, Higham JP. Primate brain size is predicted by diet but not sociality. Nat Ecol Evol. 2017;1:0112.

69. Hunter L. Carnivores of the world. Princeton (New Jersey): Princeton University Press; 2011.

70. Lyman CP, Willis JS, Malan A, Wang LCH. Hibernation and torpor in mammals and birds. New York: Academic Press; 1982.

71. Logan CJ, Clutton-Brock TH. Validating methods for estimating endocranial volume in individual red deer (Cervus elaphus). Behav Process. 2013;92:143–6.

72. van Valkenburgh B. Skeletal and dental predictors of body mass in carnivores. In: Damuth J, Macfadden BJ, editors. Body size in mammalian paleobiology: estimation and biological implications. Cambridge: Cambridge University Press; 1990. p. 181–206.

73. Ebinger P. A cytarquitectonic volumetric comparison of brains in wild and domestic sheep. Z Anat Entwickl-Gesch. 1974;144:267–302.

74. Finarelli JA. Estimation of endocranial volume through the use of external skull measures in the Carnivora (Mammalia). J Mammal. 2006;87(5):1027–36.

75. R Development Core Team. R: A language and environment for statistical computing. Version 3.2.3. 2015. http://www.R-project.org. Accessed 10 Dec 2015.

76. Paradis E, Claude J, Strimmer K. APE: analyses of phylogenetics and evolution in R language. Bioinformatics. 2004;20:289–90.

77. Orme D, Freckleton R, Thomas G, Petzoldt T, Fritz S, Isaac N, et al. caper: comparative analyses of phylogenetics and evolution in R. R package version 0.5.2. 2013. https://CRAN.R-project.org/package=caper. Accessed 25 Jan 2016.

78. Kutschera VE, Bidon T, Hailer F, Rodi JL, Fain SR, Janke A. Bears in a forest of gene trees: Phylogenetic inference is complicated by incomplete lineage sorting and gene flow. Mol Biol Evol. 2014;31(8):2004–17.

79. Noonan JP, Hofreiter M, Smith D, Priest JR, Rohland N, Rabeder G. Genomic sequencing of Pleistocene cave bears. Science. 2005;309(5734):597–9.

80. Rabeder G, Hofreiter M. Der neue Stammbaum der alpinen Höhlenbären. Die Höhle. 2004;55:58–77.

81. Rabeder G, Hofreiter M, Nagel D, Withalm G. New taxa of Alpine cave bears (Ursidae, Carnivora). Cah Sci. 2004;2:49–67.

82. Rabeder G, Debeljak I, Hofreiter M, Withalm G. Morphological responses of cave bears (Ursus spelaeus group) to high-alpine habitats. Die Höhle. 2008; 59:59–72.

83. Dabney J, Knapp M, Glocke I, Gansauge M-T, Weihmann A, Nickel B, et al. Complete mitochondrial genome sequence of a Middle Pleistocene cave bear reconstructed from ultrashort DNA fragments. PNAS. 2013;110(39): 15758–63.

84. Baca M, Mackiewicz P, Stankovic A, Popović D, Stefaniak K, Czarnogórska K, et al. Ancient DNA and dating of cave bear remains from Niedźwiedzia Cave suggest early appearance of Ursus ingressus in Sudetes. Quat Int. 2014; 339–340:217–23.

85. Stiller M, Molak M, Prost S, Rabeder G, Baryshnikov G, Rosendahl W, et al. Mitochondrial DNA diversity and evolution of the Pleistocene cave bear complex. Quat Int. 2014;339–340:224–31.

86. Nyakatura K, Bininda-Emonds OR. Updating the evolutionary history of Carnivora (Mammalia): a new species-level supertree complete with divergence time estimates. BMC Biol. 2012;10(1):1–31.

87. Giraudoux P. pgirmess: data analysis in ecology. R package version 1.6.3. 2015. https://CRAN.R-project.org/package=pgirmess. Accessed 22 Jan 2016.

88. Pohlert G. The pairwise multiple comparison of mean ranks package (PMCMR). 2014. http://CRAN.R-project.org/package=PMCMR. Accessed 22 Jan 2016.

89. Wickham H. ggplot2: elegant graphics for data analysis. New York: Springer-Verlag; 2009.

90. Maddison WP. Squared-change parsimony reconstructions of ancestral states for continuous-valued characters on a phylogenetic tree. Syst Zool. 1991;40(3):304–14.

91. Germain D, Laurin M. Evolution of ossification sequences in salamanders and urodele origins assessed through event-pairing and new methods. Evol Dev. 2009;11(2):170–90.

92. Maddison WP, Maddison DR. Mesquite: a modular system for evolutionary analysis. Version 2.75. 2011. http://mesquiteproject.org. Accessed 10 Dec 2014.

93. McLeod AI. Kendall: Kendall rank correlation and Mann-Kendall trend test. R package version 2.2. 2011. https://CRAN.R-project.org/package=Kendall. Accessed 9 May 2016.

94. Kaskan PM, Finlay BL. Encephalization and its developmental structure: how many ways can a brain get big? In: Falk D, Gibson KR, editors. Evolutionary Anatomy of the primate Cerebral Cortex. Cambridge: Cambridge University Press; 2001. p. 14–29.

95. Gibson KR, Rumbaugh D, Beran M. Bigger is better: primate brain size in relationship to cognition. In: Falk D, Gibson KR, editors. Evolutionary Anatomy of the primate Cerebral Cortex. Cambridge: Cambridge University Press; 2001. p. 79–97.

96. Hofman MA. Evolution of the human brain: when bigger is better. Front Neuroanat. 2014;8:15.

97. Roth G, Dicke U. Evolution of the brain and intelligence. Trends Cogn Sci. 2005;9(5):250–7.

98. Jerison HJ. Evolution of the brain and intelligence. New York: Academic Press; 1973.

99. Yao L, Brown JP, Stampanoni M, Marone F, Isler K, Martin RD. Evolutionary change in the brain size of bats. Brain Behav Evol. 2012;80(1):15–25.

100. Edmunds NB, Laberge F, McCann KS. A role for brain size and cognition in food webs. Ecol Lett. 2016;19(8):948–55.

101. Lefebvre L, Whittle P, Lascaris E, Finkelstein A. Feeding innovations and forebrain size in birds. Anim Behav. 1997;53(3):549–60.

102. Ratcliffe JM, Fenton MB, Shettleworth SJ. Behavioral flexibility positively correlated with relative brain volume in predatory bats. Brain Behav Evol. 2006;67(3):165–76.

103. Reader SM, Laland KN. Social intelligence, innovation, and enhanced brain size in primates. PNAS. 2002;99(7):4436–41.

104. Lefebvre L, Reader SM, Sol D. Brains, innovations and evolution in birds and primates. Brain Behav Evol. 2004;63(4):233–46.

105. Weisbecker V, Goswami A. Reassessing the relationship between brain size, life history, and metabolism at the marsupial/placental dichotomy. Zool Sci. 2014;31(9):608–12.

106. Safi K, Seid MA, Dechmann DKN. Bigger is not always better: when brains get smaller. Biol Lett. 2005;1(3):283–6.

107. Köhler M, Moyà-Solà S. Reduction of brain and sense organs in the fossil insular bovid Myotragus. Brain Behav Evol. 2004;63(3):125–40.

108. Weston EM, Lister AM. Insular dwarfism in hippos and a model for brain size reduction in Homo floresiensis. Nature. 2009;459(7243):85–8.

109. Sol D. Revisiting the cognitive buffer hypothesis for the evolution of large brains. Biol Lett. 2009;5(1):130–3.

110. Tacutu R, Craig T, Budovsky A, Wuttke D, Lehmann G, Taranukha D, et al. Human ageing genomic resources: integrated databases and tools for the biology and genetics of ageing. Nucleic Acids Res. 2013;41(D1):D1027–D33.

111. Ehrenberg K. Ein fast vollständiges Höhlenbärenneonatenskelett aus der Salzofenhöhle im Toten Gebirge. Ann Nathist Mus Wien. 1973;77:69–113.

112. Stiner MC. Mortality analysis of Pleistocene bears and its paleoanthropological relevance. J Hum Evol. 1998;34(3):303–26.

113. Radinsky L. Evolution of brain size in carnivores and ungulates. Am Nat. 1978;112(987):815–31.

114. Soul LC, Benson RBJ, Weisbecker V. Multiple regression modeling for estimating endocranial volume in extinct Mammalia. Paleobiology. 2012; 39(1):149–62.

115. Christiansen P. What size were Arctodus simus and Ursus spelaeus (Carnivora: Ursidae)? Ann Zool Fenn. 1999;36(2):93–102.

15

Microevolution of the noble crayfish (*Astacus astacus*) in the Southern Balkan Peninsula

Anastasia Laggis[1], Athanasios D. Baxevanis[1], Alexandra Charalampidou[2], Stefania Maniatsi[1], Alexander Triantafyllidis[1] and Theodore J. Abatzopoulos[1*] 🆔

Abstract

Background: The noble crayfish (*Astacus astacus*) displays a complex historical and contemporary genetic status in Europe. The species divergence has been shaped by geological events (i.e. Pleistocene glaciations) and humanly induced impacts (i.e. translocations, pollution, etc.) on its populations due to species commercial value and its niche degradation. Until now, limited genetic information has been procured for the Balkan area and especially for the southernmost distribution of this species (i.e. Greece). It is well known that the rich habitat diversity of the Balkan Peninsula offers suitable conditions for genetically diversified populations. Thus, the present manuscript revisits the phylogenetic relationships of the noble crayfish in Europe and identifies the genetic make-up and the biogeographical patterns of the species in its southern range limit.

Results: Mitochondrial markers (i.e. COI and 16S) were used in order to elucidate the genetic structure and diversity of the noble crayfish in Europe. Two of the six European haplotypic lineages, were found exclusively in Greece. These two lineages exhibited greater haplotypic richness when compared with the rest four (of "Central European" origin) while they showed high genetic diversity. Divergence time analysis identified that the majority of this divergence was captured through Pleistocene, suggesting a southern glacial refugium (Greece, southern Balkans). Furthermore, six microsatellite markers were used in order to define the factors affecting the genetic structure and demographic history of the species in Greece. The population structure analysis revealed six to nine genetic clusters and eight putative genetic barriers. Evidence of bottleneck effects in the last ~5000 years (due to climatic and geological events and human activities) is also afforded. Findings from several other research fields (e.g. life sciences, geology or even archaeology) have been utilized to perceive the genetic make-up of the noble crayfish.

Conclusions: The southernmost part of Balkans has played a major role as a glacial refugium for *A. astacus*. Such refugia have served as centres of expansion to northern regions. Recent history of the noble crayfish in southern Balkans reveals the influence of environmental (climate, geology and/or topology) and anthropogenic factors.

Keywords: Noble crayfish, mtDNA, 16S, COI, Microsatellites, Europe, Balkans, Phylogeny, Populations

Background

Quaternary climatic oscillations have influenced the flora (e.g. [1–3]) and fauna (e.g. [4–6]) of Europe. Pleistocene glaciations had substantial impact on the flora and fauna directly via major biogeographic events (e.g. displacement) and habitat alterations, while indirectly, via

fluctuations in environmental conditions [7–9]. The distribution and structure of obligate freshwater organisms frequently reflects historic, geological processes (such as tectonic activity, sea level change and glaciation) due to their dependency to the aquatic environment [10]. The long-term survival of many species depended on refugia and their current distribution, genetic structure and diversity reflects such historical processes [7–9, 11]. In Europe, the Balkan Peninsula is considered one of the major glacial refugia for many species [8]. It, also, played

* Correspondence: abatzop@bio.auth.gr
[1]Department of Genetics, Development and Molecular Biology, School of Biology, Aristotle University of Thessaloniki, 54124 Thessaloniki, Macedonia, Greece
Full list of author information is available at the end of the article

an important role in the colonization of eastern and western parts of Europe [7]. The majority of the European temperate fishes seem to have colonized the continent from the Black Sea through the Danube and Dnieper rivers [8].

More specifically, Greece with its complex geographic landscape [12], coupled with its geographic location (southernmost part of the Balkan Peninsula), offered suitable conditions for many species during glaciations (e.g. [13, 14]). In this area, the geological processes (tectonic and seismic activity) as well as climatic conditions forged a complex geographic landscape with rich habitat diversity [12]. Consequently, several restricted alluvial reaches [12] and small climatically stable areas [15] were formed, influencing the flora [1, 2] and fauna [13, 14]. During the mid- and late Holocene, the anthropogenic influence was intensified, generating the modification - to some extent - of the natural environment [16, 17].

An example of obligate freshwater species that could serve as a "model organism" for tackling historical events and biogeographical processes is the noble crayfish (*Astacus astacus*). It is a well-established European freshwater crustacean and its distribution expands from Norway (North end) to Greece (South end) and from France (West end) to Russia (East end) [18]. It plays an important role on the freshwater ecosystems, with its wide habitat usage and biological features [19]. The species is harboured in a plethora of inland freshwater habitats (streams, rivers, lakes, ponds and reservoirs [19]) and it is an opportunistic, omnivorous feeder [2, 3]. Sexual maturity is reached between 2 and 5 years of age [19, 20]. Females may be reproductively inactive for long periods (from one to several years), depending on the ambient water temperature [21]. The estimated life expectancy varies between 13 and 20 years [20], with the upper limit considered as uncommon [22], anecdotal [23] or even doubted [20]. Cukerzis (1988, cited in Keller [24]) estimated the probable longevity of the noble crayfish in Central Europe to be 7-8 years. The species has high oxygen demands [20], is sensitive to organic pollution [25] and is considered as a water quality indicator [26].

The economic and cultural value of the noble crayfish is well known in the Central and Northern Europe [19]. At the same time, several humanly induced interventions (e.g. habitat degradation, introduction of foreign species and crayfish plague) affect negatively the noble crayfish ([23] and references therein). Despite their wide distribution, the IUCN estimated a decline rate of the species between 50% and 70% [23]. Thus, the "IUCN Red List of Threatened Species" classifies the noble crayfish as a vulnerable species [23], while the "Red book of threatened fauna of Greece" classifies it at the unknown status [27]. It is worth noting that international treaties and conventions have attributed special protection to the noble crayfish (Bern Convention Appendix III, EU Habitats Directive Appendix V and directive 92/43/EEC). In order to compensate the decline or loss of populations, reintroduction of the species has been widely applied in Europe as a management tool [28]. However, variation of regulations between European countries and regional authorities based mainly on cultural traditions [19] complicates the preservation of the species.

Translocations may affect noble crayfish by influencing its range expansion, gene flow and gene pool, as identified by several studies using various molecular markers [16, 17]. It is important to notice that the majority of those translocations have disregarded the genetic structure of noble crayfish populations. The necessity of recording the genetic pool prior translocations has been recently pointed out [16, 17, 29]. To our knowledge, only one study has used microsatellite markers as monitoring tool of already translocated noble crayfish populations in the Czech Republic; the latter research was focused on the genetic variation of two translocated populations to evaluate the success of the managerial protocol applied in the last decade [30].

In the past, increasing attention was paid upon the genetic structure of the noble crayfish populations using allozymes ([31] and references therein), ISSR-PCR [32], RAPD-PCR [33] and microsatellites within rDNA-ITS1 ([34] and references therein). During the last years, three large-scale genetic analyses have been published, diagnosing the genetic diversity of the species between large geographical areas [16, 17, 29]. Microsatellite analysis differentiated Northern European (Estonia, Finland and Sweden) populations from Central European (Germany and Czech Republic) populations, with the former exhibiting lower genetic variation [17]. Furthermore, Schrimpf et al. [16, 29] revealed higher genetic diversity in the Black Sea basin populations compared to those of Southern Baltic, North and Adriatic Seas.

To our knowledge, there is currently no molecular study that encompassed samples from the southernmost limit of the species distribution (i.e. Greece). It should be noted, there are a few studies incorporating Greek crayfish to their genetic analysis, but they all refer to the stone crayfish, *Austropotamobius torrentium* [5, 35]. The few, sporadically published, studies about the noble crayfish in Greece are reporting on its geographical distribution and/or general information ([36] and references therein).

The lack of genetic information on the noble crayfish of Greece, coupled with the particular geographic features of the region, initiated the present study. This study largely focuses on *Astacus astacus* populations in the southernmost Balkan Peninsula, which has not been included in previous assessments [16, 29] although it is considered as a potentially important glacial refugium.

Mitochondrial (COI and 16S) haplotypes were used to further investigate the phylogenetic relationships of the noble crayfish in Europe. Knowing that mitochondrial DNA may have little practical value in population/conservation genetic studies of widespread organisms [37], microsatellite markers were used in order to detect potential genetic differentiation and recent demographic events that may have shaped the population structure of the noble crayfish in its southernmost distribution range (i.e. Greece). The genetic make-up and the biogeographical patterns of the Greek populations are analyzed to identify factors affecting its current structure and distribution. To accomplish the above goals (i.e. phylogeography of *A. astacus* in Europe and the genetic structure of the species in its southernmost distribution range) extensive sampling in the continental Greece was performed in an exhaustive manner. The genetic structure of the species is meant to be utilized as a baseline for management and conservation activities.

Methods

Sample collection and DNA extraction

Two hundred eighty four potential noble crayfish sites have been investigated in continental Greece. The selection of the potential sites was inferred from the bibliography ([36] and references therein), information from local residents and habitat requirements of the species (high oxygen demand and availability of shelters [20]). In order to maximise our chances to capture crayfish in areas where sporadic observations were reported from local residents, a wide search of the river catchment was performed. For instance, in Pinios river catchment (Thessaly district) a total of 53 potential sampling sites were visited and explored. Similarly, in Kalamas river catchment (Epirus district), 7 sampling sites were searched for crayfish. The thorough examination of the continental Greece resulted in the sampling of noble crayfish in 21 sites (from a total of 284 explored sites).

Most of the noble crayfish samples were collected by hand during the day. When the conditions were not favorable, LiNi traps [38] baited with meat were used during the night. The only exception was one individual in KEF site (see Table 1), which was captured while moving <2 m from the riverbank fleeing from it. In all potential sites sampled by hand, an area between 50 and 250 m was thoroughly examined (following the upward direction of the river flow) and lasted between 30 min and 1 h 30 min (sampling was performed by at least two researchers). Sampling was repeated in sites where the number of sampled individuals was less than 20 or if the presence of noble crayfish had been reported previously. Individuals were transferred to the laboratory in an isotherm with ice packs (individuals from different sites were stored in separate plastic bags). Fishing license for research purposes was retrieved from the Greek authorities. In total, 284 noble crayfish were collected from 21 sites (Table 1) and stored at –20 °C. Genomic DNA was extracted from pereiopods muscle tissue (≈ 10 mg) following the protocol of Estoup et al. [39].

Mitochondrial amplification and sequencing

Phylogenetic analysis of the noble crayfish was carried out using two mitochondrial markers: 1) a partial 16S sequence using 1471 and 1472 primers [40], and 2) a partial PCR-amplified sequence of the Cytochrome Oxidase subunit I (COI) using the universal LCO1490 and HCO2198 primers [41]. PCR reactions were adapted from Maniatsi et al. for COI [42] and Baxevanis et al. for 16S [43], with the following modifications: 0.25 μl KAPA Taq DNA Polymerase (KAPA Biosystems, South Africa), 2.5 μl 10× KAPA Taq Buffer A (KAPA Biosystems, South Africa), and 0.75 (for 16S) and 1.5 (for COI) mM MgCl2. The PCR thermal profile for COI followed Trontelj et al. [5]. For 16S the annealing temperature was slightly modified (45 °C) from Crandall and Fitzpatrick [40]. Sequencing reactions (both directions) were electrophoresed on a PRISM 3730×l DNA analyzer (Applied Biosystems, Foster City, USA) and prepared by Macrogen Inc. (Seoul, South Korea).

Mitochondrial analysis

The dataset used for the analysis comprised 37 randomly chosen noble crayfish (from a total of 284 individuals) collected from 21 sites of Greece (one or two individuals per site; Table 1) and sequences retrieved from the GenBank (19 for 16S and 32 for COI [6, 29, 35]; accession numbers DQ320033, KX370092, KF888279 to KF888295 for 16S and AY667146, KX369672, KF888296 to KF888325 for COI; Additional file 1). All produced sequences during this study have been deposited in GenBank (accession numbers KY048193 to KY048202 for 16S and KY067207 to KY067228 for COI; Additional file 1). Identity of amplified regions was confirmed using the BLAST searches. Sequences (16S and COI) were viewed in Bioedit v. 7.2.5 [44] and aligned in ClustalX v. 2.1 [45]. COI sequences were screened for pseudogenes following the procedure described in Buhay [46]. For each mitochondrial marker (COI and 16S) the heterogeneity of nucleotide frequencies, substitution and saturation and presence of phylogenetic signal were checked in DAMBE v. 5.5.29 [47]. The substitution model was determined in PartitionFinder v. 2.1.1 [48] using the Bayesian Information Criterion (BIC). Sequences (16S and COI) derived from the same individual were combined for further analysis (Additional file 1). "Concatenation" approach was followed since: 1) the two loci are linked on the mitochondria and maternally inherited [49], 2) the two genes have different resolving power [50], 3) the

Table 1 Information on sampling collection, where sites, abbreviation, habitat type, geographical coordinate, collection method, year of sampling (Year), sample size, number of mitochondrial sequences (mtDNA), number of microsatellite genotyped data (nDNA) and status are given

Site	Abbreviation	Habitat type	Water Basin[a]	Geographic coordinates	Collection method	Year	Sample size	mtDNA	nDna	Status
Arahthos	ARX	River	Arahthos	N39°15' E21°00'	Trap (recreational fisherman)	2011	1	1	-	Unknown
Kalamas	KAS	River	Kalamas	N39°39' E20°31'[b]	Trap (from Perdikaris)	2004	20	2	20	Transfer from 1) Larissa (80's[d] and 2) unknown origins (years 1990 and 1991)[e]
Tzaravina	TZA	Lake	Kalamas	N39°54' E20°30'	Hand	2011	6	2	6	Transfer from 1) Larissa (80's[d] and 2) unknown origins (years 1990 and 1991)[e]
Chani Kaber Aga	KPA	Stream	Arahthos	N39°43' E20°57'	Hand	2011	6	2	6	Unknown
Aoo1	AA1	Lake	Aoos	N39°49' E21°07'[b]	Trap (fisherman)	2009	20	2	20	Unknown
Aoo2	AA2	Lake	Aoos	N39°48' E21°06'[b]	Trap (fisherman)	2011	20	2	20	Unknown
Fragkades	FRA	Stream	Aoos	N39°50' E20°48'	Hand	2011	2	2	-	Unknown
Kalivia	KLV	Stream	Acheloos	N39°18' E21°42'	Hand	2011, 2012	16	2	16	Unknown
Neochori	NEO	Stream	Acheloos	N39°17' E21°43'	Hand	2011, 2012	11	2	11	Unknown
Karya	KRI	Stream	Ziliana[c]	N39°58' E22°25'	Hand	2011, 2012	20	2	20	Native
Skotina	SKR	Stream	Ziliana[c]	N39°59' E22°27'	Hand	2012	20	2	20	Native
Koniskos	KNS	Stream	Pinios	N39°48' E21°48'	Hand	2011, 2012	17	2	17	Transfer from Krania (last 10 to 15 years)[g]
Loggas	LOG	Lake	Pinios	N39°49' E21°55'	Hand	2011, 2012	17	2	17	Transfer from Krania (last 10 to 15 years)[g]
Begoritida/Agra	BGR	Lake	Aliakmon	N40°46' E21°46'[b]	Trap (fisherman)	2008, 2009	1	1	-	Transfer from Orhomenos and Edessa Aquaculture station[d]
Palaifyto	PLF	River	Axios	N40°47' E22°17'	Hand/Trap	2011, 2012	20	2	20	Transfer probably from Begoritida/Agra[g]
Tsivlo	TSV	Lake	Streams of North Peloponnese Beach	N38°04' E22°13'	Hand	2011, 2012	20	2	20	Transfer from unknown origin[f,g]
Doxa	DOX	Lake	Streams of North Peloponnese Beach	N37°55' E22°17'	Hand	2011, 2012	23	2	23	Unknown
Perivoli	PRV	Stream	Pinios	N39°03' E22°11'	Hand	2011	1	1	-	Native
Kefalovriso	KEF	River	Pinios	N39°53' E22°04'	Hand-land	2012	1	1	-	Transfer from unknown origin[g]
Krania	KRN	Stream	Aliakmon	N39°51' E21°18'	Hand	2012	20	2	20	Native
Pertouli	PRT	Stream	Acheloos	N39°32' E21°28'	Hand	2013	22	1	22	Unknown

[a]based on the National Commission of Water (Government Gazette 1383/8/2-9-10 and 1572/B/28-9-10).
[b]geographic coordinates obtained from Google earth.
[c]from [114].
[d]from [36].
[e]from [115].
[f]from [116].
[g]from local resident and recreational fisherman

concatenation method can perform equally or better than methods that attempt to account for sources of error introduced by incomplete lineage sorting [51], and 4) the selection of one mitochondrial region can influence the results obtained [52].

Phylogenetic inference was performed using a strict clock model and a coalescent tree prior in Beast v. 1.8 [53]. The Markov Chain Monte Carlo (MCMC) analysis run comprised 10^8 generations, sampled every 10^4 generations. In order to determine convergence and appropriate burn-in, the Effective Sample Size (ESS) values, densities plots and trace logs for each parameter were visualized in Tracer v. 1.5 [54]. The first 10^6 generations (10%) were discarded as burn-in. ESS values for all parameters were >486, larger than the threshold value of 200 identified by Tracer v. 1.5. The best fit tree was found using the Maximum clade credibility tree option in the Target tree type implemented in TreeAnnotator v. 1.8. Graphical representation of the relation of the number of haplotypes and samples with the phylogroups, were implemented in R v. 3.1.2 [55]. Pairwise within and between genetic distances of the haplotype groups for 16S and COI (Kimura 2-parameter and p-distance model), were computed in MEGA v. 6.06 [56].

Molecular dating was performed in Beast v. 1.8, using the same parameters as in the phylogenetic inference analysis (see also section "Results" for the best fit models generated for COI and 16S). To our knowledge, there are no fossil records and geophysical events in order to date the phylogroups of *Astacus astacus*. In order to estimate divergence time, different mutation rates were taken into account: 1) mitochondrial Arthropod substitution rate (2.3% pairwise sequence divergence per million years [57]), and 2) Decapod substitution rates for 16S (0.65-0.88% pairwise sequence divergence per million years) and COI (1.66-2.33% pairwise sequence divergence per million years) genes [58]. Both approaches have been previously used in crayfish [6, 35].

Microsatellite genotyping

The analysis comprised 278 samples from 16 sites [i.e. from the 21 sampled sites, 5 sites were excluded from the subsequent analysis due to low sampling size ($n < 6$); Table 1]. Six microsatellite loci (Aas8, Aas766, Aas1198, Aas2489, Aas3040 and Aas3950) were used [59, 60], with the forward primer IRD800-labelled. The PCR reaction and amplification were modified from those of Kõiv et al. [59, 60] (Additional file 2). Reagents were the same as described in mitochondrial section. In all PCR reactions 0.5 µl 10× Bovine Serum Albumin (BSA) was added.

Microsatellite analysis was performed in a semi-automated Li-COR® 4200 DNA Analyzer (Li-COR, Nebraska, USA), using 6.5% acrylamide sequencing gels (Li-COR®KB Plus™). Data was genotyped in Saga^{GT} software (Li-COR, Nebraska, USA). Allele scoring was facilitated by using the same individuals for each locus as a reference sample between gels, in combination with the molecular genetic marker (50-350 bp, Size Standard IRDye™ 800 Li-COR®). Scoring accuracy was increased by independent genotyping of the samples by two researchers.

Microsatellite analysis

Genotyping errors were assessed using Micro-Checker v. 2.2.3 [61] (95% CI and 10^4 repetitions). Mean number of alleles per locus, allele frequencies and observed, expected and unbiased expected Nei's heterozygosity [62] were calculated using Genetix v. 4.05 [63]. Genepop v. 4.2.2 [64] was used to test linkage disequilibrium, conformity to Hardy-Weinberg Equilibrium for each locus and to infer genetic differentiation for all pairs of sites. For all probability tests, Markov chain method was applied using 10^4 generations, 20 batches and 5000 iterations per batch. Significance was assessed following the sequential Bonferroni procedure [65]. Allelic richness, number of alleles and Weir and Cockerham's estimators (F_{IT}, F_{IS} and F_{ST}) were computed in FSTAT v. 2.9.3.2 [66]. R_{ST}Calc v. 2.2 [67] was used to calculate R_{ST} (10^4 permutations).

Genetic structure was inferred by the Bayesian clustering method implemented in Structure v. 2.3 [68]. The conditions performed were 20 runs for each genetic cluster (K) between 1 and 16 using a 10^5 burn-in period followed by 2 x 10^6 MCMC iterations, under an admixture model with independent allele frequencies. The number of K was determined via Structure Harvester [69]. Label Clump v. 1.1.2 [70] and Distruct v. 1.1 [71] were also employed to merge the results and generate the graphical representation of clusters, respectively. Discriminant analysis of principal components (DAPC) [72] in R package Adegenet v. 1.4-2 [73] was also used to assess the number of clusters and the relationships between populations. The number of K and the optimal number of components to be retained was assessed using the find.clusters (BIC) and optim.a.-score functions, respectively.

Monmonier's maximum difference algorithm implemented in Barrier v. 2.2 [74] was employed to identify barriers to gene flow within the data set (all sixteen sites). The procedure using the F_{ST} and D_{CE} described in [75] was followed. The tested number of probable barriers was between 1 and 10. The relationships between F_{ST}, Rousset's distance measure $F_{ST}/(1- F_{ST})$, R_{ST} and geographic distances among sites were analysed using Mantel test with 10^4 randomizations, available at Isolation By Distance Web Service v. 3.23 (IBDWS [76]).

Geographic distances were calculated in Geographic Distance Matrix Generator v. 1.2.3 [77].

Effective population size was estimated using ONe-SAMP [78], which relies on Approximate Bayesian Computation. The demographic history of noble crayfish was inferred through a hierarchical Bayesian model implemented in MsVar v. 1.3 [79], using the number of genetic clusters derived from the population structure analysis (software Structure v. 2.3). For each genetic cluster, five independent runs were conducted under an exponential model, with different seeds and hyperpriors (Additional file 3). For each data set, 10^5 steps and 9 x 10^4 thinning were used. The generation time was considered as per Pianka [80], using the formula $t_2 = (\alpha + \omega)/2$, where α is the age at maturation and ω the longevity. Since the longevity of the species is debated (see Background) two approaches were used: 1) the sexual maturity of the species (mean 3.5 years) and 2) the most probable longevity of 7.5 years (generation time 5.5). MsVar analysis was performed utilizing Aristotle University of Thessaloniki, Scientific Computing Office, IT Center, Institutional Computer Cluster resources (consuming a total of 91,561 CPU hours and 38 GB RAM capacity). 50% of each chain was discarded as burn-in. The convergence among MCMC runs was assessed visually and via the Gelman & Rubin [81] and Brooks & Gelman [82] statistics in R v. 3.1.2, using the package Coda v. 0.16-1 [83]. Point estimates of less than 1.2 were considered as a good indicator of convergence [84]. Hedge's d and mean effective size per cluster, along with their 95% CI, were calculated as described in Paz-Vinas et al. [85].

The relative probability of the demographic event was assessed using approximate Bayes Factors (BF). A generation time of 5.5 for the noble crayfish was used in the analysis. Different levels of BF were considered as in Salmona et al. [86]. BFs were computed every 50 years for a time interval between 0 and 12,000 years (Holocene). Additionally, BFs were also computed for equal length intervals (500, 100 and 50 years) covering the past 100,000, 50,000 and 7000 years, respectively. R-scripts were based on those provided by V. Sousa (personal communication) and those included in Paz-Vinas et al. [85].

Results
Mitochondrial analysis
mtDNA variation
The full dataset comprised 66 concatenated haplotypes derived from the combination of 53 unique haplotypes for COI and 17 unique haplotypes for 16S. In COI alignment no gaps were present, whereas in 16S alignment one or two gaps were observed. The aligned final dataset had a total length of 684 bp, with 334 bp for 16S and 350 bp for COI. The best fit model generated for 16S

was HKY + I [87] and for COI was HKY for codon positions 1 and 2, and HKY + Γ for codon position 3. The haplotype diversity was 0.6857 for 16S and 0.985 for COI. The number of variable and parsimony informative sites, were 17 and 11 for 16S and 49 and 32 for COI, respectively. There was no evidence for heterogeneity in base frequencies for neither mtDNA genes (16S: $\chi^2 = 1.51$, df = 165, $p > 0.05$ and COI: $\chi^2 = 9.26$, df = 207, $p > 0.05$). The substitution saturation test showed no significant saturation on either gene, since the index of substitution saturation (Iss for 16S = 0.20 and Iss for COI = 0.025) was lower than the critical value (Iss.c for 16S = 0.685 and Iss.c for COI = 0.686), with $T = 41.104$, df = 333, $p < 0.001$ for 16S and $T = 73.571$, df = 349, $p < 0.001$ for COI.

Phylogenetic inference
The tree inferred from the Bayesian phylogenetic analysis revealed six distinct genetic clades (Fig. 1). Phylogenetic clades G1 and G2 represent haplotypes found in Greece. Lineage G3 consists of haplotypes originated from Croatia, Montenegro and Germany. G4 contains haplotypes from Romania, Kosovo, Czech Republic and Austria. Phylogroup G5 comprises haplotypes belonging to several European countries (Romania, Belgium, Hungary, Bulgaria, Germany, Poland, Finland, Czech Republic and Austria). Finally, phylogenetic clade G6 is composed of Croatian, Montenegrin, German and Romanian haplotypes. It should be noted that minor discrepancies (identified in the shallow branches) have been observed between the independent gene trees produced for 16S and COI (data not shown); however, the major phylogroups were recovered in both trees.

All newly identified haplotypes were unique to Greece, and represent 36.4% of the total number of haplotypes (24 out of 66 haplotypes). The percentage of the phylogroup-specific haplotypes divided by the number of individuals (V) was greater than 58.06% in both phylogenetic lineages G1 and G2. In contrast, the percentage V for phylogroups G3 to G6 was between 4.87% and 16.32% (Fig. 2). Furthermore, the between genetic distances of the haplotype groups ranged from 0.1% to 1.9% for 16S and 0.6% to 4.1% for COI. The within genetic distances of the phylogroups were between 0.13% and 1.02% for 16S, and 0.37% and 3.5% for COI (Additional file 4).

Age estimates
Estimates of divergence times for the concatenated mtDNA dataset showed that the noble crayfish diversified during the Pleistocene. Similar time estimates were produced by the substitution rate approaches (Decapoda and Arthropoda; Table 2). Therefore, estimated time divergences using the mutation rate of Decapoda are

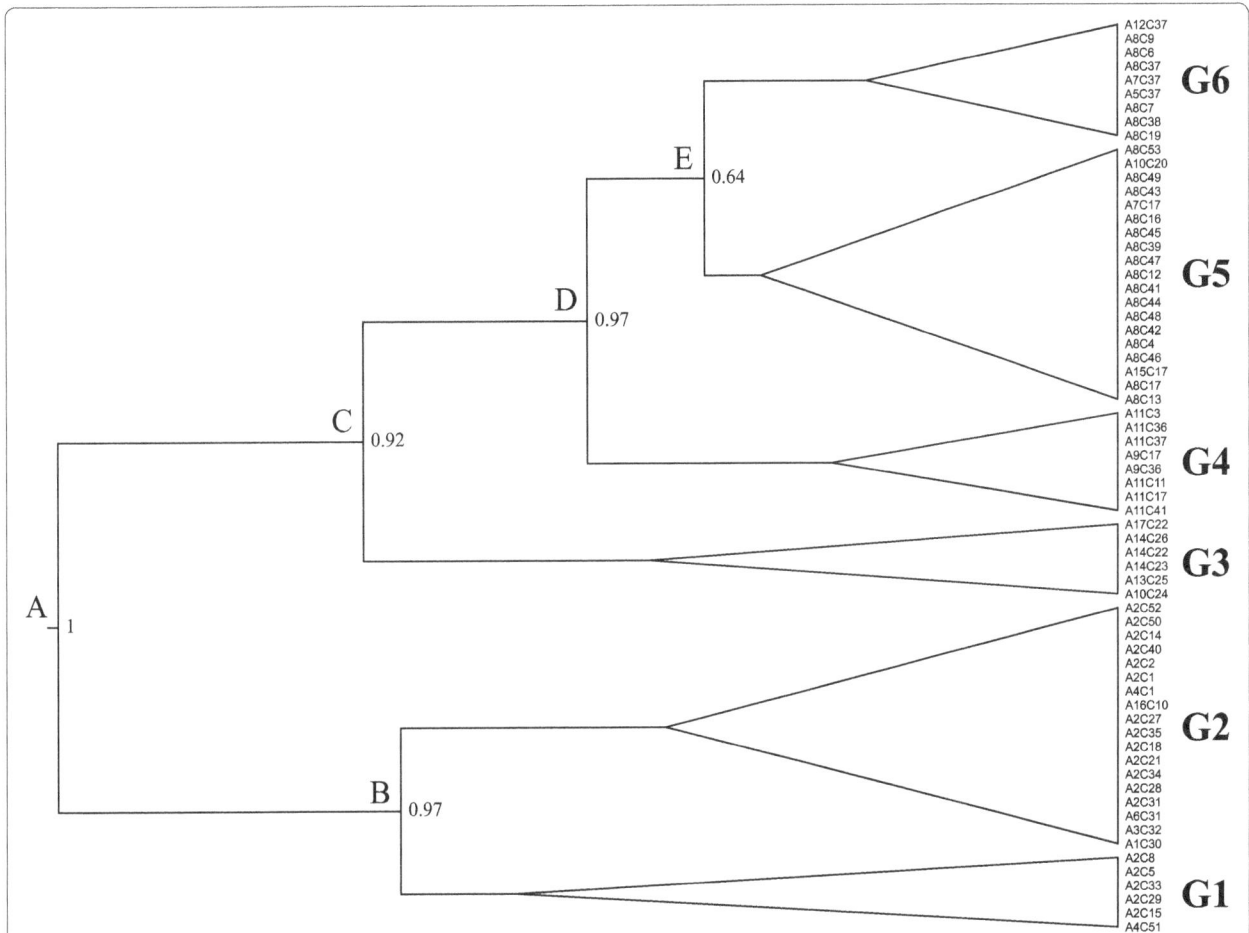

Fig. 1 Bayesian inference topology of phylogenetic relationships among 66 concatenated mtDNA haplotypes of the noble crayfish, using Beast v. 1.8. Values at nodes represent posterior probabilities. Major nodes for the noble crayfish are given (**A** to **E**). Phylogenetic clades are represented by symbols from G1 to G6, for more information see Results

further discussed. Node A, represents the separation of the southernmost distribution of *Astacus astacus* (i.e. Greece; phylogroups G1 and G2) from the remaining groups of Europe (G3 to G6) at 1.765 ± 0.0036 MYA (Table 2). The separation of the two Greek haplogroups (G1 and G2) occurred at 1.245 ± 0.0029 MYA (node B). The first group to diversify from the remaining groups of Europe is G3, at 1.284 ± 0.003 MYA (node C). Node D followed (separating G4 from groups G5 and G6) with an estimated divergence time at 0.792 ± 0.0022 MYA. Phylogenetic groups G5 and G6 shared for the last time a common ancestor at 0.556 ± 0.0016 MYA (node E). The divergence within groups (G1 to G6) was estimated at 1.765 ± 0.0036, 0.705 ± 0.0021, 0.761 ± 0.0022, 0.356 ± 0.015, 0.468 ± 0.0014 and 0.285 ± 0.0012 MYA, respectively.

Microsatellite analysis
General summary statistics
Putative null alleles were observed in 3 microsatellite loci and 5 sites (Table 3). Omitting Aas8, Aas2498 or

Aas1198 from the analysis did not have any substantial effect on the results (following the procedure described in [39]; data not shown). Therefore, all loci were included in the subsequent analysis. All microsatellite loci were polymorphic, with the number of alleles per locus ranging from 11 (Aas766) to 39 (Aas1198) (Additional file 5). A total of 111 different alleles were recognised with an average of 18.5 alleles per locus (Additional file 5). For all pairs of loci across all sites, 5 out of 240 global tests for genotypic linkage disequilibrium were statistically significant (none after Bonferroni correction) and none among locus pairs (Fisher's method). Deviation from Hardy-Weinberg equilibrium was present at 9 out of 96 tests (none after Bonferroni correction).

The mean allelic richness of each site varied from 1.86 (KRN) to 5.57 (KAS) and the mean observed heterozygosity ranged from 0.19 (KRN) to 0.77 (KAS and DOX) (Table 3; see also Additional file 5). Overall inbreeding coefficient F_{IS} for each sampled site showed heterozygote excess in several sites (AA2, LOG, NEO, KRN and

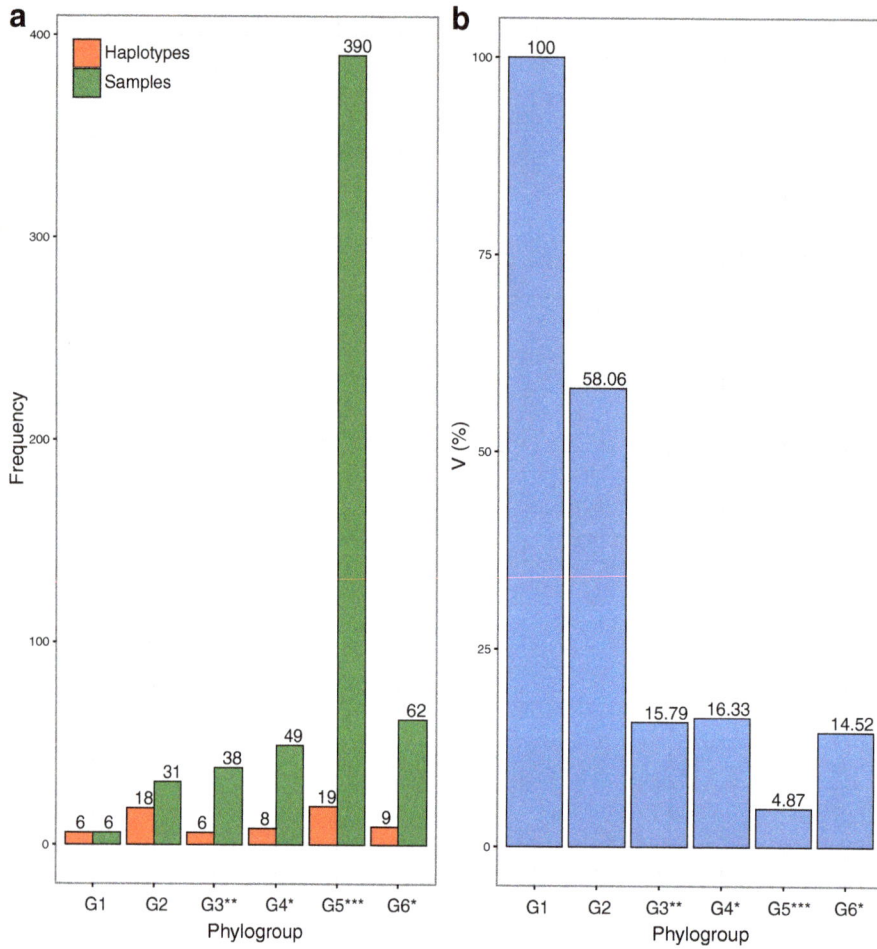

Fig. 2 Graphical representation of the composition of the phylogenetic clades (G1 to G6), based on: (**a**) the number of haplotypes (*red color*) and samples (*green color*); (**b**) the percentage of the phylogroup-specific haplotypes divided by the number of individuals (V; *blue color*). Where: *, based on [29]; **, based on [29, 35]; *** based on [6, 29, 35]

Table 2 Estimates of divergence time for the concatenated mtDNA (16S and COI) data set, based on the substitution rate of Decapoda or Arthropoda

Substitution rate	Decapoda			Arthropoda		
Phylogroup or Node	Time	SD	95% HPD	Time	SD	95% HPD
A	1.765	0.0036	(1.1433 - 2.3995)	1.807	0.0049	(0.9810 - 2.7321)
B	1.245	0.0029	(0.7819 - 1.7700)	1.276	0.0037	(0.6481 - 1.9634)
C	1.284	0.0030	(0.7610 - 1.8444)	1.317	0.0040	(0.6953 - 2.1040)
D	0.792	0.0022	(0.4428 - 1.2068)	0.809	0.0028	(0.3496 - 1.3083)
E	0.556	0.0016	(0.3042 - 0.8348)	0.569	0.0019	(0.2598 - 0.9207)
G1	1.048	0.0025	(0.6251 - 1.4989)	1.072	0.0032	(0.5545 - 1.6842)
G2	0.705	0.0021	(0.3464 - 1.1095)	0.727	0.0027	(0.3034 - 1.2180)
G3	0.761	0.0022	(0.3957 - 1.1816)	0.781	0.0028	(0.3307 - 1.2955)
G4	0.356	0.0015	(0.1254 - 0.6256)	0.364	0.0016	(0.1157 - 0.6708)
G5	0.468	0.0014	(0.2453 - 0.7109)	0.478	0.0016	(0.2188 - 0.7847)
G6	0.285	0.0011	(0.1002 - 0.4974)	0.293	0.0013	(0.0854 - 0.5367)

For each phylogenetic group (G1 - G6) and node (A-E) (defined in Fig. 1) the mean divergence time (Time), standard deviation (SD), 95% HPD lower and upper are given

Table 3 Population genetic parameters inferred from 6 microsatellite loci for 16 sites of the noble crayfish

	AA1	AA2	KAS	LOG	KNS	KLV	NEO	PLF	KRI	SKR	KRN	KPA	TZA	DOX	TSV	PRT
A	3.667	4.333	8.167	2.500	2.500	4.167	2.833	5.333	2.667	2.167	2.333	2.333	2.500	8.167	7.333	7.333
A_R	2.844	3.295	5.567	1.969	2.069	2.760	2.281	4.082	2.021	2.056	1.856	2.333	4.833	5.333	5.019	5.080
P_A	0.000	3.000	5.000	1.000	0.000	1.000	1.000	3.000	1.000	0.000	1.000	0.000	1.000	8.000	8.000	4.000
H_E	0.468	0.559	0.753	0.204	0.236	0.365	0.299	0.648	0.310	0.310	0.174	0.414	0.641	0.758	0.735	0.761
H_N	0.480	0.573	0.772	0.210	0.243	0.377	0.314	0.664	0.318	0.318	0.179	0.452	0.700	0.775	0.754	0.779
H_O	0.408	0.592	0.767	0.265	0.216	0.333	0.356	0.592	0.308	0.300	0.192	0.472	0.667	0.768	0.717	0.727
F_{IS}	0.152	−0.033	0.007	−0.271	0.114	0.118	−0.144	0.112	0.030	0.058	−0.074	−0.049	0.051	0.009	0.050	0.067
P_{H-W}	0.040	0.103	0.402	0.840	0.344	0.169	0.943	0.047	0.804	0.825	0.850	0.689	0.769	0.069	0.805	0.127
L0	Aas2489	Aas8			Aas8									Aas8		Aas1198, Aas8

Where the number of alleles (A), allelic richness (A_R), number of private alleles (P_A), expected heterozygosity (H_E), unbiased expected Nei's heterozygosity (H_N), observed heterozygosity (H_O), inbreeding coefficient (F_{IS}), exact P-value for Hardy-Weinberg equilibrium test (P_{H-W}) and loci with putative null alleles (L0) are given

KPA) due to their negative values (between −0.271 in LOG and −0.033 in AA2). Maximum value of F_{IS} was recorded at AA1 (0.152; Table 3). The total number of private alleles was 33. Sites DOX and TSV had the highest number of private alleles with 8 counts, while in AA1, KNS, SKR, KRN and KPA no private alleles were observed.

Exact tests based on allele frequencies for all pairs of sites showed that 6 out of 240 tests were significant (none after Bonferroni correction). The overall genetic heterogeneity was high ($F_{ST} = 0.4$ and $R_{ST} = 0.39$), with the estimated values of F_{ST} generally greater than R_{ST} for most of the loci (except Aas3040 and Aas1198; Additional file 6). Pairwise estimates of F_{ST} ranged from −0.0069 (KLV-NEO) to 0.7499 (KRN-SKR) and R_{ST} from −0.0325 (KLV-NEO) to 0.8778 (KRI-KRN) (Additional file 7).

Population structure

The inferred number of clusters (K) was nine for the analysis using Structure software (Additional file 8). Sites were grouped together as follows: AA1-AA2-KPA (cluster 1), KRI-SKR (cluster 2), KLV-NEO (cluster 6), KAS-TZA (cluster 8) and LOG-KNS-KRN (cluster 9), while each one of the other sites (PLF, TSV, PRT and DOX) formed different and distinct genetic clusters (cluster 3, 5, 7 and 4, respectively; Fig. 3). The proposed number of clusters for the DAPC analysis varied between 6 and 9 (Additional file 8). It is important to notice that assignment probability for each K identified by DAPC was >0.978 (Additional file 9). In both programs, K = 9 had a similar grouping and was based on the geographic proximity of the sampling sites (Fig. 3).

The majority of the genetic structure in DAPC for nine clusters was captured in the first principal component (screeplot of the eigenvalues; Fig. 4). Visual inspection of the first two principal components, revealed two distinct genetic clusters (clusters 2 and 9; Fig. 4). Similar results were observed for K between 6 and 8 (Additional file 10). No evidence of isolation by distance was

retrieved by Mantel tests (no statistically significant correlation between genetic and geographical distances was observed - Additional file 11).

Barriers

Barrier software revealed the occurrence of breaks in the genetic flow across continental Greece. Analysis of all microsatellite loci simultaneously using D_{CE} or F_{ST} identified eight probable barriers. D_{CE} approach revealed at least 93% bootstrap support for every barrier. The locus by locus analysis of F_{ST} showed that five out of six microsatellite loci (with the exception of Aas3040) agreed with the occurrence of all barriers (Additional file 12). In Aas3040 the observed differences are due to the lack of barriers between sites KRN-LOG-KNS and NEO-KLV, as well as between AA1-AA2-KPA and KAS-TZA. Since the number of data derived from the analysis was large (totally 100; Additional file 12) and the outcomes were similar, it was decided to provide the output data for eight putative barriers, 16 sites and all microsatellite loci using both F_{ST} and D_{CE} (Fig. 5). The derived genetic barriers are closely related with the presence of mountains and river systems, corroborating the findings of structure analysis for K = 9. For instance, sites KRI and SKR are delimited by Mt. Olympus, Mts. Pieria and Aliakmon river. The names of the mountains and major rivers are provided in Additional file 13.

Demographic history and effective population size

Point estimates of the genetic clusters varied between 1 and 1.09 (Additional file 14), indicating a good convergence of MCMC runs. MsVar analyses indicated a decline in all genetic clusters, with no apparent overlap of the posterior distributions of past $\log_{10}(N1)$ and present $\log_{10}(N0)$ population sizes (Additional file 15). Analyses of the global data set (random samples from different demes were merged into one dataset) gave similar results excluding the presence of a false bottleneck signal (Additional files 14 and 15; as suggested by [56]). Since

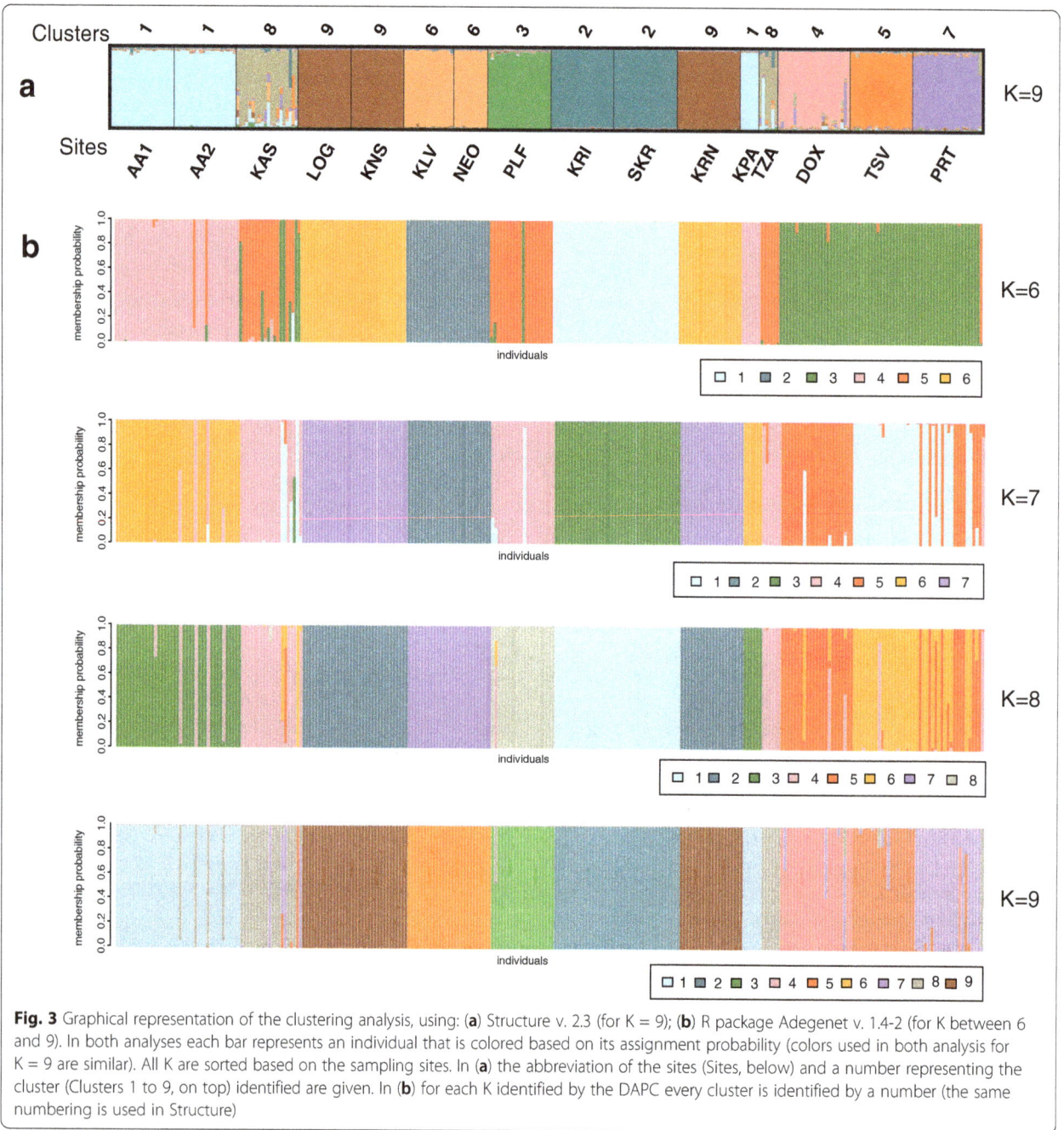

Fig. 3 Graphical representation of the clustering analysis, using: (**a**) Structure v. 2.3 (for K = 9); (**b**) R package Adegenet v. 1.4-2 (for K between 6 and 9). In both analyses each bar represents an individual that is colored based on its assignment probability (colors used in both analysis for K = 9 are similar). All K are sorted based on the sampling sites. In (**a**) the abbreviation of the sites (Sites, below) and a number representing the cluster (Clusters 1 to 9, on top) identified are given. In (**b**) for each K identified by the DAPC every cluster is identified by a number (the same numbering is used in Structure)

the descriptive statistics (Table 4) and the statistics describing the magnitude of demographic change (Table 5) are similar for either generation time used (3.5 and 5.5), it was decided to further discus the analysis assuming a generation time of 5.5. The ancestral population size $\log_{10}(N1)$ among the genetic clusters ranged from 4.97 to 5.51 (Table 4). The contemporary population size $\log_{10}(N0)$ varied between 0.5 and 1.42 (or mean N0 3.17 and 26.6, respectively) (Table 4). The current effective size estimated for each genetic cluster in ONeSAMP was small, with

values varying between 22.069 and 44.191 (Table 6). Mean effective population size (Mean $\log_{10}(N0/N1)$) of the genetic clusters decreased by 14 to 20% (average of approximately 17%, Table 5). The magnitude of the bottleneck was strong in every genetic cluster, with negative values for Hedge's d (between −6.02 and −9.91) (Table 5). Time since the population decline $\log_{10}(T)$ (Additional file 16) varied between 2.27 and 2.77 among genetic clusters (Table 4). BF analysis indicated that the decline occurred during the last around 5000 years (BF > 3, Fig. 6; see also Additional file 17).

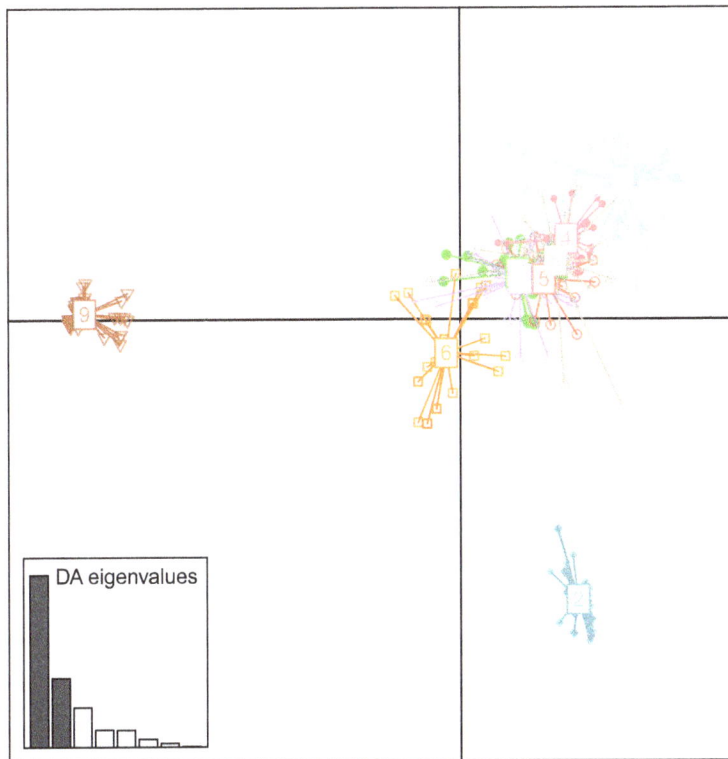

Fig. 4 DAPC scatterplot of the first two principal components for K = 9. Clusters are represented by several distinguishable colors (same as in Structure, Fig. 4). The inset shows the discriminant analysis (DA) eigenvalues

Discussion
Genetic diversity of the noble crayfish
The observed haplotypic diversity of the noble crayfish (*Astacus astacus*) in Europe (0.6857 for 16S and 0.985 for COI, respectively) is in accordance with the wide geographic range of the species in this area [18]. Six haplotypic lineages (G1 to G6, Fig. 1) were identified using a concatenated mitochondrial data set (16S and COI). The inclusion of samples from the southern Balkans (i.e. Greece) revealed the existence of two novel phylogenetic clades (G1 and G2), separated from the rest of the European haplogroups. The between pairwise genetic distances showed that the phylogroups associated to the southernmost Balkan Peninsula (i.e. Greece, G1 and G2) were clearly differentiated from the remaining European haplotype groups (G3 to G6; Additional file 4). Furthermore, a high number of unique haplotypes in Greece were observed (24 haplotypes from 37 samples; Fig. 2), indicating a high genetic diversity. It must be noted that the number of unique haplotypes and genetic diversity of the southern distribution of *A. astacus* may be biased by the distribution of the sites within each haplogroup and the number of individuals per haplogroup in the Balkan Peninsula. These findings may corroborate the taxonomical division of the noble crayfish, proposed by Karaman [88]; three subspecies are present in Europe, *A. astacus astacus*, *A. a. colchicus* and *A. a. balcanicus*. Based on our results, it is impelling to suggest that the Greek phylogroups may represent the subspecies *A. a. balcanicus*. Haplotype groups G3 to G6 (Fig. 1) are consistent with a previous study [29]. Additionally, a lower genetic diversity was observed in central and northern Europe (G4 to G6) compared to the Greek (G1 and G2) and the Croatian (G3) lineages (Fig. 2; Additional file 4), probably due to bottleneck and founder effects [16, 17, 29]. Several studies indicated that this pattern could be the result of a post-glacial range expansion from south to north regions [7, 8, 11]. Genetic distances (16S and COI; Additional file 4) were lower to those observed in *Austropotamobius torrentium* in Europe [35], probably reflecting different life-history traits. Phylogroup G5 indicates a pattern typical of a recent range expansion, based on the number of observed haplotypes. This pattern of noble crayfish distribution range is identified as a typical Southern glacial refugium [9]. Therefore, a "Mediterranean" molecular biogeographical pattern [11] and a "grasshopper" [7] or "chub" [8] model of expansion, can be deduced for the noble crayfish. The most likely colonization route of the noble crayfish towards European regions is through the Danube river (as suggested by Schrimpf et al. [29]) and observed, also, in temperate freshwater fishes [8]. In contrast, it has

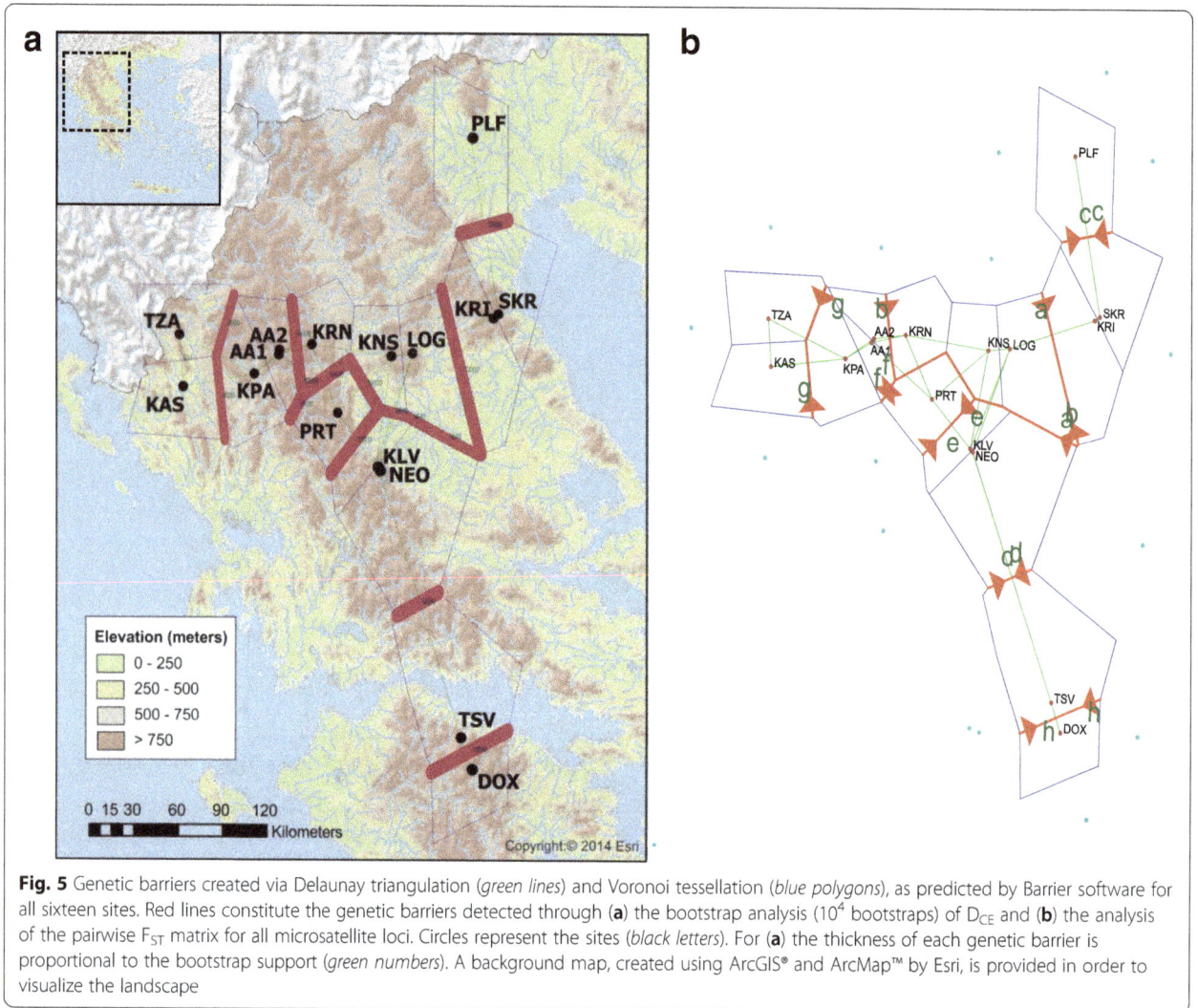

Fig. 5 Genetic barriers created via Delaunay triangulation (*green lines*) and Voronoi tessellation (*blue polygons*), as predicted by Barrier software for all sixteen sites. Red lines constitute the genetic barriers detected through (**a**) the bootstrap analysis (10^4 bootstraps) of D_{CE} and (**b**) the analysis of the pairwise F_{ST} matrix for all microsatellite loci. Circles represent the sites (*black letters*). For (**a**) the thickness of each genetic barrier is proportional to the bootstrap support (*green numbers*). A background map, created using ArcGIS® and ArcMap™ by Esri, is provided in order to visualize the landscape

been assumed that the non-endemic Plio-Pleistocene freshwater fishes of Greece originated from further north regions, e.g. the Danube river [89]. The present analysis indicates that Greece is an older glacial refugium than the eastern Black Sea (phylogenetic clade G4) and Western Balkans (phylogenetic clade G3), with the latter two regions already identified by Schrimpf et al. [29]. Nevertheless, the scenario of multiple refugia is confirmed, as suggested for the noble crayfish [17, 29] and postulated for other species (e.g. [90, 91]).

The phylogroups of the noble crayfish diverged during Early and Middle Pleistocene. Specifically, the estimated average divergence times were between ca. 0.3 and 1.8 MYA (Table 2). In the Mediterranean, major cooling

Table 4 Mean values of $\log_{10}(N0)$, $\log_{10}(N1)$ and $\log_{10}(T)$ for every genetic cluster (1 to 9) with a generation time of 3.5 and 5.5

Genetic cluster	1		2		3		4		5		6		7		8		9	
Generation time	3.5	5.5	3.5	5.5	3.5	5.5	3.5	5.5	3.5	5.5	3.5	5.5	3.5	5.5	3.5	5.5	3.5	5.5
Mean $\log_{10}(N0)$	0.885	1.006	0.522	0.661	0.889	0.781	1.289	1.425	1.298	1.273	0.837	0.582	1.315	1.233	1.411	1.201	0.509	0.501
S.D.	0.711	0.685	0.952	0.826	0.711	0.685	0.827	0.873	0.647	0.519	0.622	0.632	0.856	0.877	0.581	0.655	0.555	0.664
Mean $\log_{10}(N1)$	5.153	5.132	5.355	5.349	4.970	4.971	5.381	5.387	5.502	5.508	5.036	5.272	5.342	5.338	5.462	5.464	5.281	5.269
S.D.	0.382	0.380	0.490	0.495	0.378	0.375	0.324	0.322	0.308	0.308	0.411	0.482	0.333	0.332	0.304	0.303	0.482	0.478
Mean $\log_{10}(T)$	2.286	2.584	2.454	2.770	2.167	2.267	2.358	2.680	2.441	2.611	2.218	2.732	2.430	2.545	2.430	2.432	2.482	2.657
S.D.	0.654	0.631	0.893	0.784	0.755	0.797	0.597	0.489	0.574	0.587	0.786	0.822	0.539	0.604	0.514	0.619	0.826	0.842

Standard Deviation (SD) for every parameter is given

Table 5 Statistics describing the magnitude of the demographic change for each genetic cluster (1-9) and generation time (3.5 or 5.5)

Genetic cluster	1		2		3		4		5		6		7		8		9	
Generation time	3.5	5.5	3.5	5.5	3.5	5.5	3.5	5.5	3.5	5.5	3.5	5.5	3.5	5.5	3.5	5.5	3.5	5.5
Mean effective size $\log_{10}(N0/N1)$	-1.809	-1.734	-1.976	-1.96	-1.718	-1.764	-1.515	-1.414	-1.576	-1.555	-1.761	-1.993	-1.523	-1.576	-1.449	-1.643	-2.037	-2.030
95% CI lower	-1.809	-1.734	-1.976	-1.96	-1.718	-1.764	-1.515	-1.414	-1.576	-1.555	-1.761	-1.993	-1.523	-1.576	-1.449	-1.643	-2.037	-2.030
95% CI upper	-1.809	-1.734	-1.976	-1.96	-1.718	-1.764	-1.515	-1.414	-1.576	-1.555	-1.761	-1.993	-1.523	-1.576	-1.449	-1.643	-2.037	-2.030
S.D.	0.826	0.804	0.916	0.894	0.838	0.888	0.623	0.468	0.662	0.596	0.863	0.902	0.587	0.686	0.513	0.714	0.930	0.920
Variance	0.682	0.646	0.840	0.800	0.702	0.788	0.388	0.219	0.438	0.356	0.745	0.814	0.344	0.471	0.263	0.51	0.865	0.847
Hedges' d	-7.479	-7.447	-6.386	-6.885	-7.164	-7.584	-6.516	-6.024	-8.301	-9.915	-7.965	-8.346	-6.199	-6.19	-8.734	-8.347	-9.183	-8.241
95% CI lower	-9.439	-9.407	-8.346	-8.846	-9.124	-9.544	-8.476	-7.984	-10.261	-11.875	-9.925	-10.306	-8.159	-8.15	-10.694	-10.307	-11.143	-10.201
95% CI upper	-5.519	-5.487	-4.426	-4.925	-5.204	-5.623	-4.556	-4.063	-6.341	-7.955	-6.005	-6.386	-4.239	-4.23	-6.774	-6.387	-7.222	-6.281

Mean effective size $\log_{10}(N0/N1)$, Standard Deviation (SD), Variance, along with Hedges' d and their 95% Confidence Intervals (CI) are given. Negative values indicate significant bottlenecks

Table 6 Mean estimated contemporary size (N0) for each genetic cluster of the noble crayfish, along with their 95% Confidence Intervals (CI) (OneSAMP software)

Cluster	Mean N0	95% CI lower	95% CI upper
1	29.134	20.632	44.899
2	22.069	14.988	37.702
3	23.765	18.996	33.244
4	34.107	27.110	48.675
5	38.318	27.868	61.107
6	34.361	24.070	59.253
7	27.029	22.345	35.878
8	44.191	33.663	65.395
9	23.051	13.907	39.118

events occurred between 2.15 and 2.73 MYA ([92] and references therein), covering the upper HPD limits. During the Middle Pleistocene, two global and prominent climatic transitions seems to be the core of the induced phylogenetic divergence of the species, the "Early Middle Pleistocene Transition" (0.8 to 1.2 MYA) and the "Mid-Brunhes Event" (~0.4 MYA) [93]. Moreover, a correlation

can be observed between time of lineage divergence, increased freshwater inputs (last ca. 2.3 MYA [94]) and climatic oscillations (last 1.3 MYA [95]) of the eastern Mediterranean. It has been suggested that there was a connection between the Danube basin and the Greek river basins during the late Pliocene to Pleistocene; this was based on studies related to central European and Danubian freshwater fishes [89]. A geological event that might have influenced the noble crayfish is the opening of the Danube corridor between 0.7 and 1 MYA [96]. In Greece, the delta formation of Pinios river could have, also, affected the noble crayfish genetic structure by dividing the haplotypes into two well-defined clusters (~1.1 MYA; Faugères, 1977 in Caputo [97] and Rook and Martínez-Navarro [98]). It has, also, been suggested that during the glacial-interglacial cycles, freshwater or oligosaline conditions were prevailing in the upper Aegean areas [99]. These events may have played an important role in the expansion of the noble crayfish in northern regions.

Micro-scale evolution of the noble crayfish

The noble crayfish in Greece is confined in the upper parts of river systems, with a limited, scattered,

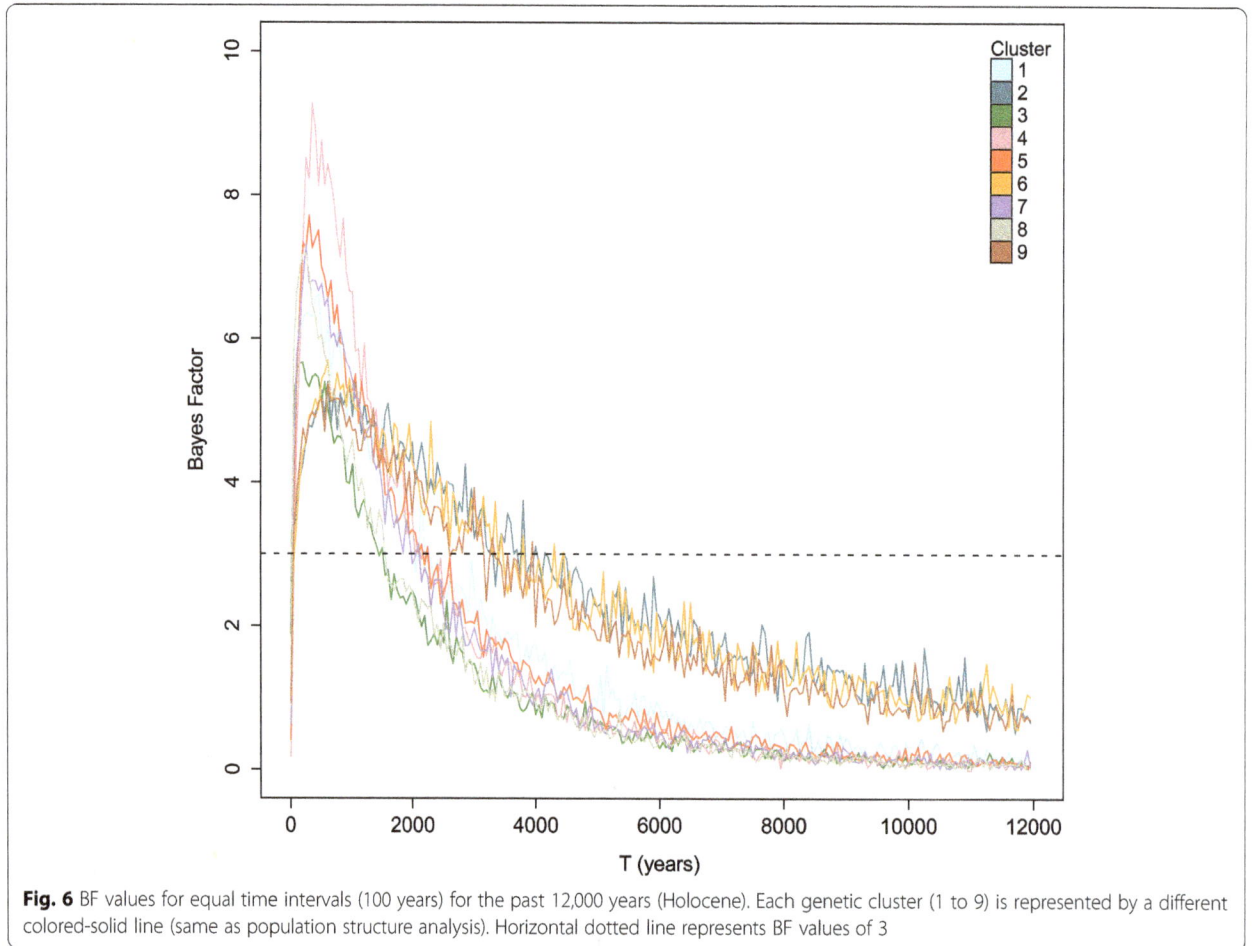

Fig. 6 BF values for equal time intervals (100 years) for the past 12,000 years (Holocene). Each genetic cluster (1 to 9) is represented by a different colored-solid line (same as population structure analysis). Horizontal dotted line represents BF values of 3

discontinuous and grouped distribution (21 crayfish sites from a total of 284 investigated sites; Additional file 12). Similar distribution pattern was observed in a former study in Greece [36], but also in other European countries [20, 100]. Anthropogenic activities (agricultural land use and settlements) are negatively affecting the species distribution in riverine systems of the Czech Republic [100]. Agriculture intensification in lowlands of the Mediterranean [101] and water quality of rivers in Greece [102] could explain the distribution pattern of the Greek noble crayfish.

The genetic heterogeneity in the southernmost distribution range of the noble crayfish (i.e. Greece) (global F_{ST} = 0.4 and between sites $-0.007 < F_{ST} < 0.75$; Additional file 6) was comparable to wider geographical studies (i.e., global F_{ST} = 0.264 [17] and $0.008 < F_{ST} < 0.723$ [29]). Structure analysis revealed a high number of genetic clusters (Fig. 3). The higher sub-structuring of Greece compared to central and northern regions of Europe [17, 29] corroborates the findings of the mitochondrial analysis.

The identified nine clusters (K = 9) revealed a relation between them and the geographic landscape of Greece (Figs. 3 and 5). Furthermore, eight putative barriers were inferred revealing that mountains played an important role in the gene flow (Fig. 5). The majority of the inferred clusters (K = 9) also revealed a close relation with their river system, suggesting geographic isolation. These findings are in accordance with field observations on wild populations of noble crayfish (indigenous to the location). Those studies indicated a sedentary nature of the noble crayfish, with low dispersal ability [20, 103]. However, discrepancies can be observed in two clusters (i.e. 1 and 9, Fig. 3), with their components belonging to two different river systems. The most likely explanation is due to human interference. Cluster 1 (sites AA1, KPA and AA2; Fig. 3) may be the result of indirect human activities, since deviation of water by the Aoos hydroelectric power plant has been reported [102]. Cluster 9 is a typical case of a recent human translocation (sites KNS and LOG translocated from site KRN; Fig. 3 and Table 1). In fact, the Geographic Distance Matrix Generator shows that in several cases it involves long-distance translocations (e.g. 53 km in a straight line, between sites KRN and LOG). The same observation was inferred for the river Kalamas, by a former study [36]. It is important to notice, that the sedentary nature of the species depends on food availability, sex, external factors and familiarity of the noble crayfish to the habitat [20, 103]. Low pairwise F_{ST} and R_{ST} values between sites from different river catchments (Additional file 6), absence of correlation between genetic and geographical distances (Additional file 11) and close genetic relation of the majority of the clusters (Fig. 4), indicated that human

translocations have shaped the genetic structure of *A. astacus* in Greece (and not, so much, relevant biogeographic events). Information on the status of the species (Table 1) disclosed that several identified distinct clusters are due to human translocation (e.g. cluster 8, Kalamas river).

Person(s) involved in translocation of the noble crayfish in Greece had not followed good conservation practices, such as similarities or dissimilarities of habitats, riverine systems or genetic make-up of the species (e.g. sites KNS, LOG and KRN, Table 1 and Fig. 3). Similar observations are also reported in other European regions [16, 17]. Another issue that must be addressed is the nature of the translocations. Unfortunately, translocations may occur without scientific guidance and/or governmental consensus (for instance sites LOG and KNS; Table 1). The absence of appropriate controls may be an issue, since overharvesting and violation of regulations from local and recreational fishermen is common practice. For example, 671 noble crayfish with highly variable lengths (approximately between 5 and 12 cm Total Length or TL) were collected in a 24 h period in Karya by a local recreational fisherman (personal observation). It should be noted that, in Greece, trading of *A. astacus* is prohibited for less than 7 cm TL (Ministerial Decision A2-3354/2007, Government Gazette 2207/B/14.11.2007), while the minimum allowed fishing length is 10 cm TL (Royal Decree 142/1971, Government Gazette 49A/1971). Similar phenomena have also been observed in Norway [104]. Prohibiting and restricting measures (e.g. restriction of fishing gear) could improve management practices.

Analysis of the demographic history in the southernmost distribution of the species (i.e. Balkan Peninsula, Greece) insinuates a severe decrease in the population size during the Holocene (last ~5000 years), resulting in small effective population sizes (Fig. 6; Tables 4, 5 and 6; Additional files 14, 15 and 16). The culminating arid conditions of the last approximately 5.5 kyr BP [105] may have played an important role in the recent evolutionary history of the noble crayfish (potential and reoccurring bottlenecks). During that period, several climatic oscillations (e.g. [106]) and variations in river flood activity (e.g. [107]) have been reported (see also Additional file 17E). Furthermore, several geological changes influencing freshwater systems of Greece have been recorded during the mid- and late Holocene [97, 108].

However, for the last ~5000 years and due to the advent of the Bronze Age (~3.1-5.3 kyr BP), it is difficult to distinguish climate-induced variations from human-mediated influence on riverine systems of Greece [109, 110]. Historical records showed that from the Bronze Age (~5.3 kyr BP) until nowadays several anthropogenic activities affected the continental Greece. Changes in settlements, hydraulic works (irrigation, e.g. aqueducts, dams, water diversions and creation of artificial lakes) and wars (e.g. [111, 112]) are some of the factors that may have directly or indirectly influenced the noble crayfish populations (potential alteration or destruction of habitats and crayfish consumption by humans). The major hydraulic/engineering project in Copais lake (Central - South Greece) during the Bronze Age by the Mynians (Mycenaean civilization [111]) illustrates the intensity of the human activity in Greece. Notably, it is known that crustaceans are consumed since the Bronze Age [113].

Conclusions

The noble crayfish revealed a high genetic diversity, with its southernmost range limit playing a crucial role. Mitochondrial analysis (16S and COI) inferred six phylogroups; two of them (corresponding to the southernmost distribution of the species) were identified for the first time. Genetic analyses revealed that the genetic diversity of A. astacus populations from Southern Balkans (Greece) is high and they form a distinct group when compared with their European counterparts. The divergence time analysis implied a potential differentiation of the species during the Early and Middle Pleistocene (between ca. 0.3 and 1.8 MYA). During that period, several Pleistocene climatic oscillations and geological processes corroborate the inferred divergence. Therefore, we consider that the area of the Southern Balkans may have played an important role in shaping the genetic make-up of A. astacus populations (old centre of expansion and/or southern glacial refugium). Microsatellite analysis focusing on the noble crayfish from its southernmost distribution (Balkan Peninsula, Greece) disclosed six to nine distinct clusters. The combination of the inferred genetic clusters with the genetic barriers and the status of each sampled site (natural or translocated) unraveled the influence of several factors. Specifically, it was deduced that the landscape of the region (mountains and river systems) and human related interventions (mostly translocations) were the main factors affecting the genetic structure of the noble crayfish. Finally, it was figured out that during mid to late Holocene, a severe decrease in the effective population size of the Greek noble crayfish took place. During that time period, several climatic, geological and anthropogenic processes occurred influencing the distribution of noble crayfish in the southernmost Balkan Peninsula (i.e. Greece). It is concluded that the noble crayfish has been primarily influenced by physical processes (climate and geology), while recently, anthropogenic implications (mainly translocations by

humans and niche degradation) may have affected its genetic and geographic structure.

Additional files

Additional file 1: List of sequences used in the mitochondrial analysis. A detailed table with the information of each sequence used is given. The information comprises the name of the site, country of origin, haplotype, GenBank Accession numbers and bibliographic references. (DOC 362 kb)

Additional file 2: The modified conditions used for the PCR and amplification of each microsatellite loci are given. (DOC 32 kb)

Additional file 3: Starting values (α, σ, β and τ) for hyperpriors (log(N0), log(N1), log(Θ) and log(T)) for each independent MCMC run. (DOC 30 kb)

Additional file 4: Pairwise within and between genetic distances of the haplotype groups (G1 to G6) for COI and 16S. (DOC 56 kb)

Additional file 5: Genetic diversity of the microsatellite loci and sampling sites. Table with the number of alleles, allelic richness, number of private alleles, expected heterozygosity, unbiased expected Nei's heterozygosity, observed heterozygosity, inbreeding coefficient and exact P-value for Hardy-Weinberg equilibrium test are listed for the microsatellite loci and the sampling sites. (DOC 140 kb)

Additional file 6: Fixation indices for all microsatellite loci. Values of the fixation indices for all microsatellite loci based on infinite allele model (F_{IT}, F_{IS} and F_{ST}) and stepwise mutation model (R_{ST}). (DOC 30 kb)

Additional file 7: Estimated pairwise F_{ST} and R_{ST} values based on sampling sites. (DOC 53 kb)

Additional file 8: Inference of number of clusters. The file contains graphical representations of a) DeltaK for each K (1 to 16) produced by Structure Harvester and b) Bayesian information criterion (BIC) for every number of clusters, using DAPC. (DOC 108 kb)

Additional file 9: Assignment probability (K between 6 and 9) and assignment probability per cluster, as identified by DAPC analysis. (DOC 35 kb)

Additional file 10: Graphical representations of the DAPC scatterplots of the first two principal components for K between 6 and 8. (DOC 325 kb)

Additional file 11: Mantel test correlation between geographical distance (kilometers) and genetic distances (given as F_{ST}, $F_{ST}/(1-F_{ST})$ or R_{ST}) for all sites. (DOC 637 kb)

Additional file 12: Graphical representation of genetic barriers (1 to 10) based on: a) sixteen sampling stations, all microsatellites loci and the genetic distance D_{CE}, b) sixteen sampling stations, all microsatellites loci and the F_{ST}, c) sixteen sampling stations, each microsatellite locus (Aas8, Aas766, Aas1198, Aas2498, Aas3040 and Aas3950) and F_{ST}, d) nine genetic clusters (of the Structure software), all microsatellites loci and the genetic distance D_{CE}, and e) nine genetic clusters (of the Structure software), all microsatellites loci and the F_{ST}. (DOC 10235 kb)

Additional file 13: Map of the sampled noble crayfish in Greece where abbreviation of the sampling sites, names of major rivers, mountains and sierra, type of genetic markers used (mitochondrial and/or nuclear; mtDNA and nDNA, respectively) and elevation are given. (DOC 2355 kb)

Additional file 14: Point estimates with their upper Confidence Interval (CI) for all parameters (log(N0), log(N1), log(Θ) and log(T)) and every genetic cluster for a generation time of 3.5 and 5.5. (DOC 48 kb)

Additional file 15: Posteriors and priors densities plots for the past (N1) and present (N0) population sizes of each genetic cluster (cluster 1 to 9) and the global data set (samples from different demes were merged into one dataset; black color). (DOC 859 kb)

Additional file 16: Posteriors and priors densities plots for time since the population decline (T) for each genetic cluster (cluster 1 to 9) and the global data set (samples from different demes were merged into one dataset; black color). (DOC 674 kb)

Additional file 17: The relative probability of different hypotheses was assessed using approximate Bayes Factors (BF). A generation time of 5.5 for the noble crayfish was used in the analysis. Hypotheses were examined for the past 12,000 and 7000 years. (DOC 2096 kb)

Abbreviations
BF: Bayes factors; BIC: Bayesian information criterion; CI: Confidence interval; COI: Cytochrome oxidase subunit I; DA: Discriminant analysis; DAPC: Discriminant analysis of principal components; ESS: Effective sample size; HPD: Highest posterior density; Iss: Index of substitution saturation; Iss.c: Index of substitution saturation critical value; IT: Information technology; K: Number of clusters; MCMC: Markov chain Monte Carlo; mtDNA: Mitochondrial DNA; PCR: Polymerase chain reaction; TL: Total length; V: Percentage of the phylogroup-specific haplotypes divided by the number of individuals

Acknowledgements
The authors would like to acknowledge the support provided by Andreas Laggis throughout the progress of this study. Also, we would like to express our gratitude to Kostas Papathanasiou for the help provided during the sampling, I. Kappas for his valuable comments during the writing of this work and V. Sousa and T. Theodosiou for their help with R. Also, we would like to thank C. Perdikaris and several fishermen for the crayfish samples provided. Finally, we kindly acknowledge the effort and time invested by two anonymous reviewers for improving the manuscript.

Funding
Not applicable.

Authors' contributions
AL conceived the experimental design, sampling, performed genetic experiments, genetic and bioinformatic analysis, wrote the paper; ADB contribute in data analysis and writing of the manuscript; AC bioinformatic analysis and technical support; SM genetic analysis; AT contributed in cross-checking of genotypes and writing; TJA supervised AL Ph.D. thesis, conceived the experimental design and analytical work and contributed to the writing of the manuscript. All authors read and approved the final manuscript.

Competing interests
The authors declare that they have no competing interests.

Author details
[1]Department of Genetics, Development and Molecular Biology, School of Biology, Aristotle University of Thessaloniki, 54124 Thessaloniki, Macedonia, Greece. [2]Scientific Computing Office, Information Technology (IT) Center, School of Sciences, Aristotle University of Thessaloniki, 541 24 Thessaloniki, Macedonia, Greece.

References
1. Tzedakis PC, Lawson IT, Frogley MR, Hewitt GM, Preece RC. Buffered tree population changes in a quaternary refugium: evolutionary implications. Science. 2002;297:2044-7.
2. Médail F, Diadema K. Glacial refugia influence plant diversity patterns in the Mediterranean Basin. J Biogeogr. 2009;36:1333-45.
3. Deffontaine V, Libois R, Kotlík P, Sommer R, Nieberding C, Paradis E, et al. Beyond the Mediterranean peninsulas: evidence of central European glacial refugia for a temperate forest mammal species, the bank vole (Clethrionomys glareolus). Mol Ecol. 2005;14:1727-39.
4. Jablonski D, Jandzik D, Mikulíček P, Džukić G, Ljubisavljević K, Tzankov N, et al. Contrasting evolutionary histories of the legless lizards slow worms (Anguis) shaped by the topography of the Balkan peninsula. BMC Evol Biol. 2016;16:99.
5. Trontelj P, Machino Y, Sket B. Phylogenetic and phylogeographic relationships in the crayfish genus Austropotamobius inferred from mitochondrial COI gene sequences. Mol Phylogenet Evol. 2005;34:212-26.
6. Jelić M, Klobučar GIV, Grandjean F, Puillandre N, Franjević D, Futo M, et al. Insights into the molecular phylogeny and historical biogeography of the white-clawed crayfish (Decapoda, Astacidae). Mol Phylogenet Evol. 2016; 103:26-40.
7. Hewitt GM. Post-glacial re-colonization of European biota. Biol J Linn Soc. 1999;68:87-112.
8. Hewitt GM. Genetic consequences of climatic oscillations in the quaternary. Philos T Roy Soc B. 2004;359:183-95.
9. Stewart JR, Lister AM, Barnes I, Dalén L. Refugia revisited: individualistic responses of species in space and time. Proc Biol Sci. 2010;277:661-71.
10. Hughes JM, Schmidt DJ, Finn DS. Genes in streams: using DNA to understand the movement of freshwater fauna and their riverine habitat. Bioscience. 2009;59:573-83.
11. Schmitt T. Molecular biogeography of Europe: Pleistocene cycles and postglacial trends. Front Zool. 2007;4:11.
12. Macklin MG, Lewin J, Woodward JC. Quaternary fluvial systems in the Mediterranean basin. In: Lewin J, Macklin MG, Woodward JC, editors. Mediterranean quaternary river environments. Rotterdam, Brookfield: A. A. Balkema; 1995. p. 1-25.
13. Alexandri P, Triantafyllidis A, Papakostas S, Chatzinikos E, Platis P, Papageorgiou N, et al. The Balkans and the colonization of Europe: the post-glacial range expansion of the wild boar, Sus scrofa. J Biogeogr. 2012;39:713-23.
14. Karaiskou N, Tsakogiannis A, Gkagkavouzis K, Papika S, Latsoudis P, Kavakiotis I, et al. Greece: a Balkan subrefuge for a remnant red deer (Cervus elaphus) population. J Hered. 2014;105:334-44.
15. Tzedakis PC. Museums and cradles of Mediterranean biodiversity. J Biogeogr. 2009;36:1033-4.
16. Schrimpf A, Schulz HK, Theissinger K, Pârvulescu L, Schulz R. The first large-scale genetic analysis of the vulnerable noble crayfish Astacus astacus reveals low haplotype diversity in central European populations. Knowl Manag Aquat Ecosyst. 2011;401:35.
17. Gross R, Palm S, Kõiv K, Prestegaard T, Jussila J, Paaver T, et al. Microsatellite markers reveal clear geographic structuring among threatened noble crayfish (Astacus astacus) populations in northern and Central Europe. Conserv Genet. 2013;14:809-21.
18. Kouba A, Petrusek A, Kozák P. Continental-wide distribution of crayfish species in Europe: update and maps. Knowl Manag Aquat Ecosyst. 2014;413:5.
19. Skurdal J, Taugbøl T. Astacus. In: Holdich DM, editor. Biology of freshwater crayfish. Oxford: Blackwell Science; 2002. p. 467-510.
20. Cukerzis J. La biologie de l'écrevisse (Astacus astacus L.). Paris: Institut National de la Recherche Agronomique Publications; 1984. (In French)
21. Skurdal J, Hessen DO, Garnås E, Vøllestad LA. Fluctuating fecundity parameters and reproductive investment in crayfish: driven by climate or chaos? Freshw Biol. 2011;56:335-41.
22. Lundberg U. Behavioural elements of the noble crayfish, Astacus astacus (Linnaeus, 1758). Crustaceana. 2004;77:137-62.
23. Edsman L, Füreder L, Gherardi F, Souty-Grosset C. Astacus astacus. IUCN Red List of Threatened Species. http://dx.doi.org/10.2305/IUCN.UK.2010- 3.RLTS. T2191A9338388 (2010). Accessed 23 Jan 2016.
24. Keller M. Ten years of trapping Astacus astacus for restocking. Freshwater Crayfish. 1998;12:518-28.
25. Pârvulescu L, Pacioglu O, Hamchevici C. The assessment of the habitat and water quality requirements of the stone crayfish (Austropotamobius torrentium) and noble crayfish (Astacus astacus) species in the rivers from the Anina Mountains (SW Romania). Knowl Manag Aquat Ecosyst. 2011;401:3.
26. Reynolds J, Souty-Grosset C, Richardson A. Ecological roles of crayfish in freshwater and terrestrial habitats. Freshwater Crayfish. 2013;19:197-218.
27. Legakis A. Invertebrates. In: Legakis A, Maragou P, editors. Red book of threatened fauna of Greece. Athens: Hellenic Zoological Society; 2009. p. 428-509. (In Greek)
28. Jussila J, Ojala K, Mannonen A. Noble crayfish (Astacus astacus) reintroduction project in the river Pyhäjoki, western Finland: a case study. Freshwater Crayfish. 2008;16:51-6.
29. Schrimpf A, Theissinger K, Dahlem J, Maguire I, Pârvulescu L, Schulz HK, et al. Phylogeography of noble crayfish (Astacus astacus) reveals multiple refugia. Freshw Biol. 2014;59:761-76.

30. Bláha M, Žurovcová M, Kouba A, Policar T, Kozák P. Founder event and its effect on genetic variation in translocated populations of noble crayfish (*Astacus astacus*). J Appl Genet. 2016;57:99–106.

31. Fevolden SE, Taugbøl T, Skurdal J. Allozymic variation among populations of noble crayfish, *Astacus astacus* L., in southern Norway: implication for management. Aquacult Fish Manage. 1994;25:927–35.

32. Schulz HK, Šmietana P, Schulz R. Assessment of DNA variations of the noble crayfish (*Astacus astacus* L.) in Germany and Poland using inter-simple sequence repeats (Issrs). Bull Fr Pêche Piscic. 2004;372–373:387–99.

33. Schulz R. Status of the noble crayfish *Astacus astacus* in Germany: monitoring protocol and the use of RAPD markers to assess the genetic structure of populations. Bull Fr Pêche Piscic. 2000;356:123–38.

34. Alaranta A, Henttonen P, Jussila J, Kokko H, Prestegaard T, Edsman L, et al. Genetic differences among noble crayfish (*Astacus astacus*) stocks in Finland, Sweden and Estonia based on the ITS1 region. Bull Fr Pêche Piscic. 2006;380–381:965–76.

35. Klobučar GIV, Podnar M, Jelić M, Franjević D, Faller M, Štambuk A, et al. Role of the Dinaric Karst (western Balkans) in shaping the phylogeographic structure of the threatened crayfish *Austropotamobius torrentium*. Freshw Biol. 2013;58:1089–105.

36. Koutrakis E, Perdikaris C, Machino Y, Savvidis G, Margaris N. Distribution, recent mortalities and conservation measures of crayfish in Hellenic fresh waters. Bull Fr Pêche Piscic. 2007;385:25–44.

37. Zhang D-X, Hewitt GM. Nuclear DNA analyses in genetic studies of populations: practice, problems and prospects. Mol Ecol. 2003;12:563–84.

38. Westman K, Pursiainen M, Vilkman R. A new folding trap model which prevents crayfish from escaping. Freshwater Crayfish. 1978;4:235–42.

39. Estoup A, Largiader CR, Perrot E, Chourrout D. Rapid one-tube DNA extraction for reliable PCR detection of fish polymorphic markers and transgenes. Mol Mar Biol Biotechnol. 1996;5:295–8.

40. Crandall KA, Fitzpatrick JF. Crayfish molecular systematics: using a combination of procedures to estimate phylogeny. Syst Biol. 1996;45:1–26.

41. Folmer O, Black M, Hoeh W, Lutz R, Vrijenhoek R. DNA primers for amplification of mitochondrial cytochrome c oxidase subunit I from diverse metazoan invertebrates. Mol Mar Biol Biotechnol. 1994;3:294–9.

42. Maniatsi S, Baxevanis AD, Abatzopoulos TJ. The intron 2 of p26 gene: a novel genetic marker for discriminating the two most commercially important *Artemia franciscana* subspecies. J Biol Res - Thessalon. 2009;11:73–82.

43. Baxevanis AD, Kappas I, Abatzopoulos TJ. Molecular phylogenetics and asexuality in the brine shrimp *Artemia*. Mol Phylogenet Evol. 2006;40:724–38.

44. Hall TA. BioEdit: a user-friendly biological sequence alignment editor and analysis program for windows 95/98/NT. Nucleic Acids. 1999;41:95–8.

45. Larkin MA, Blackshields G, Brown NP, Chenna R, Mcgettigan PA, McWilliam H, et al. Clustal W and Clustal X version 2.0. Bioinformatics. 2007;23:2947–8.

46. Buhay JE. 'COI-like' sequences are becoming problematic in molecular systematic and DNA barcoding studies. J Crustac Biol. 2009;29:96–110.

47. Xia X. DAMBE5: a comprehensive software package for data analysis in molecular biology and evolution. Mol Biol Evol. 2013;30:1720–8.

48. Lanfear R, Frandsen PB, Wright AM, Senfeld T, Calcott B. PartitionFinder 2: new methods for selecting partitioned models of evolution for molecular and morphological phylogenetic analyses. Mol Biol Evol. 2017;34:722–73.

49. Daniels SR. Reconstructing the colonisation and diversification history of the endemic freshwater crab (*Seychellum alluaudi*) in the granitic and volcanic Seychelles archipelago. Mol Phylogenet Evol. 2011;61:534–42.

50. Pedraza-Lara C, Alda F, Carranza S, Doadrio I. Mitochondrial DNA structure of the Iberian populations of the white-clawed crayfish, *Austropotamobius italicus italicus* (Faxon, 1914). Mol Phylogenet Evol. 2010;57:327–42.

51. Tonini J, Moore A, Stern D, Shcheglovitova M, Ortí G. Concatenation and species tree methods exhibit statistically indistinguishable accuracy under a range of simulated conditions. PLoS Currents Tree Of Life. 2015; doi:10.1371/currents.tol.34260cc27551a527b124ec5f6334b6be.

52. Meiklejohn KA, Danielson MJ, Faircloth BC, Glenn TC, Braun EL, Kimball RT. Incongruence among different mitochondrial regions: a case study using complete mitogenomes. Mol Phylogenet Evol. 2014;78:314–23.

53. Drummond AJ, Suchard MA, Xie D, Rambaut A. Bayesian phylogenetics with BEAUti and the BEAST 1.7. Mol Biol Evol. 2012;29:1969–73.

54. Rambaut A, Drummond AJ. Tracer version 1.5. http://beast.bio.ed.ac.uk/Tracer (2009). Accessed 21 Mar 2015.

55. R Core Team. R: A language and environment for statistical computing. http://www.r-project.org/ (2014). Accessed 25 Sep 2015.

56. Tamura K, Stecher G, Peterson D, Filipski A, Kumar S. MEGA6: molecular evolutionary genetics analysis version 6.0. Mol Biol Evol. 2013;30:2725–9.

57. Brower AV. Rapid morphological radiation and convergence among races of the butterfly *Heliconius erato* inferred from patterns of mitochondrial DNA evolution. Proc Natl Acad Sci U S A. 1994;91:6491–5.

58. Schubart CD, Diesel R, Hedges SB. Rapid evolution to terrestrial life in Jamaican crabs. Nature. 1998;393:363–5.

59. Kõiv K, Gross R, Paaver T, Kuehn R. Isolation and characterization of first microsatellite markers for the noble crayfish, *Astacus astacus*. Conserv Genet. 2008;9:1703–6.

60. Kõiv K, Gross R, Paaver T, Hurt M, Kuehn R. Isolation and characterization of 11 novel microsatellite DNA markers in the noble crayfish, *Astacus astacus*. Anim Genet 2009;40:124–124.

61. Van Oosterhout C, Hutchinson WF, Wills DPM, Shipley P. MICRO-CHECKER: software for identifying and correcting genotyping errors in microsatellite data. Mol Ecol Notes. 2004;4:535–8.

62. Nei M. Estimation of average heterozygosity and genetic distance from a small number of individuals. Genetics. 1978;89:583–90.

63. Belkhir K, Borsa P, Chikhi L, Raufaste N, Bonhomme F. GENETIX 4.05, logiciel sous Windows™ pour la génétique des populations. http://kimura.univ-montp2.fr/genetix/ (2016). Accessed 1 Jan 2016.

64. Rousset F. Genepop'007: a complete re-implementation of the genepop software for windows and Linux. Mol Ecol Resour. 2008;8:103–6.

65. Rice WR. Analyzing tables of statistical tests. Evolution. 1989;43:223–5.

66. Goudet J. FSTAT, a program to estimate and test gene diversities and fixation indices (version 2.9.3) http://www2.unil.ch/popgen/softwares/fstat.htm#top (2016). Accessed 23 Feb 2016.

67. Goodman SJ. RST calc: a collection of computer programs for calculating estimates of genetic differentiation from microsatellite data and determining their significance. Mol Ecol. 1997;6:881–5.

68. Pritchard JK, Stephens M, Donnelly P. Inference of population structure using multilocus genotype data. Genetics. 2000;155:945–59.

69. Earl DA, VonHoldt BM. STRUCTURE HARVESTER: a website and program for visualizing STRUCTURE output and implementing the Evanno method. Conserv Genet Resour. 2012;4:359–61.

70. Jakobsson M, Rosenberg NA. CLUMPP: a cluster matching and permutation program for dealing with label switching and multimodality in analysis of population structure. Bioinformatics. 2007;23:1801–6.

71. Rosenberg NA. DISTRUCT: a program for the graphical display of population structure. Mol Ecol Notes. 2004;4:137–8.

72. Jombart T, Devillard S, Balloux F. Discriminant analysis of principal components: a new method for the analysis of genetically structured populations. BMC Genet. 2010;11:94.

73. Jombart T. Adegenet: a R package for the multivariate analysis of genetic markers. Bioinformatics. 2008;24:1403–5.

74. Manni F, Guerard E, Heyer E. Geographic patterns of (genetic, morphologic, linguistic) variation: how barriers can be detected by using Monmonier's algorithm. Hum Biol. 2004;76:173–90.

75. Chambers JL, Garant D. Determinants of population genetic structure in eastern chipmunks (*Tamias striatus*): the role of landscape barriers and sex-biased dispersal. J Hered. 2010;101:413–22.

76. Jensen JL, Bohonak AJ, Kelley ST. Isolation by distance, web service. BMC Genet. 2005;6:13.

77. Ersts PJ. Geographic Distance Matrix Generator (version 1.2.3). http://biodiversityinformatics.amnh.org/open_source/gdmg/ (2017). Accessed 10 Mar 2017.

78. Tallmon DA, Koyuk A, Luikart G, Beaumont MA. ONeSAMP: a program to estimate effective population size using approximate Bayesian computation. Mol Ecol Resour. 2008;8:299–301.

79. Storz JF, Beaumont MA. Testing for genetic evidence of population expansion and contraction: an empirical analysis of microsatellite DNA variation using a hierarchical Bayesian model. Evolution. 2002;56:154–66.

80. Pianka ER. Evolutionary ecology. 2nd ed. New York, USA: Harper and Rowe Publishers; 1978.

81. Gelman A, Rubin DB. Inference from iterative simulation using multiple sequences. Stat Sci. 1992;7:457–511.

82. Brooks SPB, Gelman AG. General methods for monitoring convergence of iterative simulations. J Comput Graph Stat. 1998;7:434–55.

83. Plummer M, Best N, Cowles K, Vines K. CODA: convergence diagnosis and output analysis for MCMC. R News. 2006;6:7–11.

84. Gelman A, Hill J. Data analysis using regression and multilevel/hierarchical models. Policy anal. Cambridge: Cambridge University Press; 2007.

85. Paz-Vinas I, Quéméré E, Chikhi L, Loot G, Blanchet S. The demographic history of populations experiencing asymmetric gene flow: combining simulated and empirical data. Mol Ecol. 2013;22:3279–91.

86. Salmona J, Salamolard M, Fouillot D, Ghestemme T, Larose J, Centon J-F, et al. Signature of a pre-human population decline in the critically endangered Reunion Island endemic forest bird Coracina newtoni. PLoS One. 2012;7:e43524.

87. Hasegawa M, Kishino H, Yano T. Dating of the human-ape splitting by a molecular clock of mitochondrial DNA. J Mol Evol. 1985;22:160–74.

88. Karaman MS. Studie der Astacidae (Crustacea, Decapoda) II. Teil. Hydrobiologia. 1963;22:111–32. (In German)

89. Economidis PS, Banarescu PM. The distribution and origins of freshwater fishes in the Balkan peninsula, especially in Greece. Int Revue Ges Hydrobiol. 1991;76:257–83.

90. Salvi D, Harris DJ, Kaliontzopoulou A, Carretero MA, Pinho C. Persistence across Pleistocene ice ages in Mediterranean and extra-Mediterranean refugia: phylogeographic insights from the common wall lizard. BMC Evol Biol. 2013;13:147.

91. Thanou E, Tryfonopoulos G, Chondropoulos B, Fraguedakis-Tsolis S. Comparative phylogeography of the five Greek vole species infers the existence of multiple South Balkan subrefugia. Ital J Zool. 2012;79:363–76.

92. Rohling EJ, Foster GL, Grant KM, Marino G, Roberts AP, Tamisiea ME, et al. Sea-level and deep-sea-temperature variability over the past 5.3 million years. Nature. 2014;508:477–82.

93. Maslin MA, Brierley CM. The role of orbital forcing in the early middle Pleistocene transition. Quatern Int. 2015;389:47–55.

94. Kroon D, Alexander I, Little M, Lourens LJ, Matthewson A, Robertson AHF, et al. Oxygen isotope and sapropel stratigraphy in the eastern Mediterranean during the last 3.2 million years. In: Robertson AHF, Emeis K-C, Richter C, Camerlenghi A, editors. Proceedings of the ocean drilling program, scientific results, 160. Earth and planetary science research institute publications; 1998. p. 181–9.

95. Tzedakis PC, Hooghiemstra H, Pälike H. The last 1.35 million years at Tenaghi Philippon: revised chronostratigraphy and long-term vegetation trends. Quaternary Sci Rev. 2006;25:3416–30.

96. Fitzsimmons KE, Marković SB, Hambach U. Pleistocene environmental dynamics recorded in the loess of the middle and lower Danube basin. Quaternary Sci Rev. 2012;41:104–18.

97. Caputo R, Bravard J-P, Helly B. The Pliocene-quaternary tecto-sedimentary evolution of the Larissa plain (eastern Thessaly, Greece). Geodin Acta. 1994;7:219–31.

98. Rook L, Martínez-Navarro B. Villafranchian: the long story of a Plio-Pleistocene European large mammal biochronologic unit. Quatern Int. 2010;219:134–44.

99. Bianco PG. Potential role of the palaeohistory of the Mediterranean and Paratethys basins on the early dispersal of euro-Mediterranean freshwater fishes. Ichthyol Explor Fres. 1990;1:167–84.

100. Římalová K, Douda K, Štambergová M. Species-specific pattern of crayfish distribution within a river network relates to habitat degradation: implications for conservation. Biodivers Conserv. 2014;23:3301–17.

101. Caraveli H. A comparative analysis on intensification and extensification in Mediterranean agriculture: dilemmas for LFAs policy. J Rural Stud. 2000;16:231–42.

102. Skoulikidis NT. The environmental state of rivers in the Balkans - a review within the DPSIR framework. Sci Total Environ. 2009;407:2501–16.

103. Bohl E. Motion of individual noble crayfish Astacus astacus in different biological situation: in-situ studies using radio telemetry. Freshwater Crayfish. 1999;12:677–87.

104. Skurdal J, Garnås E, Taugbøl T. Management strategies, yield and population development of the noble crayfish Astacus astacus in lake Steinsfjorden. Bull Fr Pêche Piscic. 2002;367:845–60.

105. Finné M, Holmgren K, Sundqvist HS, Weiberg E, Lindblom M. Climate in the eastern Mediterranean, and adjacent regions, during the past 6000 years - a review. J Archaeol Sci. 2011;38:3153–73.

106. Büntgen U, Myglan VS, Ljungqvist FC, McCormick M, Di Cosmo N, Sigl M, et al. Cooling and societal change during the late antique little ice age from 536 to around 660 AD. Nat Geosci. 2016;9:231–6.

107. Luterbacher J, García-Herrera R, Akcer-On S, Allan R, Alvarez-Castro MC, Benito G, et al. A review of 2000 years of paleoclimatic evidence in the mediterranean. In: Lionello P, editor. The Climate of the Mediterranean region: From the Past to the Future. Amsterdam: Elsevier; 2012. p. 87–185.

108. Vött A, Brückner H, Schriever A, Handl M, Besonen M, van der Borg K. Holocene coastal evolution around the ancient seaport of Oiniadai, Acheloos alluvial plain, NW Greece. Coastline Rep. 2004;1:43–53.

109. van Andel TH, Zangger E, Demitrack A. Land use and soil erosion in prehistoric and historical Greece. J Field Archaeol. 1990;17:379–96.

110. Lespez L. Geomorphic responses to long-term land use changes in eastern Macedonia (Greece). Catena. 2003;51:181–208.

111. Viollet P-L. Water engineering in ancient civilizations: 5,000 years of history. Boca Raton: CRC Press; 2007.

112. Bintliff J. The complete archaeology of Greece: from hunter-gatherers to the 20th century a.D. 1st ed. Wiley-Blackwell: Chichester; 2012.

113. Mylona D. Aquatic animal resources in prehistoric Aegean. Greece J Biol Res - Thessalon. 2014;21:2.

114. Bathrellos GD, Skilodimou HD, Maroukian H, Gaki-Papanastassiou K. Late quaternary evolution of the lower reaches of Ziliana stream in south Mt. Olympus (Greece). 8th IAG International Conference on Geomorphology "Geomorphology and Sustainability". August 2013. Paris, France.

115. Association of Environmental Protection Vrontismenis. http://www.ekke.gr/estia/gr_pages/mko_po/organoseis/Grenorag/Hpiros/152.htm (2012). Accessed 4 Mar 2016. (In Greek).

116. Stavropoulou N, Papaconstantinos C, Bantaraki M, Kizilou C, Oikonomou M. The wetland of lake Tsivlou. Akrata: Center for Environmental Education Akrata; 2003. (In Greek)

16

Evolutionary radiations in the species-rich mountain genus *Saxifraga* L.

J. Ebersbach[1][*] ⓘ, J. Schnitzler[1], A. Favre[1] and A.N. Muellner-Riehl[1,2]

Abstract

Background: A large number of taxa have undergone evolutionary radiations in mountainous areas, rendering alpine systems particularly suitable to study the extrinsic and intrinsic factors that have shaped diversification patterns in plants. The species-rich genus *Saxifraga* L. is widely distributed throughout the Northern Hemisphere, with high species numbers in the regions adjacent to the Qinghai-Tibet Plateau (QTP) in particular the Hengduan Mountains and the Himalayas. Using a dataset of 297 taxa (representing at least 60% of extant *Saxifraga* species), we explored the variation of infrageneric diversification rates. In addition, we used state-dependent speciation and extinction models to test the effects of geographic distribution in the Hengduan Mountains and the entire QTP region as well as of two morphological traits (cushion habit and specialized lime-secreting glands, so-called hydathodes) on the diversification of this genus.

Results: We detected two to three rate shifts across the *Saxifraga* phylogeny and two of these shifts led to radiations within two large subclades of *Saxifraga*, sect. *Ciliatae* Haworth subsect. *Hirculoideae* Engl. & Irmsch. and sect. *Porphyrion* Tausch subsect. *Kabschia* Engl. GEOSSE analyses showed that presence in the Hengduan Mountains had a positive effect on diversification across *Saxifraga*. Influence of these mountains was strongest in *Saxifraga* sect. *Ciliatae* subsect. *Hirculoideae* given its pronounced distribution there, and thus the radiation in this group can be classified at least partially as geographic. In contrast, the evolution of the cushion life form and lime-secreting hydathodes had positive effects on diversification only in selected *Saxifraga* sections, including sect. *Porphyrion* subsect. *Kabschia*. We therefore argue that radiation in this group was likely adaptive.

Conclusions: Our study underlines the complexity of processes and factors underpinning plant radiations: Even in closely related lineages occupying the same life zone, shifts in diversification are not necessarily governed by similar factors. In conclusion, alpine plant radiations result from a complex interaction among geographical settings and/or climatic modifications providing key opportunities for diversification as well as the evolution of key innovations.

Keywords: Evolutionary radiations, alpine habitats, *Saxifraga*, diversification rates, key innovations, Hengduan Mountains

Background

Evolutionary radiations resulting in exceptionally species-rich clades are shaping current patterns of biodiversity in the plant kingdom [1, 2]. Large numbers of evolutionary radiations have been recorded in alpine plant groups [3], and references therein, contributing to the high species diversity of vascular plants found in many mountain systems [4, 5]. In fact, alpine life zones around the world are proportionally more species-rich than expected by the area they occupy [6]. In addition, alpine habitats occur at all latitudes allowing for global comparisons [7], which renders them particularly attractive to study evolutionary radiations.

Mountain systems are influenced by several extrinsic factors and processes that have the potential to trigger plant radiations. First, geological processes can cause populations to become geographically isolated, fostering allopatric speciation [8, 9]. Additionally, orogenic activity creates new environmental space, which can be colonized by pre-adapted local or immigrant lineages, thereby providing key opportunities for diversification [10–12]. Yet, viewing mountain building as sole driver for diversification is likely an oversimplification [13]. Speciation and uplift usually occur on different time scales, and in fact,

* Correspondence: ebersbach.jana@gmail.com
[1]Department of Molecular Evolution and Plant Systematics & Herbarium (LZ), Institute of Biology, Leipzig University, Johannisallee 21–23, D-04103 Leipzig, Germany
Full list of author information is available at the end of the article

many mountain plant taxa have diversified long after the uplift of their respective mountain range was initiated, as shown for the species-rich areas surrounding the Qinghai-Tibet Plateau (QTP), [14, 15] and elsewhere [3]. Thus, other mechanisms are likely to have contributed to radiations in mountains. For example, several authors have demonstrated the importance of local or global climate modifications for alpine plant radiations [16, 17].

High elevation life zones are characterized by environmental conditions that limit plant growth, such as large diurnal temperature amplitudes or strong seasonality with severe winter frosts, extended snow cover and short growing seasons [18]. Thus, successful colonization, establishment and finally radiation in this life zone might depend on the evolution of traits that allow for adaptation and provide competitive advantages in alpine environments. Novel traits (morphological, physiological, etc.) that allow species to conquer new adaptive zones are referred to as key innovations and they are often regarded as an important first step in adaptive radiations as they allow for ecomorphological divergence and thus speciation [19, 20].

Key innovation potential of selected traits can be tested using phylogenetic trees: Positive effects of a particular trait should result in substantially higher diversification rates in groups possessing that trait. Several key innovations have been identified for different montane and alpine plants such as low specific leaf area in Ericaceae Juss. [21], cushion habit in *Androsace* L. [22], and fruit type, both in *Tripterospermum* Blume [23] and in Andean bellflowers [24].

The effects of orogeny, climate change, key opportunities and key innovations can interact with each other and may differ between mountain systems, which can result in regionally restricted radiations, so-called geographic radiations [19]. Studying alpine plant groups that occur in several mountain systems may therefore contribute to disentangling the factors triggering alpine radiations. In this study, we will identify rapidly evolving clades and investigate the potential drivers for these radiations in the broadly distributed herbaceous genus *Saxifraga*.

The large arctic-alpine genus *Saxifraga* (Saxifragaceae Juss.) comprises up to 500 species [25] and is widely distributed across the Northern Hemisphere. Species diversity is concentrated in the southern European mountain ranges, the Caucasus, the Arctic as well as the QTP region [26]. The Himalayas and the Hengduan Mountains boast particularly high species numbers with up to 75 species in the subnival belt of the Hengduan Mountains alone [27]. *Saxifraga* was confirmed to be monophyletic, provided the exclusion of *Micranthes* Haworth [28, 29]. The genus comprises at least 13 sections with species of widely differing morphology and varying levels of species richness, from the monotypic section *Saxifragella* (Engl.) Gornall & Zhang to the very large section *Ciliatae* (175 species) [26]. For example, species of section *Porphyrion* (90–112 species) are

mostly cushion plants not taller than a few centimetres, whereas *Irregulares* Haworth species (10–20 species) are erect, with large, petiolate basal leaves, and reaching up to 40 cm in height [26, 30]. Several sections within *Saxifraga* have radiated within relatively short geological time-frames [31, 32], but the underlying factors behind these radiations have not been studied in detail. For these reasons, *Saxifraga* is an ideally suited study system to investigate patterns and processes connected to alpine radiations. Integrating phylogenetic, biogeographical, and morphological information, we here present a multifaceted approach to studying diversification in *Saxifraga*. We will focus on the following questions: (1) Do diversification rate shifts explain differences in clade size within *Saxifraga*? (2) Do diversification rates within *Saxifraga* vary with geographic distribution? (3) Did the evolution of the cushion life form and lime-secreting hydathodes, both of which are potential key innovations in alpine habitats, affect diversification in *Saxifraga*?

Results
Diversification rate patterns in *Saxifraga*
Independent of prior choice and assumed total species numbers, BAMM identified two to three rate shifts within *Saxifraga* (Fig. 1, Additional file 1). Two of these shifts were consistently placed within the same *Saxifraga* clades across all analyses. The first one (from here on referred to as rate shift 1) was located within sect. *Porphyrion* subsect. *Kabschia sensu* Tkach et al. [26], which originated and radiated in Europe (European Alps, Caucasus Mountains) and includes one subgroup that colonized the QTP region and diversified there. The second one (shift 2) was located within a clade that contains members of sect. *Ciliatae* subsect. *Hirculoideae*. This clade is the largest subclade of the species-rich section *Ciliatae*, which originated and radiated in the QTP region [31–33]. A third rate shift (shift 3) was present in some scenarios and placed in varying locations close to the root of the *Saxifraga* phylogeny, most often on the node separating sections *Heterisia* (A. M. Johnson) Small, *Irregulares* and *Saxifragella* (see Fig. 1) from the rest of the genus.

Analyses with BayesRate confirmed the presence of three diversification rate shifts within *Saxifraga* (Table 1). Regarding overall rate regimes, the model with three distinct diversification rates had the highest marginal likelihood: one rate shared by both clades affected by rate shifts 1 and 2, one rate shared by sections *Heterisia*, *Irregulares*, *Saxifragella* and *Pseudocymbalaria* Zhmylev (group H + I + S + P), and one rate for the rest of *Saxifraga* (Bayes factor (BF) min: 11.66, BF max 9.61; Table 1). Additionally, models specifying pure birth processes performed considerably better than those with birth-death processes, with the exception of group H + I + S + P for which both processes were almost equally likely (Table 1). Rate estimates did not differ substantially between BAMM and BayesRate

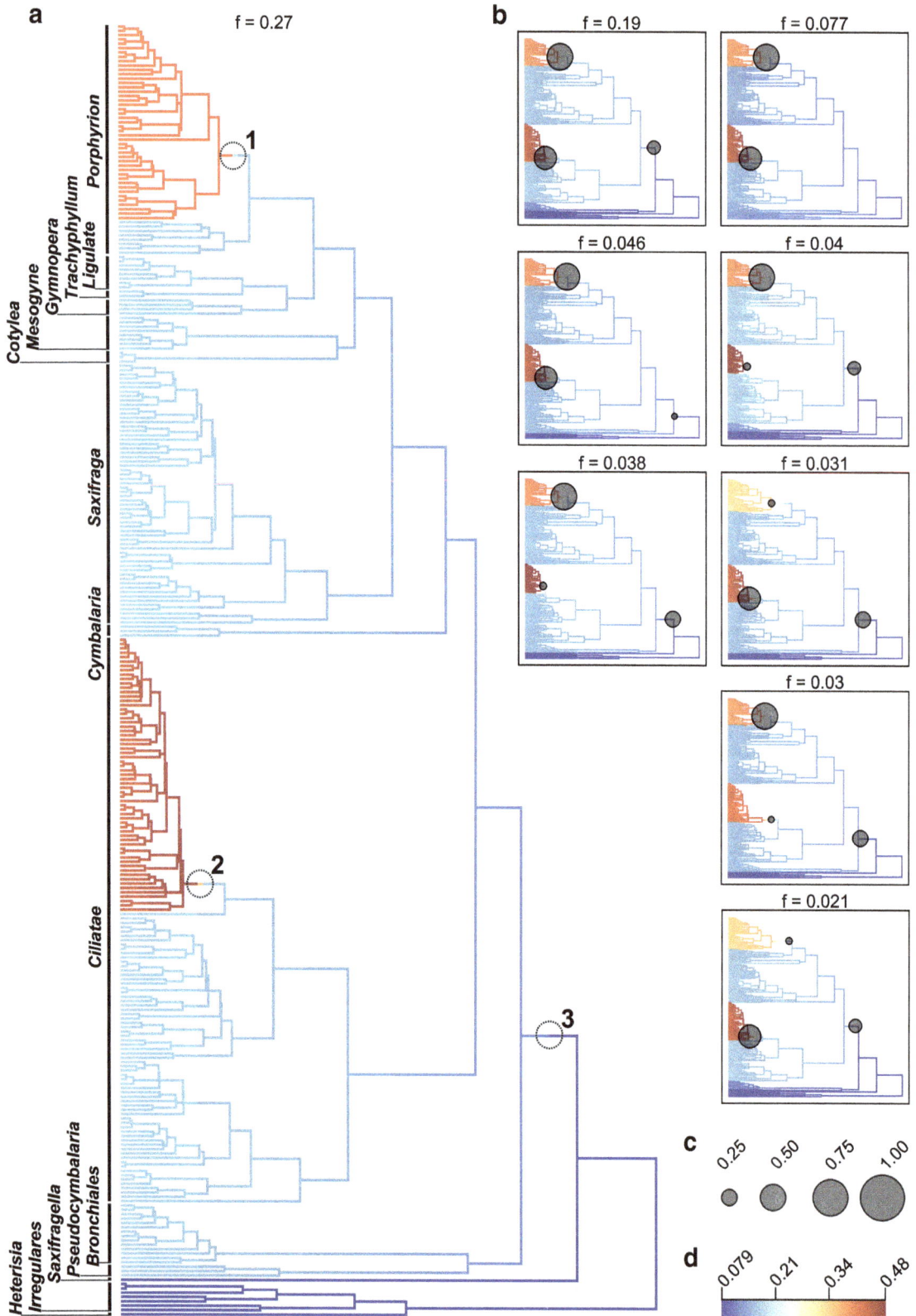

Fig. 1 (See legend on next page.)

Fig. 1 Diversification dynamics in *Saxifraga*. **a** Best shift scenario according to BAMM. Dashed circles indicate diversification rate shifts. *Saxifraga* sections are indicated according to Tkach et al. [26] and Gornall et al. [33]. **b** The eight next likely shift scenarios. **c** Size reference for marginal shift probabilities for each shift in B. **d** Colour ramp for diversification rates in species per million years. Results presented here are from the analysis with γ = 0.5 and minimum sampling fractions per section. Results from additional BAMM analyses were concordant with those presented here and are summarized in Additional file 1

(Additional file 2). While BayesRate generally produced slightly higher net diversification rate estimates (Additional file 2), the considerable overlap of the 95% HPD intervals indicates an overall agreement between the two approaches. Importantly, the differences between *Ciliatae* subsect. *Hirculoideae* and the other clades were found to be very similar (e.g., an increase over the background rate of 2.6- and 2.4-fold, respectively).

Diversification rate estimates were very similar between minimum and maximum species number analyses (Table 2). Under pure birth processes, net diversification rates in clades *Ciliatae* subsect. *Hirculoideae* and *Porphyrion* subsect. *Kabschia* were about 2.5 times higher than the background rate (net diversification rates min, max: *Hirculoideae* and *Kabschia* clade 0.455, 0.471 species/Myr; background rate 0.171, 0.184 species/Myr; Table 2). Sections *Heterisia*, *Irregulares*, *Saxifragella* and *Pseudocymbalaria* on the other hand diversified at a rate that was ca. 3 times lower than the *Saxifraga* background rate (min, max: 0.054, 0.071 species/Myr).

Geography-associated diversification

The GEOSSE model with the lowest overall AIC value specified a pure birth process in the Hengduan Mountains region (area A) and differential speciation, extinction and transition rates between areas (Additional file 3.1). Two similar models, also with differential speciation and transition rates between areas, also had good model fit (ΔAIC < 6; Additional file 3.1). Initial parameter estimates using maximum likelihood inference of these three best scoring models were highly congruent (Additional file 3.2) and we decided to use the best overall model to run MCMC parameter estimation. Net diversification rates were ca. 3.5 times higher in the Hengduan Mountains region than in the rest of the world (Fig. 2, Table 3). The empirical ΔAIC for the Hengduan Mountain dataset was well separated from the distribution of simulated ΔAIC values indicating a robust signal in the empirical data for region-dependent diversification (Additional file 4).

Analyses regarding the entire QTP region yielded similar results with respect to the most likely diversification model

Table 1 Comparison of different diversification rate scenarios for *Saxifraga* using BayesRate

Total no. of shifts	Ciliatae subsect. Hirculoideae	Porphyrion subsect. Kabschia	Group H + I + S + P	Back-ground	min LogL	min BF	max LogL	max BF
3	PB*		PB	PB	−665.42	–	−693.57	–
3	PB*		BD	PB	−666.13	1.42	−694.82	2.50
3	PB*		BD	BD	−668.67	6.50	−697.40	7.66
3	BD*		BD	BD	−670.55	10.25	−698.69	10.23
2	PB*		-	PB	−671.25	11.66	−698.38	9.61
2	PB*		-	BD	−673.20	15.55	−700.35	13.56
2	PB	PB	-	PB	−674.16	17.48	−701.91	16.67
2	BD*		-	BD	−675.16	19.49	−702.43	17.72
2	BD	BD	-	BD	−679.17	27.50	−706.92	26.70
1	PB	-	-	PB	−682.72	34.59	−708.13	29.10
1	BD	-	-	BD	−686.32	41.80	−711.20	35.26
1	-	PB	-	PB	−700.92	71.00	−725.73	64.32
1	-	PB	-	PB	−703.61	76.37	−728.10	69.05
0	-	-	-	PB	−722.36	113.88	−748.80	110.46
0	-	-	-	BD	−723.53	116.22	−749.73	112.31

Various scenarios for shifts in three *Saxifraga* clades under pure birth (PB) and birth-death (BD) processes were tested. Asterisks indicate scenarios in which selected clades were constrained to have the same net diversification rates. Log marginal likelihoods (LogL) and results from Bayes factor (BF) tests are given for analyses with minimum and maximum species

Table 2 Net diversification rate estimates

	min mean [HPD]	max mean [HPD]
Ciliatae subsect. *Hirculoideae*	0.455 [0.349–0.575]	0.471 [0.357–0.593]
Porphyrion subsect. *Kabschia*	0.455 [0.349–0.575]	0.471 [0.357–0.593]
Group H + I + S + P	0.054 [0.020–0.090]	0.071 [0.032–0.116]
Background rate	0.171 [0.133–0.211]	0.184 [0.144–0.226]

Mean net diversification (speciation - extinction) rates and 95% highest posterior density (HPD) intervals for the different *Saxifraga* clades from the BayesRate analysis

and parameter estimates (Additional files 5.1 and 2). However, there was no substantial difference in model fit between simulated and empirical data suggesting that asymmetrical transition rates alone, not differential diversification, could have led to the observed distribution pattern (Additional file 4).

Tests for key innovations

When tested across the entire *Saxifraga* phylogeny, neither the evolution of lime-secreting hydathodes nor the cushion exhibited significant patterns of differential speciation or extinction rates. However, as pointed out by Beaulieu & Donoghue [34] and Beaulieu & O'Meara [35], it can be problematic to use BiSSE to investigate the effects of certain

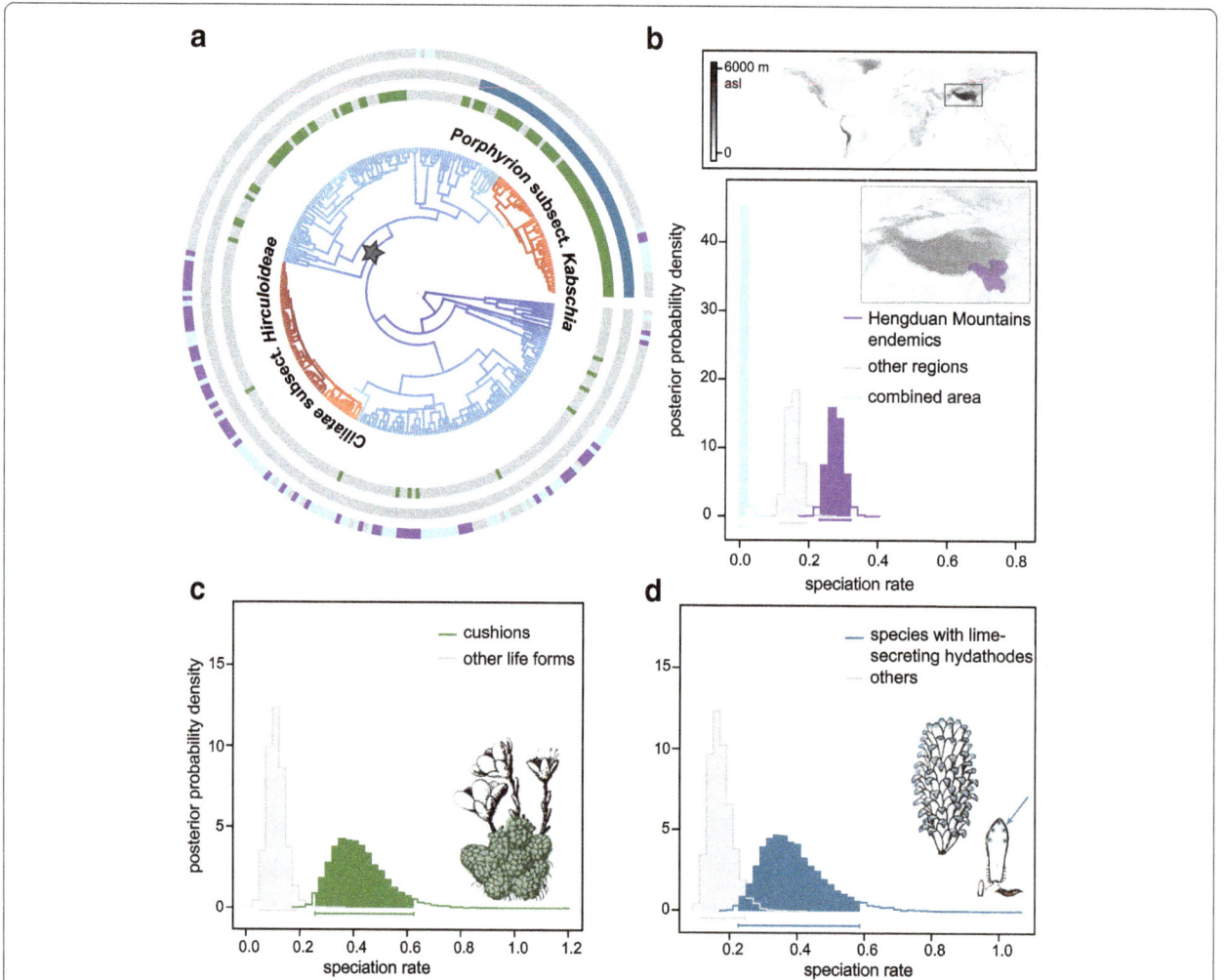

Fig. 2 State-dependent diversification in *Saxifraga*. **a** *Saxifraga* phylogeny with estimated diversification rate regimes according to BAMM (Fig. 1, warmer colours indicate higher rates). Phylogenetic distributions of species' traits are indicated in green (cushion habit) and blue (lime-secreting hydathodes), the geographical distribution (presence in the Hengduan Mountains) is shown in purple. **b** Speciation rates in the Hengduan Mountain hotspot vs. other regions. **c** Speciation rates of cushion saxifrages vs. other life forms. Analysis was constrained to parts of the *Saxifraga* phylogeny indicated by the star in **a** Drawing depicts habit of *S. lilacina* [87], cushion indicated by green shading. **d** Speciation rates of lineages carrying lime-secreting hydathodes vs. others. The analysis was constrained to parts of the *Saxifraga* phylogeny indicated by the star in **a** Drawings depict stem of *S. imbricata*, with lime-secreting hydathodes on each leaf tip and a single cauline leaf of *S. lilacina* bearing five lime-secreting hydathodes [87]

Table 3 Parameter estimates for best scoring GEOSSE model for state-dependent diversification of *Saxifraga* in the Hengduan Mountains region

	λ mean [HPD]	μ mean [HPD]	d mean [HPD]	r mean [HPD]
Hengduan Mountains	0.277	–	0.464	0.277
	[0.231–0.321]		[0.316–0.623]	[0.231–0.321]
Other areas	0.156	0.076	0.004	0.080
	[0.118–0.193]	[0.021–0.133]	[0.001–0.008]	[0.049–0.11]
Joined area	0.006			
	[0.000–0.018]			

Mean parameter estimates and 95% highest probability density (HPD) intervals are given for speciation rates (λ), extinction rate (μ), transition rates to joined area (d) and net diversification rates (r; λ – μ) of *Saxifraga* in the Hengduan Mountains region

traits on phylogenies with multiple rate shifts. Both of the traits that we tested are limited to specific clades of *Saxifraga* (Fig. 2). Importantly, they are almost entirely absent in *Ciliatae* and in *Ciliatae* subsect. *Hirculoideae* for which rate shift 2 was detected. Restricting the analysis to a smaller part of the tree comprising sections *Cymbalaria* Griseb., *Saxifraga*, *Cotylea* Tausch, *Mesogyne* Sternb., *Gymnopera* D. Don, *Trachyphyllum* (Gaudin) W. D. J. Koch, *Ligulatae* Haworth and *Porphyrion* yielded a different result: Plants with lime-secreting hydathodes and cushion saxifrages had substantially higher diversification rates (Fig. 2, Table 4). In addition, trait simulations showed that the empirical ΔAIC were substantially different from simulated ΔAIC values revealing a strong signal for differential diversification rates under these two traits for this part of the *Saxifraga* phylogeny (Additional file 4).

Discussion

As a species-rich group with an affinity to high altitude habitats, *Saxifraga* is well suited to study underlying patterns and factors related to alpine radiations and the diversification process in general. Our study shows that multiple rate shifts led to radiations within *Saxifraga* and that specific combinations of extrinsic and intrinsic factors drove these radiations. To our knowledge, this is the most extensive study on diversification rates with regard to the species-rich, yet still poorly understood, alpine areas surrounding the QTP to date.

Diversification rate patterns in *Saxifraga*

We found a slight increase in the inferred number of rate shifts when increasing the hyperprior on the expected number of rate shifts (Additional file 1A), but prior choice did not affect model selection (scenario with maximum posterior probability; Additional file 1B & C). This was further corroborated in a recent study by Mitchell & Rabosky [36]. While both model selection via Bayes factors, as well as pure-birth models of diversification would also be available in BAMM, BayesRate additionally allows for speciation and extinction rates to be linked between clades. In our case, this revealed that speciation and extinction rates of clades *Ciliatae* subsect. *Hirculoideae* and *Porphyrion* subsect. *Kabschia* were not significantly different (i.e., a model with same rates was preferred over one with independent rates for both clades).

We found evidence for three distinct diversification rate shifts within *Saxifraga*. Two of these rate shifts led to substantially increased diversification rates within *Saxifraga* sections *Ciliatae* and *Porphyrion* which had previously been suggested as having experienced rapid radiations [31, 32].

Table 4 Parameter estimates for best scoring models for state-dependent diversification for cushion life form and lime-secreting hydathodes in *Saxifraga*

	λ mean [HPD]	μ mean [HPD]	q mean [HPD]	r mean [HPD]
Cushions	0.420	0.219	0.056	0.201
	[0.247–0.614]	[0–0.452]	[0.024–0.096]	[0.094–0.312]
Other life forms	0.114	0.065	0.055	0.048
	[0.051–0.179]	[0–0.188]	[0.002–0.119]	[–0.074–0.143]
Hydathodes	0.393	0.159	0.003	0.234
	[0.226–0.587]	[0–0.395]	[0–0.009]	[0.118–0.356]
Non–hydathodes	0.173	0.063	0.006	0.110
	[0.113–0.244]	[0–0.150]	[0–0.015]	[0.057–0.163]

Mean parameter estimates and 95% highest probability density (HPD) intervals are given for speciation rates (λ), extinction rates (μ); transition rates (q) and net diversification rates (r; λ – μ)

214   A Comprehensive Approach to Evolutionary Biology

Therefore, diversification rate shifts at least partially explain clade size differences in *Saxifraga*.

Only a few diversification rate estimates have been published for other taxa with pronounced radiations in the QTP region. The rate estimates reported here (mean estimates min, max: 0.455, 0.486 species/Myr) were similar to those found in other alpine radiations from the QTP region such as in *Gentiana* L. (0.37 species/Myr) [37] and in several subgenera of *Delphinium* L. (subg. *Delphinastrum* (DC.) Peterm. and *Oligophyllon* Dimitrova 0.37–0.81 species/Myr, subg. *Aconitum* 0.42–1.11 species/Myr, all assuming low extinction) [38]. In contrast, diversification rates were very low (0.066 species/Myr) in *Rhododendron* L. despite the taxon being famously species-rich in the QTP region [21]. In addition, the rate estimates reported here do not match the exceptionally high rates that have been reported for several Andean plant groups (e.g., *Lupinus* L.: 1.56–5.21 species/Myr [39]; *Gentianella* Moench 1.48–3.21 species/Myr [40]) despite the fact that the Andes and the Hengduan Mountains share several environmental features (e.g., young age, high physiographic heterogeneity). This suggests that comparisons of alpine radiations across the globe might also need to incorporate regional dynamics (e.g., geologic and glaciation history, biogeographic connectivity).

An additional rate shift was present towards the root of the phylogeny, separating sections *Heterisia*, *Irregulares*, *Saxifragella* and *Pseudocymbalaria* (group H + I + S + P) with very low net diversification rates from the rest of *Saxifraga*. Pure birth and birth-death processes were almost equally likely in this clade, so this depauperate clade may result from very low speciation, relatively high extinction or a combination thereof. In general, pure birth processes had superior model fit in *Saxifraga* and this matches very low extinction rate estimates that have been found for several other alpine plant groups [21, 37]. However, extinction rates are difficult to estimate from molecular phylogenies and rate estimates are often low with large confidence intervals [35, 41, 42] but see [43].

Drivers of diversification within *Saxifraga*
Similar to *Saxifraga*, multiple infrageneric rate shifts yielding substantial increases in diversification rates were shown in other species-rich alpine or subalpine genera such as *Rhododendron* L. [21], *Lupinus* L. [39] and *Hypericum* L. [16], suggesting that some plant groups might be particularly prone to radiate. This could be due to intrinsic or extrinsic factors or factor combinations that are shared by all infrageneric clades involved in these radiations. In *Saxifraga*, this seems unlikely as clades affected by rate shifts are morphologically and phylogenetically distinct with contrasting distribution patterns and biogeographic histories [26, 31]. Rather, it appears plausible that each rate shift was triggered by a specific combination of intrinsic and extrinsic factors including geographical distribution and key innovations as will be discussed in the next section.

Influence of the Hengduan Mountains region on diversification
Occurrence in the Hengduan Mountains region had a positive effect on *Saxifraga* diversification. Section *Ciliatae* subsect. *Hirculoideae* affected by rate shift 2 is distributed almost entirely in the QTP region with a large proportion of species occurring in the Hengduan Mountains [30, 44]. While some *Ciliatae* taxa recently colonized other areas including East Asia, Northern Asia and even North America, they did not diversify in those areas [31]. Thus, local key opportunities in the Hengduan Mountains region at the time of diversification may have triggered the diversification rate increase in this group. While rapid radiations have been documented for several plant groups distributed in these mountains [3], ours is the first study to formally compare diversification rates of taxa distributed in the Hengduan Mountains region with those of close relatives of other areas.

The Hengduan Mountains stand out from other mountain ranges in the QTP region due to several characteristics. In contrast to the plateau, parts of which had reached 4000 m altitude as early as 40 MY ago [14, 45] and the Himalayas, which were uplifted to significant altitudes during the Miocene [15], the Hengduan Mountains are considered to be relatively young (Miocene, late Pliocene) [46, 47]. These different geological histories have led to strongly contrasting biogeographic patterns throughout the region. While recent *in situ* diversification was disproportionally more important for the species assembly in the Hengduan Mountains, Himalayan biodiversity was largely influenced by immigration [48]. The Hengduan Mountains and the eastern Himalayas have been under the influence of the monsoon system (i.e., greater summer rainfall) since their orogeny, whereas the QTP proper and the western Himalayas are almost entirely beyond the reach of the summer monsoon. In addition, extinctions during Pleistocene glaciations were likely low due to north-south orientation of the valleys of the Hengduan Mountains region [48]. Finally, the Hengduan Mountains have been described as highly heterogeneous in terms of topography, featuring deeply dissected landscapes with steep elevational gradients [3].

Recently, the interplay of high physiographic heterogeneity and episodes of rapid climate oscillations (promoting the "species pump" effect) has been highlighted as a potential driver of alpine diversification [19, 49]. Acting together, these two factors may enhance diversification in mountains by repeatedly modifying dispersal barriers, thus promoting allopatric speciation while buffering extinction through the availability of nearby refugia resulting in a large number of species with an island-like distribution and a high chance of survival. Among others, global climate oscillations have been reconstructed for the Early Pliocene as well as for the Pleistocene [50, 51]. This time corresponds roughly to the period that was suggested as origin of the majority of species within subsect. *Hirculoideae* (<5 Myr ago) [31]. Therefore, both high habitat heterogeneity and climate oscillations were present during the early stages of the radiation of *Saxifraga* in the Hengduan Mountains. Extant *Saxifraga* distribution patterns fit the expected results of the "species pump" effect and long-term survival: There are at least 35 *Saxifraga* species endemic to the subnival belt of the Hengduan Mountains, many of them with highly restricted distribution areas [30, 44].

However, while subsect. *Hirculoideae* showed a substantial shift in its diversification rate, the remaining subgroups of sect. *Ciliatae* that also predominantly occur in the QTP region (and particularly in the Hengduan Mountains) did not. This strongly suggests that the presence in these mountains is not sufficient to explain the diversification rate pattern in this section and that geographic radiations might require additional factors other than extrinsic key opportunities. Subsection *Hirculoideae* is not characterized by any obvious morphological or physiological traits that could serve as clade-specific key innovations. Another driver that has been found to be associated to plant radiations and shifts in diversification rates are whole genome duplications [52]. A survey of polyploidy in the flora of the Hengduan Mountains showed that, contrary to expectations, the frequency of polyploidy in the flora was low (22%) [53]. For *Saxifraga*, the authors found that the proportion of polyploid species was higher than average (44%) but still lower that what would have been expected judging from other high mountain floras (45–85%) [53]. However, chromosome numbers were only available for nine *Saxifraga* species distributed in the Hengduan Mountains, so further work will be required to examine this pattern in more detail. Regarding hybridization, another mechanism often connected to rapid speciation in plants, we were only able to find one report of a potential trace of such an event in *S. egregia* of sect. *Ciliatae* [54], however more data is needed to clarify this. Finally, niche shifts or changes in niche width could also be an important factor involved in diversification rate shifts, as demonstrated, for instance, by Matuszak et al. [23] and Favre et al. [37] who found that increased diversification rates corresponded to niche shifts in two species-rich genera of Gentianinae G. Don (i.e., *Gentiana* and *Tripterospermum*). For *Saxifraga*, Rubio de Casas et al. [55] showed that the cliff and rock face habitat was most common but that a switch to tundra habitats (which exhibited high within-habitat diversification) occurred within a large unresolved clade

containing species of sections *Ciliatae*, *Saxifraga* and *Porphyrion*. Since the exact location of this shift within the phylogeny remained unclear, ecological niches will have to be studied in more detail for these sections to investigate this pattern more closely.

Key Innovations in alpine environments

Saxifraga rate shift 1 was observed in *Porphyrion* subsect. *Kabschia*, which is largely distributed in European and Asian mountain systems, with only one subclade occurring in the QTP region, where the subsection is more diverse in the Himalayas than in the Hengduan Mountains [56]. Due to this wide and uneven distribution pattern, it is unlikely that key opportunities connected to geographic distribution have triggered the rate shift in this group. Rather, morphological traits likely acted as key innovations causing the observed rate shift in *Porphyrion* subsect. *Kabschia*. First, the cushion life form, which is widespread throughout the genus, was associated with differential rates of diversification in certain clades of *Saxifraga*. Cushion plants are widespread in arctic and alpine habitats and this life form has been shown to be an adaptation to cold climates through reduction in water loss and desiccation [57]. Similar to *Saxifraga*, the cushion life form was found to foster lineage diversification in the species-rich alpine genus *Androsace* [22, 58]. The results derived from both genera corroborate the findings of Boucher et al. [57] who showed that the cushion life form arose more than 100 times independently across Angiosperms and concluded that it is likely to be a convergent key innovation for occupancy of extremely cold environments.

Rather than a single key innovation driving radiations, it has been suggested that several sequentially acquired traits may act together as drivers of lineage diversification [59]. Beside the cushion habit, lime-secreting hydathodes, which are restricted to sections *Ligulatae* and *Porphyrion*, were also associated with higher diversification rates in some clades of *Saxifraga*. Hydathodes, in general, are associated with guttation, or water discharge from the leaf interior, and occur widely in *Saxifraga* [60]. Lime-secreting hydathodes which are morphologically distinct from simple hydathodes serve as a means for excreting excess calcium salts leaving characteristic white encrustations on the leaves [61]. This mechanism is particularly interesting considering that many *Saxifraga* species, in particular those of sections *Porphyrion*, *Ligulatae*, *Trachyphyllum* and *Gymnopera*, occur in habitats with little soil, including cliff ledges and rock crevices, and display strong substrate specificity with regards to calcareous or siliceous substrates. Conti et al. [61] suggested that lime-secreting hydathodes represent the derived state of hydathodes in *Saxifraga* and current phylogenetic placement of the aforementioned sections suggests that they evolved either once in the ancestor of these four sections and were subsequently lost twice (in sections

Mesogyne and *Trachyphyllum*), or less likely, that lime-secretion evolved twice independently in *Porphyrion* and in *Ligulatae*. Following the first scenario, it would be likely that the ancestor of these four sections was calcicole. However, at this point substrate preference data are not complete for species of sect. *Porphyrion*. While there is a general trend that species possessing lime-secreting hydathodes occur on base-rich, calcareous soils or rock, there are several exceptions to this pattern: Several species with lime-secreting pores are not bound to base-rich substrates (e.g., *S. juniperifolia*, *S. cotyledon*, *S. florulenta*) and species that do not possess lime-secreting hydathodes can occur alongside others that do on calcareous soils (e.g., *S. exarata* subsp. *moschata*) [60]. Thus, this mechanism and its implications clearly deserve closer attention, in particular considering its positive correlation with diversification in *Saxifraga* as shown in this study.

In general, it is important to point out that while we have achieved substantial taxon coverage of *Saxifraga*, there are some species missing from our dataset, in particular from sections for which rate shifts were inferred (e.g., section *Irregulares*: sampling proportion: 0.35–0.7%, section *Porphyrion*: 0.47–0.58%, Additional file 6). Sampling bias can affect the detection of rate shifts as well as the location of identified shifts [62, 63] and caution when interpreting the results is therefore advised. However, we took this into account as thoroughly as possible by supplying section-specific sampling proportions during all analyses and running minimum and maximum species number analyses. Furthermore, most sections of *Saxifraga* are morphologically distinct and molecular work so far confirmed the current taxonomic delineation [26]. Thus, it is unlikely that these missing taxa would affect our results.

Conclusion

Our study highlights the complexity of plant radiations. Even in closely related lineages occupying the same life zone, shifts in diversification rates are not necessarily governed by the same factors. In particular, radiation within one subgroup of *Saxifraga* (section *Ciliatae* subsect. *Hirculoideae*) was driven by extrinsic opportunities associated with the Hengduan Mountains region as well as additional factors, possibly niche shifts. A second infrageneric radiation (in sect. *Porphyrion* subsect. *Kabschia*) was likely independent of these processes. Instead, the cushion habit and the acquisition of lime-secreting hydathodes emerged as likely key innovations indicating that the radiation in this clade might have been adaptive (*sensu* Simões et al. [19]). The case of *Saxifraga* thus shows that alpine plant radiations have complex underlying causes, which are to be viewed in both, geographical and biotic contexts. Future studies of (alpine) plant radiations should

therefore attempt to identify combinations of intrinsic and extrinsic factors relevant to the diversification process.

Methods
Dataset

For this study, we used the same Saxifragaceae data set as in Ebersbach et al. [31], consisting of 420 taxa and an aligned length of 4225 bp of markers ITS, *trnL–trnF*, and *matK*. Prior to analyses, we modified the posterior distribution of post-burnin BEAST trees from Ebersbach et al. [31] by removing all taxa not belonging to *Saxifraga* as well as several duplicate taxa. In addition, three taxa (*S. haplophylloides*, *S. diffusicallosa* and *S. maxionggouensis*) from the dataset of Gao et al. [32] were added to all trees using the bind.tip function of the *phytools* package [64] in R [65]. This function allows to add tips to a phylogeny while rescaling the tree to conserve its ultrametric properties. According to Gao et al. [32] these three taxa are all part of subsect. *Hirculoideae* which form a well-defined subclade of *Saxifraga* section *Ciliatae*, but their exact position is unclear due to low resolution. We therefore placed these taxa randomly within that subclade (also using randomly generated node heights) to mirror the existing uncertainty. The final dataset consisted of 297 *Saxifraga* taxa. Analyses were run on the maximum clade credibility (MCC) tree generated by TreeAnnotator v1.8.2 [66] as well as on a set of 100 trees randomly sampled from all trees in the posterior distribution.

Diversification rates

We used BAMM (v.2.5.0) [67] and the R package *BAMMtools* [68] as well as the programme BayesRate (v.1.6.3, [43]) to assess diversification rate heterogeneity across the *Saxifraga* phylogeny. BAMM identifies distinct configurations of rate shifts without the need for an *a priori* specification of their number and location. Any sampling bias is likely to affect the accuracy of rate estimates [69], so we specified section-specific sampling fractions to account for incomplete taxon sampling. However, given that published species numbers for some sections differ widely (Additional file 6, e.g., [26, 30, 70, 71]), we calculated sampling fractions for minimum and maximum published species numbers per section and ran all analyses with both values.

We ran four MCMC chains with 20 million generations each, saving the output every 1000th generation. The first 10% of samples were discarded as burn-in and the effective sample sizes for the number of shifts and the log likelihood were calculated to assess convergence. It has been suggested that the likelihood function of BAMM might be incorrect and that the programme suffers from strong prior sensitivity and unreliable rate estimates [72]. However, Rabosky et al. (in press) [73] recently reported that they were unable to reproduce the prior sensitivity and that the shortcomings indicated by Moore et al. [72] were likely due to a misapplication of the programme and poorly designed test data. We performed two additional analyses to address these issues: First, we ran the analyses under a range of settings for the prior distribution on the number of rate shifts (γ = 10, 2, 1, 0.5, 0.1) to assess the potential prior sensitivity. Second, we used the programme BayesRate to further evaluate the various rate shift scenarios. BayesRate was specifically designed for hypothesis-based testing of diversification regimes (i.e., rate shifts through time, rate differences between particular clades, etc.) across a given phylogeny or set of phylogenies. Importantly, BayesRate employs a different likelihood function [42] than BAMM, and thus provides an independent test of the diversification model (number of rate shifts) and the associated rates. Finally, BayesRate allows for a straightforward incorporation of phylogenetic (topological and temporal) uncertainty, while there is currently no direct way of accounting for phylogenetic uncertainty in BAMM. We used BayesRate to test the most likely number of rate regimes and to compare the likelihood of pure birth (yule) and birth-death via thermodynamic integration. After identifying the best-fit model via a Bayes factor test [74], we estimated the model parameters across all 100 randomly sampled phylogenetic trees to account for phylogenetic uncertainty.

Key innovations
We tested the effect of two potential key innovations (cushion habit and the presence of lime-secreting hydathodes) on *Saxifraga* diversification. Trait information was assembled from several floras as well as specific *Saxifraga* literature [30, 60, 75–77]. We supplied state-specific sampling proportions to account for missing species and sampling bias (sampling proportions: cushion plants: 0.8, non-cushions 0.5; plants with lime-secreting hydathodes: 0.85; plants without lime-secreting hydathodes: 0.55). To asses trait-associated diversification, we fitted the Multiple State Speciation and Extinction (MuSSE) model [78] on 100 phylogenetic trees in a Bayesian framework (as outlined in [79], script available at https://github.com/dsilvestro/mcmc-diversitree). MuSSE is a multistate extension of the BiSSE (Binary State Speciation and Extinction) model [80], which simultaneously estimates the rates of speciation (λ) and extinction (μ) under two character states (0 and 1) as well as transition rates between these states (q_{01} and q_{10}).

Recently, serious concerns about the statistical power and interpretations of state-dependent speciation and extinction (SSE) models have been raised [81–83]. The statistical power of BISSE and other SSE models is relatively low when using small phylogenies with less than 300 tips and/or trait states that are unevenly represented in the phylogeny (high tip ratio). Furthermore, SSE models have been shown to be affected by high Type 1 error rates that stem from not accounting for independent shifts in character states that are not associated to diversification *per se* [83]. In order to account for this, we employed a simulation scheme suggested by Rabosky & Goldberg [83] exploiting the Maximum Likelihood framework of BiSSE in the R package *diversitree* [78]. The empirically estimated mean transition rates q_{10} and q_{01} were used to simulate 100 sets of traits across the *Saxifraga* phylogeny using the R package *phytools* [64]. The simulated traits were then used to fit two models each, the unconstrained diversification model ($\lambda_0 \neq \lambda_1$, $\mu_0 \neq \mu_1$, $q_0 \neq q_1$) and the best scoring model for each trait. We used the distribution of ΔAIC values (difference in AIC between both models) to investigate the discriminative power of these models and compared this with the results from our empirical analysis.

Geographical distribution
GeoSSE, another member of the SSE model class, is specifically designed to investigate the association of geographical distribution patterns with lineage diversification [84]. It estimates rates of speciation in the target area (area A) and the remaining distribution (area B) as well as in the joint range (area AB). Range contraction is modelled as local extinction (AB -> B, AB -> A) while transition rates between each area and the joint area (A -> AB, B -> AB) correspond to range expansion. We used GeoSSE as implemented in the R package *diversitree* [78] to test for an association between diversification in *Saxifraga* and occurrence in the Hengduan Mountains region. Distribution data were retrieved from various floras (30, 60, 75, 76; sampling proportion of *Saxifraga* from Hengduan Mountains region: 0.5, in other areas: 0.7, widespread species in combined area: 0.7). In order to test the robustness of our results we performed an additional analysis testing for an effect of occurrence in the QTP region in general, here broadly defined as the QTP itself, the Hengduan Mountains, the Himalayas, the Karakorum Range, Pamir, the Tianshan, and the Altai Mountains (sampling proportion of *Saxifraga* from QTP region: 0.65, in other areas: 0.8, widespread species in combined area: 0.9). Area delineation of the Hengduan Mountains region was done in accordance with Boufford [85]. In total, we compared 36 constrained GeoSSE models (e.g., equal speciation, extinction, and/or transition rates) to a fully unconstrained model (differential speciation, extinction and transition rates) using the AIC and Akaike weights [86]. Consecutively, parameter estimation was performed via an MCMC analysis of 50,000 generations for the model with the lowest AIC. The first 500 steps were discarded as burn-in and trait simulations were run as described above.

Additional files

Additional file 1: Results of prior sensitivity tests and species number comparisons in BAMM (PDF 1143 kb)

Additional file 2: Diversification rate estimates for *Saxifraga* from BayesRate and BAMM (DOCX 32 kb)

Additional file 3: 1. Best scoring GEOSSE models for state-dependent diversification of *Saxifraga* in Hengduan Mountains. 2. Initial parameter estimates of GEOSSE models for state-dependent diversification of *Saxifraga* in Hengduan Mountains (DOCX 39 kb)

Additional file 4: Results for model fitting of simulated traits in *Saxifraga (PDF 416 kb)*

Additional file 5: 1. Best scoring GEOSSE models for state-dependent diversification of *Saxifraga* in QTP region. 2. Parameter estimates for best scoring GEOSSE model for state-dependent diversification of *Saxifraga* in QTP region (DOCX 35 kb)

Additional file 6: Minimum and maximum species numbers for all *Saxifraga* sections (DOCX 29 kb)

Abbreviations

QTP: Qinghai-Tibet Plateau; H + I + S + P: Clade consisting of sections *Heterisia, Irregulares, Saxifragella, Pseudocymbalaria*; BF: Bayes factor; PB: Pure birth (or Yule) diversification process; BD: Birth-death diversification process; HPD: Highest posterior density; SSE model: State-dependent speciation and extinction model

Acknowledgements
We would like to thank Ingo Michalak (formerly Leipzig University) for methodological assistance during early stages of this study as well as Natalia Tkach (Martin Luther University Halle) for continued advice and constructive discussions. Furthermore, we would like to thank Hang Sun and Ende Liu (both Kunming Institute of Botany) for their assistance with Chinese samples and data retrieval from foreign language sources.

Funding
Financial support for this study was provided by the German Science Foundation (project no. MU 2934/3–1) and by the research funding programme 'LOEWE _ Landes-Offensive zur Entwicklung Wissenschaftlich-oekonomischer Exzellenz' of Hesse's Ministry of Higher Education, Research, and the Arts to Alexandra N. Muellner-Riehl.

Authors' contributions
AF, ANM-R and JE designed the study. JE and JS analysed the data and JE, JS and AF interpreted the results. JE wrote the first draft of the manuscript and all authors contributed to writing the final version.

Authors' information
Jana Ebersbach is a PhD candidate at Leipzig University. She is mainly studying the taxonomy, biogeography and diversification history of *Saxifraga* with a particular focus on the influence of past climatic and geological changes in the region of the Qinghai–Tibet Plateau.

Competing interests
The authors declare that they have no competing interests.

Author details
[1]Department of Molecular Evolution and Plant Systematics & Herbarium (LZ), Institute of Biology, Leipzig University, Johannisallee 21–23, D-04103 Leipzig, Germany. [2]German Centre for Integrative Biodiversity Research (iDiv) Halle-Jena-Leipzig, Deutscher Platz 5e, D-04103 Leipzig, Germany.

References
1. Linder HP. Plant species radiations: where, when, why? Philos Trans R Soc Lond Ser B Biol Sci. 2008;363:3097–105.
2. Zanne AE, Tank DC, Cornwell WK, Eastman JM, Smith SA, FitzJohn RG, et al. Three keys to the radiation of angiosperms into freezing environments. Nature. 2014;506:89–92. doi:10.1038/nature12872.
3. Hughes CE, Atchison GW. The ubiquity of alpine plant radiations: from the Andes to the Hengduan Mountains. New Phytol. 2015;207:275–82.
4. Kohler T, Maselli D, editors. Mountains and climate change: From understanding to action. Bern: CDE; 2009.
5. Barthlott W, Mutke J, Rafiqpoor D, Kier G, Kreft H. Global centers of vascular plant diversity. Nova Acta Leopold. 2005;92:61–83.
6. Spehn EM, Rudmann-Maurer K, Körner C. Mountain biodiversity. Plant Ecol Divers. 2011;4:301–2. doi:10.1080/17550874.2012.698660.
7. Körner C. Mountain Biodiversity, its causes and function. Ambio Spec Rep. 2004:11–7.
8. Luebert F, Muller LAH. Effects of mountain formation and uplift on biological diversity. Front Genet. 2015;6:54. doi:10.3389/fgene.2015.00054.
9. Wen J, Zhang J, Nie Z, Zhong Y, Sun H. Evolutionary diversifications of plants on the Qinghai-Tibetan Plateau. Front Genet. 2014; doi:10.3389/fgene.2014.00004.
10. Hughes C, Eastwood R. Island radiation on a continental scale: exceptional rates of plant diversification after uplift of the Andes. Proc Natl Acad Sci U S A. 2006;103:10334–9.
11. Winkworth R. Evolution of the New Zealand mountain flora: Origins, diversification and dispersal. Org Divers Evol. 2005;5:237–47. doi:10.1016/j.ode.2004.12.001.
12. Uribe-Convers S, Tank DC. Shifts in diversification rates linked to biogeographic movement into new areas: An example of a recent radiation in the Andes. Am J Bot. 2015;102:1854–69. doi:10.3732/ajb.1500229.
13. Hoorn C, Mosbrugger V, Mulch A, Antonelli A. Biodiversity from mountain building. Nat Geosci. 2013;6:154. doi:10.1038/ngeo1742.
14. Renner SS. Available data point to a 4-km-high Tibetan Plateau by 40 Ma, but 100 molecular-clock papers have linked supposed recent uplift to young node ages. J Biogeogr. 2016;
15. Favre A, Päckert M, Pauls SU, Jähnig SC, Uhl D, Michalak I, et al. The role of the uplift of the Qinghai-Tibetan Plateau for the evolution of Tibetan biotas. Biol Rev Camb Philos Soc. 2015;90:236–53. doi:10.1111/brv.12107.
16. Nürk NM, Uribe-Convers S, Gehrke B, Tank DC, Blattner FR. Oligocene niche shift, Miocene diversification – cold tolerance and accelerated speciation rates in the St. John's Worts (Hypericum, Hypericaceae). BMC Evol Biol. 2015;15:190. doi:10.1186/s12862-015-0359-4.
17. Wang L, Schneider H, Zhang X, Xiang Q. The rise of the Himalaya enforced the diversification of SE Asian ferns by altering the monsoon regimes. BMC Plant Biol. 2012;12:210. doi:10.1186/1471-2229-12-210.
18. Nagy L, Grabherr G. The biology of alpine habitats. Oxford: Oxford University Press; 2009.
19. Simões M, Breitkreuz L, Alvarado M, Baca S, Cooper JC, Heins L, et al. The evolving theory of evolutionary radiations. Trends Ecol Evol. 2016;31:27–34. doi:10.1016/j.tree.2015.10.007.
20. Hodges SA, Arnold ML. Spurring plant diversification: Are floral nectar spurs a key innovation? Proc R Soc B. 1995;262:343–8.
21. Schwery O, Onstein RE, Bouchenak-Khelladi Y, Xing Y, Carter RJ, Linder HP. As old as the mountains: the radiations of the Ericaceae. New Phytol. 2015;207:355–67. doi:10.1111/nph.13234.
22. Boucher FC, Thuiller W, Roquet C, Douzet R, Aubert S, Alvarez N, et al. Reconstructing the origins of high-alpine niches and cushion life form in the genus Androsace s.l. (Primulaceae). Evolution. 2012;66:1255–68. doi:10.1111/j.1558-5646.2011.01483.x.
23. Matuszak S, Favre A, Schnitzler J, Muellner-Riehl AN. Key innovations and climatic niche divergence as drivers of diversification in subtropical Gentianinae in southeastern and eastern Asia. Am J Bot. 2016;103:899–911. doi:10.3732/ajb.1500352.
24. Lagomarsino LP, Condamine FL, Antonelli A, Mulch A, Davis CC. The abiotic and biotic drivers of rapid diversification in Andean bellflowers (Campanulaceae). New Phytol. 2016;210:1430–42. doi:10.1111/nph.13920.
25. Vargas P. A phylogenetic study of *Saxifraga* sect.*Saxifraga* (Saxifragaceae) based on nrDNA ITS sequences. Plant Syst Evol. 2000;223:59–70.

26. Tkach N, Röser M, Miehe G, Muellner-Riehl AN, Ebersbach J, Favre A, Hoffmann MH. Molecular phylogenetics, morphology and a revised classification of the complex genus *Saxifraga* (Saxifragaceae). Taxon. 2015: 1159–87. doi: 10.12705/646.4.

27. Xu B, Li Z, Sun H. Plant diversity and floristic characters of the alpine subnival belt flora in the Hengduan Mountains, SW China. J Syst Evol. 2014; 52:271–9. doi:10.1111/jse.12037.

28. Soltis DE, Kuzoff RK, Conti E, Gornall RJ, Ferguson IK. matK and rbcL gene sequence data indicate that *Saxifraga* (Saxifragaceae) is polyphyletic. Am J Bot. 1996;83:371–82.

29. Prieto JAF, Arjona JM, Sanna M, Perez R, Cires E. Phylogeny and systematics of Micranthes (Saxifragaceae): an appraisal in European territories. J Plant Res. 2013;126:605–11. doi:10.1007/s10265-013-0566-2.

30. Pan J, Gornall RJ, Ohba H. *Saxifraga*. In: Wu C, Raven PH, editors. Flora of China: Brassicaceae through Saxifragaceae. St. Louis: Science Press; Missouri Botanical Garden Press; 2001. p. 280–344.

31. Ebersbach J, Muellner-Riehl AN, Michalak I, Tkach N, Hoffmann MH, Röser M, et al. In and out of the Qinghai-Tibet Plateau: divergence time estimation and historical biogeography of the large arctic-alpine genus *Saxifraga* L. J Biogeogr. 2017;44:900–10. doi:10.1111/jbi.12899.

32. Gao Q, Li Y, Gornall RJ, Zhang Z, Zhang F, Xing R, et al. Phylogeny and speciation in *Saxifraga* sect. *Ciliatae* (Saxifragaceae): Evidence from *psbA-trnH*, *trnL-F* and ITS sequences. Taxon. 2015;64:703–13. doi:10.12705/644.3.

33. Gornall RJ. An outline of a revised classification of *Saxifraga* L. Bot J Linn Soc. 1987:273–92.

34. Beaulieu JM, Donoghue MJ. Fruit evolution and diversification in campanulid angiosperms. Evolution. 2013;67:3132–44. doi:10.1111/evo.12180.

35. Beaulieu JM, O'Meara BC. Extinction can be estimated from moderately sized molecular phylogenies. Evolution. 2015;69:1036–43. doi:10.1111/evo.12614.

36. Mitchell JS, Rabosky DL. Bayesian model selection with BAMM: effects of the model prior on the inferred number of diversification shifts. Methods Ecol Evol. 2017;8:37–46.

37. Favre A, Michalak I, Chen C, Wang J, Pringle JS, Matuszak S, et al. Out-of-Tibet: The spatio-temporal evolution of Gentiana (Gentianaceae). J Biogeogr. 2016. doi:10.1111/jbi.12840.

38. Jabbour F, Renner SS. A phylogeny of Delphinieae (Ranunculaceae) shows that *Aconitum* is nested within *Delphinium* and that Late Miocene transitions to long life cycles in the Himalayas and Southwest China coincide with bursts in diversification. Mol Phylogenet Evol. 2012;62:928–42. doi:10.1016/j.ympev.2011.12.005.

39. Drummond CS, Eastwood RJ, Miotto STS, Hughes CE. Multiple continental radiations and correlates of diversification in *Lupinus* (Leguminosae): Testing for key innovation with incomplete taxon sampling. Syst Biol. 2012;61:443–60. doi:10.1093/sysbio/syr126.

40. von Hagen KB, Kadereit JW. The phylogeny of *Gentianella* (Gentianaceae) and its colonization of the southern hemisphere as revealed by nuclear and chloroplast DNA sequence variation. Org Divers Evol. 2001;1:61–79. doi:10.1078/1439-6092-00005.

41. Rabosky DL. Extinction rates should not be estimated from molecular phylogenies. Evolution. 2010;64:1816–24. doi:10.1111/j.1558-5646.2009.00926.x.

42. Nee S, Holmes EC, May RM, Harvey PH. Extinction rates can be estimated from molecular phylogenies. Philos Trans R Soc Lond Ser B Biol Sci. 1994; 344:77–82. doi:10.1098/rstb.1994.0054.

43. Silvestro D, Schnitzler J, Zizka G. A Bayesian framework to estimate diversification rates and their variation through time and space. BMC Evol Biol. 2011;11:311. doi:10.1186/1471-2148-11-311.

44. Xu B, Li Z, Sun H. Seed plants of the alpine subnival belt from the Hengduan Mountains. Southwest China: Science Press; 2014.

45. Lippert PC, van Hinsbergen DJ, Dupont-Nivet G. Early Cretaceous to present latitude of the central proto-Tibetan Plateau: A paleomagnetic synthesis with implications for Cenozoic tectonics, paleogeography, and climate of Asia. In: Nie J, Horton BK, Hoke GD, editors. Toward an Improved Understanding of Uplift Mechanisms and the Elevation History of the Tibetan Plateau. Boulder: Geological Society of America; 2014. p. 1–22. doi:10.1130/2014.2507(01).

46. Wang P, Scherler D, Liu-Zeng J, Mey J, Avouac J, Zhang Y, et al. Tectonic control of Yarlung Tsangpo Gorge revealed by a buried canyon in Southern Tibet. Science. 2014;346:978–81. doi:10.1126/science.1259041.

47. Sun B, Wu J, Liu Y, Ding S, Li X, Xie S, et al. Reconstructing Neogene vegetation and climates to infer tectonic uplift in western Yunnan, China. Palaeogeogr Palaeoclimatol Palaeoecol. 2011;304:328–36. doi:10.1016/j.palaeo.2010.09.023.

48. Xing Y, Ree RH. Uplift-driven diversification in the Hengduan Mountains, a temperate biodiversity hotspot. Proc Natl Acad Sci U S A. 2017; doi:10.1073/pnas.1616063114.

49. Mosbrugger V, Favre A, Muellner-Riehl AN, Päckert M, Mulch A. Cenozoic Evolution of Geo-Biodiversity in the Tibeto-Himalayan Region. In: Mountains, Climate, and Biodiversity.

50. Zachos JC, Dickens GR, Zeebe RE. An early Cenozoic perspective on greenhouse warming and carbon-cycle dynamics. Nature. 2008;451:279–83. doi:10.1038/nature06588.

51. Zachos JC, Pagani M, Sloan L, Thomas E, Billups K. Trends, rhythms, and aberrations in global climate 65 Ma to present. Science. 2001;292:686–93. doi:10.1126/science.1059412.

52. Tank DC, Eastman JM, Pennell MW, Soltis PS, Soltis DE, Hinchliff CE, et al. Nested radiations and the pulse of angiosperm diversification: increased diversification rates often follow whole genome duplications. New Phytol. 2015;207:454–67. doi:10.1111/nph.13491.

53. Nie Z, Wen J, Gu Z, Boufford DE, Sun H. Polyploidy in the flora of the Hengduan Mountains Hotspot, Southwestern China. Ann Mo Bot Gard. 2005;92:275–306.

54. Zhang F, Li Y, Gao Q, Lei S, Khan G, Yang H, Chen S. Development and characterization of polymorphic microsatellite loci for *Saxifraga egregia* (Saxifragaceae). Appl Plant Sci. 2015. doi: 10.3732/apps.1500037.

55. Rubio de Casas R, Mort ME, Soltis DE. The influence of habitat on the evolution of plants: a case study across Saxifragales. Ann Bot. 2016. doi: 10.1093/aob/mcw160.

56. Bürgel J. Hybridisation in *Saxifraga* subsection *Kabschia* (Saxifragaceae) from the Central Himalaya. Phyton. 2007;47:191–204.

57. Boucher FC, Lavergne S, Basile M, Choler P, Aubert S. Evolution and biogeography of the cushion life form in angiosperms. Perspect Plant Ecol Evol Syst. 2016;20:22–31. doi:10.1016/j.ppees.2016.03.002.

58. Roquet C, Boucher FC, Thuiller W, Lavergne S. Replicated radiations of the alpine genus Androsace (Primulaceae) driven by range expansion and convergent key innovations. J Biogeogr. 2013;40:1874–86. doi:10.1111/jbi.12135.

59. Donoghue MJ. Key innovations, convergence, and success: macroevolutionary lessons from plant phylogeny. Paleobiology. 2005;2:77–93.

60. Webb DA, Gornall RJ. A Manual of Saxifrages and their cultivation. 1st ed. Portland: Timber Press, Incorporated; 1989.

61. Conti E, Soltis DE, Hardig TM, Schneider J. Phylogenetic relationships of the silver saxifrages (*Saxifraga*, sect. *Ligulatae* Haworth): implications for the evolution of substrate specificity, life histories, and biogeography. Mol Phylogenet Evol. 1999;13:536–55. doi: 10.1006/mpev.1999.0673 .

62. Spriggs EL, Clement WL, Sweeney PW, Madrinan S, Edwards EJ, Donoghue MJ. Temperate radiations and dying embers of a tropical past: the diversification of *Viburnum*. New Phytol. 2015;207:340–54. doi:10.1111/nph.13305.

63. Bouchenak-Khelladi Y, Onstein RE, Xing Y, Schwery O, Linder HP. On the complexity of triggering evolutionary radiations. New Phytol. 2015;207:313–26. doi:10.1111/nph.13331.

64. Revell LJ. phytools: An R package for phylogenetic comparative biology (and other things). Methods Ecol Evol. 2012;3:217–23. doi:10.1111/j.2041-210X.2011.00169.x.

65. R Core Team. R: A Language and Environment for Statistical. Vienna, R Foundation for Statistical Computing; 2016.

66. Drummond AJ, Suchard MA, Xie D, Rambaut A. Bayesian phylogenetics with BEAUti and the BEAST 1.7. Mol Biol Evol. 2012;29:1969–73. doi:10.1093/molbev/mss075.

67. Rabosky DL. Automatic detection of key innovations, rate shifts, and diversity-dependence on phylogenetic trees. PLoS One. 2014;9:e89543. doi: 10.1371/journal.pone.0089543.

68. Rabosky DL, Grundler M, Anderson C, Title P, Shi JJ, BROWN JW, et al. BAMMtools: An R package for the analysis of evolutionary dynamics on phylogenetic trees. Methods Ecol Evol. 2014;5:701–7. doi:10.1111/2041-210X.12199.

69. Rabosky DL, Santini F, Eastman J, Smith SA, Sidlauskas B, Chang J, et al. Rates of speciation and morphological evolution are correlated across the largest vertebrate radiation. Nat Commun. 2013;4:1958. doi:10.1038/ncomms2958.

70. DeChaine EG, Anderson SA, McNew JM, Wendling BM. On the evolutionary and biogeographic history of *Saxifraga* sect. *Trachyphyllum* (Gaud.) Koch (Saxifragaceae Juss.). PLoS One. 2013;8:e69814. doi:10.1371/journal.pone.0069814.

71. Zhang Z, Chen S, Gornall RJ. Morphology and anatomy of the exine in *Saxifraga* (Saxifragaceae). Phytotaxa. 2015;212:105. doi:10.11646/phytotaxa.212.2.1.

72. Moore BR, Hohna S, May MR, Rannala B, Huelsenbeck JP. Critically evaluating the theory and performance of Bayesian analysis of macroevolutionary mixtures. Proc Natl Acad Sci U S A. 2016;113:9569–74. doi:10.1073/pnas.1518659113.

73. Rabosky DL, Mitchell JS, Chang J. Is BAMM flawed? Theoretical and practical concerns in the analysis of multi-rate diversification models. Syst Biol. 2017. doi:10.1093/sysbio/syx037.

74. Kass RE, Raftery AE. Bayes Factors. J Am Stat Assoc. 1995;90:773–95. doi:10.1080/01621459.1995.10476572.

75. Brouillet L, Elvander PE. *Saxifraga*. In: Flora of North America Editorial Committee, editor. Flora of North America; 2009. p. 43–166.

76. Akiyama S, Gornall RJ. *Saxifraga*. In: Watson A, et al., editors. Flora of Nepal; 2012. p. 254–303.

77. Horný R, Webr KM, Byam-Grounds J. Porophyllum saxifrages. Stamford: Byam-Grounds Publications; 1986.

78. FitzJohn RG. Diversitree: Comparative phylogenetic analyses of diversification in R. Methods Ecol Evol. 2012;3:1084–92. doi:10.1111/j.2041-210X.2012.00234.x.

79. Burin G, Kissling WD, Guimaraes PR Jr, Sekercioglu CH, Quental TB. Omnivory in birds is a macroevolutionary sink. Nat Commun. 2016;7:11250. doi:10.1038/ncomms11250.

80. Maddison WP, Midford P, Otto SP. Estimating a binary character's effect on speciation and extinction. Syst Biol. 2007;56:701–10.

81. Maddison WP, FitzJohn RG. The unsolved challenge to phylogenetic correlation tests for categorical characters. Syst Biol. 2015;64:127–36. doi:10.1093/sysbio/syu070.

82. Davis MP, Midford PE, Maddison W. Exploring power and parameter estimation of the BiSSE method for analyzing species diversification. BMC Evol Biol. 2013;13:38. doi:10.1186/1471-2148-13-38.

83. Rabosky DL, Goldberg EE. Model inadequacy and mistaken inferences of trait-dependent speciation. Syst Biol. 2015;64:340–55. doi:10.1093/sysbio/syu131.

84. Goldberg EE, Lancaster LT, Ree RH. Phylogenetic inference of reciprocal effects between geographic range evolution and diversification. Syst Biol. 2011;60:451–65. doi:10.1093/sysbio/syr046.

85. Boufford DE. Biodiversity Hotspot: China's Hengduan Mountains. Arnoldia. 2014;72:24–35.

86. Akaike H. On the Likelihood of a Time Series Model. Underst Stat. 1978;27:217. doi:10.2307/2988185.

87. Engler H, Irmscher E. Saxifragaceae – *Saxifraga*. In: Engler H, editor. Das Pflanzenreich. Regni vegetabilis conspectus, vol. 1919. Leipzig: Engelmann; 1916. p. 449–709.

Permissions

The contributors of this book come from diverse backgrounds, making this book a truly international effort. This book will bring forth new frontiers with its revolutionizing research information and detailed analysis of the nascent developments around the world.

We would like to thank all the contributing authors for lending their expertise to make the book truly unique. They have played a crucial role in the development of this book. Without their invaluable contributions this book wouldn't have been possible. They have made vital efforts to compile up to date information on the varied aspects of this subject to make this book a valuable addition to the collection of many professionals and students.

This book was conceptualized with the vision of imparting up-to-date information and advanced data in this field. To ensure the same, a matchless editorial board was set up. Every individual on the board went through rigorous rounds of assessment to prove their worth. After which they invested a large part of their time researching and compiling the most relevant data for our readers.

The editorial board has been involved in producing this book since its inception. They have spent rigorous hours researching and exploring the diverse topics which have resulted in the successful publishing of this book. They have passed on their knowledge of decades through this book. To expedite this challenging task, the publisher supported the team at every step. A small team of assistant editors was also appointed to further simplify the editing procedure and attain best results for the readers.

Apart from the editorial board, the designing team has also invested a significant amount of their time in understanding the subject and creating the most relevant covers. They scrutinized every image to scout for the most suitable representation of the subject and create an appropriate cover for the book.

The publishing team has been an ardent support to the editorial, designing and production team. Their endless efforts to recruit the best for this project, has resulted in the accomplishment of this book. They are a veteran in the field of academics and their pool of knowledge is as vast as their experience in printing. Their expertise and guidance has proved useful at every step. Their uncompromising quality standards have made this book an exceptional effort. Their encouragement from time to time has been an inspiration for everyone.

The publisher and the editorial board hope that this book will prove to be a valuable piece of knowledge for researchers, students, practitioners and scholars across the globe.

List of Contributors

Katharina Weiss, Gudrun Herzner and Erhard Strohm
Evolutionary Ecology Group, Institute of Zoology, University of Regensburg, Universitätsstr. 31, 93053 Regensburg, Germany.

Yasuko Akiyama-Oda
Laboratory of Evolutionary Cell and Developmental Biology, JT Biohistory Research Hall, 1-1 Murasaki-cho, Takatsuki 569-1125, Osaka, Japan.
Department of Microbiology and Infection Control, Osaka Medical College, Takatsuki, Osaka, Japan.

Hiroki Oda
Laboratory of Evolutionary Cell and Developmental Biology, JT Biohistory Research Hall, 1-1 Murasaki-cho, Takatsuki 569-1125, Osaka, Japan.
Department of Biological Sciences, Graduate School of Science, Osaka University, Osaka, Japan.

Mizuki Sasaki
Laboratory of Evolutionary Cell and Developmental Biology, JT Biohistory Research Hall, 1-1 Murasaki-cho, Takatsuki 569-1125, Osaka, Japan.
Current address: Department of Parasitology, Asahikawa Medical University, 2-1-1-1 Midorigaoka-higashi, Asahikawa 078-8510, Hokkaido, Japan.

Le Liu
College of Geoscience and Surveying Engineering, China University of Mining and Technology (Beijing), Beijing 100083, China.
Key Laboratory of Orogenic Belts and Crustal Evolution, Department of Geology, Peking University, Beijing 100871, China.

Deming Wang and Jinzhuang Xue
Key Laboratory of Orogenic Belts and Crustal Evolution, Department of Geology, Peking University, Beijing 100871, China.

Meicen Meng
Science Press, China Science Publishing & Media Ltd., Beijing 100717, China.

Ahmad Al-Salam
Department of Laboratory Medicine and Pathobiology, Faculty of Medicine, University of Toronto, 1 King's College Circle, Toronto, ON M5S 1A8, Canada

David M. Irwin
Department of Laboratory Medicine and Pathobiology, Faculty of Medicine, University of Toronto, 1 King's College Circle, Toronto, ON M5S 1A8, Canada
Banting and Best Diabetes Centre, University of Toronto, Toronto, ON, Canada.

Marleen M. P. Cobben
Department of Animal Ecology, Netherlands Institute of Ecology (NIOO-KNAW), 6700, AB, Wageningen, The Netherlands. Theoretical Evolutionary Ecology Group, Institute for Animal Ecology and Tropical Biology, University of Würzburg, Emil-Fischerstr. 32, 97074 Würzburg, Germany.

Oliver Mitesser
Theoretical Evolutionary Ecology Group, Institute for Animal Ecology and Tropical Biology, University of Würzburg, Emil-Fischerstr. 32, 97074 Würzburg, Germany.

Alexander Kubisch
Theoretical Evolutionary Ecology Group, Institute for Animal Ecology and Tropical Biology, University of Würzburg, Emil-Fischerstr. 32, 97074 Würzburg, Germany.
Institute for Landscape and Plant Ecology, University of Hohenheim, August-von-Hartmann-Str. 3, 70599 Stuttgart, Germany.

Chaoyang Zhao
Department of Botany and Plant Sciences, University of California, Riverside, Riverside, CA 92521, USA.

Paul D. Nabity
Department of Botany and Plant Sciences, University of California, Riverside, 900 University Avenue, Batchelor Hall room 2140, Riverside, CA 92521, USA.

Qian Zhao, Dongna Ma and Minsheng You
State Key Laboratory for Ecological Pest Control of Fujian/Taiwan Crops and College of Life Science, Fujian Agriculture and Forestry University, Fuzhou 350002, China.

Institute of Applied Ecology, Fujian Agriculture and Forestry University, Fuzhou 350002, China.
Fujian-Taiwan Joint Centre for Ecological Control of Crop Pests, Fujian Agriculture and Forestry University, Fuzhou350002, China.
Key Laboratory of Integrated Pest Management for Fujian-Taiwan Crops, Ministry of Agriculture, Fuzhou 350002, China.

Liette Vasseur
State Key Laboratory for Ecological Pest Control of Fujian/Taiwan Crops and College of Life Science, Fujian Agriculture and Forestry University, Fuzhou 350002, China.
Institute of Applied Ecology, Fujian Agriculture and Forestry University, Fuzhou 350002, China.
Department of Biological Sciences, Brock University, 1812 Sir Isaac Brock Way, St. Catharines, ON L2S 3A1, Canada.

Manjusha Chintalapati, Michael Dannemann and Kay Prüfer
Max Planck Institute for Evolutionary Anthropology, 04103 Leipzig, Germany

Shixia Xu, Xiaohui Sun, Xu Niu, Zepeng Zhang, Ran Tian, Wenhua Ren, Kaiya Zhou and Guang Yang
Jiangsu Key Laboratory for Biodiversity and Biotechnology, College of Life Sciences, Nanjing Normal University, 1 Wenyuan Road, Nanjing 210023, China

Hiroaki Nakano and Hideyuki Miyazawa
Shimoda Marine Research Center, University of Tsukuba, 5-10-1, Shimoda, Shizuoka 415-0025, Japan.

Akiteru Maeno and Toshihiko Shiroishi
Mammalian Genetics Laboratory, National Institute of Genetics, 1111 Yata, Mishima, Shizuoka, 411-8540, Japan.

Keiichi Kakui
Faculty of Science, Hokkaido University, N10 W8, Kita-ku, Sapporo, Hokkaido 060-0810, Japan.

Ryo Koyanagi and Miyuki Kanda
DNA Sequencing Section, Okinawa Institute of Science and Technology Graduate University, Onna, Okinawa 904-0495, Japan.

Noriyuki Satoh
Marine Genomics Unit, Okinawa Institute of Science and Technology Graduate University, Onna, Okinawa 904-0495, Japan.

Hisanori Kohtsuka
Misaki Marine Biological Station, The University of Tokyo, 1024 Koajiro, Misaki, Miura, Kanagawa 238-0225, Japan.

Akihito Omori
Misaki Marine Biological Station, The University of Tokyo, 1024 Koajiro, Misaki, Miura, Kanagawa 238-0225, Japan.
Present address: Sado Marine Biological Station, Faculty of Science, Niigata University, Sado, Niigata 952-2135, Japan.

Yoko Arakaki, Hiroko Kawai-Toyooka and Hisayoshi Nozaki
Department of Biological Sciences, Graduate School of Science, University of Tokyo, 7-3-1 Hongo, Bunkyo-ku, Tokyo 113-0033, Japan.
Department of Life Science and Technology, School of Life Science and Technology, Tokyo Institute of Technology, 2-12-1 Ookayama, Meguro-ku, Tokyo 152-8550, Japan.

Shin-ya Miyagishima and Takayuki Fujiwara
Department of Cell Genetics, National Institute of Genetics, 1111 Yata, Mishima, Shizuoka 411-8540, Japan.

Kaoru Kawafune
Department of Biological Sciences, Graduate School of Science, University of Tokyo, 7-3-1 Hongo, Bunkyo-ku, Tokyo 113-0033, Japan. Department of Life Science and Technology, School of Life Science and Technology, Tokyo Institute of Technology, 2-12-1 Ookayama, Meguro-ku, Tokyo 152-8550, Japan.

Jonathan Featherston
Evolutionary Studies Institute, University of the .Witwatersrand, Johannesburg 2000, South Africa.
Agricultural Research Council, Biotechnology Platform, Pretoria 0040, South Africa.

Pierre M. Durand
Evolutionary Studies Institute, University of the Witwatersrand, Johannesburg 2000, South Africa.
Department of Ecology and Evolutionary Biology, University of Arizona, Tucson, AZ 85721, USA.

Martin S. Fischer
Department of Human Evolution, Max Planck
Institute for Evolutionary Anthropology, Leipzig,
Germany.

Patrick Arnold
Department of Human Evolution, Max Planck
Institute for Evolutionary Anthropology, Leipzig,
Germany.
Institut für Spezielle Zoologie und
Evolutionsbiologie mit Phyletischem Museum,
Friedrich-Schiller-Universität Jena, Jena, Germany.

Borja Esteve-Altava
Structure & Motion Lab, Department of Comparative
Biomedical Sciences, Royal Veterinary College,
Hatfield, UK.

Vinh Sy Le and Cuong Cao Dang
University of Engineering and Technology, Vietnam
National University Hanoi, Hanoi, Vietnam

Quang Si Le
School of Pharmacy and Biomedical Sciences,
University of Portsmouth, Winston Churchill
Avenue Portsmouth, Portsmouth, PO1 2UP, UK

Kristof Veitschegger
Palaeontological Institute and Museum, University
of Zurich, Karl Schmid-Strasse 4, 8006 Zürich,
Switzerland

**Anastasia Laggis, Athanasios D. Baxevanis,
Stefania Maniatsi, Alexander Triantafyllidis and
Theodore J. Abatzopoulos**
Department of Genetics, Development and
Molecular Biology, School of Biology, Aristotle
University of Thessaloniki, 54124 Thessaloniki,
Macedonia, Greece.

Alexandra Charalampidou
Scientific Computing Office, Information
Technology (IT) Center, School of Sciences, Aristotle
University of Thessaloniki, 541 24 Thessaloniki,
Macedonia, Greece.

J. Ebersbach, J. Schnitzler and A. Favre
Department of Molecular Evolution and Plant
Systematics & Herbarium (LZ), Institute of Biology,
Leipzig University, Johannisallee 21–23, D-04103
Leipzig, Germany.

A.N. Muellner-Riehl
Department of Molecular Evolution and Plant
Systematics & Herbarium (LZ), Institute of Biology,
Leipzig University, Johannisallee 21–23, D-04103
Leipzig, Germany.
German Centre for Integrative Biodiversity
Research (iDiv)H alle-Jena-Leipzig, Deutscher Platz
5e, D-04103 Leipzig, Germany.

Index

www.ingramcontent.com/pod-product-compliance
Lightning Source LLC
Chambersburg PA
CBHW082057190326
41458CB00010B/3520